[redacted]

Reef Diagenesis

Edited by
J. H. Schroeder and B. H. Purser

With 187 Figures

Springer-Verlag
Berlin Heidelberg New York
London Paris Tokyo

Professor JOHANNES H. SCHROEDER
Ph.D. George Washington University, Washington

Technische Universität Berlin
Institut für Geologie und Paläontologie
Hardenbergstraße 42
1000 Berlin 12, FR Germany

Professor Dr. BRUCE H. PURSER

Laboratoire de Pétrologie
Sédimentaire et Paléontologie
Université Paris-Sud Orsay
Bâtiment 504
91405 Orsay Cedex, France

Cover illustration. Zoned sparry calcite cement in Miocene corals from Khor Eit, NE Sudan (Cathodoluminescence microphotograph Schroeder/Zinkernagel).

Inner front cover. Left: Marine aragonite needle cement in Recent algal cup reefs, Bermuda (SEM Schroeder). *Right:* Calcified algal filament and micrite, both high Mg calcite, with aragonite needle cement, all marine; Recent algal cup reefs, Bermuda (SEM Schroeder).

Inner end cover. Left: Marine aragonite needle cement in subaerially exposed boulders from Pleistocene algal reefs, Bermuda (SEM Schroeder). *Right:* Low Mg calcite microsparite on aragonite substrate in dead Holocene corals, Poe Beach, New Caledonia (SEM Aïssaoui).

ISBN 3-540-16594-0 Springer-Verlag Berlin Heidelberg New York
ISBN 0-387-16594-0 Springer-Verlag New York Heidelberg Berlin

Library of Congress Cataloging-in-Publication Data. Reef diagenesis. Includes index. 1. Diagenesis. 2. Reefs. 3. Sedimentation and deposition. I. Schroeder, Johannes H., 1939– . II. Purser, B. H. QE571.R39 1986 551.4′24 86-15604.

Typesetting: Fotosatz GmbH, Beerfelden
Offsetprinting and bookbinding: Konrad Triltsch, Graphischer Betrieb, Würzburg
2132/3130-543210

Contents

II Mesozoic Reefs

III Paleozoic Reefs

List of Contributors

You will find the addresses at the beginning of the respective contribution

Reef Diagenesis: Introduction

J. H. SCHROEDER[1] and B. H. PURSER[2]

1 Introduction

A symposium convened during the Vth International Coral Reef Congress in Papeete, Tahiti, 1985, encouraged the editors to assemble this volume of case studies by participating and, especially, by nonparticipating scientists. An attempt was made to include case studies from various regions and geological periods, carried out on various scales from regional to ultrastructural. We hope to present an overall view of reef diagenesis.

Although the volume focuses on reef diagenesis, fields also to be considered are biology, paleontology, and sedimentary facies distribution, as they provide the context and, to some extent, encompass the determinants of diagenetic processes. The scope has been limited to reef diagenesis because we feel that reefs have relatively clearly defined geometries, which facilitate the evaluation of diagenetic trends and the definition of diagenetic models. On the other hand, their many different components make reefs somewhat more complex than other deposits, and this creates difficulties in deciphering diagenetic histories; the study of reefs, therefore, is not the simplest manner of solving the many problems relating to carbonate diagenesis.

An additional reason for evaluating reef diagenesis is the reservoir potential of these carbonate bodies. To illustrate the point, in the recent collection of 35 case studies of carbonate reservoirs (Roehl and Choquette 1985), reefs were involved in 15. The emphasis on porosity development in many studies of the present volume is therefore not of mere academic interest.

Most scientists concerned with reef diagenesis probably aim at presenting the perfect example showing the three-dimensional distribution of diagenetic phenomena and porosity, from which a clear understanding of the successive diagenetic environments and processes can be derived. The "model" can then be applied to comparable cases. Although only few models are offered in this volume, each case study with its particular scope and scale is a step toward this ambitious goal. Therefore the sum of these contributions, rather than one single study, suggests possible directions and pertinent questions for future research.

[1] Institut für Geologie und Paläontologie, Technische Universität Berlin, Hardenbergstr. 42, 1000 Berlin 12, FRGermany

[2] Laboratoire de Pétrologie Sédimentaire et Paléontologie. Bâtiment 504, Université de Paris-Sud, 91405 Orsay Cedex, France

Reef Diagenesis
Edited by J. H. Schroeder and B. H. Purser

2 Terminology, Concepts, and Historical Aspects

In order to establish a minimal basis of communication, it may be useful to recall some general definitions, to refer to some general concepts, and to provide a brief historical perspective.

2.1 Reefs

The AGI glossary (Bates and Jackson 1980) generally serves as a good starting point. Its definition of an "organic reef" reads: "A ridgelike or moundlike structure, layered or massive, built by sedentary calcareous organisms, esp. corals, and consisting mostly of their remains; it is wave-resistant and stands above the surrounding comtemporaneously deposited sediment". Many would debate one or the other point of this definition, and indeed, the literature includes many such discussions, which would certainly lead far beyond the scope of this introduction.

As a next step toward the subject, beyond mere definition, two recent reviews of James (1983, 1984) provide convenient points of (re-)entry into the study of reefs; the first is broad and well illustrated, the second concise; both have many helpful references. From there, the way invariably leads back to neo-classical papers such as Heckel (1974), Ladd (1969), and Nelson et al. (1962). Even today a step back to the classics such as Walther (1888) involves more than a nostalgic scientific experience.

A primary concern of many reef workers has been morphology and internal framework as well as type and function of frame builders. This approach provides a basis for many concepts and classifications reviewed by Heckel (1974). As the frame builders evolve through geological time, so the reefs vary in external morphology and internal structure. However, reefs exist in which no frame builders appear; Dunham (1970) called them "stratigraphic reefs" in order to distinguish them from "ecologic reefs" with frame builders.

Another approach views reefs as one in a system of facies belts comprising a composite profile extending from coast to basin. Wilson (1975) defined the standard facies, among them the "organic reef of platform margin" whose "ecologic character varies in response to water energy, steepness of slope, organic productivity, amount of framework construction, binding or trapping, frequency of subaerial exposure and consequent cementation". He distinguished three types, the downslope mud and debris accumulation, the knoll reef, and the frame-constructed reef rim. Within their respective facies context he traced them throughout geologic history. On a complementary note, Longman (1981) draws attention to the important relation between reef growth, sea-level fluctuation and subsidence.

On a somewhat smaller scale, there is a third approach based essentially on the morphology of the individual reef body. Maxwell (1968) distinguished reef patterns based on shape, morphological zonation, and central topography which reflect the respective "hydrologic, bathymetric and biological balance". Reefs evolve from an "embryonic colony" through various patterns in response to changes in this balance. In this context an important French contribution comes to mind which includes many excellent examples of physiographic/ecologic zonations of reefs; a very useful reference, although not necessarily the most representative of this work, is the bi-lingual contribution to coral reef terminology (Clausade et al. 1971).

A fourth approach, with roots in the work of Walther (1888) and of Ginsburg and Lowenstam (1958), focuses on the sedimentological processes involved (Schroeder and Zankl 1974). Reefs are considered the result of successively effective reef-forming and reef-destroying processes: construction by various organisms, bioerosive and mechanical destruction, internal sedimentation, and cementation. In a given reef, or any portion thereof, the sequence of processes may be highly varied. Ultimately, all skeletal structures may be replaced by cemented internal sediment which fills successive borings of various bioerosive organisms. The processes involved are partly genetic and partly diagenetic.

All four approaches are useful in the sedimentological and reservoir analysis of reefs, although on different scales; all provide an essential background for the study of reef diagenesis.

2.2 Diagenesis

The definition of Bates and Jackson (1980) again serves as a starting point; for diagenesis it reads: "All the chemical, physical, and biological changes undergone by a sediment after its initial deposition, and during and after its lithification, exclusive of surficial alteration (weathering) and metamorphism". An important general review of the term and its use has been presented by Dunoyer de Segonzac (1968).

Reef diagenesis is treated in two recent review papers by MacIntyre (1984) and Haley (1984). The former is concerned with preburial processes, both biological (encrustation, bioerosion, and soft tissue destruction) and geological (mechanical destruction and cementation). Halley (1984) deals with diagenetic alteration of original aragonite and high magnesium calcite by freshwater, brine, mixed water, deep-sea water, and subsurface water, and in some respects expands the process approach presented by Matthews (1974). Even this very brief summary of these reviews shows that in reef diagenesis, as in reef formation, one is faced with an interaction of physical, chemical, biological, and geological processes.

When tracing the history of research on reef diagenesis, one invariably turns to Cullis' (1904) outstanding study of the Funafuti cores with its wealth of observation: it remains essential reading for any student of reef diagenesis. One should not forget, however, that a number of valuable observations concerning living and raised reefs were collected long before Cullis, in the late nineteenth century (see Walther 1894 for review).

Interest in the subject decreased at the beginning of our century, but rose again to considerable heights in the 1960's; this increase was preceded by a number of important individual studies, among them that of Newell (1955), who postulated submarine cementation for the Permian Guadelupe Reefs, and that of

Ladd and Schlanger (1960), who described drusy calcite cements in short cores from living reefs of Eniwetok which they could not explain with the knowledge available at that time.

There were three major developments in our knowledge of reef diagenesis in the late 1960's: (1) The pene-simultaneous discovery of submarine cement in reefs of Bermuda (Ginsburg et al. 1967), Barbados (MacIntyre et al. 1968), Jamaica (Land and Goreau 1970) and other locations. (2) Intensive work on raised Pleistocene reefs, notably that of Matthews and his group on Barbados (ref. see Matthews 1974). (3) Simultaneous research on older fossil reefs, especially of Devonian age of which Krebs (1969) presented early results. Subsequently the work has flourished: Greater bathymetric depths were reached at Recent reefs (James and Ginsburg 1979; Land and Moore 1980). New tools were applied (stable isotopes, cathodoluminescence, fluid inclusions) and older tools were considerably improved (trace elements, scanning electron microscopy). An ever-increasing flow of studies on Recent and fossil reefs has been published. We have now attained the point where one should evaluate the current state and future of research in this field, take note of the answers which have already been found, and formulate the questions which remain. We hope that this volume will help in the assessment.

Reef diagenesis, however, must be seen in the context of carbonate diagenesis in general, a field which has been characterized by rapid development since the 1960's: The publications edited by Pray and Murray (1965), Füchtbauer (1969) and Bricker (1971) mark the early increase in interest. The textbook of Bathurst (1971/1975) twice provided important perspectives for teaching and research. More recently, the papers assembled by Schneiderman and Harris (1985) afforded a timely and welcome report on the state of the art. The reviews, textbooks, and studies by many scientists have helped in the realization of the present progress report.

Acknowledgments. We wish to express our gratitude to all our co-authors for their patience in revising their contributions, and to some of them for reviewing early drafts of other manuscripts. We thank F. Lyons for linguistic assistance concerning various drafts of several manuscripts. It is also a pleasure to acknowledge the cooperation of W. Engel of Springer Verlag who showed great concern and understanding throughout the stages of planning and production.

References

Bates RL, Jackson JA (1980) Glossary of geology (2nd ed). American Geological Institute, Falls Church, Va, 751 pp

Bathurst RGC (1971/1975) Carbonate sediments and their diagenesis (1st ed/2nd ed). Elsevier, Amsterdam, 620/658

Bricker OP (1971) Carbonate cements. Johns Hopkins University Studies in Geology 19:1 – 376

Clausade M, Gravier N, Picard J, Pichon M, Roman M-L, Thomassin B, Vasseur P, Vivien M, Weydert P (1971) Morphologie des récifs coralliens de la région de Tuléar (Madagascar): Eléments de terminologie récifale. Téthys supplement 2:1 – 74

Cullis CG (1904) The mineralogical changes observed in the cores of the Funafuti borings. In: The Atoll of Funafuti. R Soc London 392 – 420

Füchtbauer H (ed) (1969) Lithification of carbonate sediments. Sedimentology 12:1 – 322
Dunham RJ (1970) Stratigraphic reefs versus ecologic reefs. Am Assoc Petrol Geol Bull 54:1931 – 1932
Dunoyer de Segonzac G (1968) The birth and development of the concept of diagenesis. Earth Sci Rev 4:153 – 201
Ginsburg RN, Lowenstam HA (1958) The influence of marine bottom communities on the depositional environment of sediments. J Geol 66:310 – 318
Ginsburg RN, Shinn EA, Schroeder JH (1967) Submarine cementation and internal sedimentation within Bermuda reefs. Geol Soc Am Spec Pap 115:78 – 79
Halley RB (1984) Post-depositional diagenesis of metastable carbonates: An overview. Palaeontogr Am 54:245 – 253
Heckel PH (1974) Carbonate buildups in the geologic record: A review. Soc Econ Paleontol Miner Spec Publ 18:90 – 154
James NP (1983) Reef environment. Am Assoc Petrol Geol Mem 33:345 – 462
James NP (1984) Reefs. In: Walker RG (ed) Facies models (2nd ed). Geosc Canada Repr Ser 1:229 – 244
James NP, Ginsburg RN (1979) The seaward margin of Belize barrier and atoll reefs. Int Assoc Sedimentol Spec Publ 3:1 – 191
Krebs W (1969) Early void-filling cementation in Devonian forereef limestones (Germany). Sedimentology 12:279 – 299
Ladd HS (1969) Existing reefs – geological aspects. Proc North American Paleontological Convention 1273 – 1300
Ladd HS, Schlanger SO (1960) Drilling operation on Eniwetok. US Geol Surv Prof Pap 260 – Y: 863 – 905
Land LS, Goreau TF (1970) Submarine lithification of Jamaican reefs. J Sediment Petrol 40:457 – 462
Land LS, Moore CH (1980) Lithification, micritization and syndepositional diagenesis of biolithites on the Jamaican island slope. J Sediment Petrol 50:357 – 370
Longman MW (1981) A process approach to recognizing facies and reef complexes. Soc Econ Paleontol Miner Spec Publ 30:9 – 40
MacIntyre IG (1984) Preburial and shallow-subsurface alteration of modern scleractinian corals. – Palaeontogr Am 54:229 – 244
MacIntyre IG, Mountjoy EW, D'Anglejan BF (1968) An occurrence of submarine cementation of carbonate sediments of the west coast of Barbados WI. J Sediment Petrol 38:660 – 664
Matthews RK (1974) A process approach to diagenesis of reefs and reef associated limestones. Soc Econ Paleontol Miner Spec Publ 18:234 – 256
Maxwell WGH (1968) Atlas of the Great Barrier Reef. Elsevier, Amsterdam, 249
Nelson HF, Brown CW, Brineman JH (1962) Skeletal limestone classification. Am Assoc Petrol Geol Mem 1:224 – 252
Newell ND (1955) Depositional fabric of Permian reef limestones. J Geol 63:301 – 309
Pray LC, Murray RC (eds) (1965) Dolomitization and limestone diagenesis. Soc Econ Paleontol Miner Spec Publ 13:1 – 180
Roehl PO, Choquette PW (eds) (1985) Carbonate petroleum reservoirs. Springer, New York, pp 1 – 622
Schneidermann N, Harris PM (eds) (1985) Carbonate cements. Soc Econ Paleontol Miner Spec Publ 36:1 – 379
Schroeder JH, Zankl H (1974) Dynamic reef formation: A sedimentological concept based on studies of Recent Bermuda and Bahama reefs. Proc II Internat Symposium Coral reefs 2:413 – 428
Walther J (1888) Die Korallenriffe der Sinaihalbinsel. Abhandlungen der mathematisch-physischen Classe der Königlichen Sächsischen Gesellschaft der Wissenschaften 14:438 – 505
Walther J (1894) Lithogenesis der Gegenwart. Fischer, Jena, pp 535 – 1055
Wilson JL (1975) Carbonate facies in geologic history. Springer, New York, pp 471

I Cenozoic Reefs

Regional Distribution of Submarine Cements Within an Epicontinental Reef System: Central Great Barrier Reef, Australia*

J. F. MARSHALL[1]

1 Introduction

Since their rediscovery nearly 20 years ago (Ginsburg et al. 1967, Macintyre et al. 1968), submarine cements have been widely reported in a variety of reefs throughout the world. This has led to the recognition that submarine cementation was an important process in many ancient reefs (e.g., Davies 1977, Lohman and Meyers 1977, James and Kobluk 1978, Mazzullo and Cys 1979, Walls et al. 1979, Mazzullo 1980, Playford 1980). The variety of synsedimentary cements reported from these older deposits have textures and fabrics similar to modern submarine cements, e.g., the comparison between former aragonite botryoid cements in Permian phylloid algal mounds (Mazzullo and Cys 1979) and modern submarine botryoidal aragonite in Holocene reef limestones (Ginsburg and James 1976). However, it is also apparent that other modern and ancient reefs do not show evidence of submarine cementation or that certain parts of a given reef are relatively uncemented compared to others (e.g., James et al. 1976; Playford 1980; Marshall 1983a).

The only previous regional analysis of the distribution of submarine cements within modern epicontental reef complexes, i.e., Belize (James et al. 1976), has indicated that cementation is restricted to the seaward margins of shelf-edge reefs, and that reefs landwards of the edge of the shelf are uncemented. This restriction in the distribution of submarine cements has been related to environmental factors, principally the relatively high energy conditions required to pump substantial volumes of seawater through the reefs' pore system in order to precipitate significant quantities of carbonate cement (e.g., Ginsburg et al. 1971, James et al. 1976, Focke and Gebelein 1978). The degree of cementation has also been related to specific reef facies where the grain size (or pore size) of cavity-filling internal sediment is a determining factor (Ginsburg et al. 1971, James et al. 1976, Marshall 1983a).

Comparison with the Belize reefs might suggest that those ancient reefs showing evidence of synsedimentary cementation were originally shelf-edge phenomena, whereas those that do not contain submarine cements were more nearshore reefs. This implies that the distribution of cements can be used as a paleogeographical tool. However, while the Belize model, with its semi-continuous line of outer reefs, might be applicable to ancient barrier reef complexes, such as the Upper Devonian reefs of Western Australia and Alberta,

* Published with the permission of the Director, Bureau of Mineral Resources.
[1] Bureau of Mineral Resources, GPO Box 378, Canberra City, ACT, 2601, Australia

Reef Diagenesis
Edited by J. H. Schroeder and B. H. Purser

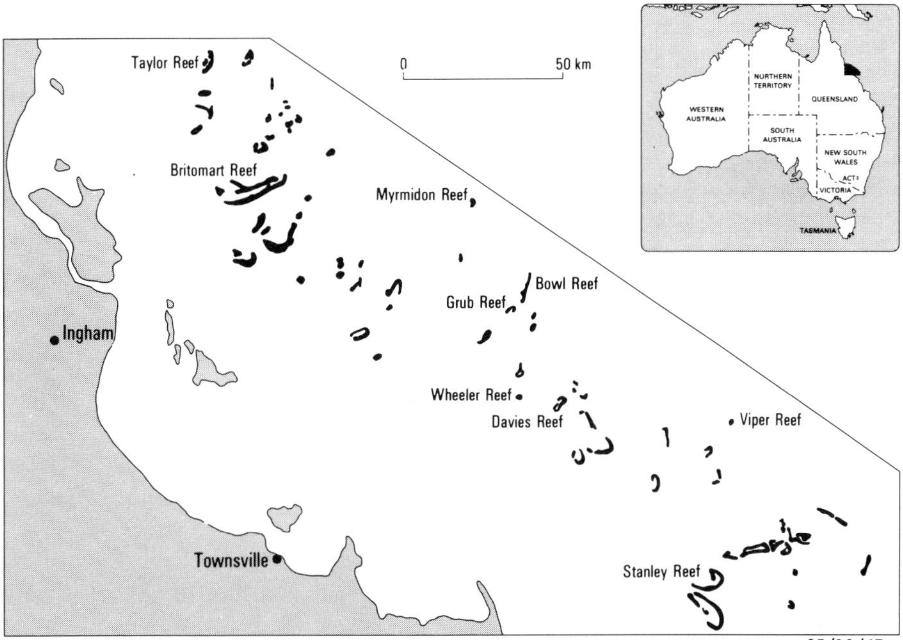

Fig. 1. Map showing the positions of reefs that have been drilled in the central Great Barrier Reef

it would not necessarily be a suitable model for many other types of reef. Non-barrier or platform-type reefs may show different cement distributions. The model clearly needs to be more refined, particularly in an epicontinental reef complex that is more open than the barrier reef type.

The 120-km-wide continental shelf off Townsville (Fig. 1) is an excellent field laboratory to study cross-shelf variations in submarine cementation and the possible factors controlling this variation. The outer half of the shelf forms a reef tract consisting of numerous platform reefs of varying size and shape. Unlike Belize there is no semi-continuous barrier system. While the outermost reefs receive the brunt of the dominant southeast swell, the reef complex is sufficiently open to allow relatively high energy conditions to impinge on some of the more inner reefs. This is usually the situation where reefs are not protected to any great degree by others. Thus Stanley Reef, situated near the inner edge of the reef tract, has only several smaller reefs protecting it from the southeast swell, whereas Wheeler Reef is protected by three large reefs (Fig. 1).

Fifteen holes, ranging in depth between 7.8 and 30.0 m, have been drilled on five reefs in the region (Figs. 1 and 2) by the Bureau of Mineral Resources (BMR). In addition, several holes have been drilled on other reefs in the area; e.g., Britomart Reef (Johnson et al. 1984), Davies Reef (Grimes 1982), Grub and Taylor Reefs (Davies and Hopley 1983), as well as some of the nearshore fringing reefs (Hopley et al. 1983). Data from the BMR holes include detailed drilling logs, facies analysis, petrographic descriptions, and radiocarbon age determinations. These data have been used in this paper to analyze the distribution of submarine cements and its relationship, if any, to variations in reef growth.

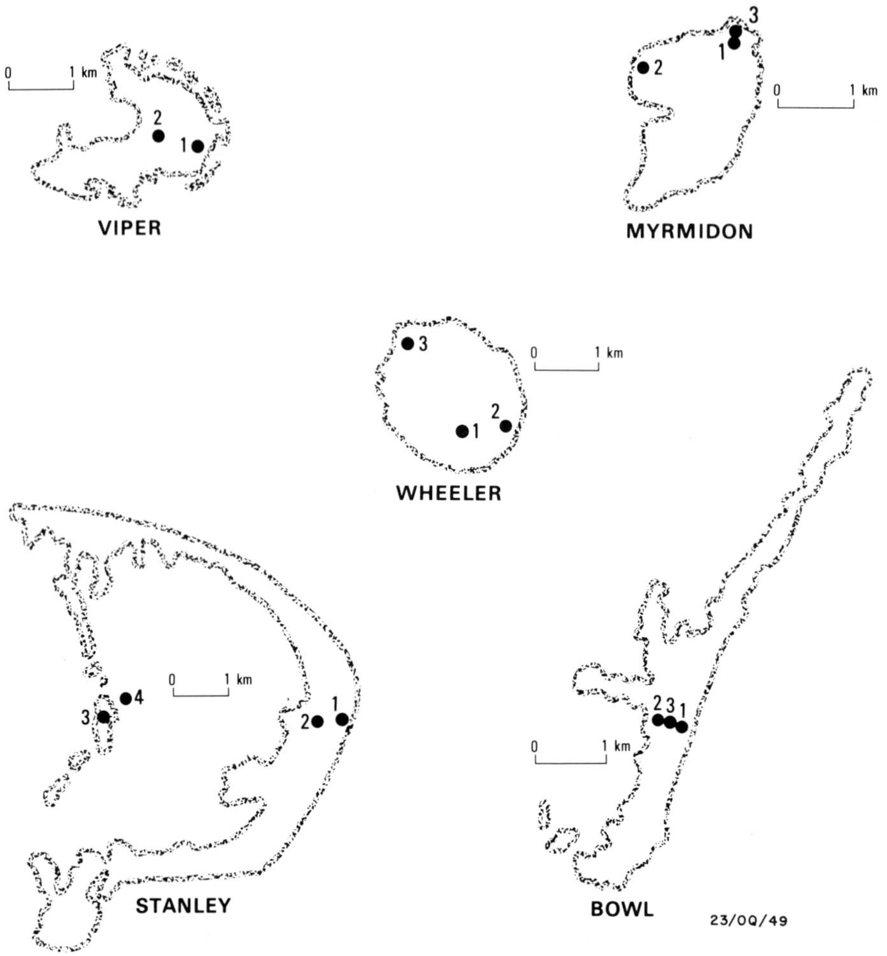

Fig. 2. Well locations on reefs drilled by BMR, showing their position relative to each other and to windward (*right*) and leeward (*left*) margins

2 Cross-Shelf Variations in Reef Facies

Reef facies in the Great Barrier Reef can be grouped broadly into (1) a framework assemblage and (2) a detrital assemblage (Marshall and Davies 1984). The framework assemblage consists of two main elements: (a) a branching coral facies and (b) a coral head facies. Other less extensive framework facies include a *Millepora* facies and an algal pavement facies; the latter is not particularly apparent in the central Great Barrier Reef. The detrital assemblage may consist of several facies that are essentially differentiated on the basis of grain size and environment of deposition (Marshall and Davies 1984). While very coarse-grained (gravel) and fine-grained (mud) facies do occur within the reefs (e.g., Grimes 1982, Marshall and Davies 1984, Johnson et al. 1984), there appears to be

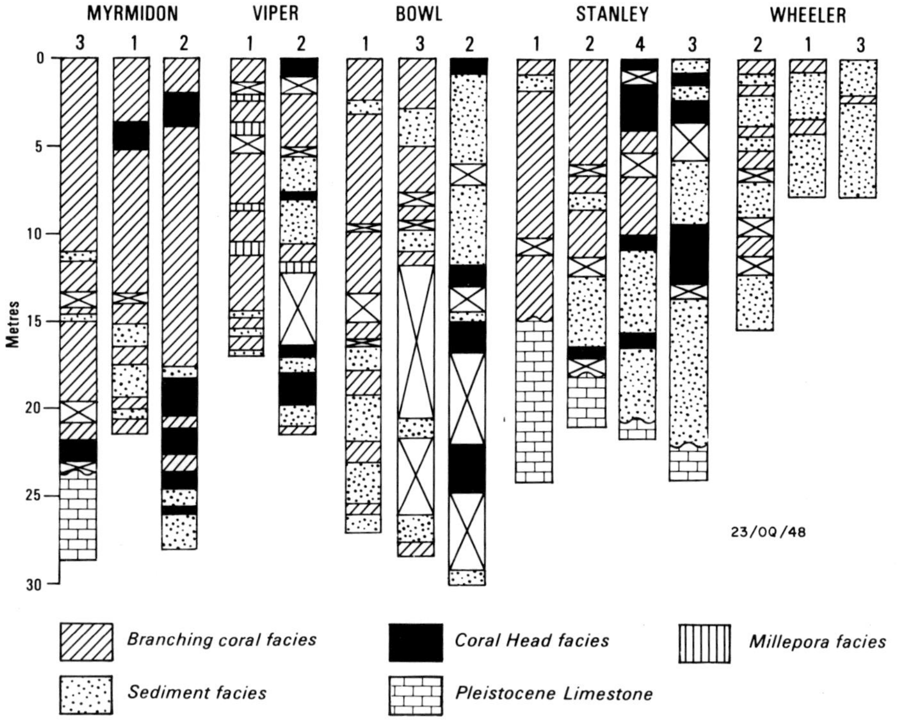

Fig. 3. Well logs from the five reefs discussed in the text, showing the distribution and cross-shelf variation of the facies

a predominance of sand-size unconsolidated sediments within the detrital assemblage.

One of the most obvious results of drilling in the central Great Barrier Reef is the decrease in the amount of framework across the shelf (Fig. 3). Reefs on the outer edge of the shelf, such as Myrmidon and Viper Reefs, are framework-dominated. This is particularly the case of their windward margins (holes 1 and 3 on Myrmidon and hole 1 on Viper; Fig. 3) where the framework, exclusive of cavities, accounts for 80 to 95% of the reef mass. Even the leeward margins of these reefs are framework-dominated. As one proceeds back across the reef tract, the amount of unconsolidated sediment in the reefs begins to increase significantly, but more particularly on the leeward margins (e.g., Bowl and Stanley Reefs; Fig. 3). On the inner edge of the reef tract (e.g., Wheeler Reef) the framework assemblage accounts for only 18 to 30% of the reef structure, the remainder being unconsolidated sediment. This paucity of framework in the inner reefs is also apparent in drill holes on Britomart and Davies Reefs (Johnson et al. 1984, Grimes 1982). Results from nearshore, fringing reefs indicate that many have a detrital core which, in some reefs, is terrigenous (Hopley et al. 1983).

The branching coral facies is the dominant framework assemblage in most reefs (Fig. 3). At Myrmidon Reef it is very distinctive, mainly because of the

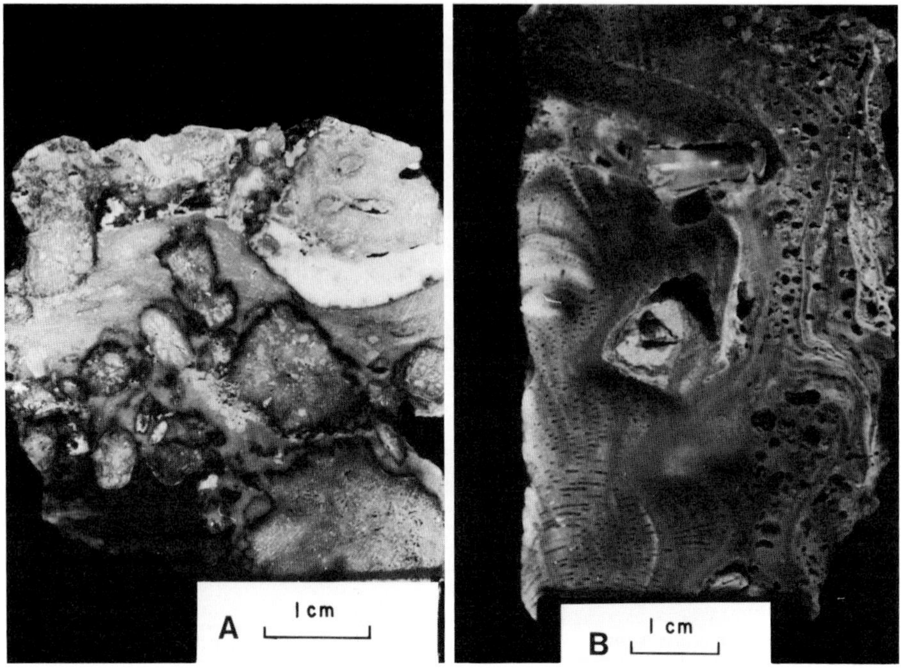

Fig. 4. A Section of core (Myrmidon Reef, hole 1) showing cross-sections of delicate branching corals, which have been bored and filled, and extensive interskeletal fill by cemented silt-size internal sediment. Note the dark iron oxide(?) coating around the periphery of the corals. **B** *Millepora* encrusted with coralline algae (Viper Reef, hole 1) that has been bored by bivalves (*upper centre*) and sponges (*right*)

delicate nature of many of the branching corals and the extensive fill by cemented internal sediments within the numerous 1 to 10 cm high cavities between the thicket-like branches (Fig. 4A). Most of the corals are species of *Acropora* whose branch diameter generally is less than 2 cm. Most branching corals have a thin (<1 mm) brown coating of unknown origin which, in cut sections of the core, outlines the coral branches against the extensive infill of the interskeletal cavities. A similar situation occurs at Viper Reef, where the branching coral facies, while being a slightly more massive framework, shows considerable filling and cementation within interskeletal cavities.

The branching coral facies in those reefs removed from the edge of the shelf, such as Bowl, Stanley and Wheeler Reefs, tends to form a more open framework than the shelf-edge reefs, and corals, such as *Acropora,* tend to be more massive with shorter, but thicker, branches. This type of framework can be very cavernous, with some holes attaining 3 m in height. In some drillhole sections cavities are volumetrically more abundant than framework (e.g., Bowl Reef holes 2 and 3; Fig. 3). This type of branching coral framework, particularly when compared to the outer reefs, has significantly less cemented interskeletal infill, probably a direct result of its open nature.

The coral head facies tends to have a fairly scattered distribution, although it is better developed beneath the leeward margins of the reefs, where it is often interspersed within the sediment facies (Fig. 3). This differs from the situation at One Tree Reef in the southern Great Barrier Reef, where the coral head facies dominates the windward margin of that reef (Marshall and Davies 1982). *Porites* is the dominant coral within this facies, and in places it forms single heads up to 2–3 m thick.

At Viper Reef a *Millepora* facies is interspersed with the branching coral facies (Figs. 3, 4B). *Millepora* is commonly encrusted by coralline algae and lithified crusts, and although not a significant framework-builder in the other reefs, it can be fairly common on the present-day reef slopes of these reefs.

The detrital assemblage which, from the drill logs, seems to be as important as the branching coral facies, consists mainly of gravel- and sand-size skeletal fragments. Very little unconsolidated lime mud is evident in the reefs, although Grimes (1982) and Johnson et al. (1984) have reported muddy matrices within their rudstone units in Davies and Britomart Reefs and Hopley et al. (1983) found similar units within the fringing reef at Rattlesnake Island.

The sediment facies consists of the broken skeletons of reef organisms, in particular corals, coralline algae, molluscs, *Halimeda,* foraminifera, and echinoids. At Wheeler Reef unconsolidated sediments predominate over framework to such an extent that it appears that very little in-situ framework exists within much of this reef (Fig. 3). Only on the present reef slope is there any abundance of frame-building organisms. This suggests that in this type of reef environment the framework is ultimately degraded, but the actual cause of degradation is unknown.

3 Submarine Cements and Their Distribution

Both aragonite and Mg-calcite, the only cement mineralogies within the Holocene section of these reefs, show a diversity of textures and fabrics. Aragonite cement commonly occurs as acicular, mesh and micrite forms, whereas Mg-calcite cements predominantly occur as micrite, peloids and bladed spar. These cement textures and fabrics are the same as those reported from other reefs, both in the Caribbean and elsewhere in the Great Barrier Reef (e.g., Macintyre 1977, James and Ginsburg 1979, Land and Moore 1980, Marshall and Davies 1981, Marshall 1983 a). In the central Great Barrier Reef, all these types of cement may be present as both intraskeletal and interskeletal fillings.

Mg-calcite and aragonite micrite are microcrystalline cements (<4 µm) that occur predominantly within the interstices of silt-size internal sediment fillings. Aragonite mesh cement (Fig. 5A) consists of randomly orientated, needle-like crystals, 10–40 µm long and less than 6 µm wide, which have nucleated on silt-size particles within mudstones and packstones to form a fairly open mesh within the intergranular pores.

Peloids are a particular form of Mg-calcite micrite (0.5–2.0 µm size) that form spherical or subspherical aggregates that usually range in diameter between 20 and 60 µm (Fig. 5B). In the past, peloids have been variably described as fecal pellets, calcified algal filaments, detrital grains, and submarine cement (see

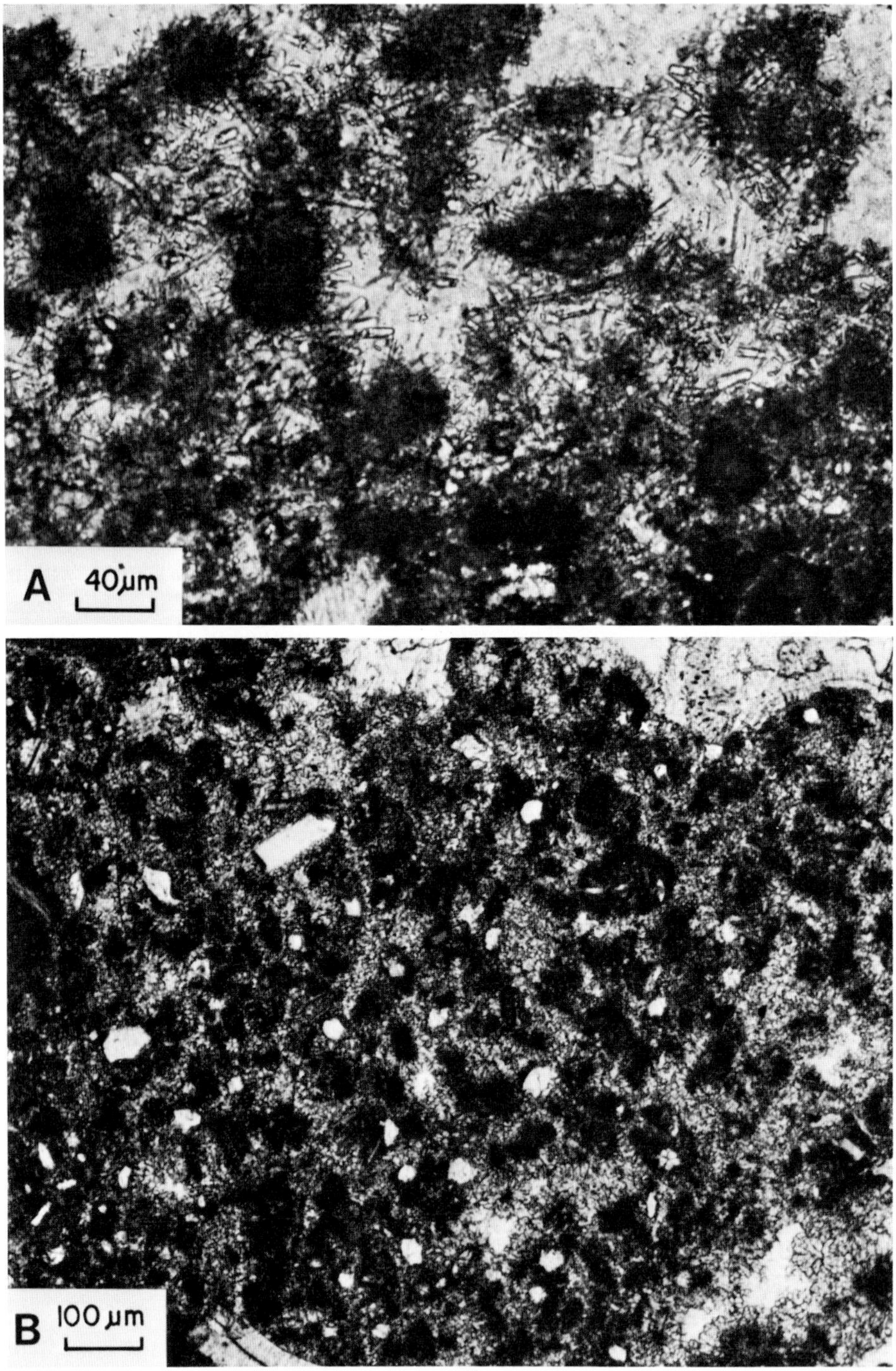

Fig. 5. A Photomicrograph showing needles of aragonite mesh cement developed on silt-size nuclei that partly fill interparticle pore space. Plane polarized light. **B** Photomicrograph of silt-size peloids consisting of microcrystalline Mg-calcite with rims of crystalline Mg-calcite cement that have filled interpeloidal pores. Plane polarized light

Macintyre 1985 for discussion). It is Macintyre's (1977, 1985) contention that peloids are a precipitate and are, as such, a particular form of submarine cement. This origin has been supported by Marshall and Davies (1981) and Marshall (1983 a). However, disagreement is centered on whether peloids are a product of repeated nucleation (Macintyre 1984, 1985) or are a relatively spontaneous nucleation (Marshall 1983 a). Both mechanisms agree that growth ceases after a critical size range is reached. Peloids are often rimmed by a later stage of more crystalline Mg-calcite cement. These rim cements are of the order of 5 to 40 μm in length, and may completely occlude the remaining pore space in relatively dense peloidal fillings (Fig. 5 B).

In the central Great Barrier Reef, both acicular aragonite and bladed Mg-calcite spar cements appear to form at an early and at a relatively late stage during the lithification process. The early cements are found lining or filling coral pores and other skeletal chambers that may be followed by micrite or mesh cemented internal sediments. The later cements are more coarsely crystalline than their predecessors, and they line or fill either secondary borings or the remaining space within partly filled (usually geopetal) cavities.

3.1 Myrmidon Reef

At Myrmidon Reef the interskeletal filling of the numerous 1 to 10 cm high cavities is of two types: (1) a skeletal wackestone or packstone (Fig. 6 A), and (2) a finely laminated mudstone (Fig. 6 B – D). The wackestone/packstone, whose constituents ranges from pebble- to silt size, mainly fills the larger (>2 cm) cavities. The larger skeletal fragments include corals, coralline algae, molluscs, foraminifera, *Halimeda* and echinoids. Both the coarser grains and the silt fraction (predominantly sponge chips) are cemented to a varying degree by all the cement types mentioned previously. However, Mg-calcite cements predominate; bladed spar occurs as rims about 30 μm thick around sand-size grains and as linings, up to 1.4 mm thick, on the walls of smaller cavities within the infill (Fig. 7 C).

The laminated mudstone, which was deposited after the wackestone/packstone, is similar to that described by James and Ginsburg (1979) from the fore-reef slope of Belize. The laminae range from 0.1 to 3.0 mm in thickness and they appear to have been produced as a result of a combination of contrasting textures and composition, the degree of packing, and differences in cement mineralogy (Fig. 6 C). Both aragonite (mesh and micrite) and Mg-calcite (micrite and peloids) cements are present in the mudstones. Commonly, aragonite cement occurs near the basal portions of the fill and Mg-calcite towards the top, but they may be frequently interlayered. A normal sequence would consist of silt-size internal sediments cemented by aragonite followed by silt-size internal sediments cemented by Mg-calcite (with interlayering between the two). The final fill often consists of Mg-calcite peloids with relatively little internal sediment particles. The peloids commonly have 10 – 30 μm pore-filling rim cements (Fig. 5 B). A distinctive type of laminated mudstone are *laminated peloidal mudstones* (Fig. 6 D). Up to 80% of the bulk of these mudstones consist of Mg-calcite peloids, the remainder being

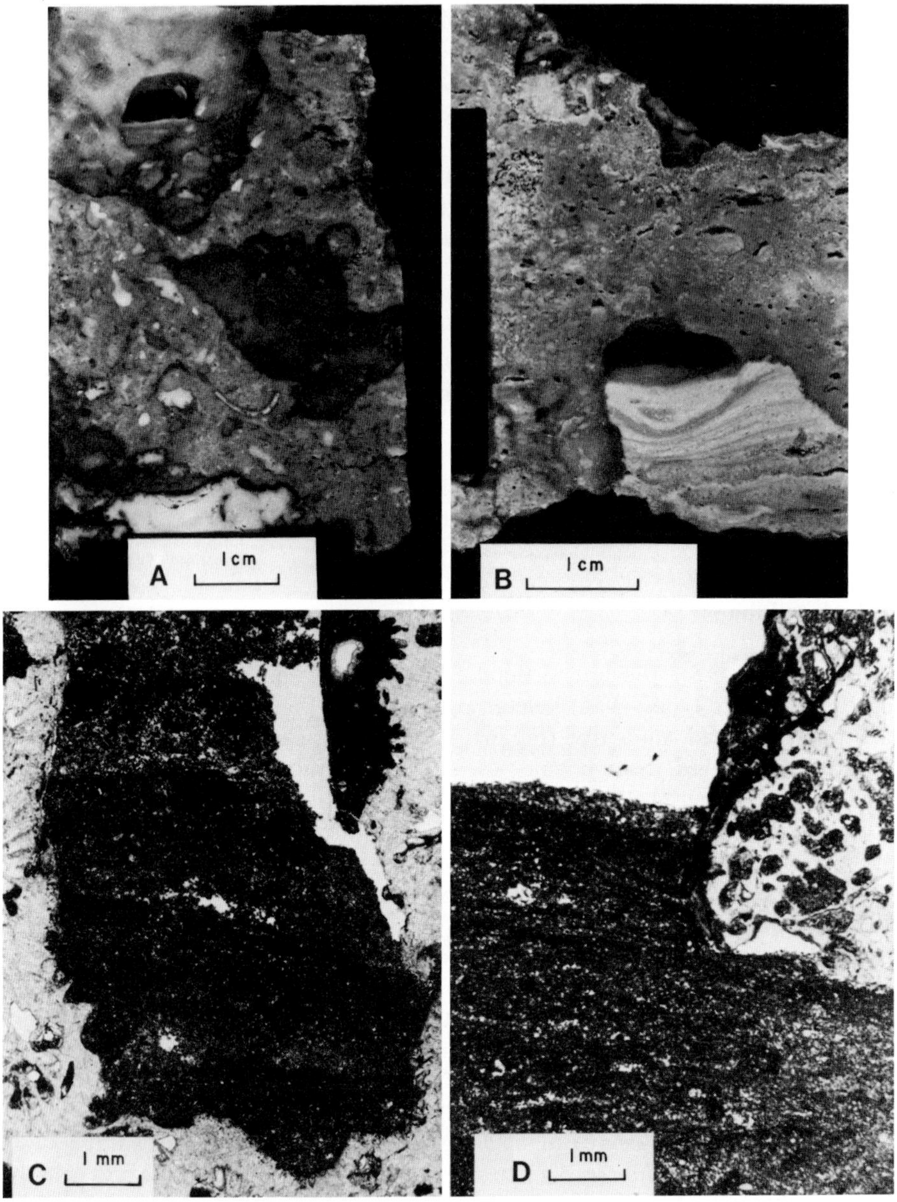

Fig. 6. A Cavity-filling packstone between branching coral framework (Myrmidon Reef, hole 1). Upper coral has been bored and filled or partly filled (geopetal) with laminated mudstone. **B** Laminated mudstone that has filled a cavity and part of a large boring (Myrmidon Reef, hole 2). The lighter-coloured layers are cemented by aragonite, whereas the darker, thinner layers are cemented by Mg-calcite. **C** Photomicrograph of a laminated mudstone filling (Myrmidon Reef, hole 1). Laminations in this example are the result of packing and the degree of cementation of the predominantly silt-size infill. Plane polarized light. **D** Photomicrograph of a laminated peloidal mudstone partly filling a small cavity between the branching coral framework (Myrmidon Reef, hole 1). Plane polarized light

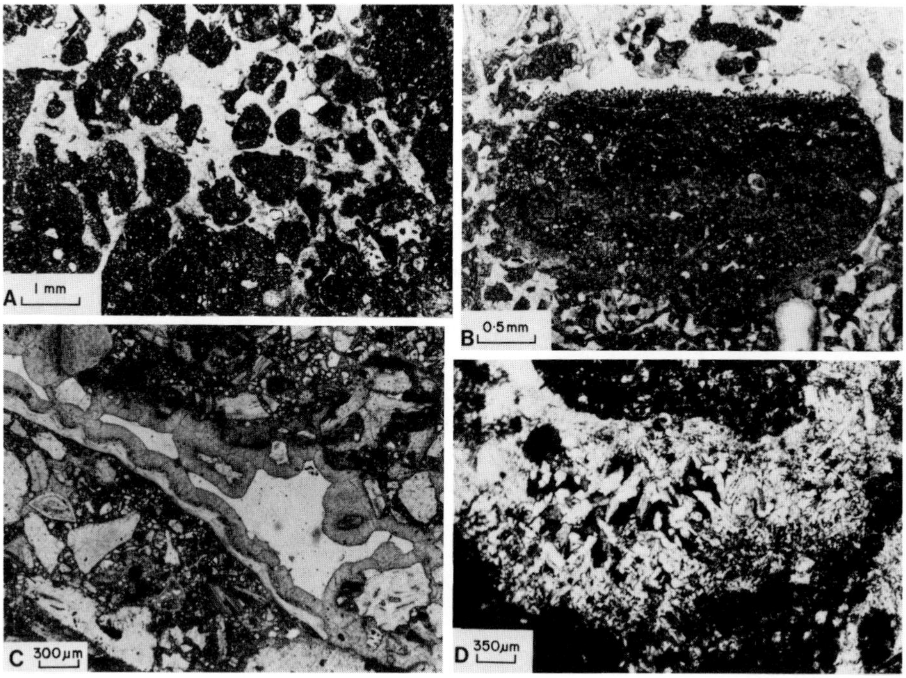

Fig. 7. A Photomicrograph showing extensive sponge boring and filling by Mg-calcite micrite cemented internal sediment, which have almost obliterated the original coral structure (Myrmidon Reef, hole 2). Plane polarized light. **B** Photomicrograph of a large boring with silt-size internal sediment that is cemented alternatively by aragonite and Mg-calcite (Myrmidon Reef, hole 1). Plane polarized light. **C** Photomicrograph of interskeletal packstone/wackestone showing lining of cavities by relatively thick rims of late stage bladed Mg-calcite spar cement (Myrmidon Reef, hole 3). Plane polarized light. **D** Photomicrograph of a secondary boring that is partly filled with a late-stage, coarsely acicular aragonite cement (Myrmidon Reef, hole 2). Crossed polarizers

silt and some sand-size internal sediments. Laminations in this type of fill have developed as a result of the size, sorting, porosity and the degree of cementation of the peloids in the various layers. Peloids of 40 – 200 μm diameter having 30-μm-size rim cements that have filled almost all original porosity alternate with 10 – 70 μm diameter peloids with only 4 – 6 μm rim cements and high interpeloidal porosity. No aragonite cement is present in this type of mudstone.

At Myrmidon Reef there has been significant alteration of the original framework as a result of extensive boring, and filling of both skeletal pores and borings by silt-size internal sediments and cements (Fig. 7A). The branching coral framework has been bored mainly by sponges and, to a lesser extent, by polychaetes and bivalves. Sponge borings have produced cavities measuring 0.05 to 5.00 mm in diameter. While many borings have only one type of fill, others shows successive generations, each generation having a different texture or mineralogy (Fig. 7B). The normal succession is basically that described by Marshall (1983a) with (1) aragonite silt; (2) silt-size internal sediments in Mg-calcite or aragonite micrite; (3) peloids with crystalline rim cements; and (4) either coarsely acicular aragonite or relatively wide (>100 μm) rims or mammelons of bladed Mg-calcite spar. Some of the larger borings contain geopetal infills of lime mud exhibiting alternating Mg-calcite and aragonite cements, similar to the laminated mudstones (Fig. 7B).

In places, second-generation borings, which affect both the original framework and filled primary borings, and partly filled cavities are the sites for the precipitation of coarsely acicular aragonite cement and/or thick rims of bladed Mg-calcite spar (Fig. 7C, D). The coarsely acicular aragonite forms individual crystals up to 300 μm in length and 30 μm in width (Fig. 7D). In cross or oblique section the aragonite cement resembles "sparite" and appears to be similar to coarsely crystalline aragonite or aragonite "spar" described by Land and Moore (1980) from the fore-reef slope of Jamaica. At Myrmidon Reef this type of cement definitely has not formed by replacement, as it grows into cavities.

Coral pores and other skeletal chambers, such as encrusting bryozoans, foraminifera and vermetid gastropods, all exhibit frequent filling by silt-size internal sediment and cements. Coral pores include all types of submarine cement mentioned previously. Some corals also contain fillings of fine "blocky" aragonite cement similar to that described by James and Ginsburg (1979) and Marshall (1983a).

3.2 Viper Reef

The extent of submarine cementation at Viper Reef is slightly less than at Myrmidon Reef. Interskeletal cavity fillings are dominated by lithified crusts, and boring and pore fill is less abundant. A feature not evident at Myrmidon Reef, but which is fairly common at Viper Reef, relates to the development of lithified crusts. These Mg-calcite, peloid-dominated crusts are medium-grey in colour and vary in thickness between 1 and 50 mm (Fig. 8A). The lithified crusts are similar to those described previously by Land and Goreau (1970), Macintyre (1977), James and Ginsburg (1979), Land and Moore (1980), and Marshall (1983a). The lithified crusts essentially consist of two morphologies: (1) smooth and (2) columnar. The smooth crusts (Fig. 8A) are similar to those described by Land (1971) with very little internal structure such as laminations (e.g., Marshall 1983a). While there is sometimes a hint of lamination or mottling, the majority are homogeneous. The columnar crusts are similar to those described by Marshall (1983a) and Land's (1971) "knobby crusts". They are usually only 1 – 2 mm high, but in places have formed build-ups of 1 cm thickness. The columnar crusts have sometimes developed on top of smooth crusts. At the base of both types of crust there is often a dark band, up to 2 mm thick, which is

Fig. 8. A Section of core showing smooth lithified crust developed on top of an extensively cement filled coral. Note thin dark coating at base of lithified crust (Viper Reef, hole 1). **B** Core section of a framework of encrusted branching corals that has extensive filling by cemented internal sediments within cavities (Viper Reef, hole 1). **C** Core section from the windward margin of Stanley Reef (hole 2) showing extensive alteration of the branching coral framework and the development of columnar lithified crusts (*upper left* and *lower right*). **D** Slightly domal, laminated lithified crust (Stanley Reef, hole 2)

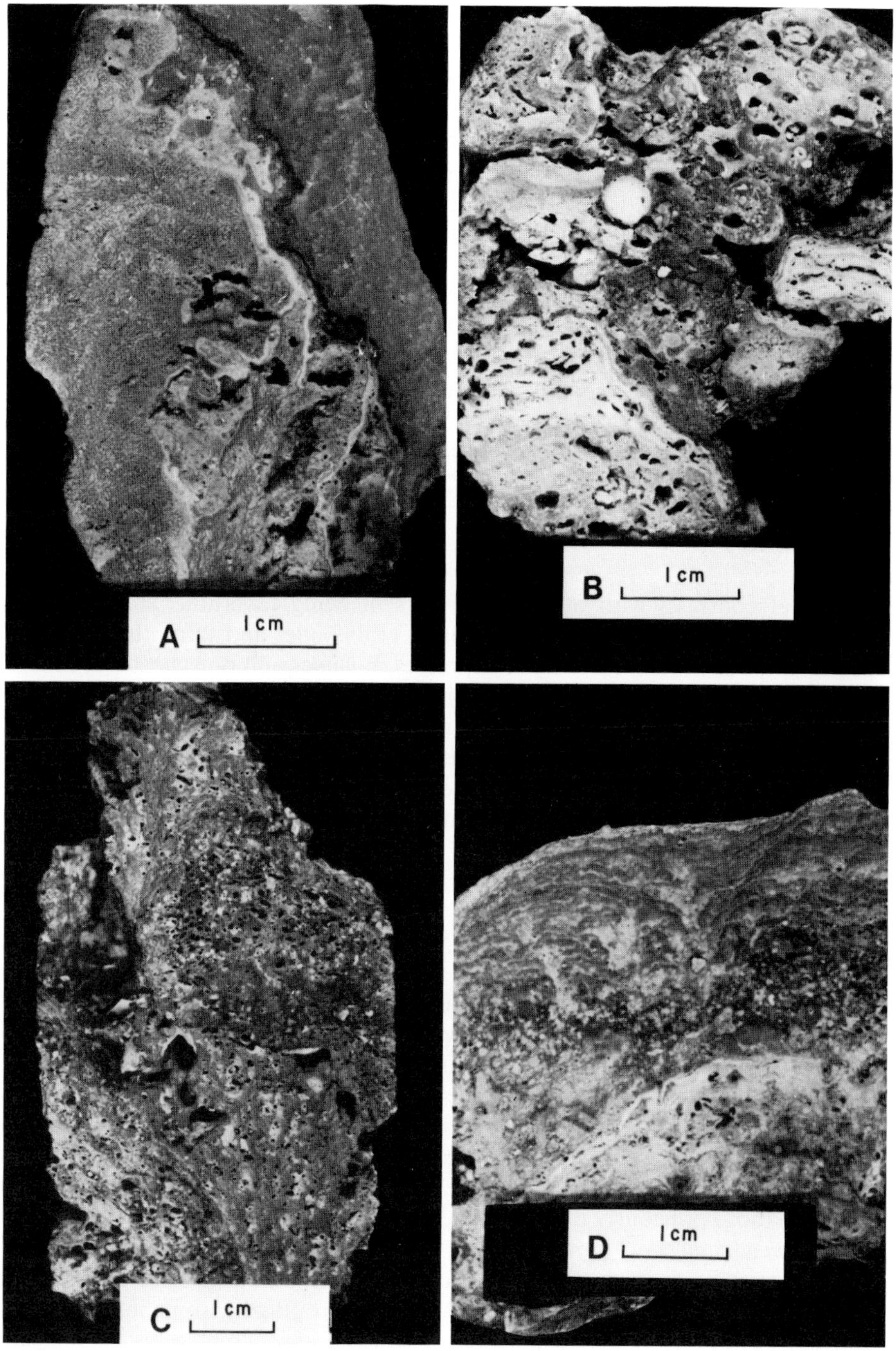

Fig. 8

probably an iron/manganese coating similar to those described by James and Ginsburg (1979) and Land and Moore (1980).

The distribution of lithified crusts shows them to be ubiquitous in hole 1, where they line cavities in both the branching coral and *Millepora* facies. However, in the more leeward hole (No. 2), lithified crusts are present only within the branching coral facies between 3.5 and 5.0 m (Fig. 3).

The framework elements are modified by boring, internal sediment filling and cementation (Fig. 8 B), but this is not as intense as at Myrmidon Reef. Boring organisms include sponges, barnacles, bivalves and polychaetes, with sponges being predominant. Sponge borings range in diameter from 0.5 to 10 mm, and while many are empty, others are filled or partly filled (geopetal) with silt-size internal sediments and cements (predominantly Mg-calcite). Some of the larger sponge borings contain laminated mud fills that have alternate aragonite and Mg-calcite cements. Coral pores and chambers of other skeletal organisms are similarly variably filled with internal sediment and cement.

3.3 Bowl Reef

Bowl Reef contrasts with other outer reefs in being cavernous and relatively devoid of submarine cement. Cavity-filling grainstones and thin lithified crusts are present, but are rare. Borings by sponges, bivalves and polychaetes are more often empty than filled. Cores penetrating massive corals indicate that the barnacles present are nestlers rather than borers (Bromley 1978). The base of the barnacle-created cavities is small and then increases in size upwards. There is no obvious access from the exterior of the coral, which would explain why these distinctly striated cavities, (3 – 8 mm in diameter and up to 30 mm in length) generally remain unfilled.

3.4 Stanley Reef

Situated towards the inner edge of the reef tract, Stanley Reef shows contrasting amounts of submarine cementation between windward and leeward margins. The two holes drilled on the windward margin (holes 1 and 2; Fig. 2), while containing some interskeletal cemented internal sediments, have many cavities lined or partly filled with lithified crusts. These Mg-calcite, peloidal-rich crusts vary in thickness between 0.1 and 5.0 cm. Three types of lithified crust are present: (1) smooth (2) laminated and (3) columnar. The smooth and columnar crusts are similar to those previously described at Viper Reef. The smooth crusts generally are relatively thin (1 – 3 mm) and tend to line cavities or coat coral branches, whereas the columnar crusts can be up to 5 cm thick. The latter often are mottled where roofed-over cavities between the columns are filled with lighter-coloured lime mudstone (Fig. 8 C). The laminated lithified crusts, which generally vary in thickness between 1 and 4 cm, are similar to crusts described from Jamaica (Land and Moore 1980) and the southern Great Barrier Reef (Marshall 1983 a). They commonly have a prominent dark coating at their base, comparable to the iron-manganese coatings described by James and Ginsburg (1979) and Land and Moore (1980). The crusts often are domal in shape with wavy, often discontinuous laminations (Fig. 8 D). Columnar crusts may develop on top of them.

The corals in the windward margin holes show extensive boring, particularly by sponges and, to a lesser extent, bivalves. The periphery of massive corals and the interior of branching corals often are riddled by sponge borings measuring 0.5 to 10.0 mm in diameter. Most borings are filled entirely or partly (geopetal) with silt-size internal sediments and a variety of aragonite and Mg-calcite cements. Some borings contain internal sediments that remain uncemented.

In contrast, the leeward margin holes (3 and 4; Fig. 2) are relatively devoid of cements. Some 1 – 5-mm-thick columnar lithified crusts have coated cavities, but there is virtually no interskeletal cementation in these holes. Similarly, while both sponge and bivalve borings are reasonably numerous, the majority are either empty or filled with uncemented internal sediments.

3.5 Wheeler Reef

At Wheeler Reef, apart from the rare, thin lithified crust and some cemented boring infill, there is no sign of cementation within the drill holes. Most of the recovered framework shows extensive boring by sponges, bivalves and polychaetes, but almost all borings are empty. Some encrustation and cementation is apparent within the top 10 cm of hole 2 on the windward margin.

3.6 Regional Distribution

The preceding data indicate that, in general, the degree of submarine cementation decreases from the outer to the inner shelf. Both Myrmidon and Viper Reefs, at the very edge of the shelf, contain considerable quantities of inter- and intraskeletal cements and cemented internal sediment fillings. At Myrmidon Reef the amount of cemented fill is so abundant in places that it is volumetrically equal to the framework. The larger reefs, such as Bowl near the shelf edge and Stanley towards the inner edge, do contain submarine cements, particularly within their windward margins, but the amount of cementation has decreased appreciably relative to the outer reefs. Protected inner tract reefs, such as Wheeler, show virtually no evidence of interskeletal cementation; most cements are restricted to a limited amount of passive fill. A similar situation exists at Britomart Reef (Johnson et al. 1984), where cements are limited to minor aragonite within coral pores.

In general, all reefs or parts of reefs which have a substantial amount of submarine cement show a similar style of facies development. This consists of a branching framework that contains numerous, relatively small (1 – 20 cm) interskeletal cavities. Most of these cavities have become the sites of fine-grained internal sediment deposition. In contrast, reefs or parts of reefs that have a more open framework or are dominated by detrital facies show relatively little cementation. This relationship between facies and cements has been noted previously from other regions (James et al. 1976, Marshall 1983a). However, while specific facies appear to exert a strong control on cementation, it is equally apparent that these same facies show variations in the degree of cementation across the shelf.

On closer inspection it is apparent that this cross-shelf trend is not necessarily uniform, an obvious example being the difference in cementation between Bowl and Stanley Reefs. Bowl, which is located towards the outer edge of the reef tract, shows less cementation than Stanley. While facies control could explain this anomaly (Bowl appears to be more cavernous than Stanley), closer examination reveals that parts of each reef, particularly beneath their windward margins, show similar branching coral facies, but it is only at Stanley that it has become filled by internal sediment and cemented.

Another explanation for this departure from the cross-shelf trend could be variations in the depth of the respective reef tops during the Holocene transgression, when most of the cements were being precipitated within the framework. Analysis of reef growth and sea level during the Holocene (Davies et al. 1985) indicates that Bowl Reef was some 15 to 22 m below sea level during the period 6000 – 8000 years B.P. and did not catch up to sea level until 3500 years B.P. (Fig. 9). This contrasts with the windward margin of Stanley Reef which reached

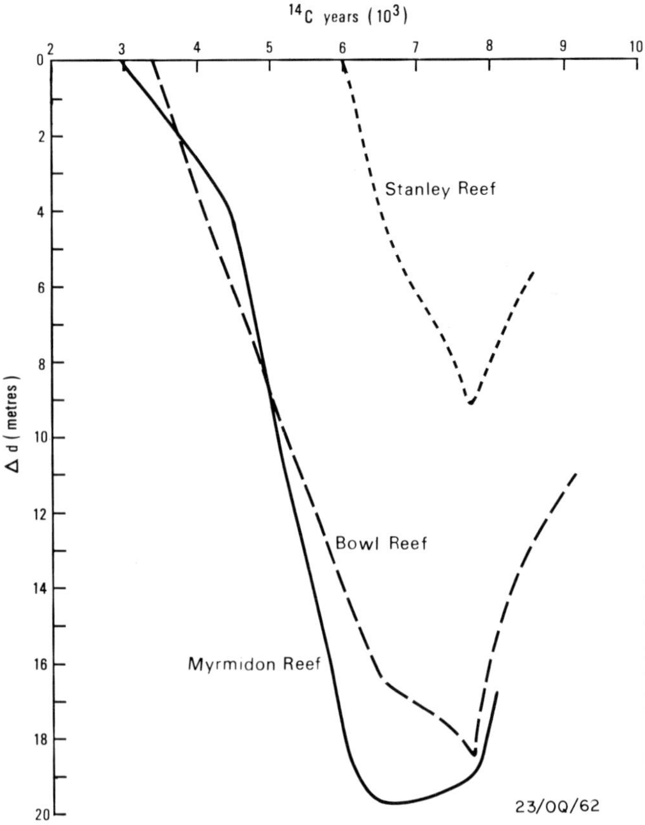

Fig. 9. Diagram showing the variation in depth between the top of the reef and sea level during the Holocene. Depth variation was plotted from the sea level curve of Thom and Roy (1983) and representative reef growth curves (see Davies et al. 1985) for the windward margin of each reef

a maximum of 9 m below sea level during the same period (Fig. 9), and which reached the stabilized sea level soon after (6000 years B.P.). Both Myrmidon and Viper Reefs were also some 15 to 20 m below sea level between 6000 and 8000 years B.P. (Fig. 9), but it has been postulated (Marshall 1985) that upwelling, resulting from internal tides and Kelvin waves, along the shelf edge in this area provided a suitable pumping mechanism for the precipitation of a substantial amount of submarine cement within these reefs. It is suggested, however, that the upwelling did not extend far enough across the outer shelf to affect Bowl Reef. This is consistent with the present-day oceanographic situation (Wolanski and Pickard 1983).

4 Discussion

While the actual mechanism for the precipitation of submarine cements is not fully understood, particularly with respect to biological as opposed to physio-chemical processes (Marshall 1983 b), it is accepted that high energy conditions are required in order to pump large volumes of seawater through the porous system of the reef so as to precipitate significant quantities of calcium carbonate. This should be a prime controlling factor with respect to the distribution of submarine cements, both within individual reefs and within entire reef systems. Previously noted cross-shelf trends of submarine cementation have been restricted to one region, i.e., Belize (James et al. 1976). A preliminary analysis of cores from the northern Great Barrier Reef indicates that a situation similar to Belize exists: the outer ribbon reefs are well cemented, but there appears to be relatively little submarine cement in the inner and mid-shelf reefs. However, the situation in the central Great Barrier Reef is different from that at Belize and the northern Great Barrier Reef. Here reefs do not form a semi-continuous outer barrier, but constitute a reasonably dense patchwork of platform reefs scattered over the outer and mid-shelf region (Fig. 1).

In the central Great Barrier Reef, although high energy conditions are not limited to the outermost reefs, it is not this present-day hydrologic regime that has influenced cementation within the bulk of the Holocene reefs. Rather, conditions during the vertical stage of reef growth have determined cementation. This is based on the contention that submarine cementation is contemporaneous with reef growth, as indicated by numerous studies (e.g., Land and Goreau 1970, Ginsburg et al. 1971, James et al. 1976, Macintyre 1977, Focke and Gebelein 1978, Land and Moore 1980, Marshall 1983 a).

The decrease in submarine cementation from the outer to the inner reefs can be attributed to several inter-related factors such as decrease in energy and increase in the proportion of unconsolidated sediments relative to framework. In addition the varying degree of submarine cementation between the reefs can be explained by a *reef growth factor,* which relates to the position of the reef with respect to sea level during the transgressive or vertical stage of reef growth. Reefs that maintained themselves close to sea level, or at least above wave base, during this stage experienced continually higher energy conditions than reefs which were drowned during the same period, and which caught up subsequent to sea level stabilization. The examples of these two styles of growth are Stanley and Bowl Reefs respectively. However, because this reef growth factor concerns only *shallow* high energy conditions relating to wave or tidal pumping, further complications can arise. Reefs such as Myrmidon and Viper have growth histories similar to that of Bowl Reef; during the transgressive phase these reefs were progressively drowned, and did not begin to catch up to sea level until this became stabilized. It would appear that some shelf-edge process, possibly upwelling, wave surge from long-period oceanic swells or an as yet unexplained mechanism, caused sufficient quantities of seawater to be pumped through the outer reefs' framework, which ultimately led to the formation of significant quantities of submarine cement.

These observations indicate that in trying to predict the distribution of submarine cements, either within a single reef or within an entire reef system, knowledge of the growth history of the reef is necessary. Because reefs have different growth histories, any analysis of the energy-related factors that produce submarine cementation and, ultimately, the regional distribution of submarine cements, tends to be complicated. This complication can be compounded in ancient reefs where the growth history, particularly with respect to sea level, usually is unknown. While the simple Belize model may be applicable to ancient barrier reef systems, it does not take into account the possibility that inner reefs may have been able to maintain themselves at a higher elevation than the outer barrier reefs during the transgressive stage, and thus be affected by submarine cementation. This situation is evident to some extent in the northern Great Barrier Reef. Furthermore, well-lithified mounds could have been cemented in a relatively deep submarine environment if water movement was related to oceanographic conditions other than wave or tidal pumping. For these reasons it may be unwise to place too much importance on the distribution of submarine cements in ancient reefs as a method for determining outer as opposed to inner shelf environments or shallow as opposed to deep reef environments.

5 Conclusions

Analysis of cores from 15 shallow holes drilled on five Holocene reefs within the central Great Barrier Reef show the following:

1. Reefs situated on the outer edge are framework dominated, but the amount of framework decreases towards the inner edge of the reef tract where reefs are sediment dominated.
2. Co-incident with this change in facies there is a decrease in the degree of submarine cementation within the framework facies across the shelf. However, submarine cementation is not restricted to shelf-edge reefs. Unlike Belize, there is a relatively substantial amount of cementation within the windward margin of some inner reefs.
3. While it can be maintained that high energy conditions are an important factor in the cementation process, these conditions may be deeper than the

often invoked mechanisms of wave and tidal pumping. On the outer shelf of the central Great Barrier Reef one possible mechanism involves currents created by upwelling.

4. In ancient reefs it is doubtful whether the regional distribution of submarine cements can be used with reliability as a predictive tool for determining depositional environments.

References

Bromley RG (1978) Bioerosion of Bermuda reefs. Palaeogeogr Palaeoclimatol Palaeoecol 23:169–197

Davies GR (1977) Former magnesian calcite and aragonite cements in Upper Paleozoic reefs of the Canadian Arctic: a summary. Geology (Boulder) 5:11–15

Davies PJ, Hopley D (1983) Growth facies and growth rates of Holocene reefs in the Great Barrier Reef. BMR J Aust Geol Geophys 8:237–251

Davies PJ, Marshall JF, Hopley D (1985) Relations between sea level and reef growth in the central and northern Great Barrier Reef. Proc 5th Int Coral Reef Symp, Tahiti 3:95–103

Focke JW, Gebelein CD (1978) Marine lithification of reef rock and rhodolites at a forle-reef slope locality (−50 m) off Bermuda. Geol Mijnbouw 57:163–171

Ginsburg RN, James NP (1976) Submarine botryoidal aragonite in Holocene reef limestone, Belize. Geology (Boulder) 4:431–436

Ginsburg RN, Shinn EA, Schroeder JH (1967) Submarine cementation and internal sedimentation within Bermuda reefs. Geol Soc Am Spec Pap 115:78–79

Ginsburg RN, Marszalek DS, Schneidermann N (1971) Ultrastructure of carbonate cements in a Holocene algal reef of Bermuda. J Sediment Petrol 41:472–482

Grimes D (1982) Drill-site report, AIMS Davies No 1. Geol Surv Qld Record 1982/18 (unpubl)

Hopley D, Slocombe AM, Muir F, Grant C (1983) Nearshore fringing reefs in north Queensland. Coral Reefs 1:151–160

James NP, Ginsburg RN (1979) The seaward margin of Belize barrier and atoll reefs. Int Assoc Sediment Spec Publ 3:1–191

James NP, Kobluk DR (1978) Lower Cambrian patch reefs and associated sediments: southern Labrador, Canada. Sedimentology 25:1–35

James NP, Ginsburg RN, Marszalek DS, Choquette PW (1976) Facies and fabric specificity of early subsea cements in shallow Belize (British Honduras) reefs. J Sediment Petrol 46:523–544

Johnson D, Cuff C, Rhodes E (1984) Holocene reef sequences and geochemistry, Britomart Reef, central Great Barrier Reef, Australia. Sedimentology 31:515–529

Land LS (1971) Submarine lithification of Jamaican reefs. In: Bricker OP (ed) Carbonate cements. Johns Hopkins Stud Geol 19:59–62

Land LS, Goreau TF (1970) Submarine lithification of Jamaican reefs. J Sediment Petrol 40:457–462

Land LS, Moore CH (1980) Lithification, micritization and syndepositional diagenesis of biolithites on the Jamaican island slope. J Sediment Petrol 50:357–370

Lohmann KC, Meyers WJ (1977) Microdolomite inclusions in cloudy prismatic calcites – a proposed criterion for former high-magnesium calcites. J Sediment Petrol 47:1078–1088

Macintyre IG (1977) Distribution of submarine cements in a modern Caribbean fringing reef, Galeta Point, Panama. J Sediment Petrol 47:503–516

Macintyre IG (1984) Extensive submarine lithification in a cave in the Belize barrier reef platform. J Sediment Petrol 54:11–25

Macintyre IG (1985) Submarine cements – the peloidal question. In: Schneidermann N, Harris PM (eds) Carbonate cements. Soc Econ Paleontol Miner Spec Publ 36:109–116

Macintyre IG, Mountjoy EW, D'Anglejan BF (1968) An occurrence of submarine cementation of carbonate sediments off the west coast of Barbados, W.I. J Sediment Petrol 38:660–664

Marshall JF (1983 a) Submarine cementation in a high-energy platform reef: One Tree Reef, southern Great Barrier Reef. J Sediment Petrol 53:1133 – 1149
Marshall JF (1983 b) Marine lithification in coral reefs. In: Barnes DJ (ed) Perspectives on coral reefs. Clouston, Canberra, pp 231 – 239
Marshall JF (1985) Cross-shelf and facies related variations in submarine cementation in the central Great Barrier Reef. Proc 5th Int Coral Reef Symp, Tahiti 3:221 – 226
Marshall JF, Davies PJ (1981) Submarine lithification on windward reef slopes: Capricorn-Bunker Group, southern Great Barrier Reef. J Sediment Petrol 51:953 – 960
Marshall JF, Davies PJ (1982) Internal structure and Holocene evolution of One Tree Reef, southern Great Barrier Reef. Coral Reefs 1:21 – 28
Marshall JF, Davies PJ (1984) Facies variation and Holocene reef growth in the southern Great Barrier Reef. In: Thom BG (ed) Coastal geomorphology in Australia. Academic Press, Sydney, pp 123 – 134
Mazzullo SJ (1980) Calcite pseudospar replacive of marine acicular aragonite and implications for aragonite cement diagenesis. J Sediment Petrol 50:409 – 423
Mazzullo SJ, Cys JM (1979) Marine aragonite sea-floor growths and cements in Permian phylloid algal mounds, Sacramento Mountains, New Mexico. J Sediment Petrol 49:917 – 936
Playford PE (1980) Devonian "Great Barrier Reef" of Canning Basin, Western Australia. Am Assoc Petrol Geol Bull 64:814 – 840
Thom BG, Roy PS (1983) Sea level change in New South Wales over the past 15 000 years. In: Hopley D (ed) Australian sea levels in the last 15 000 years. A review. Monograph Series, Geogr Dept James Cook Univ Occasional Paper 3, pp 64 – 84
Walls RA, Mountjoy EW, Fritz P (1979) Isotopic composition and diagenetic history of carbonate cements in Devonian Golden Spike reef, Alberta, Canada. Geol Soc Am Bull 90:963 – 982
Wolanski E, Pickard GL (1983) Upwelling by internal tides and Kelvin waves at the continental shelf break on the Great Barrier Reef. Aust J Mar Freshwater Res 34:65 – 80

Model of Reef Diagenesis: Mururoa Atoll, French Polynesia

D. M. AISSAOUI[1], D. BUIGUES[2], and B. H. PURSER[1]

1 Introduction

Located within the Tuamotu Archipelago (Fig. 1), the Mururoa Atoll is characterized by an elongate morphology (25 × 9 km) which probably reflects the composite nature of its volcanic basement (Gachon 1985) situated at minimum depths of 180 m below lagoonal sea level and at depths varying between 350 and 450 m below peripheral islands (Buigues 1982). Radiometric datings (Labeyrie et al. 1969) indicate that volcanic activity ceased about 7 million years ago, approximately at the end of the Miocene. The considerable thickness of reefoid carbonates has been determined essentially by sustained Neogene subsidence of the Darwin ridge (Brousse 1985).

The sedimentary pile, consisting of pure carbonate down to depths of at least 450 m in peripheral wells, includes an annular lense of dolomite (Fig. 1 B). This important diagenetic body has a maximum thickness of about 320 m in peripheral wells, thinning to zero within the center of the atoll. Recent wells deviated laterally through the flanks of the atoll suggest that this body also disappears towards the oceanic limits of the platform (Purser and Aïssaoui 1985). The lensoid dolomite mass, for reasons discussed in the following text, extends from near volcanic basement to minimum depths of −130 m.

1.1 Previous Work

Initial investigations concerning surface morphology and ecology of the living reefs were carried out by Deneufbourg (1969) and by Chevalier et al. (1969). Following completion of the first well ("Colette") in 1964, systematic sampling by Fontes et al. (1969) demonstrated important differences between the isotopic (O and C) values of the dolomites and the overlying limestones, as discussed subsequently. The first detailed petrographic description of the sedimentary sequence at Mururoa by Repellin (1977) was limited to a single well ("Colette") located on the N periphery of the atoll.

A two-dimensional evaluation of the sedimentary and diagenetic facies at Mururoa Atoll by Buigues (1982) included wells located both on the periphery and in the lagoon. It demonstrated the lenticular nature and lateral variations

[1] Laboratoire de Pétrologie Sédimentaire et Paléontologie, Bâtiment 504, Université de Paris-Sud, 91405 Orsay Cedex, France
[2] Commissariat à l'Energie Atomique, 91680 Bruyères-le-Châtel, France

Reef Diagenesis
Edited by J. H. Schroeder and B. H. Purser

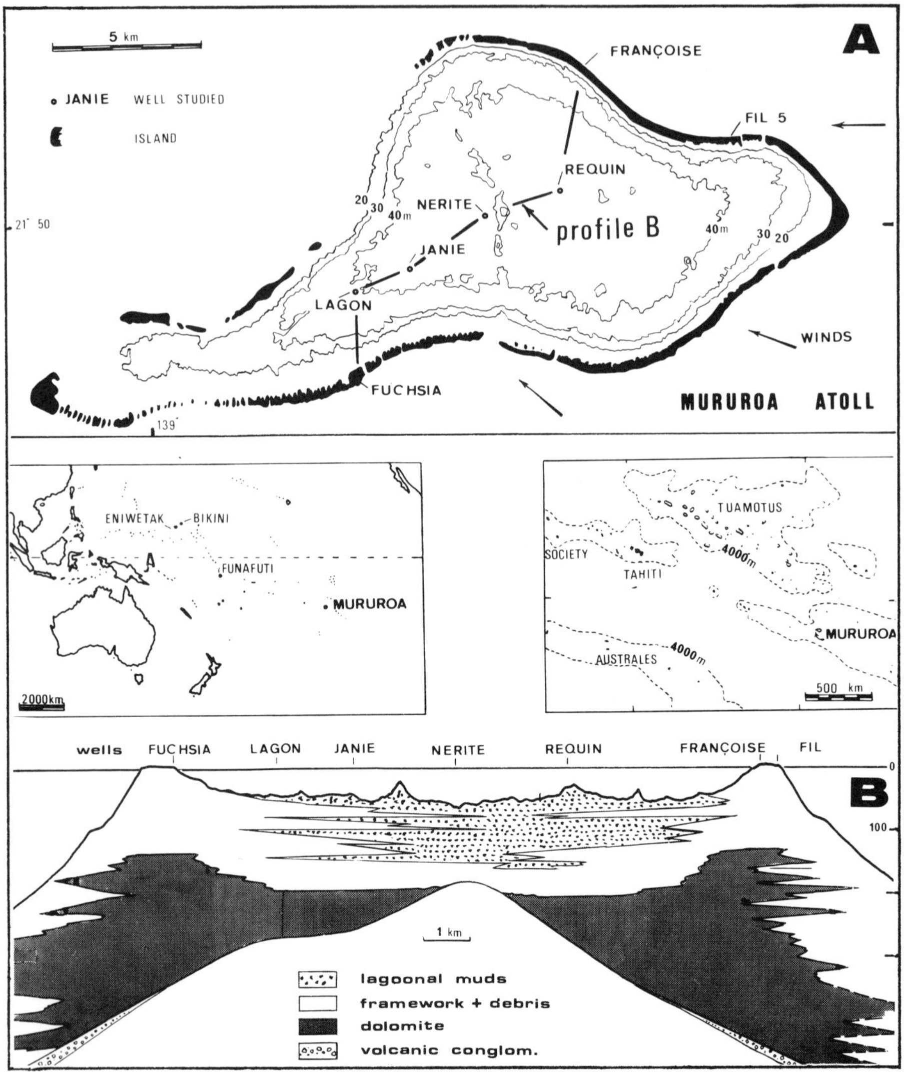

Fig. 1. A Localisation and morphology of Mururoa Atoll; note the distribution of wells studied. **B** Section across Mururoa Atoll showing the distribution of the main sedimentary unit with respect to the lenticular dolomite body. Note that lagoonal muds (i.e., atoll morphology) appear relatively late in the evolution of the build-up

within the dolomite body. A more recent study by Trichet et al. (1984) stresses the presence of a dolomite "front" separating limestones (above) from dolomite, at depths of −125 m, giving the impression that this contact is stratiform; published data (Buigues 1982) clearly show, however, that the dolomite top descends to −204 m below the lagoon before disappearing within the center of the atoll.

1.2 Objectives of this Contribution

Summaries by Buigues (1985), Aïssaoui and Purser (1985) and by Purser and Aïssaoui (1985) included results from recent wells deviated seawards from peripheral islands and which penetrated thick sequences of talus material highly lithified by spectacular submarine cements. Together with new information, these results are synthesized to give an overall view of the diagenesis affecting an oceanic platform. This is favored by exceptional well control (all depths given are below mean sea level).

2 The Sedimentary Framework and Evolution of Mururoa Atoll

Major facies have been defined (Buigues 1985) whose distribution expresses morphological changes through time (Fig. 1 B). Basal volcanic conglomerates (Gachon 1985) are overlain by a thick series of muddy coralgal facies composed of fragile branched (*Leptoceris*) and lamellar corals, and coralline algae. Together with coral rudstones, these sediments express free marine circulation around the emergent volcanic island. Perivolcanic sedimentation ceased at about −180 m and was succeeded by more varied sediments, including massive coral framestone built mainly by *Porites, Acropora* and less abundant *Meandrina* and coralline algae. These open marine constructions, although best developed in peripheral wells, also occur below the present lagoon, suggesting continued marine circulation unmodified by peripheral barriers.

The increased development of muddy sediments above depths of about −120 m (Fig. 1 B), notably within wells situated in the E parts of the present lagoon, indicates morphological differentiation into lagoon and peripheral barrier. This is confirmed by the contemporaneous growth of framestones penetrated in peripheral wells where thick-branched corals, often in growth position, extend up to the present surface. Thus, the modern atoll-type morphology appears to have developed only during the later stages of the evolution. The presence of at least five karst surfaces, each extending across the atoll, demonstrates that this evolution was interrupted by repeated emergence.

3 Cementation at Mururoa Atoll: Calcite and Aragonite Cements

Because these cements have been described and illustrated by Aïssaoui and Purser (1985), they will be reviewed briefly in order to present the complete picture of diagenesis at Mururoa. The essential properties of these cements are summarized on Table 1.

Table 1. Geochemical properties of the principal diagenetic phases[a]

		Major and Trace Elements							Isotopes	
	Calcites	mol % $CaCO_3$	mol % $MgCO_3$	ppm Mg	ppm Sr	ppm Na	ppm Fe	ppm Mn	$\delta^{13}C$ PDB	$\delta^{18}O$ PDB
Mg calcite	Type 1 ("Fil 5" −228 m)	83.7 to 93.4	7.0 to 16.9		0−1900	0−2200	0−1830	0−700	+3.68 to +3.87	+0.92 to +3.58
Mg calcite	Type 2 ("Fil 5" −276.5 m)	86.2 to 90.3	9.9 to 13.7		0−1340	0−800	0−400	0−800	not measured	
Mg calcite	Type 3 ("Fil 5" −177.4 m)	88.0 to 93.8	6.2 to 12.0		0−720	0−860	0−850	0−840	not measured	
	Low Mg fibrous calcite ("Fr" −103.4 m)	97.3 to 99.5		320 to 5000	770 to 1500	0−5700	0−1000	0−1100	−3.92	−4.86
	Low Mg Calcite spar ("Fr" −11.5 m)	98 to 99.6		800 to 5000	0−2000	0−700	0−1600	0−1200	−3.13	−4.15
Aragonite	acicular ("Fr" −7.2 m)	98.2 to 99.0		0−400	9000 to 10000	800 to 3000	0	0−1000	not measured	
Aragonite	botryoidal ("Fil 5" −228 m)	98.2 to 99.7		0−350	8700 to 10000	0−2700	0−770	0−510	+4.88 to +4.93	+1.05 to +1.23
	Dolomites									
	Type 1 ("Fil 5" −248.6 m)	56.9 to 63.7	36.2 to 43.0		0−1400	0−1500	0−200	0−570	+0.9 to +2.7	+2.7 to +3.7
	Type 2 ("Requin" −224 m)	52.0 to 59.8	40.1 to 47.8		0−600	0−700	0−1000	0	average +1.7	average +3.2
	("Janie" −222 m)	49.0 to 52.7	47.2 to 50.5		0	0−1300	0−570	0−670	+2.7 to +3.7	+3.8 to +5.1
	Type 3 ("Fr" −142.5 m)	54.6 to 58.1	41.3 to 45.3		0−400	0−1600	0−700	0−440	average +3.2	average +4.4
	("Fr" −432.6 m)	52.3 to 58.5	41.4 to 47.7		0−470	0−1200	0−1800	0−1400		

[a] All measurements have been made with CAMEBAX microprobe. Those listed are based on a series of analyses made on a given representative sample (e.g., "Fil 5" −228 m).

3.1 Calcite Cements

3.1.1 High Magnesian Calcite

Type 1 (Figs. 2, 3A). In thin section, it consists of elongate fibers, each composed of a series of radially oriented, secondary crystals. Fibers are grouped into bundles having undulatory extinction. Type 1 magnesian calcites are often overlain by contemporaneous internal, marine sediments. These cements are very common at Mururoa especially within peripheral wells, "Françoise" and deviated "Fil 5", where they form isopachous crusts up to 10 mm in thickness.

The marine origin of Type 1 Mg-calcite cements is suggested both by their mineralogy, their frequent association with marine internal sediment, and their isotopic signal (Table 1).

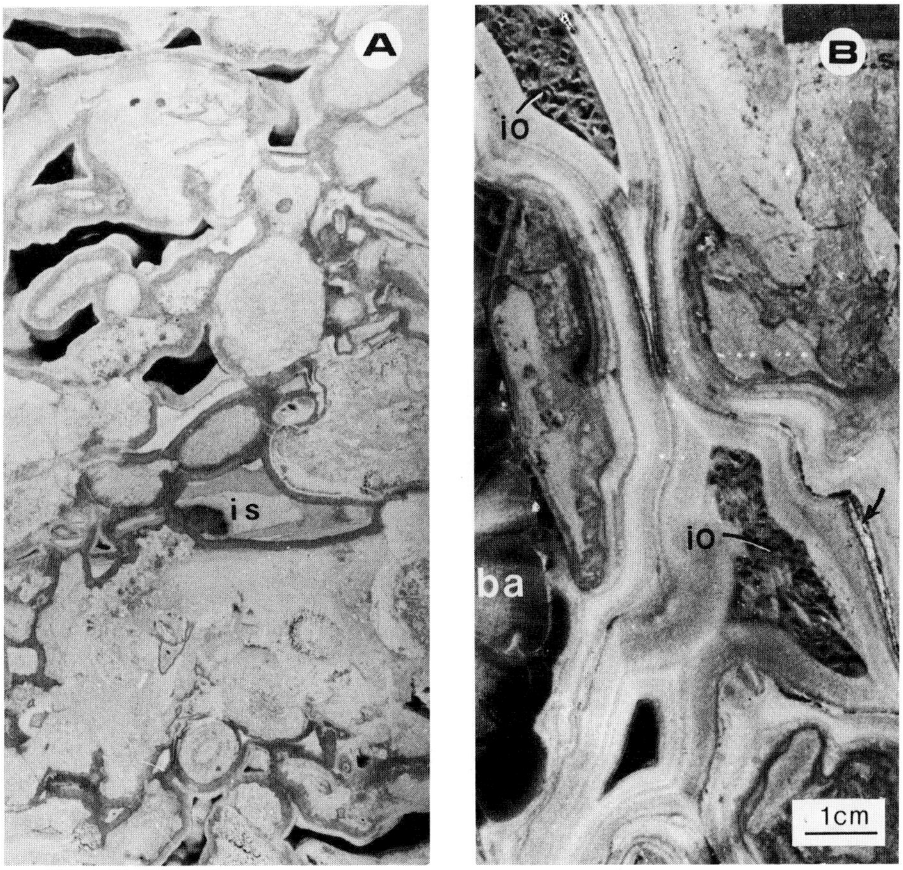

Fig. 2 A, B. Polished core surfaces showing well-developed marine cements. **A** Type 1 Mg-calcite cement overlain by internal sediment (*is*) (peripheral well, "Françoise"). **B** Botryoidal aragonite (*ba*) overlain by type 1 fibrous Mg-calcite. This thick Mg-calcite cement is locally dissolved (*arrows*) and is overlain by encrusting organisms (*io*); (flank well "Fil 5")

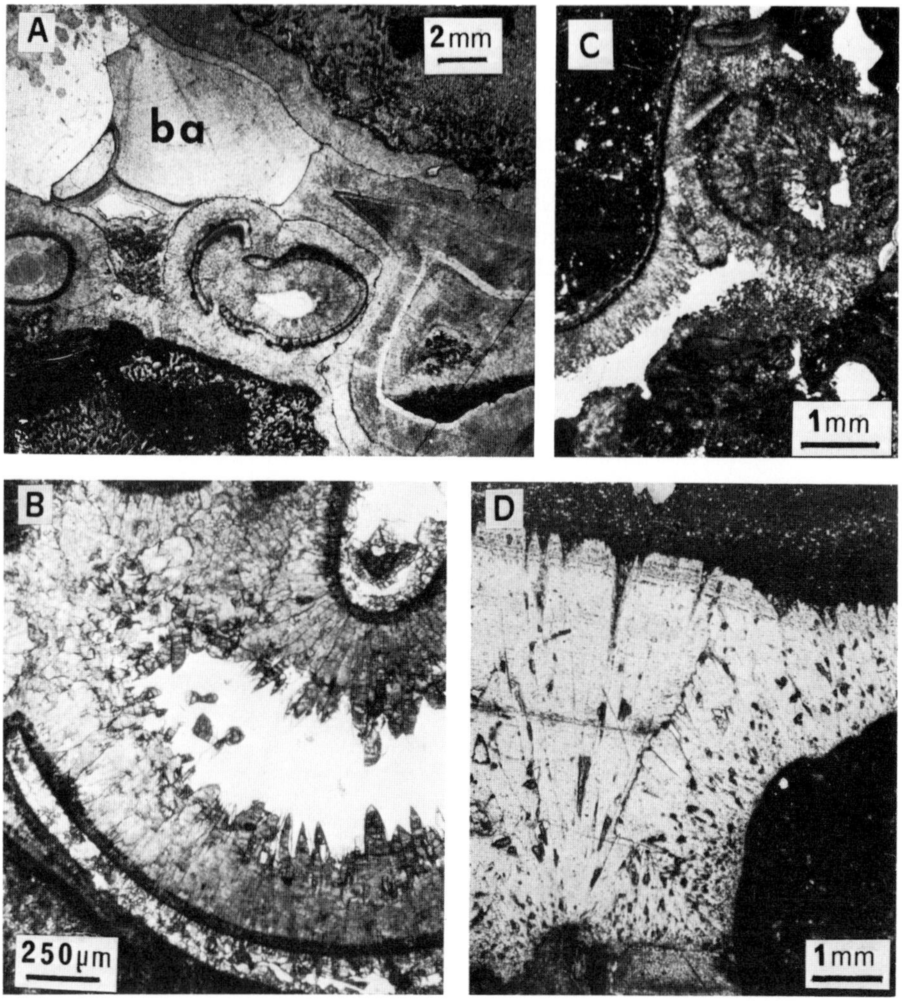

Fig. 3 A – D. Principal types of cement. **A** Mamelon of botryoidal aragonite (*ba*) preceded and overlain by type 1, Mg-calcite ("Fil 5", −228 m). **B** Type 2 fibrous Mg-calcite within dissolved gastropod ("Françoise" −231.15 m). **C** Acicular aragonite cement ("Fil 5" −69 m). **D** Fibrous, low Mg-calcite characterized by elongate crystals. Note abrasion on top of cement overlain by internal sediment. ("Françoise" −103 m)

Type 2 (Fig. 3 B). This cement is composed of well-defined subparallel crystals forming an isopachous layer never exceeding 400 μm in thickness. Contrary to Type 1 cement, this Mg-calcite cement seems to form a distinct, single layer. Individual crystals are fibrous to bladed with a predominance of planar crystal faces. Similar cements have been described as "blade spar" by James and Ginsburg (1979) from the Belize barrier reef. Type 2 cement may exhibit undulose extinction, especially when the limits between adjacent crystals are irregular. However, most fibers exhibit unit extinction.

Type 2 magnesian calcite, like Type 1, is restricted to the peripheral wells, principally "Françoise" and "Fil 5". However, it is less common than Type 1 and affects thin, preferentially coarse grainstone intervals in these wells.

Type 3. This calcite consists of well-defined, closely packed bladed crystals having somewhat elongate form, whose average length is 150 µm. When sufficiently large, elongate crystals present undulatory extinction. Morphology of Type 3 cement is similar to a form of Mg-calcite spar reported from Belize and qualified by the term "stubby" (James and Ginsburg 1979). Under the SEM, each crystal seems to be composed of subcrystals, although boundaries between the latter are not well defined.

3.1.2 Low Magnesian Calcite

Fibrous Low Magnesian Calcite (Fig. 3D). Highly elongate (up to 10 mm) these crystals exhibit planar faces with well-developed cleavages. Moreover, this low Mg-calcite is characterized by a dark brown color in hand specimen due to numerous inclusions, and by a slight pleochroism in thin section. Although forming fans, these cements do not have undulatory extinction and each crystal is optically independent. They are occasionally covered by nonmarine internal sediment.

Mg and Sr concentrations (Table 1) and negative isotopic signatures strongly suggest a subaerial origin. A near-surface origin under meteoric influences is suggested also by the association with pedogenetic and karstic phenomenon.

Sparry Low Magnesian Calcite. Crystals varying in size from 20 – 200 µm are readily distinguished from the fibrous calcite cements not only because of their equant fabric but also because of their disorganized, irregular distribution both through the sedimentary pile and within a given thin section. Unlike the fibrous calcites, these sparitic cements have few inclusions and their trace-element concentrations are very low (Table 1). These cements are related to the series of karst surfaces which extend across the atoll (Buigues 1982). This origin is confirmed by a negative isotopic signal (Table 1).

3.2 Aragonite Cements

3.2.1 Acicular Aragonite (Fig. 3C)

The length of fibers varies from 50 to 300 µm, and their width from 2 to 5 µm; they are generally more loosely packed than fibrous magnesian calcite. Frequent within corals, acicular aragonite may also form as submarine cement within certain grainstones. Distribution of fibers within the primary pore network is irregular although individual crystals may form a continuous layer.

3.2.2 Botryoidal Aragonite (Figs. 2B, 3A)

A given botryoid is composed of a fan-shaped, densely packed mass of fibers exhibiting typical undulose extinction. Individual fibers are composed of twinned, euhedral subcrystals. Botryoidal aragonite at Mururoa generally alternates with layers of fibrous Mg-calcites (Type 1) within a given cavity (Fig. 2B), reflecting variations in pure-fluid composition or temperature. That it has formed from marine waters is suggested by its positive isotopic signature (Table 1). This characteristic cement type has been recognized only within the laterally deviated wells, at depths between 120 and 130 m. Described from other localities by Shinn (1969), James and Ginsburg (1979), and by Aïssaoui (1985), it is characterized by a botryoidal form, individual mamelons attaining a radius of 3 cm at Mururoa (Fig. 2B). Aragonite botryoids are scattered through the sediment generally within residual cavities of primary origin.

Aragonite cements are limited in volume with respect to the more abundant submarine, calcite cements. In common with the latter, they are best developed in peripheral wells. These conclusions are remarkably consistent with those noted for Belize barrier reef (James and Ginsburg 1979).

3.3 Distribution of the Cements

Beneath the present lagoon, the greater part of the sedimentary pile is characterized by a very low degree of cementation; friable textures are due mainly to lack of cement rather than dissolution. Fine sediment may be totally uncemented, even at depths exceeding 100 m below the floor of the present lagoon. When lithified, sediments of the central part of Mururoa atoll are cemented almost exclusively by sparry calcite forming thin, dense, more or less stratiform horizons extending across the atoll. These spar-cemented sediments are apparently related to periods of subaerial exposure. Beneath the peripheral islands and external slope of the present atoll, almost the entire sequence is highly cemented. Within certain rudstones the volume of cement may exceed that of the sedimentary components (Fig. 2B) and there is notable increase in the degree of cementation from the central parts towards the periphery of the atoll, a zone of maximum cementation (Aïssaoui and Purser 1985). Similar distribution has been noted in the Holocene barrier reef in SE Florida (Lighty 1985).

4 Dissolution

In addition to several karst surfaces clearly visible in cores, individual grains and cements often show varying degrees of dissolution.

4.1 Dissolution of Sedimentary Components

At Mururoa, the selective dissolution of carbonate components begins within the topmost meters of the sedimentary sequence in all wells. When the dissolution is minor, it is generally fabric-selective and affects preferentially the aragonitic grains such as corals and gastropods. Within the dolomite body, preservation of aragonite is rare and aragonitic components are totally dissolved, while the calcitic grains have resisted dissolution. However, these tendencies are only general, and selective dissolution may be more subtle. Thus, within the same thin section, certain molluscs or foraminifera are totally dissolved while others of the same species are not. Similarly, aragonitic grains may be preserved despite partial dolomitization. Interestingly, these "exceptions" are generally related to significant amounts of cementation. In other words, dissolution of grains depends partly on the degree of protective cementation which slows the circulation of the aggressive fluid especially on a microscopic scale, mineralogy of the components seemingly being a secondary parameter.

4.2 Dissolution of Cements

This occurs mainly within the highly cemented intervals in the peripheral wells "Fil 5" (−141, 147, 188, 228 m), and occasionally within "Françoise" (−136.8 m). It preferentially affects the multilayered Type 1 Mg-calcite (Fig. 4), but does not seem to alter other cement types; moreover, at "Fil 5" (−228 m), botryoidal aragonite is both preceded and followed by two distinct Type 1 Mg-calcite cement generations, both of which are partly dissolved. Under the

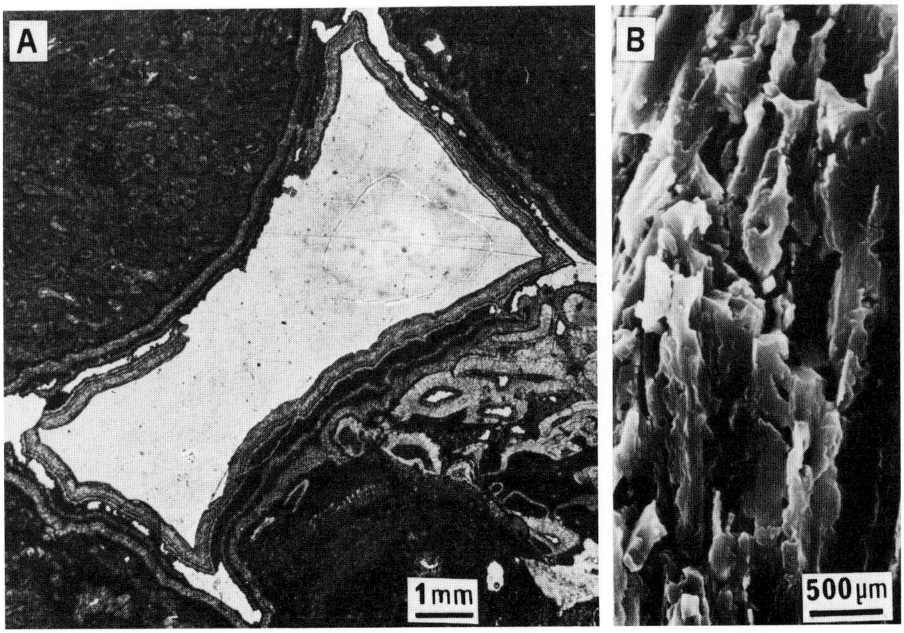

Fig. 4 A, B. Dissolution of cement. **A** Dissolution of type 1 fibrous Mg-calcite (thin section, "Fil 5" −141 m). **B** SEM view of dissolved type 1 cement. Note that dissolution develops between fibers ("Fil 5" −156.8 m)

SEM, dissolution of the cement fibers is fairly evident and results in the creation of small voids along the boundaries of adjacent fibers (Fig. 4B) and is more or less related to cement and crystal growth discontinuities, these sites probably serving as microscopic drains for solvents.

4.3 Dissolution of Dolomite

In addition to the dissolution of aragonitic and calcitic skeletons, there are many secondary voids whose forms are clearly rhombohedral. Because these voids occasionally cut into skeletal debris, notably corals, it is clear that the precursor was a crystal which had replaced a sedimentary particle, and thus is more likely to have been dolomite rather than calcite (Purser and Aïssaoui 1985). Dolomite dissolution occurs in all wells between the surface and the present top of the dolomite body and may constitute between 5 and 40% of the prosity in a given sample. Dissolution indicates that the present top of the dolomite body is a residual limit, and that original dolomite distribution cannot be established by X-ray analyses alone.

4.4 Lateral Distribution of Dissolution Phenomena

The intensity of dissolution is difficult to appreciate because of a considerable microporosity. Present to varying degrees in all wells, including those deviated towards the outer flanks, it is less developed within the outer parts of the platform. This is due partly to the protective effects of the cements, particularly well developed near the periphery of the atoll.

5 Dolomites and Dolomitization at Mururoa Atoll

The dolomites at Mururoa, as already noted, form a lenticular mass whose maximum thickness of about 320 m is situated below the peripheral islands. The body thins and ultimately disappears both in wells ("Nerite") located in the central parts of the present lagoon, and towards the peripheral flanks ("Fil 5") of the platform (Fig. 1B). The top of the dolomites never rises above −130 m, although numerous rhombohedral pore spaces (dedolomite) and scattered rhombohedra indicate minor dolomitization up to the surface. Interpretation of the mechanisms favoring dolomitization at Mururoa must clearly take into account these geometric parameters. Dolomite distribution is complicated by the fact that the system, at Mururoa, is multiple; there are repeated phases of dolomitization and the overall geometry of the dolomite body is probably a composite effect.

Finally, the dolomites at Mururoa, in common with their Caribbean counterparts (Land 1973, Pierson 1982, Ward and Halley 1985), and elsewhere, are associated with dissolution. The often intimate relationships between dolomitization and meteoric diagenesis obviously influence interpretation of the dolomitization mechanisms.

5.1 Dolomite Types

Because the four basic types of dolomite have already been described by Purser and Aïssaoui (1985), only a brief description is given (Fig. 5). The following are pure dolomites.

Types 1 and 2 are fine-grained dolomites whose 5 – 50 μm-sized rhombohedra are often difficult to recognize in thin section. Type 1 is a dark dolomicrite whose individual crystals are not visible in thin section, while type 2 crystals, visible in thin section, have dark centers and clear rims. They are fabric-preserving and seem to form mainly by replacement. These highly porous, chalky dolomites predominate within the central parts of the platform.

Type 3 is a crystalline dolomite which occurs as a void-lining cement or as clear overgrowths on replacement types 1 and 2. This dolomite is best developed towards the periphery of the platform, forming a hard crystalline rock.

Type 4 comprises relatively large (100 – 250 μm) crystals, often scattered within a finer dolomitic matrix. It appears to form by recrystallization of an earlier dolomite, possibly during periods of renewed dolomitization. All dolomite types are associated with dissolution phenomena. Their mineralogical and geochemical properties are given on Table 1.

5.2 Geochemistry of Mururoa Dolomites (Table 1)

5.2.1 Major and Trace Elements

Microprobe analyses show variations in major elements (Ca, Mg) as well as trace elements (Sr, Na, Fe, and Mn), both within single crystals and on a regional scale. Microcrystalline "replacement" dolomite (type 1) and large, type 3 dolospar crystals lining voids show no reproducible differences in major element concentrations. In other words, local inter- or intra-crystalline variations reflect the considerable heterogeneity of Mururoa dolomites as in many other dolomites (Land 1980). These variations are even more complex if one considers the trace elements. Sr is present in all analyzed samples but, for no obvious reason, its concentration varies from 0 to 1000 ppm. This is also the case for Na (0 to 3200 ppm). Fe and, to a lesser degree, Mn are locally important in deep intervals of peripheral wells ("Françoise" 298, 332.5, 372, 411.5 m, "Fuchsia" – 335, 361 m, "Zoé" – 303 m), iron seemingly being related to the slightly colored dolomites situated near the base of the sequence (Buigues 1982).

On a regional scale, variations, although not particularly striking, are as follows:

– $MgCO_3$ content is variable within the peripheral wells "Françoise" and "Fuchsia", especially near the top of the dolomite body; downwards, there is a noticeable increase of $MgCO_3$, a trend particularly expressed at "Françoise";

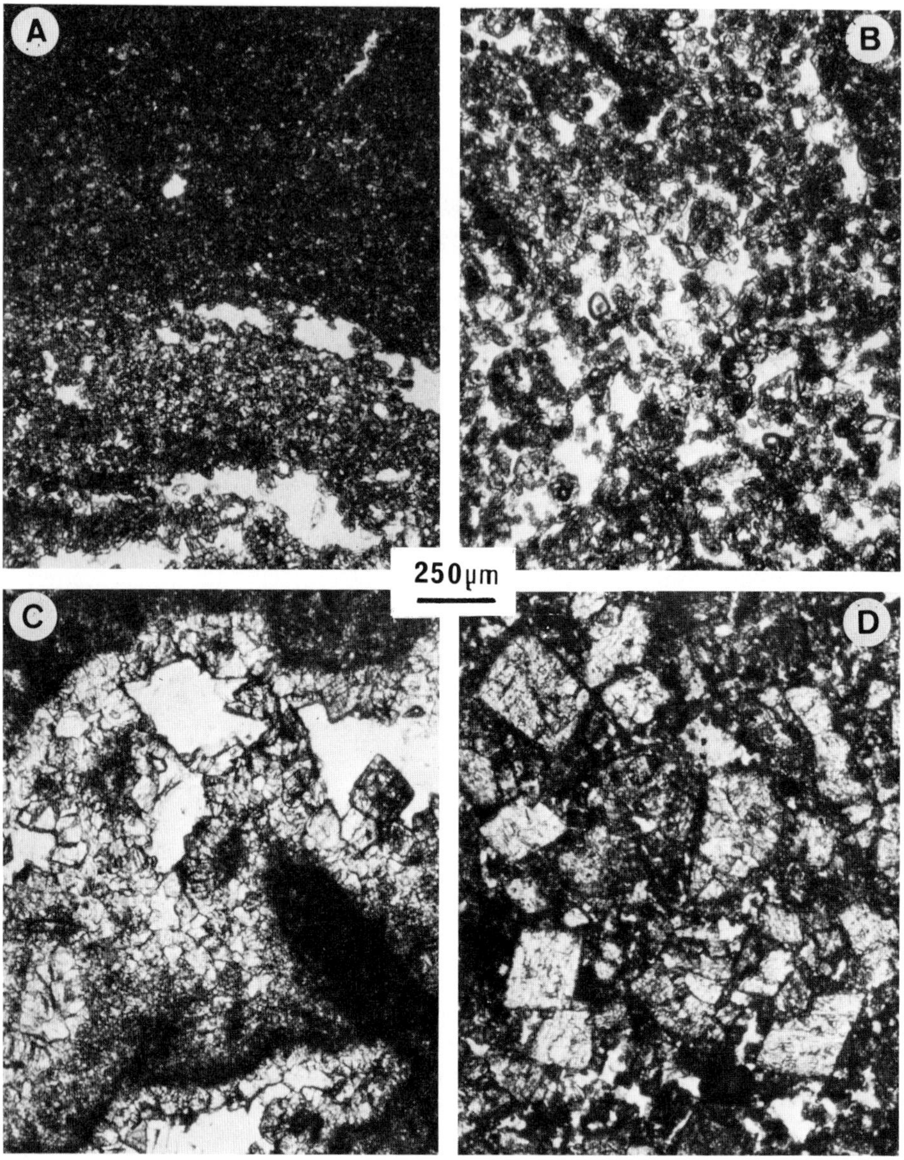

Fig. 5A–D. Principal dolomite petrotypes; all are 100% dolomite. **A** Type 1 (*dark*) and type 2, essentially "replacement" dolomite. Note dissolution void ("Janie" –198.3 m). **B** Type 2 dolomite showing small rhombohedra with cloudy centers and limpid peripheries ("Janie" – 205 m). **C** Type 3 dolosparite cement lining dissolution cavity ("Fuchsia" –180 m). **D** Type 4 recrystallized dolomite showing two generations of rhombohedra ("Fuchsia" –339 m)

– There is a clear increase in the $MgCO_3$ content from the external flanks of the atoll towards the peripheral wells "Françoise" and "Fuchsia": at "Fil 5", a laterally deviated well, dolomites are clearly calcian, $MgCO_3$ ranging from 36 to a maximum of 43 mol %, while at "Françoise", $MgCO_3$ ranges from 43 to 49 mol %. It is possible that external dolomites in "Fil 5" correspond to an initial phase in which unstable, magnesian-poor dolomite is precipitated.

5.2.2 Stable Isotopes (Fig. 6 and Table 1)

Although selective sampling of individual dolomite types has not been done, the isotopic qualities of a mixture of types 1 and 2 in samples from the central wells "Nérite" and "Lagon 2" are slightly lighter than those of samples rich in type 3 dolosparite from the peripheral well "Colette" (Fig. 6). The average values $\delta^{18}O$ of +3.3‰ PDB and $\delta^{13}C$ of +1.7‰ PDB within the lagoonal wells are comparable to those obtained by Pierson (1982) from Pleistocene Bahamian dolomites and especially to those of Ward and Halley (1985) from Pleistocene dolomites from NE Yucatan. However, interestingly, the dolomites from the peripheral well "Colette" are somewhat enriched in heavier isotopes both with respect to the "lagoonal" dolomites at Mururoa and to average "Caribbean" values of Land (1973), Supko (1977), Pierson (1982), and Ward and Halley (1985). The average $\delta^{18}O$ value of +4.7‰ from the peripheral localities at Mururoa, however, is very close to that of the dolomites from Baffin Bay, Texas, interpreted by Behrens and Land (1972) as marine in origin.

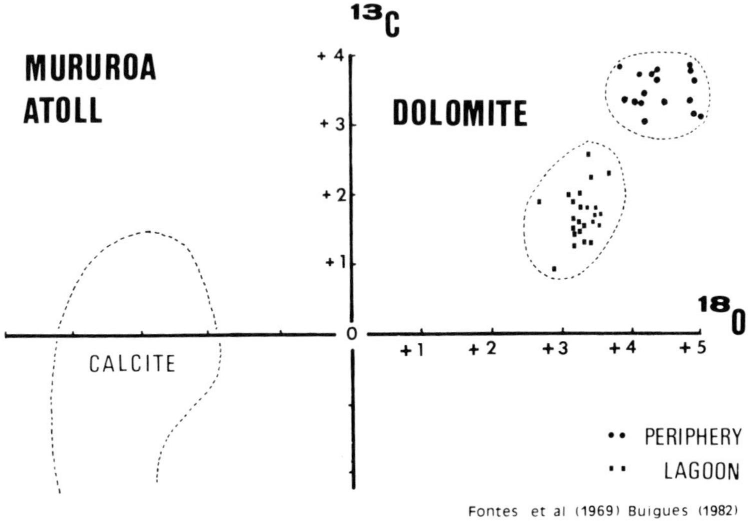

Fig. 6. Stable isotope signatures for Mururoa dolomites and limestones. Note separation between slightly heavier peripheral and somewhat lighter lagoonal dolomites

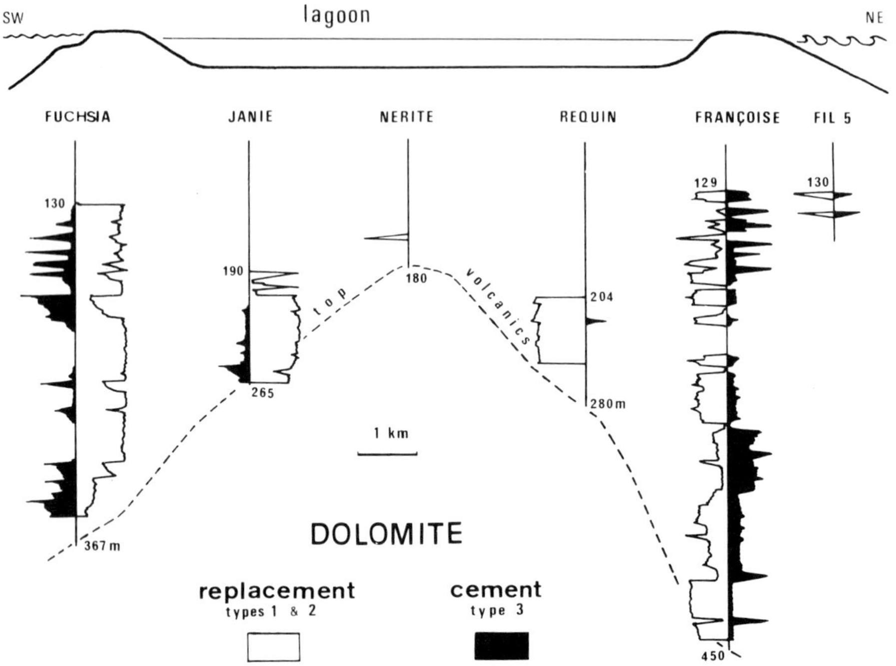

Fig. 7. Cross-section showing the distribution of "replacement" (types 1 and 2) and dolosparite (type 3) cement, this latter best developed in peripheral wells

5.3 Distribution of the Principal Dolomite Petrotypes

Lateral variations are evident. As shown in Figure 7, wells such as "Requin" and "Janie", situated within the inner parts of the atoll, have mainly types 1 and 2 replacement dolomite; type 3 cement is rare or absent. This latter (type 3) is best developed in the well "Françoise", where virtually all dolomite samples comprise both replacement and cement fabrics. This peripheral concentration of type 3 dolomite cement may vary laterally around the platform. Best developed in "Françoise" situated on the NE (windward) side of the atoll, this dolomite is less abundant in the well "Fuchsia" located in a more leeward situation on the S side of the platform.

The major variation therefore concerns an important centrifugal transition from nondolomitized limestone in the center (well "Nérite") to a porous dolomite facies composed mainly of replacement dolomite, to a peripheral, more crystalline (and somewhat thicker) mass of dolomite in which dolomite cement is well developed (Fig. 7).

5.4 Limestone-Dolomite Transitions and Their Significance

5.4.1 Vertical Transitions in Cores

The vertical transition between the dolomite body and the overlying limestone varies from well to well. In wells located within the platform, such as "Janie" and "Requin", the contact is sharp. In "Requin", for example, one passes from 100% dolomite to 100% limestone in less than 1 m (−203 m), this rapid diagenetic change coinciding with a marked change in the sedimentary facies; the top 2 – 3 m of dolomite lack bioclasts and include dissolution cavities (karst) and numerous stromatolitic laminations, while the overlying limestones have abundant marine fauna. These diagenetic and sedimentary breaks suggest that dolomitization may have extended to the surface of a brackish water body, at least near the central parts of the platform.

In peripheral wells such as " Françoise", the vertical transition from dolomite to limestone is progressive and does not coincide with a karst surface or with any marked variation in sedimentary facies. This would suggest that dolomitization did not extend up to the surface of the sediment. Peripheral dolomites include a series of interbedded limestones and other partly dolomitized levels, never observed within the inner parts of the platform.

5.4.2 Dolomitization of Fine-Grained Limestones

The dolomite invariably consists of isolated and/or grouped, interpenetrating rhombohedra whose average size is less than 10 μm (Fig. 8). SEM study reveals that the smallest rhombohedra are located in interstitial voids, suggesting that they have precipitated *between* calcite grains (Fig. 8 A). In other words, dolomite develops initially as a *cement within* the microscopic voids (Fig. 8 B). Similar process has been reported from Andros Island, protodolomite *cementing* aragonite mud (Gebelein et al. 1980).

When adjacent dolomite rhombohedra interpenetrate, they show preferential enlargement to form dolomitic patches within the calcitic sediment (Fig. 8 C). On polished samples, the patches of dolomite often have sinuous boundaries and seemingly develope to the detriment of the adjacent calcite crystals (Fig. 8 C, D). This crystal evolution closely resembles "cellular phagocytosis", a digestive biological process! Interestingly, this process characterizes both dolomitic zones, i.e., groups of dolomite rhombohedra and isolated dolomite rhombohedra (Fig. 8 E). On polished or on broken surfaces, dolomite is seen to grow first around calcite, and the engulfed grains may either be totally digested or remain as relics within dolomite rhombohedra (Fig. 8 E). In this process, calcite-dolomite transformation includes a *dissolution* phase involving a micron-sized front clearly visible under SEM.

5.4.3 Dolomitization of Coarse-Grained Limestones

In coarse grainstones and rudstones, generally highly cemented by fibrous marine calcites, dolomitization affects selectively the varied sedimentary compo-

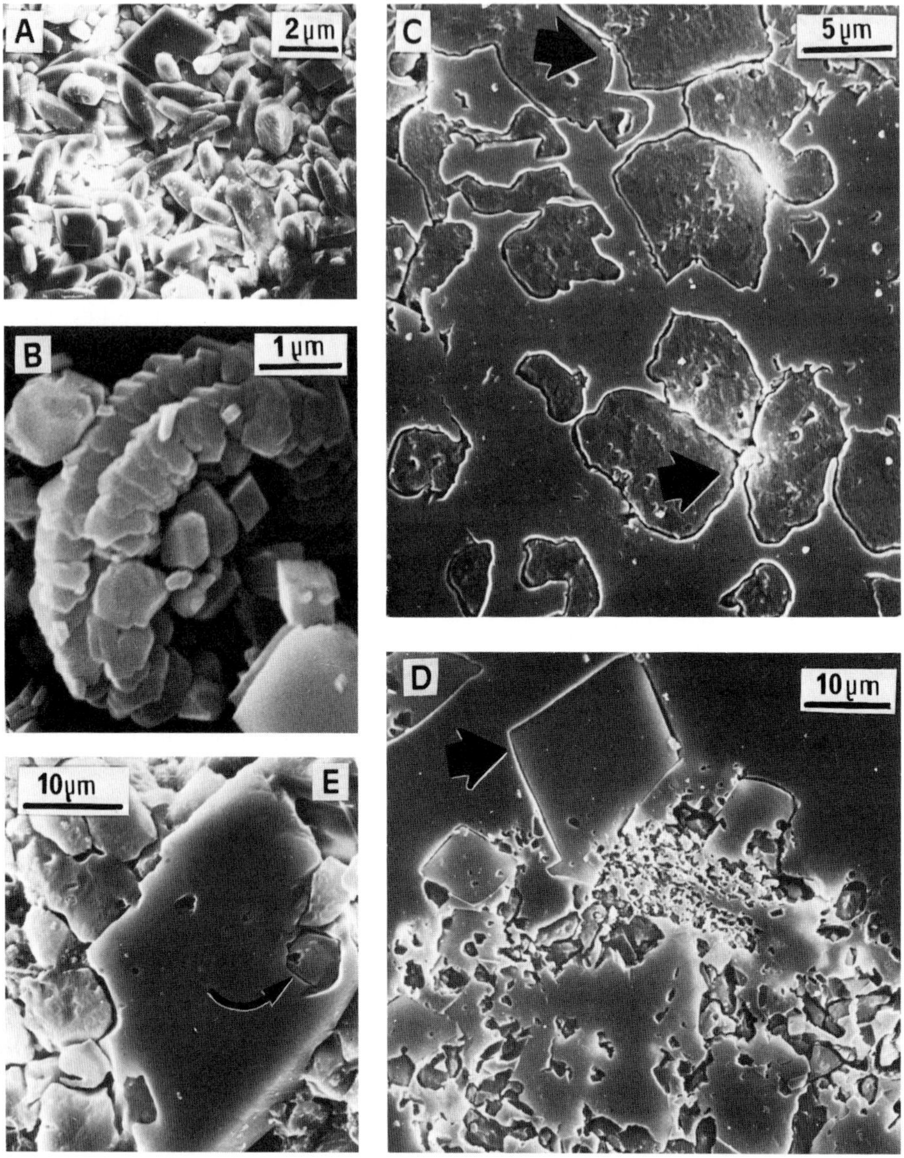

Fig. 8 A–E. Dolomitization of fine-grained limestone (SEM views). **A** Dolomite rhombohedra forming in interstitial pores. Note that this initial growth occurs in open voids ("Fil 5" – 248.6 m). **B** Dolomite rhombohedra precipitated as a cement within coccolith ("Fil 5" – 239 m). **C** Dolomite patches (*lower part*) developing to detriment of calcite. Note that initial dolomitization develops between calcite crystals (*arrows*) ("Françoise" – 269.5 m). **D** Dolomite rhombohedra developed both towards voids as a cement (*arrows*) and within the calcite grain as a "replacement" ("Françoise" – 128 m). **E** Development of individual dolomite rhombohedron which engulfs calcite grains (*arrow*) ("Fil 5" – 239 m)

nents *and* their cements. Within the peripheral well "Françoise", these phenomena are spectacular and all stages of dolomitization are clearly visible.

Dolomitization of sedimentary particles may be partial and its selectivity indicates that both initial mineralogy *and* the quality of the pore network within the components influences dolomitization. Aragonitic components such as corals or gastropods are totally dolomitized below depths of −120 m in peripheral wells and below 190 m in central wells. This transformation occurs in two distinct stages. The first involves selective dissolution of the biogenic aragonite creating a void. The second concerns precipitation of a dolosparitic cement which rarely fills the pre-existing void. Direct aragonite-dolomite transformation has not been observed.

5.4.4 Dolomitization of Cements

A frequent aspect of the partial dolomitization of grainy limestones at "Françoise" concerns the multiple alternation of calcite and dolomite cements within given voids (Fig. 9), this phenomenon having been reported from Mururoa (Buigues 1982) and other reefal series (Bourrouilh 1982; Ward and Halley 1985). When occurring within millimetric voids, these alternations are perfectly continuous and would appear to result from the alternate precipitation of calcite and dolomite. However, when occurring within centimetric or larger cavities, the dolomite may form a *discontinuous* lamination at the bottom *and/or* at the top of the fibrous calcite when the cavity is lined by a single cement layer, or between each individual cement layer when the cavity is lined by several distinct cement layers (Fig. 9 A).

In addition to their discontinuous aspect, dolomite "layers" seemingly emphasize discontinuities or, more probably, heterogeneities within the calcite cement. This is confirmed, for example, by frequent lateral transitions between the dolomite and small voids localized between layers of calcitic cement; like the dolomite, these voids also vary laterally in width. Because the dolomite crystals do not exhibit dissolution traces (plane faces, sharp edges), these planar voids probably result from calcite dissolution. Consequently, it is more logical to interpret these particular dolomites as cements formed in planar voids created by partial dissolution of an earlier calcitic cement (Figs. 4, 10). On a local scale when a pair of dolomite lamina enclose a calcite layer, the dolomite lamina are interconnected by dolomite crystals via minute spaces in the intervening calcite (Figs. 9 B, F, 10). On close examination, dolomite is seen to envelop the calcite, indicating that dolomite develops initially between calcite crystals (Figs. 10, 11). All transformations may be contemporaneous and may produce a wide spectrum of cement microsequences representing intermediate fabrics of a single dolomitization phase. Interestingly, Schroeder (1985) has described a similar process for calcite cement microsequences in the Miocene corals from NE Sudan.

Although the above processes appear to be very frequent, alternations of calcite and dolomite cements also exist which clearly result from a true change in fluid composition.

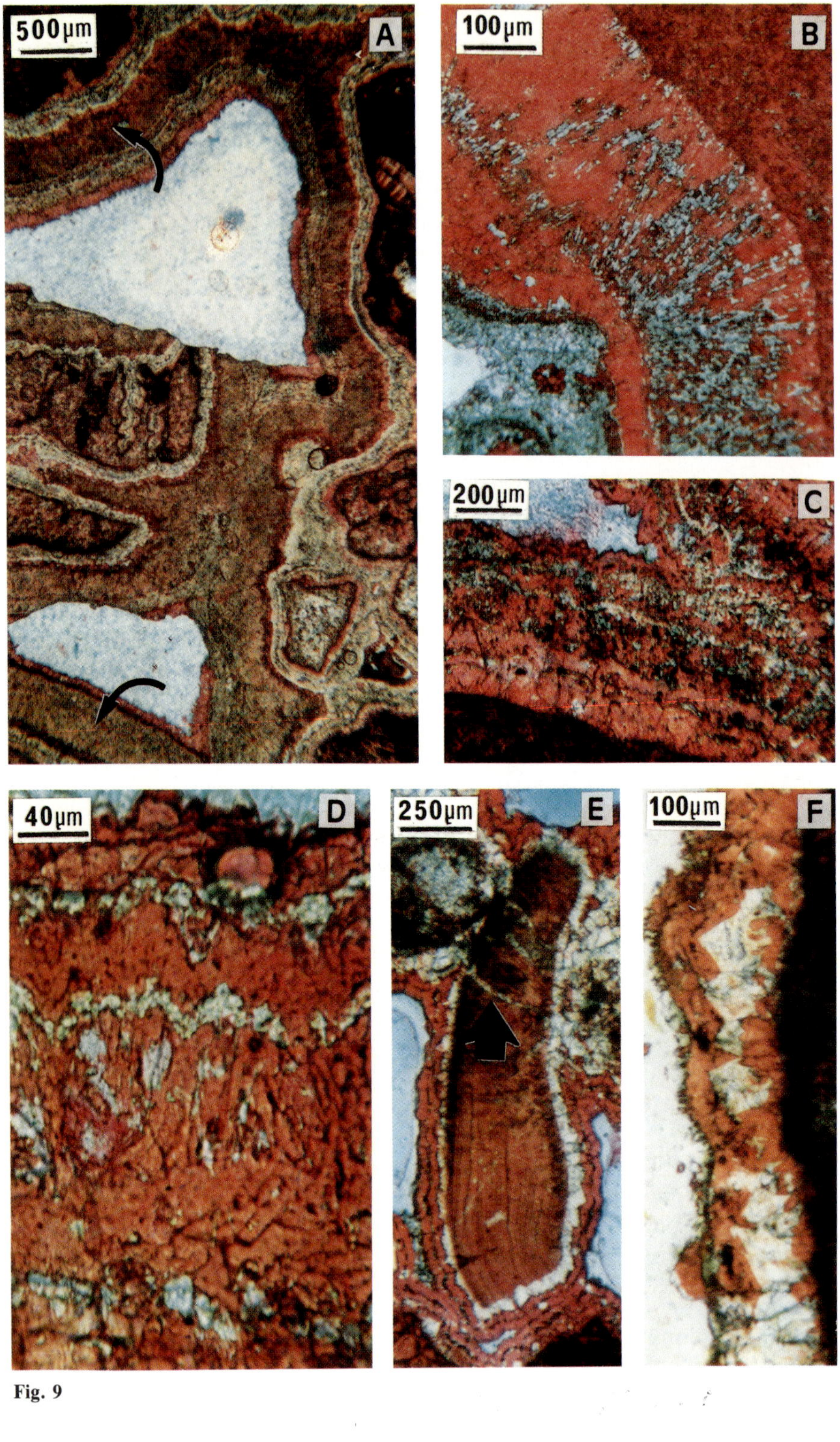

Fig. 9

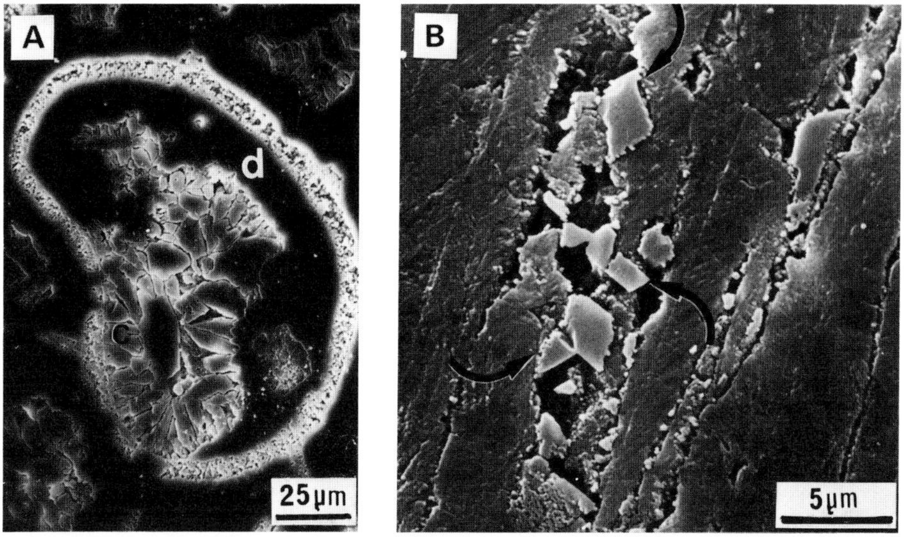

Fig. 10 A, B. Dolomitization of cements (SEM views). **A** Calcite-dolomite microsequence within skeletal cavity resulting from the partial dolomitization of the initial calcite cement. This microsequence clearly does not reflect alternating calcite and dolomite precipitation ("Françoise" −269.65 m). **B** Initial stage of dolomitization of fibrous Mg calcite cement showing development of rhombohedra (*arrows*) in microdissolution cavities located between calcite crystals ("Françoise" −134 m)

5.5 Origin of the Dolomite

5.5.1 Small-Scale Transformation (Fig. 11)

Dolomitization begins by the formation of micron-sized rhombohedra on free surfaces and, as such, is a cement. Crystals grow into primary voids and, especially, into micron-sized dissolution cavities which form contemporaneously with rhombohedral growth. Thus, dolomitization is essentially a process involving dissolution-precipitation. At this stage, dolomite is rarely recognized in thin section and only SEM examination reveals its presence. This intimate association of aragonite and calcite dissolution with dolomite growth implies that parental waters be aggressive vis-à-vis calcite and therefore cannot be normal, near surface, seawater. Further dolomite growth leads to the development of patches

Fig. 9 A – F. Dolomitization of cements; red is alizarine-stained calcite white is dolomite, blue is porosity. **A** Dolomite "layer" (*arrow near bottom*) passing laterally to calcite (*arrow near top*) suggesting that the dolomite is an alteration of the type 1 Mg-calcite cement ("Françoise" −143 m). **B** Partial dolomitization of fibrous Mg-calcite cement. Compare with SEM view Fig. 10B) ("Françoise" −231.15 m). **C** Alternation of dolomite and calcite showing discontinuities in dolomite laminations ("Françoise" −134 m). **D** Detail of Fig. 10C. **E** General view demonstrating that dolomite forms within the calcite cement inducing dolomite-calcite alternations, and within fractures (*arrows*) ("Françoise" −131 m). **F** Detail of dolomite rhombohedra formed within the calcite cement layer; Note their discontinuous arrangement ("Françoise" −164.5)

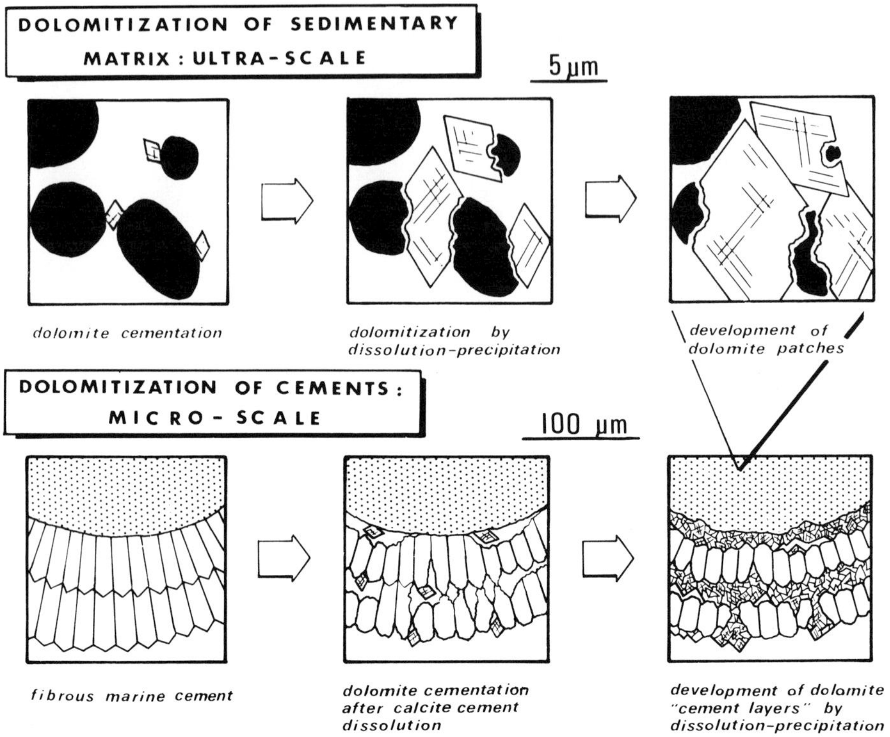

Fig. 11. Schematic model of limestone-dolomite transformation

extending either into sedimentary matrix or to syn-sedimentary cements, easily visible in thin section.

5.5.2 Large-Scale Model (Fig. 12)

The ubiquitous evidence of meteoric waters throughout the sedimentary column has persuaded previous workers (Fontes et al. 1969, Repellin 1977, Buigues 1982, Purser and Aïssaoui 1985) that dolomitization at Mururoa Atoll is related to the mixing of marine and meteoric waters. This hypothesis is confirmed in this chapter.

There is only one plausible source for the magnesium required for massive dolomitization: the sea. That dolomitizing waters *approached* normal marine composition is suggested by the isotopic qualities of the dolomite (Fontes, personal communication). However, both the $\delta^{13}C$ and the $\delta^{18}O$ values for Mururoa dolomites vary somewhat according to their position within the dolomite body (Buigues 1982 and Fig. 6).

These gradients are clearly expressed by the petrographic differences already noted; the inner platform is characterized by the predominance of fine, chalky type 1 and 2 dolomites while the peripheral wells have much crystalline, type 3

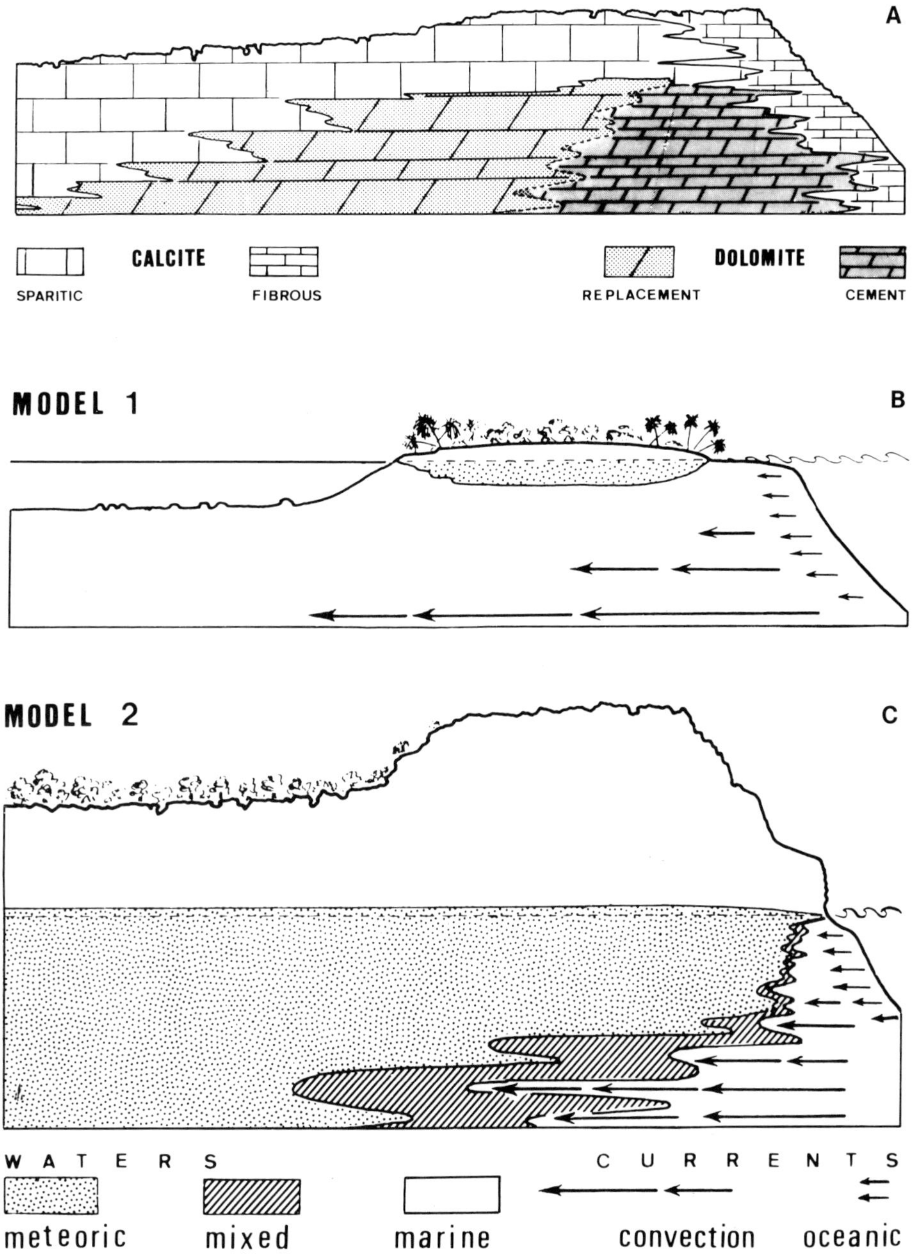

Fig. 12A–C. Models of diagenesis at Mururoa Atoll. **A** Schematic profile illustrating the lateral relationships between the main diagenetic units. Note that both dolomite and calcite cementation are best developed towards the periphery of the platform, the central parts of which are highly porous. **B** Model 1 with *thin* aquifer relating to low relief and small size. Currents responsible for mixing marine and meteoric waters pass mainly below this aquifer and dolomitization is minor. **C** Model 2 with *thick* aquifer relating to large-scale emergence of the platform. Thermal convection currents thus affect the basal parts of this aquifer causing mixing of oceanic and meteoric waters, favoring dolomitization dolomite cement (Fig. 7). This latter, both because of its peripheral distribution and its isotopic properties, expresses somewhat more oceanic conditions of formation. Petrographically identical dolomite cements have been demonstrated by Ward and Halley (1985, Fig. 5 B) from reefal carbonates in NE Yucatàn, where they are also interpreted as having formed from marine waters.

That meteoric waters have also contributed to the favorable properties of the dolomitizing fluids is indicated by the intimate association of dolomite and dissolution phenomena. Virtually all dolomite at Mururoa shows traces of leaching and in slightly dolomitized limestones, dolomite distribution often is concentrated within secondary voids formed by the dissolution of aragonite and calcite skeletons.

Finally, the disappearance of massive dolomite within wells deviated through the oceanic flanks of the platform implies that *pure* oceanic waters alone do not favor dolomitization. The overall diagenetic evolution at Mururoa, which generally begins with submarine cementation, notably in the coarser peripheral facies, was followed by widespread meteoric dissolution and karstification; only subsequent to, and most probably during the meteoric phase, did massive dolomitization occur.

In conclusion, it seems difficult to escape the fact that dolomitizing waters, although essentially marine in composition, were either continuously, or periodically, capable of dissolving aragonite and calcite. As such, they cannot have been of purely marine composition. Indeed, it may well be argued that *dissolution* of $CaCO_3$ is a prerequisite for dolomitization.

5.5.3 The Hydrodynamic System Favoring Dolomitization

A mixed, but predominantly marine dolomitizing water, suggested both by isotopic and petrographic data, requires that the platform be emergent during dolomitization. In that Mg is the essential element concerned, it is most logical to expect that waters will move *inwards* towards the center of the sedimentary pile. That this is indeed the case is suggested by the relatively high degree of both dolomite and magnesian calcite cementation towards the periphery of the platform (Fig. 12 A).

Oceanic waters may be pumped towards the center of an oceanic platform where they mix progressively with the freshwater lense. At least two mechanisms are possible:

– Waters may be driven by oceanic currents, notably during storms, as shown by Buddemeier and Oberdorfer in this Volume.

– Thermal convection resulting from differences in temperature profiles between the platform center and that of the adjacent oceanic column may result in the lateral influx of oceanic water. Studies at Eniwetak (Saller 1984, Samaden et al. 1985) and at Mururoa (Rougerie and Wauthy 1985) demonstrate a *lateral* thermal gradient, at depths of 300 m, from 15 °C in the oceanic waters to more than 25 °C within the sedimentary pile. This difference increases with depth. In that the warmer waters are less dense than the cooler oceanic water, there is a disequilibrium favoring centripetal inflow. This water eventually rises towards the lagoonal floor. Both calculations and measurement at Mururoa (personal communication Caristan, CEA) indicate that the inflow of oceanic water, while depending partly on position of the most permeable horizons, is less effective within the uppermost part of the sedimentary pile. This may be important. It suggests that in order to have a mixing of incoming marine waters with a meteoric aquifer, the aquifer must have a minimum thickness, i.e., the emerged platform (or island) must have a certain relief and surface area; low, km-sized islands with thin aquifers will not favor large-scale mixing, most incoming oceanic waters passing below the freshwater lense (Fig. 12B). However, larger platforms with considerable relief above sea level will have thicker aquifers, the basal parts of which may be affected by thermal convection or other currents, the resultant mixing favoring dolomitization (Fig. 12C).

The concept of "minimum size aquifers" finds some confirmation in the distribution of dolomites, both in the Pacific and in the Caribbean region. In the Pacific, there is little or no dolomite associated with the thin, freshwater lenses on modern atolls such as Mururoa, Eniwetak, and Ouvea (Loyalty Group), and neither below small Bahamian islands such as Joulters Key (Harris 1978) or, seemingly, the Florida Keys. Larger Bahamian islands such as Eleuthera (Pierson 1982) and in somewhat older rocks in Florida (Hanshaw et al. 1971) have important dolomite bodies.

This "minimum size" concept may explain the marked decrease of dolomite at depths of less than 130 m at Mururoa. During any one the several periods of emersion marked by platform-wide karst surfaces, an important freshwater, dolomitizing aquifer must have existed. This would also have been the case during Quaternary sea-level lows when the lagoon was dry or partly filled with fresh or brackish water and the "atoll", in fact, was a major island platform. In other words, emergent platforms, but *not* atolls, are the favored sites for mixed-water dolomitization.

Finally, it is recalled that the dolomite body at Mururoa thins, and ultimately disappears, within the most central parts of the platform, reflecting the absence of dolomitizing fluids. The presence of sparitic calcite and well-developed karst surfaces, but the total absence of fibrous marine cements, indicates that these central parts of the platform supported a freshwater lense. Lack of favorable mixing was due to lack of marine water input.

6 Conclusions: The Diagenetic Model

The sedimentary build-up at Mururoa has evolved progressively through fringing reef, platform, to atoll morphologies, this evolution being interrupted by multiple periods of emergence during which freshwater aquifers were developed. Thus, the potential for an intense and varied diagenesis has existed during much of the history of Mururoa.

Calcite diagenesis has involved both marine and meteoric processes, the former being more effective in terms of porosity destruction. The periphery of the buildup is highly cemented (Fig. 12A) mainly by fibrous magnesian calcite and the sequence in peripheral and laterally deviated wells is dense. Sparitic, nonmarine cements increase towards the center of the atoll, where their effect nevertheless is moderate; the carbonates within the central parts of the pile are very porous and permeable.

Dolomitization, which strongly affects the lower two-thirds of the sedimentary pile, is highly variable. Always associated with the dissolution of sedimentary particles and/or matrix, dolomitization is related to the presence of an important aquifer. Because the magnesium required for dolomitization must come from the sea, it is most logical to expect a centripetal hydrodynamic regime. This is confirmed both by isotopic and petrographic variations, and by the overall geometry of the dolomite body. The dolomites thin, and ultimately disappear towards the center of the platform. They are highly cemented by dolosparite in peripheral wells, while more porous "replacement" dolomites predominate towards the inner parts of the dolomite body.

In summary, this oceanic platform exhibits a significant increase both in calcitic and in dolomitic diagenesis towards its periphery. Although these diagenetic trends may be somewhat exaggerated under these dynamic oceanic conditions, the platform nevertheless demonstrates the interplay between marine and meteoric waters. Because these fluids are the principal agents of diagenesis in most ancient carbonate sequences, the diagenetic model of Mururoa Atoll may help in predicting diagenetic changes in ancient carbonate basins.

Acknowledgments. The authors express their thanks to the French Commissariat à l'Energie Atomique (CEA) for permission to publish this research and for paying the costs of the colour-plate. We have pleasure in thanking, in particular, MM. Goutière, Caristan, Gachon, and other colleagues from the CEA for helpful discussions. The French oil Company ELF prepared thin-sections. Isotope analyses were carried out at the Laboratoire de Géochimie Isotopique of the Université de Paris Sud-Orsay. The authors thank their colleague J. C. Fontes for his help and for useful discussion of results.

References

Aissaoui DM (1985) Botryoidal aragonite and its diagenesis. Sedimentology 32:345 – 361

Aissaoui DM, Purser BH (1985) Reef diagenesis: cementation at Mururoa Atoll, French Polynesia. Proc 5th Int Coral Reef Congr, Tahiti 3:257 – 262

Behrens EW, Land LS (1972) Subtidal Holocene dolomite Baffin Bay, Texas. J Sediment Petrol 42:155 – 166

Bourrouilh-le Jan FG (1982) Les étapes de la dolomitisation. Géochimie et sédimentologie d'un atoll surélevé du SW Pacifique et de sa couverture bauxitique: Rennell, Iles Salomon. Livre jubilaire G. Lucas, Mém Géol, Univ Dijon, pp 3 – 20

Brousse R (1985) The age of the islands in the Pacific Ocean: volcanism and coral reef build-up. Proc 5th Int Coral Reef Congr, Tahiti 3

Buigues D (1982) Sédimentation et diagenèse des formations carbonatées de l'atoll de Mururoa, Polynésie française. Thesis Univ Paris Sud-Orsay

Buigues D (1985) Principal facies and their distribution at Mururoa Atoll, French Polynesia. Proc 5th Int Coral Reef Congr, Tahiti 3:249 – 255

Chevalier JP, Denizot M, Mougin JL, Plessis Y, Salvat B (1969) Etude géomorphologique de l'Atoll de Mururoa (Tuamotu). Cah Pac 12:1 – 144

Cullis CG (1904) The mineralogic changes observed in the core of the Funafuti borings. In: Bonney TG (ed) The Atoll of Funafuti. R Soc Lond, pp 392 – 420

Deneufbourg G (1969) Les forages de Mururoa. Cah Pac 13:23 – 44

Evamy BD (1967) Dedolomitization and the development of rhombohedral pores in limestones. J Sediment Petrol 37:1204 – 1215

Fontes JC, Kulbicki G, Letolle R (1969) Les sondages de l'Atoll de Mururoa: aperçu géochimique et isotopique de la série carbonatée. Cah Pac 13:69 – 74

Gachon A (1985) Phases d'abrasion d'un volcan et installation corallienne. Contribution de la morphologie du soubassement volcanique (Atoll de Mururoa, Polynésie Française). Proc 5th Int Coral Reef Congr, Tahiti 3:185 – 191

Gebelein CD, Steinen RP, Garett P, Hoffman EJ, Queen JM, Plummer LN (1980) Subsurface dolomitization beneath the tidal flats of Central West Andros Island, Bahamas. In: Zenger DH, Dunham JB, Ethington RL (eds) Concepts and models of dolomitization. Soc Econ Paleontol Miner Spec Publ 28:31 – 49

Hanshaw BB, Back W, Deike RG (1971) A geochemical hypothesis for dolomitization by groundwater. Econ Geol 66:710 – 724

Harris PM (1978) Holocene marine-cemented sands, Joulter Ooid Shoal, Bahamas. Trans Gulf Coast Assoc Geol Soc 28:175 – 183

James NP, Ginsburg RN, Marszalek DS, Choquette PW (1976) Facies and fabric specificity of early subsea cements in shallow Belize (British Honduras) reefs. J Sediment Petrol 46:523 – 544

James NP, Ginsburg RN (1979) The seaward margin of Belize barrier and atoll reefs. Int Assoc Sediment Spec Publ 3:1 – 191

Kendall AC (1977) Fascicular-optic calcite: a replacement of bundled acicular carbonate cements. J Sediment Petrol 47:1056 – 1062

Kinsman DJJ (1969) Interpretation of Sr^{2+} concentrations in carbonate minerals and rocks. J Sediment Petrol 39:486 – 508

Labeyrie J, Lalou C, Delebrias G (1969) Etude des transgressions marines sur l'Atoll de Mururoa par les datations des différents niveaux de corail. Cah Pac 13:203 – 207

Land LS (1973) Contemporaneous dolomitization of Middle Pleistocene reefs by meteoric water, North Jamaica. Bull Mar Sci 23:64 – 92

Land LS (1980) The isotopic and trace element geochemistry of dolomite: the state of the art. In: Zenger DH, Dunham HB, Ethington AL (eds) Concepts and models of dolomitization. Soc Econ Paleontol Miner Spec Publ 28:87 – 110

Lighty RG (1985) Preservation of internal reef porosity and diagenetic sealing of submerged early Holocene barrier reef, Southeast Florida shelf. In: Schneidermann N, Harris PM (eds) Carbonate cements. Soc Econ Paleontol Miner Spec Publ 36:123 – 151

Loreau JP (1982) Sédiments aragonitiques et leur genèse. Mém Mus Natl Hist Nat Ser C, XLVII

Pierson BJ (1982) Cyclic sedimentation, limestone diagenesis and dolomitization in Upper Cenozoic carbonates of the Southeastern Bahamas, PhD Thesis, Univ Miami, Florida

Purser BH (1985) Dedolomite porosity and reservoir properties of Middle Jurassic carbonates in the Paris Basin, France. In: Roehl PD, Choquette PW (eds) Carbonate petroleum reservoirs. Springer, Berlin Heidelberg New York, pp 341 – 356

Purser BH, Aissaoui DM (1985) Reef diagenesis: dolomitization and dedolomitization at Mururoa Atoll, French Polynesia. Proc 5th Int Coral Reef Congr, Tahiti 3:263 – 269

Repellin P (1977) Contribution à l'étude pétrologique d'un récif corallien: le sondage "Colette", Atoll de Mururoa (Polynésie Française). Cah Pac 20

Rougerie F, Wauthy B (1985) Atoll-oasis functioning and "endo-upwelling" concept. 5th Int Coral Reef Congr, Tahiti 1985

Ruzie G (1985) Apport des techniques géophysiques à l'étude des carbonates dans les atolls: application à l'atoll de Mururoa (Polynésie Française). 5th Int Coral Reef Congr, 2nd Symp, Tahiti 1985

Saller AH (1984) Diagenesis of Cenozoic limestones on Enewetak atoll. PhD Thesis, Univ of Louisiana

Samaden G, Dallot P, Roche R (1985) Atoll d'Eniwetok. Système géothermique insulaire à l'état naturel. La Houille Blanche 2:143 – 151

Sandberg PA (1985) Aragonite cements and their occurrence in ancient limestones. In: Schneidermann N, Harris PM (eds) Carbonate cements. Soc Econ Paleontol Miner Spec Publ 36:33 – 57

Schroeder JH (1985) Sparry calcite cement in Miocene corals from Khor Eit; NE Sudan. Proc 5th Int Coral Reef Congr, Tahiti 3:283 – 288
Shinn EA (1969) Submarine lithification of Holocene carbonate cements in the Persian Gulf. Sedimentology 12:109 – 144
Supko PR (1977) Subsurface dolomites, San Salvador, Bahamas. J Sediment Petrol 47:1063 – 1077
Trichet J, Repellin P, Ouestriere P (1984) Stratigraphy and subsidence of the Mururoa Atoll (French Polynesia). Mar Geol 56:241 – 257
Ward WC, Halley RB (1985) Dolomitization in a mixing zone of near-seawater composition, Late Pleistocene, Northeastern Yucatan Peninsula. J Sediment Petrol 55:407 – 420

The Primary Surface Area of Corals and Variations in Their Susceptibility to Diagenesis

B. R. CONSTANTZ[1]

1 Introduction

Reef-related carbonate sediments are well known for their extreme diagenetic susceptibility in meteoric environments. The rate at which mineralogically unstable aragonite and compositionally unstable magnesium calcites dissolve in meteoric reservoirs is regulated partly by the amount of sediment surface area that is available for reaction with the bulk aquifer. Most common reef allochems are composed of submicron crystals (Sorby 1879) which have disproportionately large surface areas (Fig. 1, Chave and Schmalz 1966). The total surface area of common polycrystalline, carbonate skeletal allochems, is often orders of magnitude higher than monocrystalline carbonate grains of equivalent dimensions (Walter and Morse 1984). Reef-related carbonate allochems thus have very high diagenetic rates that result from the high surface area of their biogenic, polycrystalline constituents.

My principal objective here is to relate the total surface area of carbonate skeletal allochems to their diagenetic susceptibility in meteoric diagenetic environments. The results presented are twofold. First, I will present evidence about the mechanisms of crystal dissolution in reef sediments during meteoric diagenesis, stressing pertinent ultrastructural data from scleractinian corals. This is important because if dissolution is controlled only by diffusion, rather than by

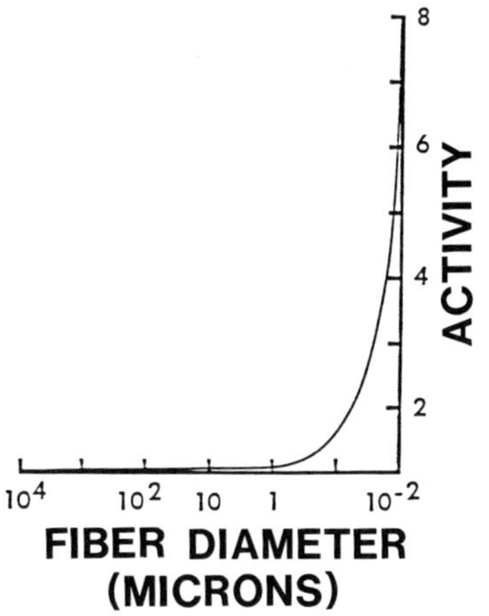

Fig. 1. The solubility of calcite as a function of grain size. (After Chave and Schmalz 1966)

[1] Earth Sciences Board, University of California, Santa Cruz, CA 95064, USA.

Reef Diagenesis
Edited by J. H. Schroeder and B. H. Purser

surface reaction, the high surface areas of carbonate allochems would not influence diagenetic rate because intragranular surfaces would be sheltered from reaction by limited diffusion pathways (Pingitore 1982, Walter and Morse 1984).

Second, I will demonstrate that whole scleractinian coral skeletons have varied primary surface areas but are compositionally and mineralogically equivalent. While no other study has measured the primary surface area of modern skeletal allochems and then followed these same allochem types through a progressive diagenetic sequence and noted their relative susceptibility to reaction, the relationship between the diameter of carbonate crystals and their reaction rate is well established (Chave and Schmalz 1966, Berner and Morse 1974, Walters and Morse 1984). The differing rates of diagenetic reaction of coral skeletons in reef frameworks are observed commonly in outcrop, although no study has specifically explained why.

2 Origin of Samples

Barbados Island is a fore-arc accretionary complex of imbricate thrust sheets of Tertiary turbidites and pelagic sediments located 100 km seaward of the Lesser Antilles Island Arc. Tectonic uplift of Barbados is estimated to have been about 28 cm/1000 years through the Pleistocene (Mesollela 1968). A thin succession of Pleistocene reef tracts cap most of the island and are well known for their meteoric diagenetic histories (Matthews 1968).

Fossil coral skeletons from the successively older Pleistocene coral reefs on Barbados provide material for this investigation because they enable the evaluation of surface area as a diagenetic parameter in a controlled situation. First, because rates of mineralogic transformation of modern reef-related sediments are greatest at the inception of meteoric water diagenesis, relative differences in diagenetic susceptibility between carbonate particles should be best expressed during the initial steps of equilibration; this also avoids the added complexity of multiple diagenetic generations. Second, while magnesium calcites can vary between 0 and 43.5 mol% $MgCO_3$, aragonites of many important corals species have relatively minor variations in $SrCO_3$ incorporated into skeletal aragonite (Constantz 1985) and thus are compositionally and mineralogically equivalent. Third, corals have a diverse range of total surface areas (Table 2) that should amplify and separate the role of reactive surface areas in diagenetic equilibration. Lastly, corals are abundant reef framework builders and allochem contributors.

3 Methods

Forty species of modern corals were collected in Barbados, the Grenedines, Jamaica, the U.S. Virgin Islands, and the Bahamas. All Pleistocene species collected in Barbados live today, justifying a precise comparison with their

Table 1. Methods of laboratory study

Parameter	Method of study
Mineralogy	
composition	X-ray diffraction
	Scanning electron microscope
	some samples stained with Feigl's solution
composition and crystal orientation	Transmission electron microscope
Morphology	Scanning electron microscope
	broken section
	polished and etched section
Permeability	Nitrogen permeometer
	2.5 cm diam. cores
Total surface area	BET[a] gas absorption method
	50, 400 and 8000 μm fractions
	(deKanel and Morse 1979)

[a] BET = *B*runauer, *E*mmet and *T*eller 1938.

modern counterparts. Weathered outcrops were sampled with a corer to depths of at least 20 cm below the weathered surface. Over half the samples were collected from freshly cut quarry walls and new road cuts; additional well cores were also examined.

The analytical methods of this study are summarized in Table 1.

4 Variations in Modern Coral Skeleton

4.1 The Structure of Modern Coral Skeleton

4.1.1 Skeletal Structure (Sensu Stricto)

The whole skeleton, or corallum, of scleractinian corals is an exoskeleton: it is entirely outside the coral's body. The corallum is composed of corallites that are constructed by individual coral polyps. Each corallite is a round polycrystalline, aragonitic tube that is connected to other corallites with varied aragonitic structures, or coenosteum. Vertical partitions, or septa, are arranged radially within the corallite with sixfold symmetry (Fig. 2).

The entire corallum is composed of radiating fans, or sclerodermites, of polycrystalline aragonite fiber bundles, or fasciculi. At the center of every sclerodermite are seed nuclei packets, or 'centers of calcification'. Every structure in the corallum is organized in this basic plan. Linearly accreting sclerodermites form trabeculae which comprise the septa and are of extremely varied organization. Fig. 2 illustrates these structures schematically.

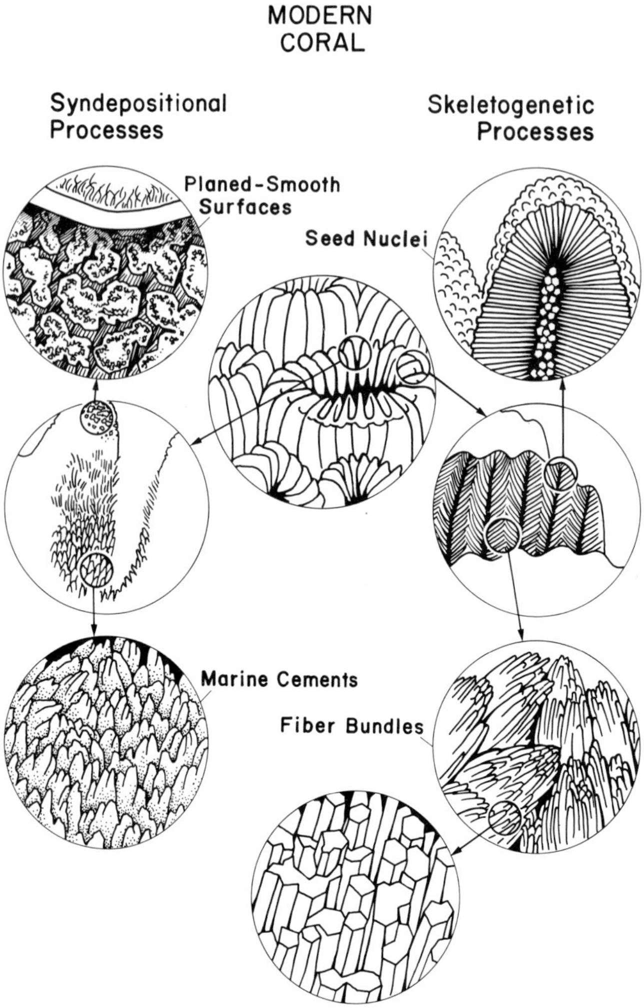

Fig. 2. Schematic illustration of modern coral skeleton architecture, showing the relationship between the skeletal elements (not to scale). Seed nuclei packets (*upper right*) form the center of the trabecular axis. Fiber bundles (*lower right*) nucleate on the seed nuclei packets. Fiber bundles are planed-smooth (*upper left*) by the coral on finished portions of the corallum. As the corallite extends, marine cements grow in vacated parts of the corallite

4.1.2 Syndepositional Processes

Two common syndepositional processes in all living coral skeletons are secondary precipitates that occur on corallum surfaces which have been planed-smooth by the coral (Fig. 3A) (Constantz 1985), and common marine cements (Fig. 3BCD) (Hubbard and Swart 1982, Lightly 1985). Figure 2 illustrates these skeletal components schematically. When corals plane-off growing fasciculi, halting their growth, a smooth layer of submicron aragonite micrite precipitates on these surfaces.

The other syndepositional product is abiotic, marine cement precipitated on vacated corallum surfaces which is usually fibrous, aragonite (Fig. 3 BC). A second, less important marine cement, also abiotic in origin, occurs on both unvacated and vacated portions of the corallum, as micritic crystals, finely disseminated on the corallum surface (Fig. 3 C).

4.2 Results: Variations in Skeletal Structure and Surface Area

4.2.1 Seed Nuclei Packets

The dimensions of individual seed nuclei are only several unit cells across. The distribution and relative percentage of seed nuclei packets is very specific to each scleractinian supertaxon. For instance, the distribution of seed nuclei packets at the sclerodermite centers (Fig. 4 A) are well established systematic characters for scleractinians (Wells 1956). With a few exceptions, the total percentage of skeletal volume represented by the seed nuclei packets is only a few percent (Fig. 4 F). Both the microstructure and mineralogy of scleractinian seed nuclei are unusual; specific information about them will be reported elsewhere.

4.2.2 Polycrystalline Fiber Bundles

Of greatest importance by volume, are the polycrystalline aragonite fiber bundles that are organized into linearly accreting spherulitic fans, or sclerodermites (Wells 1956), which form the skeletal trabeculae (Fig. 4). These aragonite fiber bundles grow rapidly, lifting the coral as they grow. Twinning and growth dislocations are common imperfections in the crystal lattice of aragonite fibers. Figure 5 A is a high magnification scanning electron micrograph of a rapidly growing polycrystalline fiber bundle, illustrating how abundant twin planes and growth dislocations can develop as primary features in the aragonite fibers. Wenk et al. (1983), and Bruni and Wenk (1985) have reported the same features in other modern aragonites. Figure 4 BCDE show examples of variability in organization and crystal size among three important coral species that represent the three major suborders of the Scleractinian order; all are important in the reef framework of modern Caribbean reefs and Pleistocene Barbadian reefs. Table 2 includes TEM measurements of common aragonite fiber diameters.

4.2.3 Surface Area Measurements

Table 2 includes a list of common crystalline diameters encountered in aragonite fibers of modern skeletons of several species, and the reactive surface area at 50 μm, 400 μm, and 8000 μm size fractions. Figure 6 is a plot of measured total surface area against aragonite fiber diameter of eight important species. In most species of the suborder Faviina and to a lesser degree in suborder Astrocoeniina, tight packing of aragonite fibers restricts intercrystalline porosity by cutting off

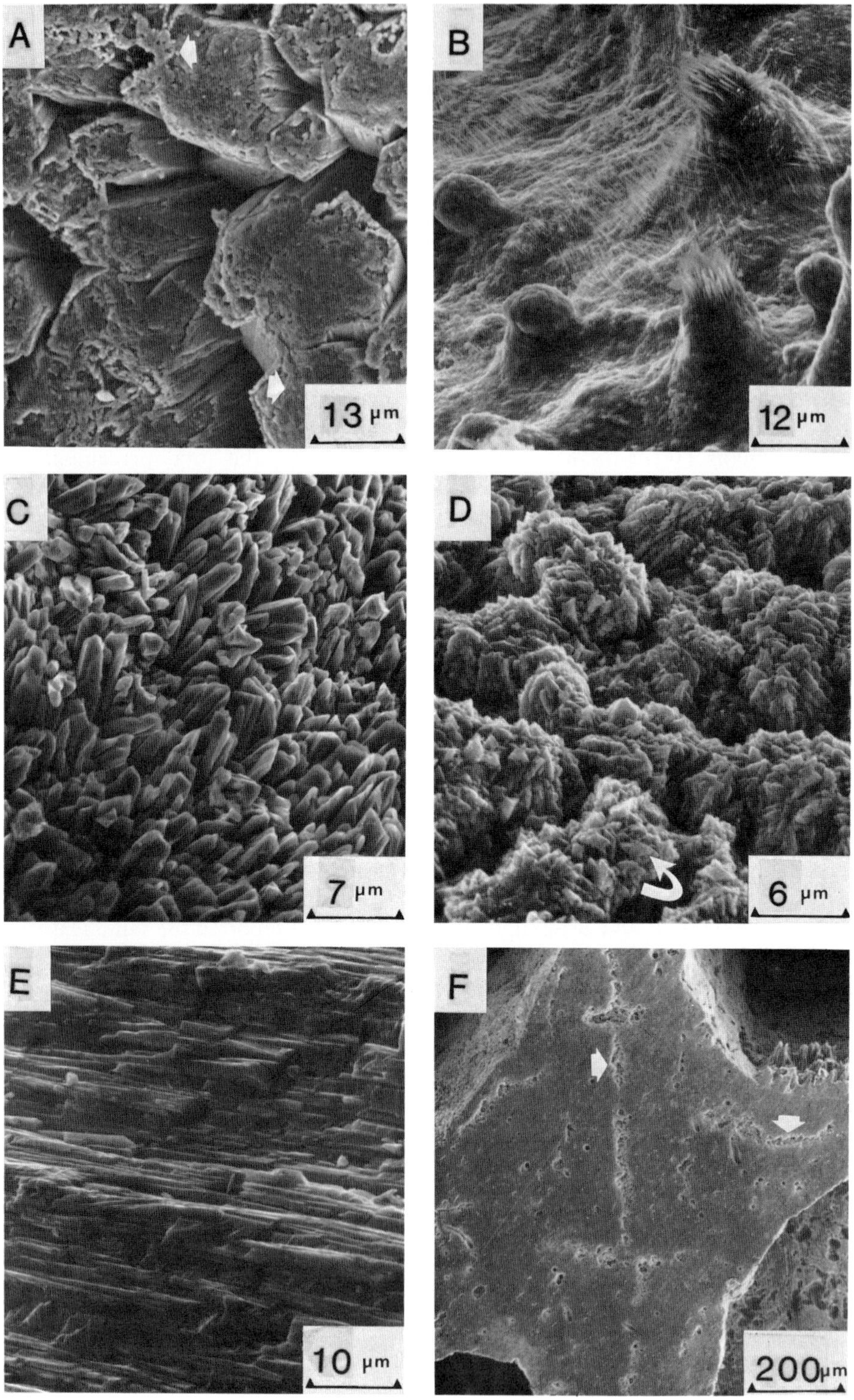

Fig. 3

long grain contacts that reduces total surface area. By contrast, members of the suborder Fungiina have relatively loosely packed crystals with more open grain boundaries, giving the skeleton higher amounts of total surface area; even though the wider crystalline diameters lower the surface area/volume ratio of these crystals, the relatively open crystal packing gives them a higher total surface area. Faviina corals have the lowest total surface are because of relatively larger crystalline diameters and tight crystal packing and annealling. Astrocoeniina corals have crystals with relatively high surface area/volume ratios but have tight crystal packing that reduces, but does not eliminate long intercrystalline pores. These patterns are the result of the differing skeletogenetic mechanisms among these supertaxa.

4.2.4 Porosity and Permeability Measurements

Porosity and permeability measurements for several coral species important in Barbados Pleistocene rocks are included in Table 2. There is no clear relationship between skeletal porosity or permeability and total surface area. In general, fast growing species usually have the lowest skeletal porosity and slow growing species usually have the highest skeletal porosity. Skeletal permeability is determined by each taxon's skeletal architecture and shows no general trend.

5 Variation in Diagenesis Resulting from Variations in Surface Area

5.1 Carbonate Crystal Dissolution and Growth Kinetics in Meteoric Water

The influence of total surface area on the instability of carbonate sediments is dependent on the surface reaction mechanisms operating during diagenesis. If diagenetic reactions were diffusion-controlled only, the total surface area of the sediments would not substantially influence reaction rates because most internal surfaces would be sheltered from reaction.

Fig. 3. A (modern) *Colpophyllia natans.* Smoothed (micritized) polycrystalline aragonite fiber bundles. *White arrows* point to micritic aragonite precipitates deposited on the planed-smooth surfaces. **B** (modern) *Diploria labyrinthiformis.* Aragonite cements in vacant portions of corallum. These fibrous cements have nucleated both on the planed-smooth surfaces and skeletal dentations. On the skeletal dentations, where seed nuclei are at the surface of the skeleton, cements resemble skeletal aragonite fiber bundles. **C** (modern) *Montastrea cavernosa.* Well-developed aragonite marine cements in basal portions of the corallum. **D** (modern) *Porites porites.* Micritic, perfect rhombohedral crystals finely disseminated within the skeletal fiber bundles. These crystals occur both on parts of the corallum adjacent to the calcicoblastic epithelium and on vacant portions of the corallum. **E** (268 000 y) *Colpophyllia natans.* Fractured polycrystalline aragonite fiber bundle. This specimen was collected from a true vadose locality, where there was no evidence of phreatic conditions. **F** (127 000 y) *Siderastrea radians.* Section through septal junction showing preferentially leached seed nuclei packets

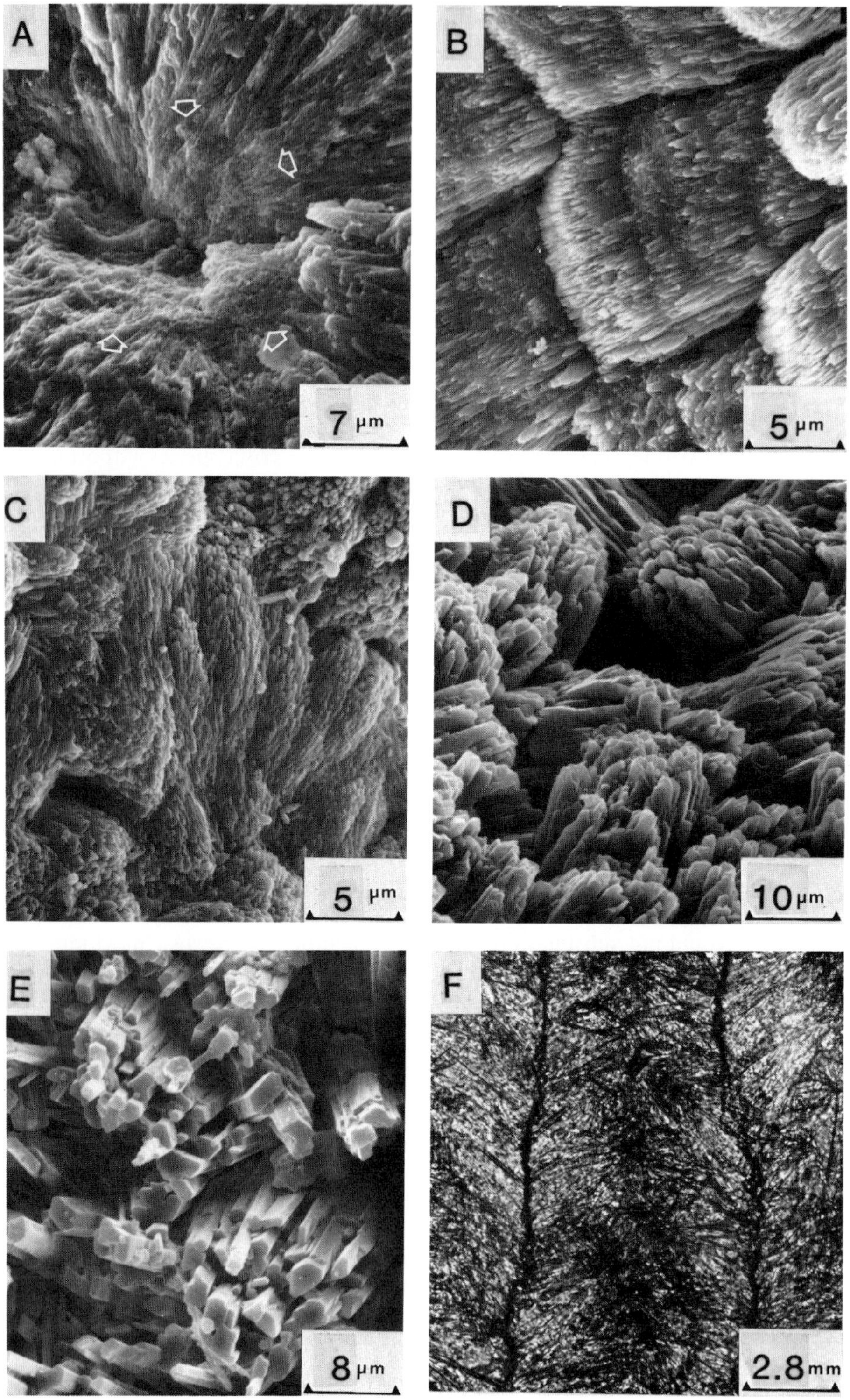

Fig. 4

Two main processes commonly control rates of dissolution or growth of a solid mineral phase in aqueous systems (Berner 1978). (1) Under *surface-reaction control,* reaction rates are limited at the surface of a mineral, typically by detachment or attachment of chemical species at the crystal surface. The concentration of solutes immediately adjacent to the grain will be the same as the bulk solution. (2) Under *diffusion control,* transport of ions in solution to the crystal surface limits reaction rate. The solution immediately adjacent to the crystal will be more or less in equilibrium with the crystal, and gradients will exist between this zone and the bulk solution. Berner and Morse (1974) have shown that different mechanisms of dissolution, and hence rate controlling processes, are dependent on degree of undersaturation.

When the rate of dissolution of a crystal is *diffusion-controlled,* the crystal becomes rounded and dissolution is more or less uniform on all surfaces. When the rate of dissolution is controlled by surface reaction, dissolution is selective; etch pits form around crystal defects, different crystallographic surfaces preferentially dissolve at different rates, and there is no general rounding of crystals. Berner and Morse (1974) demonstrated this by examining experimentally dissolved calcite crystals. Therefore, dissolution effects of meteoric water on Pleistocene coral skeletons can provide criteria for determining the precise mechanism of dissolution. If diffusion is limiting dissolution, strong gradients exist between reaction sites and the bulk reservoir. The degree of skeletal alteration via dissolution is then expected to be greatest on the outsides of polycrystalline skeletal grains, where exchange of dissolved Ca^{2+} and HCO_3^- with the bulk aquifer is greatest; in this way allochems are rounded. Moving into a polycrystalline skeletal grain, amounts of dissolution are expected to decline progressively because of impaired diffusion pathways to the bulk aquifer.

Under *surface reaction-controlled dissolution,* the whole aquifer approaches saturation for calcite and no strong diffusion gradients appear. Because the reservoir everywhere approaches saturation, dissolution rates in small intragranular skeletal pores in communication with the outside are comparable to those on the outer surfaces of polycrystalline skeletons; dissolution thus is dependent on the reactivity of crystal surfaces and rates of dissolution are uniform throughout entire polycrystalline grains. Evidence for surface reaction-con-

Fig. 4. A (modern) *Porites porites.* Fracture through sclerodermite center. Seed nuclei packet is denoted by white arrows. Note the granular apparence of seed nuclei. Each calcium carbonate nuclei is only several unit cells in dimension. Polycrystalline aragonite fiber bundles, radiating outward from the seed nuclei packet, nucleate directly from seed nuclei. **B** (modern) *Acropora cervicornis.* Polycrystalline aragonite fiber bundles. Individual Astrocoeniina aragonite fibers are tightly packed together and very narrow. **C** (modern) *Agaricia agaricites.* Polycrystalline aragonite fiber bundles. Individuel Fungiina aragonite fibers are moderately well packed and relatively wide. **D** (modern) *Colpophyllia natans.* Polycrystalline aragonite fiber bundles. Faviina aragonite fibers have very tight annealing at grain boundaries and are relatively wide. **E** (modern) *Diploria labyrinthiformis.* Newly nucleated polycrystalline aragonite fiber bundles. Individual fibers are diverging from principal growth axes, starting new fiber bundles. **F** (modern) *Siderastrea radians.* Seed nuclei packets are centers of spherulitic nucleation for aragonite fiber bundles. Vertical dark lines are tissue from the calicoblastic epithelium that has become trapped between extending trabeculae. The trabeculae are denoted by the chains of seed nuclei packets

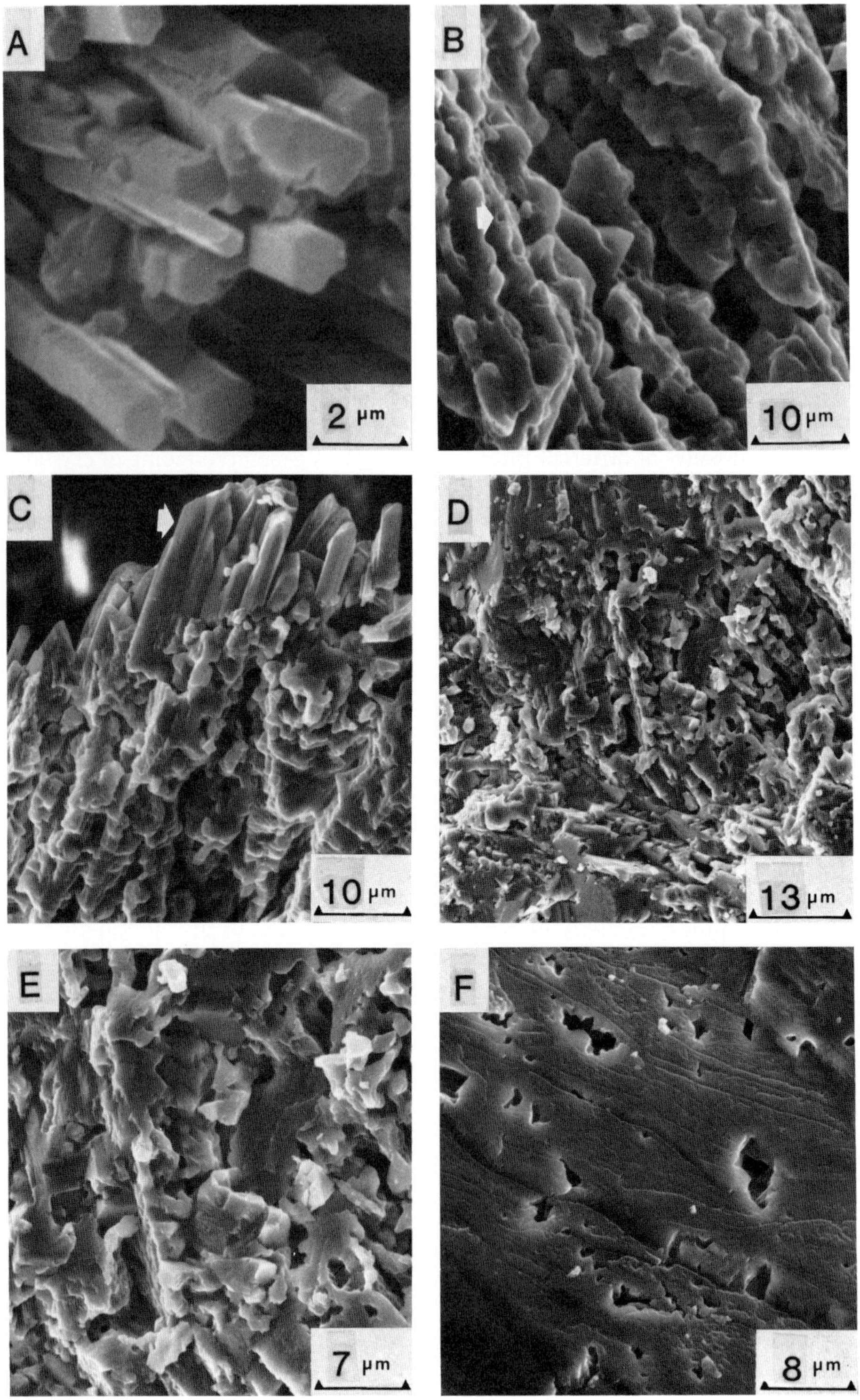

Fig. 5

trol in dissolution includes selective etching of pits around crystal defects and preferential dissolution of some crystallographic surfaces and crystal boundaries.

Before turning to ultrastructural criteria demonstrating mechanisms of dissolution, it is important to establish the position of these mechanisms in models demonstrating carbonate diagenesis. One type, *the bulk solution disequilibrium model,* assumes that diagenetic stabilization proceeds in some restricted reaction zone (microenvironment, "thin film", immobile water, etc.) which is not in equilibrium with the bulk solution of the aquifer (Veizer 1978, Pingitore 1982). In essence, this type of model assumes an open diagenetic system in which a strong diffusion-limited gradient exists between reaction sites and the bulk aquifer. This is interpreted to indicate that dissolution-reprecipitation during aragonite to calcite transformation is very efficient and restricted to some narrow zone. Under these conditions, the bulk aquifer solution is assumed to be undersaturated and the reactive site, to which dissolution-reprecipitation is restricted, is assumed to approach saturation.

An alternative model, *the bulk solution equilibrium model,* is based on evidence suggesting that the bulk solution of meteoric ground water in Barbados and Barbuda Pleistocene reefs is approximately in equilibrium with the carbonate phases in the sediments (Morrow and Mayer 1978a, b). Under this model, rapid exchange with bulk aquifer solution maintains the reactive site at equilibrium with the bulk solution and allows carbonate dissolution to be dissociated from reprecipitation, implying that the bulk aquifer is at or near saturation. Shallow, limestone, meteoric aquifers have Ca^{++} and HCO_3^- as major dissolved ions, and are generally close to equilibrium with calcite (Plummer et al. 1979). Therefore, it is uncommon in limestone to have a meteoric aquifer where dissolution reaction sites and the bulk aquifer are in disequilibrium because both should be very close to the saturation state for calcite. Although, if aquifer recharge is rapid, the bulk reservoir may never reach saturation; this occurs in the vadose zone, where meteoric water percolates through the sediments rapidly.

Fig. 5. A (modern) *Diploria labyrinthiformis.* High magnification view (x 10500) of growing polycrystalline aragonite fiber bundle. Growth dislocations become abundant features of these aragonite fibers. **B** (127000 y) *Montastrea cavernosa.* Partially dissolved aragonite fibers with selective dissolution features that have about the same distribution of growth dislocations in the primary aragonite. **C** (127000 y) *Montastrea cavernosa.* Partially dissolved aragonite fibers extend to edge of septal wall where aragonite marine cements are unaltered. The algal-enhanced, rapid growth of the skeletal fibers give skeletal aragonite fibers a much higher purportion of growth dislocations than the slow growing marine cements, the skeletal aragonite fibers are significantly more susceptible to selective dissolution than the marine cements. **D, E** (127000 y) *Montastrea cavernosa.* LMC crystal growing to entomb partially dissolved aragonite fiber framework. The LMC can be recognized by its darker color that results from better electrical grounding because it is a continuous crystal. The aragonite fibers can be recognized by their lighter color that results from poor electrical grounding because they are not in coherent contact with the LMC. **F** (127000 y) *Diploria labyrinthiformis.* External surface of LMC crystal that entombed aragonite fiber framework; note growth inclusions

Table 2. Skeletal physical properties

Suborder/species	Fiber diameter (μm)	BET surface area ($cm^2\ g^{-1}$)			Porosity (%)	Permeability (darcy)
		50 μm	400 μm	8000 μm		
Astrocoeniia						
Acropora palmata	0.05 – 0.15	770	400	170	48	0.45
Acropora cervicornis	0.05 – 0.15	700	350	90	40	0.40
Faviina						
Diploria labyrinthiformis	0.4 – 0.6	580	430	120	56	1.50
Diploria stigosa	0.4 – 0.6	–	–	–	68	1.50
Diploria clivosa	0.4 – 0.6	–	–	–	61	1.31
Colpophyllia natans	0.5 – 0.7	–	–	–	82	0.45
Montastrea annularis	0.2 – 0.4	420	380	150	69	0.12
Montastrea cavernosa	0.2 – 0.4	640	290	70	52	0.76
Fungiina						
Agaricia agaricites	0.5 – 0.6	840	360	80	50	0.01
Siderastrea radians	0.4 – 0.6	450	360	60	52	0.29
Porites porites	0.4 – 0.6	780	640	120	71	0.87
Porites asteroides	0.4 – 0.6	–	–	–	70	0.80

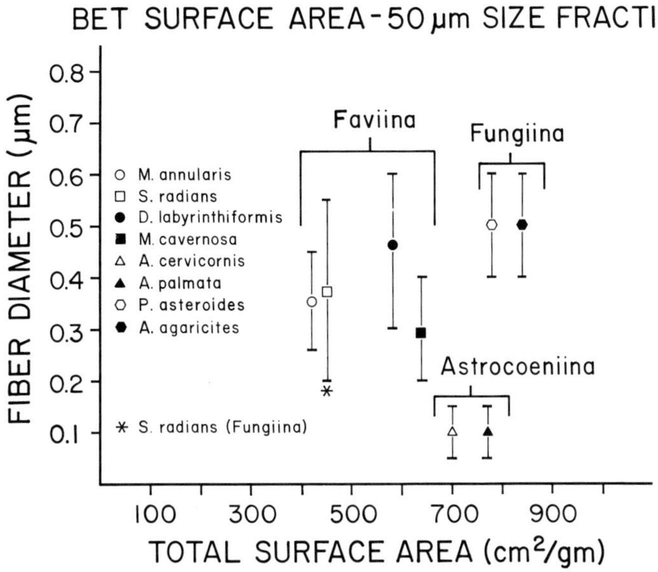

Fig. 6. Plot of typically encountered fiber diameters versus BET surface area at 50 µm size fraction

5.2 Results: Style and Variation in Diagenetic Development in Corals

5.2.1 Diagenetic Reaction Styles

Dissolution of Skeletal Aragonites and Precipitation of Low Magnesium Calcite Cements. Electron microscope examination of Pleistocene corals indicated both surface reaction-controlled dissolution of aragonite, occasionally followed by subsequent LMC cement filling (reprecipitation), and neomorphism of partially dissolved aragonite to LMC. Figure 5B illustrates surface reaction-controlled dissolution of aragonite fibers as etched pits along the fibers and preferential narrowing along specific axes. Dissolution pit boundaries have a similar distribution to twin planes and growth dislocations seen in the primary skeletal aragonite (Fig. 5A).

Figure 5C demonstrates that syntaxial growths of inorganic aragonite cement on the corallum are relatively unaffected by dissolution while their skeletal substrate crystals are profoundly affected by dissolution, reflecting the crystallographic instability of the biogenically precipitated aragonite.

In many examples, partial dissolution of aragonite fibers may occur throughout portions of the corallum. This gives a chalky texture without subsequent filling by LMC cement and indicates that dissolution and reprecipitation are not always coincident, nor is dissolution-reprecipitation necessarily an "efficient mechanism".

Figure 3F illustrates preferentially dissolved packets of seed nuclei at the centers of coral septae with no visible conduits to the outer surface (James 1974, Pingitore 1976). This indicates that diffusion was not a limiting parameter in the

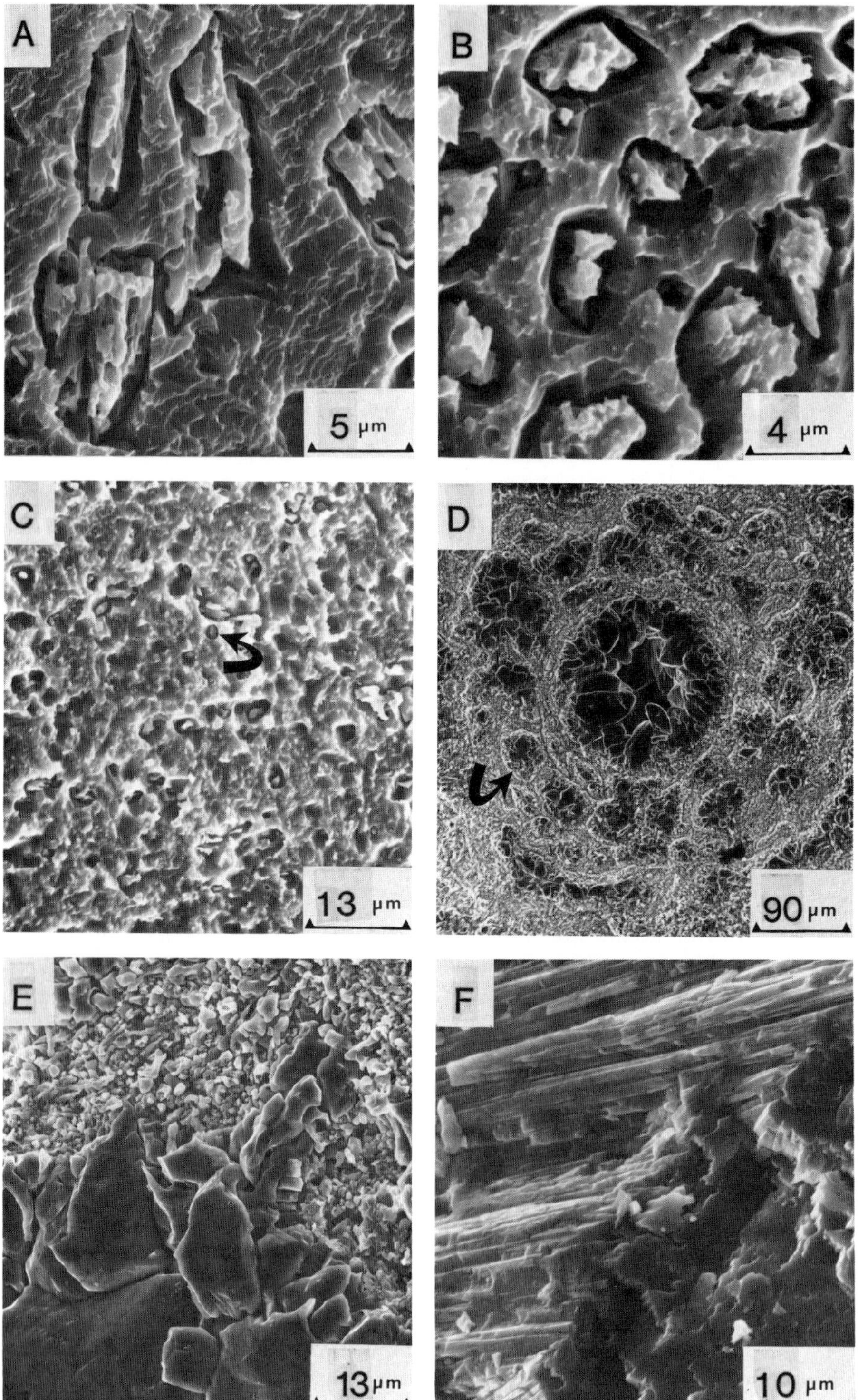

Fig. 7

dissolution of this skeleton because dissolution occurs on the inside and moves outwards.

Figure 5 CDE illustrate partially dissolved aragonite fibers being filled by LMC cement whose crystals locally incorporate aragonite fibers as solid inclusions. Figure 7 ABC illustrate examples of partially dissolved aragonite fiber relics (Sandberg and Hudson 1983) within LMC crystals. Figure 5 F illustrates the surface of a infilling LMC crystal with many fluid inclusions, reflecting its dispersed mode of crystal growth.

Neomorphic Transformations. In addition to LMC infilling, partially dissolved aragonite fibers may be affected by wet mineralogic transformation (neomorphism) across a thin irregular front (Fig. 7 DEF). Figure 7 C illustrates a boring that forms the canal that is cutting the skeleton which is infilled with LMC phreatic cements. There is a neomorphic halo around the duct, indicating that neomorphism of partially dissolved aragonite fibers is closely related to intragranular water movement. This neomorphic LMC never has entombed aragonite fibers, while LMC cements infilling partially dissolved aragonite fiber frameworks always do. Figure 7 EF are additional examples of direct neomorphism of partially dissolved aragonite fibers to LMC cement. They show that the aragonite fiber framework is partially dissolved to micrite-sized fiber fragments that show evidence of their reaction-controlled dissolution history. Porosity between partially dissolved aragonite fibers is very low in Fig. 7 DEF when compared with Fig. 5 BCDE. Neomorphism is observed most often in tightly bound fabrics of partially dissolved aragonite fiber needles. Loose aragonite fiber frameworks are markedly dissolved and have high amounts of intercrystalline porosity between aragonite fibers; these are most often infilled by LMC cements. This implies that neomorphism occurs across a narrow front of a few hundred angstroms in situations where interstitial waters in skeletal micropores are nearly *static* or very restricted (Kinsman 1969; Pingitore 1976), whereas, aragonite fibers become incorporated into LMC mosaics as solid inclusions in less restricted zones where their dissolution is not required for the precipitation of LMC cement.

5.2.2 Surface-Area and the Diagenetic Susceptibility of Corals

Vadose Environments. While most of the Pleistocene reefs on Barbados have been affected both by phreatic and vadose processes due to Pleistocene sea level

Fig. 7. A (268 000 y) *Siderastrea radians.* Relic aragonite fiber bundles entombed in LMC crystals; fiber bundles viewed perpendicular to c-axes. **B** viewed parallel to c-axes. **C** (127 000 y) *Porites porites.* Extensively dissolved aragonite fiber relics entombed in LMC crystal. Due to moderately loose crystal packing, Fungiina fibers typically experience extreme amounts of dissolution prior to LMC infilling, explaining the chalky texture which characterizes them at outcrops. *Black arrow* points to a relic aragonite fiber. **D** (127 000 y) *Diploria strigosa.* LMC cement infilling boring with neomorphic halos. Neomorphic LMC appears darker because its continuous larger crystals have better electrical grounding. Black arrow points to a neomorphic halo, that are incorporated into neomorphic LMC at inception of neomorphism. **E** (127 000 y) *Diploria strigosa.* Neomorphic LMC front progressing through uncemented and partially cemented aragonite fiber framework in polished, etched section. **F** (127 000 y) *Colpophyllia natans.* Neomorphic LMC progressing through unaltered aragonite fibers in broken section fluctuation and differential tectonics on the Barbadian High, Matthews and his students at Brown University have established that specific locations clearly were dominated by phreatic histories and others by vadose histories (see Matthews 1968). At vadose localities lacking evidence of phreatic conditions sampling below weathered surfaces yielded largely unaltered corals, irrespective of age. Apart from vadose cements and small amounts of partial rounding dissolution on the outer surfaces, corals from the localities exhibited internal skeletal preservation indistinguishable from that of modern corals (Fig. 3E).

Many "vadose" localities have some evidence of phreatic conditions, although they may now occupy a vadose setting; evidence for this includes skeletal strontium values below 1000 ppm and large amounts of large equant blocky crystals of low magnesium void-filling calcite.

Phreatic Environments. My studies of the diagenetic state of corals sampled on Barbados in outcrop and well cores suggest that diagenetic rate is much higher in localities with a predominately phreatic diagenetic history than at sites where vadose conditions predominated. Furthermore, at any one locality, particular species are alway most altered. At phreatic localities and "vadose" localities with phreatic histories, coral species of the suborder Astrocoeniina show the greatest degree of alteration. However, most coral species of the suborder Fungiina are usually altered at intermediate levels and species of the suborders Faviina and Caryophylliina are usually intact even when an adjacent Astrocoeniina coral is completely dissolved.

I commonly observe entire coralla where all the aragonite fibers are partially dissolved, and there is no infilling of LMC and no neomorphic LMC. Because of the greatly restricted pore throats in primary coral skeletons (10^3-10^4 nm) and the selective etching of aragonite fibers along crystallographic imperfections, partial dissolution of fibers must have occurred at Ca^{2+} and HCO_3^- concentrations approaching saturation. This condition occurs both within the central parts of "solid" coral skeletons and in parts of grains at the surface of intergranular pores, indicating that saturated conditions that determine surface reaction are not particular circumstances that occur only in restricted micropore reaction zones.

The different fiber diameters typical of each coral species profoundly affect the degree of dissolution that takes place between coralla of each species under the same bulk fluid saturation levels. Because species of the suborder Astrocoeniina have aragonite fiber diameters below 0.1 μm, the range where solubility begins to increase enormously, these corals are expected to be most affected during diagenesis. This is confirmed by the observed diagenetic histories of *Acropora palmata* and *A. cervicornis* in the field; both species exhibit total dissolution or total dissolution of the internal axial corallites (Constantz 1984) and neomorphism to LMC of the outer rim of radial corallites. Modern *A. palmata* and *A. cervicornis* typically have outer radial corallites completely filled with marine cements on basal parts of the skeleton. The preserved outer rinds of these two species in Pleistocene limestones represent neomorphosed parts of the skeleton that were completely filled by marine cement and are thus less susceptible to subsequent dissolution and LMC cement filling because of the wider crystal diameters and the less flawed crystal lattice of the cements. This results in a low porosity fabric that neomorphoses rather than being partially dissolved.

Species of the suborders Faviina, Fungiina, and Caryophylliina are well preserved because they never completely dissolve before LMC cement fills the partially dissolved aragonite fibers or because they neomorphose to LMC. (No Dendrophyllids were collected on Barbados and are a minor part of the fauna of modern Caribbean coral reefs). These suborders are represented by species with significantly wider crystal diameters which are several orders of magnitude less reactive (Fig. 1) (Chave and Schmalz 1966) than the Astrocoeniina aragonite fibers (Table 2). The Faviina species, *Diploria labyrinthiformis, D. strigosa, D. clivosa, Monatstrea cavernosa, M. annnularis, and Colpophyllia natans* are always the very well preserved. All Faviids studied have relatively tight crystal packing which combined with wide aragonite fiber diameters, gives these skeletons low amounts of total surface area (Fig. 6).

Species of the suborder Fungiina studied, *Agaricia agaricites, Siderastrea radians, Porites porites,* and *P. asteroides* generally have loose crystal packing with the exception of *S. radians* which has tight crystal packing and annealing, typical of the Faviids. The family Sidereastreidae share many other characteristics with the Faviids and this seeming discontinuity in ultrastructure style may be more closely related to the uncertainty of scleractinian systematics (Wells 1956) than to ultrastructural heterogeneity within supertaxa (Constantz 1985). It is not suggested that these patterns will prevail throughout entire suborders. *S. radians* preserves very well and behaves like the Faviids studied during diagenesis; this corroborates with its Faviid-like, low reactive surface area (Table 2). The other Fungiids are characterized by loose crystal packing and high amounts of intragranular porosity. These Fungiids are very susceptible to "chalkification" (James 1974) where all the skeletal aragonite fibers are partially dissolved, often with no subsequent cementation or neomorphism. Extensive dissolution of aragonite fibers in the Fungiina typically occurs before void filling cements incorporate them as solid inclusions. Figure 7C shows extremely dissolved aragonite fiber relics in *P. porites*; note that their contribution to crystal volume is very low. No examples of neomorphism were observed in the Fungiina species.

The only Caryophylliina species, *Eusmillia fastigata,* was always exceptionally well preserved as either all aragonite or all neomorphic LMC; this species has tight primary crystal packing and very large aragonite crystals.

6 Interpretations

Differences in diagenetic rate and process between these species appears to be dependent on two properties of their primary skeletons; (1) aragonite fiber width, (2) packing of aragonite fibers. The Astrocoeniina are altered pervasively by meteoric waters compared to other taxa indicating that their very soluble aragonite fibers are dissolving rapidly. Even with relatively low amounts of intragranular porosity, crystal boundaries in direct contact are preferentially dissolved, thus opening the secluded intragranular channels between aragonite fibers. Because of the relatively insignificant portions of lattice comprising each aragonite fiber (about 200 unit cells in width) their high reactivity makes them susceptible to rapid dissolution.

Faviina corals, on the other hand, are well preserved relative to other taxa, this reflecting their wide aragonite fiber diameters and low intragranular skeletal porosity which reduces the amounts of reactive surface area and shields the bulk aragonite from dissolution. Dissolution occurs preferentially along aragonite fiber boundaries, but the greater fiber diameter considerably reduces the volume of material dissolved, leaving relatively thick, partially dissolved aragonite fibers. This results in entire coralla exhibiting partially dissolved aragonite fibers, with the characteristic selective etching of surface reaction-controlled dissolution, plus aragonite fibers entombed in LMC void-filling cement, and neomorphic LMC. Neomorphism to relic-free LMC mosaics is quite common in the Faviina-type corals, where this phenomenon occurs along a narrow front (a few hundred nanometers in width) where permeability is very high (Fig. 7DEF).

Fungiina corals have relatively wide crystals and long interfiber spaces. This gives the skeleton a relatively high reactive surface area, even though the individual wide fibers have comparatively low surface areas. These corals undergo pervasive dissolution to completely chalkified frameworks of partially dissolved aragonite fibers. When these friable, chalky coralla are infilled with cements, the aragonite relic inclusions are very reduced in diameter and are relatively sparse (Fig. 7C) in LMC crystal mosaics.

The Caryophylliina species, *E. fastigata,* with large primary aragonite crystals that are tightly packed has low amounts of intracrystalline porosity, favoring neomorphism.

7 Discussion

7.1 Reactive Surface Area of Carbonate Sediments

During meteoric diagenesis, both the outer surfaces of whole allochems and intragranular surfaces are potential reactive surfaces. Because intragranular surfaces of polycrystalline allochems are sheltered, it may seem logical to assume that, during dissolution, these internal surfaces are relatively unimportant reaction sites. When strong gradients in Ca^{2+} and HCO_3^- concentration exist between sites of dissolution on intragranular surfaces and the bulk solution, intragranular dissolution is limited by the rates at which solutes diffuse away from the reaction zone since these internal surfaces are relatively isolated (Pingitore 1982). However, the development of strong dissolution gradients in Barbados aquifers is an ephemeral phenomenon that seldom interrupts a predominant state of bulk saturation, and there are no diffusion gradients (see below). Moreover, the frequent observation of uniform dissolution throughout polycrystalline skeletons suggests that diffusion-limited regimes, that would impair dissolution of micropore surfaces within grains, are not important in meteoric, phreatic environments on Barbados.

Walter and Morse (1984) measured total surface area and reactive surface area on Fungiina coral skeleton, echinoid test, and *Halimeda* at 81 μm, 153 μm, 215 μm, 275 μm, and 513 μm size fractions. Their results demonstrate that, during undersaturated conditions, the intragranular surfaces of these allochems are relatively unavailable for reaction with the bulk solution; dissolution was significantly lower than expected from total measured surface areas (Brunauer et al. 1938, de Kandel and Morse 1979). This conclusion agrees with those of Berner and Morse (1974); at relatively undersaturated levels, dissolution is diffusion-limited and restricted intragranular surfaces may be expected to be relatively sheltered from reaction by limited diffusion pathways compared to surfaces on the outside of allochems.

In natural, meteoric, limestone aquifers, the bulk phreatic waters are commonly at, or near, saturation for calcite and dissolution is surface reaction-limited. Solution within both small intragranular pores and the bulk solution of stable limestone aquifers approach saturation for calcite; diffusion of Ca^{2+} and HCO_3^- does not limit dissolution. Only at the edges of a limestone meteoric reservoir, where foreign undersaturated waters are introduced and begin to become saturated, can strong saturation gradients exist. Furthermore, when the saturated reservoir is flushed with foreign undersaturated meteoric waters, diffusion gradients are established but soon decay. However, since the bulk phreatic waters of limestone reservoirs are usually saturated for calcite, rates of surface reaction-controlled dissolution are similar in both grain micropores with very low permeability and on the outside of grains, in situations of high permeability. This implies that dissolution reaction rates within grain micropores are no more dependent on the removal via diffusion of Ca^{2+} and HCO_3^- than reactions at outer surfaces, where the bulk solution is similarly saturated. When dissolution is limited by total surface area (Berner and Morse 1974), as just described, the enhanced solubility of small grains with radii less than 1.0 μm begins to be important relative to the solubility of large masses of stable lattice. Chave and Schmalz (1966) note that particles of calcite with diameters less than 0.1 μm are quite unstable in water and that a major increase in solubility is expected of grain radii between 0.1 and 0.01 μm (Fig. 1). Table 2 and Fig. 6 show that aragonite fiber diameters for corals are in the range of enhanced solubility, particularly the Astrocoeniina which fall in the range of the major solubility increase. Seed crystal packages appear to dissolve most quickly because of the seed nuclei's small crystal size and resultant high surface energy.

Walter and Morse (1984) have shown that BET surface areas of three types of skeletal carbonates are all significantly greater than those calculated for equidimensional solid particles of equivalent size. Even though the diameters of typical coral aragonite fibers are arranged into disorganized spherulitic fans, the negative relationship between size fraction and measured BET surface area indicate that part of the intragranular aragonite fiber surfaces are not in communication with the external surfaces of the skeleton (Table 1) and play no part in any initial surface reaction. As grain size increases, the proportion of internal surface area in communication with the external grain surface decreases because of the progressive isolation of grain interiors. Walter and Morse (1984) found that the BET surface areas of coral grains decreased by a factor of 2, from 81 μm to 513 μm. The amount of surface area decrease predicted for solid equidimensional particles would be a factor of 6 over this same size interval, indicating that some, but not most, of the intragranular micropores are cut off as grain size increases. Table 2 shows that measured BET surface areas decrease by a factor of less than 2 from the 50 μm fraction to the 400 μm size fraction, a similar range and value as indicated by Walter and Morse (1984). This indicates that even though coral skeleton dissolves as an essentially nonporous material at high degrees of undersaturation, under conditions approaching saturation, dissolution within polycrystalline grains is pervasive. This interpretation corroborates well the geological observations.

7.2 Reactive Surfaces and Temporal Variations in Meteoric Waters over Geological Time

The study of Walter and Morse (1984) and this study are similar in one respect; both measured BET surface area of skeletal carbonates for comparison with their chemical reactivity. In another respect, however, the two studies are very different, although their results are quite complementary. Walter and Morse (1984) measured rates of dissolution at a fixed level of undersaturation over a period of a few minutes; this study measured rates of dissolution (and later reprecipitation or neomorphism) at unknown and fluctuating degrees of saturation over a period of several hundred thousand years. A major difference between the two studies is that the present meteoric groundwater on Barbados varies both spatially and temporally in degree of saturation with respect to carbonate phases (Steinen et al. 1978); similar variations have almost certainly prevailed over the last several hundred thousand years. Steinen et al. (1978) calculated average groundwater seepage velocities, using average rates of recharge, rock porosity, and water table thickness, of 43 – 143 m yr^{-1}. They point out that average rates of aquifer flow are significantly less than instantaneous rates following periodic flushing by abnormal rainfall and storm-driven tides. They conclude that during most of the year, the rate of meteoric water flux is very low and that single events with instantaneous rates of several thousand m yr^{-1} give even the near-static averages somewhat high values. In other words, during most of the year, the bulk solution within the Pleistocene reefs on Barbados is essentially static. Periodic storms and storm-driven tides flush the reservoir and initiate a new static period that evolves toward saturation. Water held in the reservoir during these static periods can rapidly approach saturation for calcite because of the high reaction rates with the metastable carbonate sediments present.

The history of the Barbadian meteoric water reservoir, has been dominated by long periods in which the bulk aquifer of the Pleistocene reefs has approached saturation for calcite, during which no Ca^{2+} or HCO_3^- gradients were present in the reservoir, and surface reaction-controlled dissolution of carbonate allochems dominated, including in the smallest micropores. Because of the near-saturation state of the interstitial waters, both surface reaction-controlled dissolution and

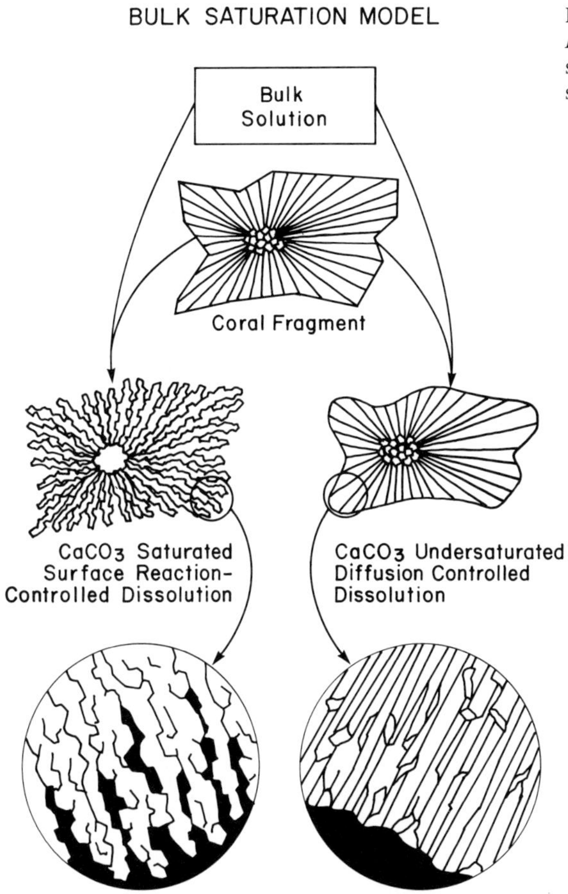

Fig. 8. Schematic illustration of the *Bulk Saturation Model,* showing how saturation state controls diagenetic style

LMC crystal growth could be temporally contemporaneous, though not necessarily restricted to the same site. During episodic periods of rapid meteoric water recharge in the aquifer, steep gradients may exist between dissolution sites within the micropore surfaces of the sediment and the bulk aquifer. Because of the tortuous pathways of rapid meteoric water flux through the aquifer, the dissolution in intragranular micropores, compared to the outer pore surfaces, then becomes a process that is diffusion-limited. In such a model, here termed *the bulk saturation model,* both diffusion-controlled regimes and surface reaction-controlled regimes occur at different times within a meteoric carbonate reservoir, depending on the saturation state of the bulk aquifer solution with respect to Ca^{2+} and HCO_3^-. The bulk saturation model is illustrated schematically in Fig. 8. This model explains and accommodates existing studies that support differing points of view (Morrow and Mayer 1978ab, Veizer 1978, Pingitore 1982).

8 Conclusions

8.1 Surface Reaction Processes in Meteoric Water Diagenesis of Reefs

Pleistocene reefs on Barbados have undergone meteoric diagenesis over the last several hundred thousand years under conditions where the bulk groundwaters approached saturation for calcite, promoting *reaction surface-controlled dissolution* of coral skeletons throughout the reservoir. Evidence for this conclusion is illustrated by preferential etching of the finest crystallographic defects throughout entire coralla and rates of dissolution that are equivalent throughout coral skeletons, both at the outer surfaces and within the allochem interiors. This condition is most prevalent in meteoric, phreatic environments.

Occasionally, rapid recharge of the aquifer flushes saturated waters, replacing them with highly undersaturated meteoric waters that produce steep diffusion gradients of Ca^{2+} and HCO_3^- between dissolution reaction sites and the bulk aquifer. This condition is most prevalent in meteoric vadose environments, but also occurs periodically in phreatic environments. This promotes a *diffusion-controlled regime,* where dissolution reactions are primarily on the surfaces of large pores between grains and not in intragranular micropores where exchange with the bulk reservoir is limited by diffusion (Pingitore 1982). Reef diagenesis under natural meteoric water conditions is best described by a model of *bulk solution saturation* because it describes an entire range of reaction styles evolving through the history of a reservoir.

8.2 Surface Area of Carbonate Skeletal Allochems and Diagenetic Reactivity

Both subcrystal diameter and intraskeletal permeability affect the solubility of skeletal carbonates during dissolution in meteoric waters near saturation for calcite. Aragonite coral skeletons with subcrystal diameters below 0.1 μm are very susceptible to diagenesis, even when relatively low amounts of intraskeletal porosity are present, because dissolution occurs preferentially along grain boundaries and channels are opened between fibers. Skeletons with subcrystal aragonite fibers wider than 0.1 μm are similarly susceptible, but to a much lesser degree, rarely reaching total dissolution. Amounts of intraskeletal porosity greatly influence rates of meteoric dissolution. However, whole skeletal porosity and permeability are not important factors affecting the diagenetic susceptibility of skeletons in meteoric waters near saturation for calcite, though they would clearly be significant in a diffusion-controlled dissolution regime. Many biogenic aragonites are further unstable because of high densities of crystallographic imperfections that favor preferential dissolution. Both processes of void-filling cementation and neomorphism occur in partially dissolved subcrystal frameworks, the two end products being distinguished by the presence of subcrystal relics in void-filling cements and absence of relics in neomorphic, low magnesium calcites. Neomorphism is more prolific in low permeability frameworks and LMC cementation in high permeability frameworks.

Acknowledgments. Funded by National Science Foundation Grant #EAR-8313414, a doctoral dissertation grant for the author, sponsored by L. F. Laporte. This article is a chapter in the author's dissertation. I thank R. E. Garrison, L. F. Laporte, W.-C. Dullo, and especially B. H. Purser, and J. H. Schroeder, for their critical reviews of this manuscript. I spoke on this topic at the Fifth International Coral Reef Congress in Papeete, Tahiti in May 1985 and received many interesting questions and comments. R. K. Trench originally suggested Barbados as a suitable locality to study differential diagenesis between corals. Wayne Hunte, director of the Bellairs Research Institute on Barbados, was very accommodating during my field work, as were the Barbadian National Water Works, who helped me obtain well cores, and many farmers, ranchers, fishermen, rum refineries, government officials, and citiziens of Barbados. Gulfranz Ahmed, of the Stanford Petroleum Research Institute, generously instructed me in use of their permeometer. I have gained much information in discussions with L. M. Walters, and J. W. Morse, who performed the BET surface area measurements. Technical assistance from the following individuals is greatly appreciated: Annemarie Meike, Stephen Wessels, Jan Nowell, Patsy Bolt, Gene Gonzales, Chris Flannigan, Robbie Marshall, Michael Moore, Pat Conroy, Syd Temple, Keith Craig, Jeremy Woodley, Bridgette McDonald, and Wayne Hunte.

References

Berner RA (1978) Rate control of mineral dissolution under earth surface conditions. Am J Sci 278:1235 – 1252

Berner RA, Morse JW (1974) Dissolution kinetics of calcium carbonate in sea water IV. Theory of calcite dissolution. Am J Sci 274:108 – 134

Brunauer S, Emmet PH, Teller E (1938) Adsorption of gases in multimolecular layers. J Am Chem Soc 60:309 – 319

Bruni FS, Wenk H-R (1985) Replacement of aragonite by calcite in sediments from the San Cassiano Formation (Italy). J Sediment Petrol 55:159 – 170

Chave KE, Schmalz RF (1966) Carbonate-seawater reactions. Geochim Cosmochim Acta 30: 1037 – 1048

Constantz BR (1984) Comparative function of microarchitecture in *Acropora cervicornis* and *Acropora palmata.* Palaeontogr Am 54:48 – 52

Constantz BR (1985) The skeletal ultrastructure of Scleractinian Corals. Thesis, Univ California, Santa Cruz

De Kanel J, Morse JW (1979) A simple technique for surface area determination. J Phys E Sci Instrum 12:272 – 273

Hubbard JAEB, Swart PK (1982) Sequence and style in scleractinian oral preservation in reefs and associated facies. Palaeogeogr Palaeoclimatol Palaeoecol 37:165 – 219

James NP (1974) Diagenesis of scleractinian corals in the subaerial vadose environment. J Sediment Petrol 48:785 – 799

Kinsman DJS (1969) Interpretation of Sr^{++} concentrations in carbonate minerals and rocks. J Sediment Petrol 39:486 – 508

Lightly RG (1985) Preservation of internal reef porosity and diagenetic sealing of submerged early Holocene Barrier Reef, Southeast Florida Shelf. In: Schneiderman N, Harris PM (eds) Carbonate Cements. Soc Econ Paleontol Miner Spec Publ 36:123 – 151

Matthews RK (1968) Carbonate diagenesis: equilibration of sedimentary mineralogy to subaerial environment; coral cap of Barbados, West Indies. J Sediment Petrol 38:1110 – 1119

Mesolella KJ (1968) The uplifted reefs of Barbados, physical stratigraphy, facies relationships, and absolute chronology, Thesis, Brown Univ

Morrow DW, Mayers IR (1978a) Simulation of reef diagenesis – a model based on strontium depletion. Can J Earth Sci 15:376 – 396

Morrow DW, Mayers IR (1978b) Model of limestone diagenesis – a model based on strontium depletion: Reply. Can J Earth Sci 15:1685 – 1687

Morse JW (1974) Dissolution kinetics of calcium carbonate in sea water III. A new method for the study of carbonate reaction kinetics. Am J Sci 274:97 – 107

Pingitore NE (1976) Vadose and phreatic diagenesis: processes, products and their recognition in corals. J Sediment Petrol 46:985 – 1006

Pingitore NE (1982) The role of diffusion in carbonate diagenesis. J Sediment Petrol 52:27 – 39

Plummer LN, Parkhurst DL, Wigley TML (1979) Critical review of the kinetics of calcite dissolution and precipitation. In: Jenne EA (ed) Chemical Modeling in Aqueous Systems. Am Chem Soc Symp Ser 93:537 – 573

Sandberg PA, Hudson JD (1983) Aragonite relic preservation in Jurassic calcite-replaced bivalves. Sedimentology 30:879 – 892

Steinen RP, Matthews RK, Sealy HA (1978) Temporal variations in geometry and chemistry of the freshwater phreatic lens: the coastal carbonate aquifer of Christ Church, Barbados, West Indies. J Sediment Petrol 48:733 – 742

Sorby HC (1879) Anniversary address of the president: structure and origin of limestones. Geol Soc Lond Proc 35:56 – 95

Veizer J (1978) Simulation of limestone diagenesis – a model based on strontium depletion: Discussion. Can J Earth Sci 15:1683 – 1685

Walter LM, Morse JW (1984) Reactive surface area of skeletal carbonates during dissolution: effect of grain size. J Sediment Petrol 54:1081 – 1090

Wells JW (1956) Scleractinia. In: Moore RC (ed) Treatise on invertebrate paleontology, Pt F. Geol Soc Am, pp 328 – 443

Wenk H-R, Barber DJ, Reeder RJ (1983) Microstructures in carbonates. In: Reeder RJ (ed) Carbonates: mineralogy and chemistry. Miner Rev Miner Soc Am 11:301 – 367

Variation in Diagenetic Sequences: An Example from Pleistocene Coral Reefs, Red Sea, Saudi Arabia

W.-C. DULLO[1]

1 Introduction

The coastal plain of the Red Sea is characterized by young Tertiary and Quarternary deposits of continental or marine origin. The latter frequently form terraces of different elevation. Other parts of the coast include steep cliffs of Precambrian basement.

The Pleistocene marine terraces are characterized by fringing reefs comparable to their Recent counterparts. Because of the great similarity between Recent and Pleistocene carbonates, together they provide an ideal site for the study of variations of meteoric diagenesis.

The series of Pleistocene terraces are characterized by a progressive diagenetic sequence which influences the fossil record of the various biota. James (1974) has described the coexistence of different diagenetic alterations in scleractinian corals from the Pleistocene of Barbados and Scherer (1975) has illustrated the possible pathways of these diagenetic processes in Pleistocene corals from Florida and the Bahamas. Gvirtzman and Friedman (1977) were the first to demonstrate a progressive diagenetic sequence for the corals in the Gulf of Aqaba region. Dullo (1983, 1984) has noted the different stability of the various biota in the Pleistocene terraces of the Red Sea. This preservation potential reflects different mineralogies and calcification patterns. Finally, Schroeder (1984) has presented a petrogenetogram for scleractinians expressing spatial variations in diagenetic sequences.

The present paper is based on two cross-sections in the Pleistocene carbonates of the Red Sea coast which are characterized by a variation in the rate of diagenetic alteration. Particular attention is given to the scleractinians whose diagenetic evolution is influenced both by external factors of the diagenetic environments and by internal skeletal parameters such as the arrangement of the crystal fabric and the size of constituent aragonite needles.

2 Material and Methods

Samples from the Pleistocene terraces are very porous. Therefore they have been impregnated with resin under vacuum for producing thin sections. The same method was used for Recent unlithified skeletal sand and cemented carbonates.

[1] Institut für Palaeontologie, Universität Erlangen, Loewenichstraße 28, 8520 Erlangen, FR-Germany

Reef Diagenesis
Edited by J. H. Schroeder and B. H. Purser

The quantitative composition of the biota and nonskeletal grains was established by point counting (grain solid: Flügel 1982), using 65 thin sections, 10×15 cm in size.

Three scleractinian species were collected from the different terraces and onlapping reefs: *Stylophora pistillata* (Esper) (Astrocoeniina, 20 specimens), *Porites lutea* Milne Edwards and Haime (Fungiina, 20 specimens), and *Favia laxa* (Klunzinger) (Faviina, 14 specimens). The microstructure and diagenetic alteration of the scleractinians were studied by means of thin sections and SEM, using fresh fractures, cleaned with distilled water and pressured air. The carbonate mineralogy was analyzed by X-ray diffraction.

3 Geological Setting, Facies, and Biotic Composition of the Terraces

3.1 Section I

Section I runs along the southern cliff of Sharm al Harr, a drowned valley which was formed during the last glacial sea level low (Gvirtzman et al. 1977). Even the Recent fringing reef in front of the profile follows this old erosional pattern (Fig. 1). The lower terrace unit here reaches a maximum elevation of 5 m above present sea level.

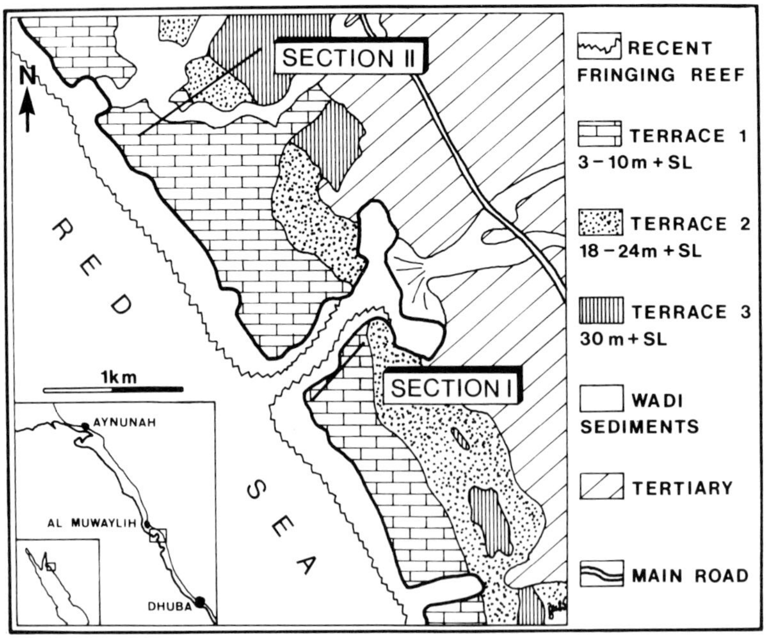

Fig. 1. Map of the Sharm al Harr area, showing location of sections studied

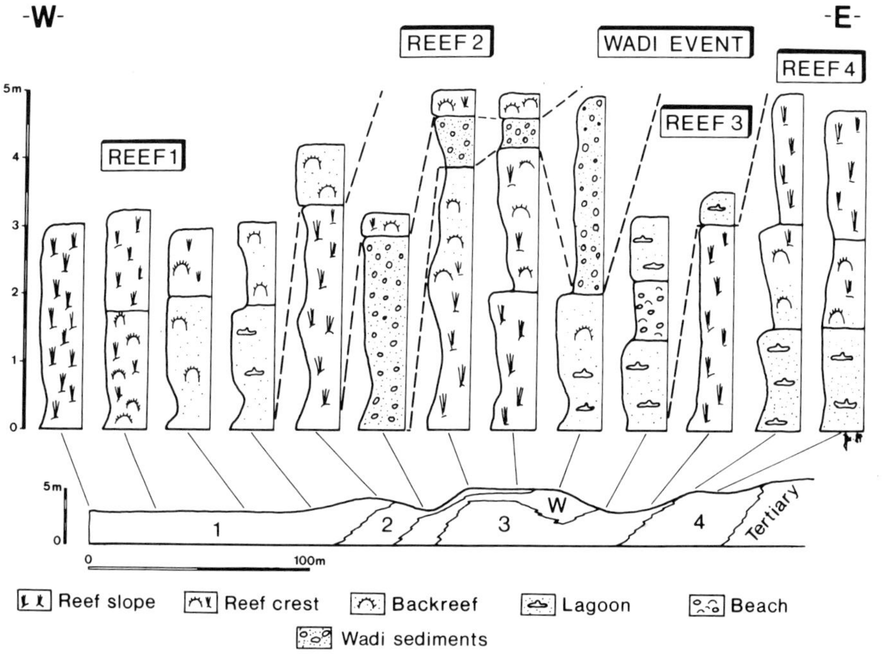

Fig. 2. Cross-section through the lower terrace south of Sharm al Harr. Facies evolution within vertical sections reveals transgressive sequences, belonging to four reef cycles

Thirteen vertical sections have been measured from the seaward margin in the west to the landward edge in the east. Each vertical section is characterized by a transgressive sequence, beginning with a basal landward facies and ending in seaward facies types. However, some sections exhibit an inverse sequence, the result of four onlapping reef cycles. These cycles are interrupted by an erosion of a wadi which cuts through reef 3 and on which onlaps reef 2 (Fig. 2).

The four cycles seem to reflect reef formation comparable to that proposed in the model by Mergner and Schuhmacher (1974), demonstrated locally (Al Sayari et al. 1983) in the Gulf of Aqaba. The formation of the four reefs within the lower terrace at Sharm al Harr is the result of at least four sea level changes, resulting in repeating onlap sequences. The fluctuation was in the order of few meters. Such fluctuations are known from the last highstand of sea level, which has occurred between 82 000 and 125 000 years BP. (Mesolella 1969, Matthews 1973, 1984, Gvirtzman and Friedman 1977). The existence of these small-scale sea level changes is reflected by the diagenetic features and fossil preservation in the different reef cycles (Fig. 3).

The four cycles are characterized by a progressive diagenetic modification of the biotic composition. Diagenesis is dominated by leaching and minor cementation of the resulting porosity. The quantitative distribution of biota, nonskeletal grains, cement, matrix, and porosity (Fig. 3) shows that algae, aragonitic molluscs, foraminifers, and unknown biogenic allochems are strongly affected by leaching and increased porosity. Although scleractinians also are

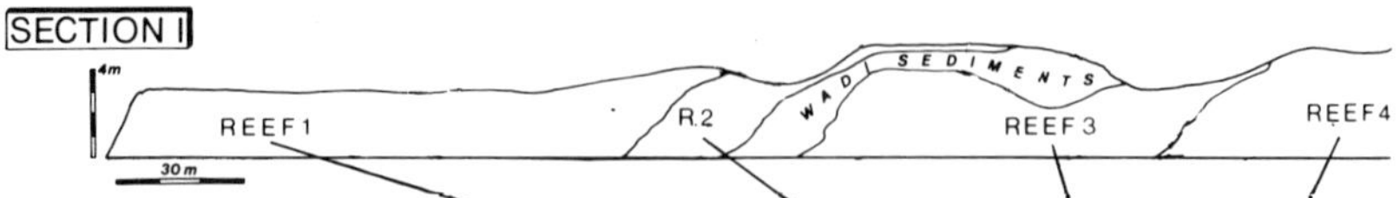

	REEF 1	REEF 2	REEF 3	REEF 4
Scleractinians	14.1	13.3	13.7	13.7
Calcareous Algae	16.6	13.8	10.4	8.8
Foraminifers	5.3	4.9	1.8	2.2
Echinoderms	1.5	0.9	1.0	2.8
Aragonitic Molluscs	2.5	1.8	1.2	1.0
Calcitic Molluscs	1.5	2.6	1.1	1.6
Ostracods, Bryozoans	1.8	0.8	0.8	1.4
Unknown Biogen. Allochems	7.4	7.5	5.7	3.2
Cement	10.8	11.0	14.9	15.2
Micrite	21.7	24.4	25.2	25.1
Peloids	3.4	2.7	2.6	3.8
Silicates	3.5	5.4	11.2	4.4
Porosity	9.9	10.9	10.4	16.8

	RECENT	TERRACE 1	TERRACE 2	TERRACE 3
Scleractinians	16.8	15.1	9.4	4.1
Calcareous Algae	19.0	20.7	15.6	10.4
Foraminifers	8.3	2.0	1.2	0.6
Echinoderms	3.2	1.7	1.7	1.7
Aragonitic Molluscs	7.0	5.0	1.8	1.7
Calcitic Molluscs	1.6	0.9	1.0	1.0
Ostracods, Bryozoans	2.4	0.8	0.8	0.9
Unknown Biogen. Alloch.	9.2	8.9	2.0	2.0
Cement	5.1	10.0	32.8	47.7
Micrite	13.3	18.0	18.4	14.5
Peloids	3.0	2.0	2.7	4.4
Silicates	2.3	0.2	2.2	5.4
Porosity	8.8	14.7	10.4	5.6

Fig. 3. Quantitative composition of biota, nonskeletal grains, cement, matrix, and porosity of the different reef cycles within section I and of the three terraces in section II in comparison to the Recent environment in front of the sections (area %)

affected by freshwater diagenesis, because of neomorphism they are less reduced in number than the other biota.

The alteration of marine cements by fresh water is also progressive. Reef 1 exhibits up to 56% of marine cements (point counter in thin section), which decrease to 18% in reef 2 and to 7% in reef 3; in reef 4 they comprise only 4% of the total cement. This evolution indicates repeated emergence and meteoric diagenesis prior to the formation of the overlying reef cycle.

3.2 Section II

Section II traverses three morphological terrace units in contrast to section I (Fig. 1). Due to the lack of deep erosion along wadis, it was impossible to measure vertical sections and to map the different reef cycles. However, at least between terrace one and two an onlap was observed. The origin of the morphological step between terrace two and three, whether only an erosive step within the same reef formation or true onlap, is not clear.

The three terrace units probably express major sea level highs which occurred between 85 000 – 125 000 years (lower), 200 000 – 250 000 years (middle), and more than 250 000 years (third), as indicated by comparison with similar terrace systems on the Sudanese coast (Berry et al. 1966) and the Sinai Peninsula where an absolute age was obtained by $^{230}Th/^{234}U$ datation (Gvirtzman and Friedman 1977).

Diagenetic features are dominated by leaching, but concomittant cementation prevails in terrace two and three. The quantitative composition of the sediments of terrace one is similar to that of the Recent reef (Fig. 3). This is also confirmed by the frequency of unaltered marine cements, which comprise up to 80% of all cements.

Terraces two and three are characterized by a complete alteration of marine cements. Terrace two still includes many algae and scleractinians, whereas terrace three is poor in preserved fossils (Fig. 3), many of which have been dissolved leaving molds, now filled with freshwater cement. Frequently there are vugs now filled with cement, in which, of course, fossils have been present also prior to dissolution. In contrast to the younger terraces, the upper parts of terrace three are characterized by well cemented limestones with little porosity (Fig. 3).

4 Diagenetic Features of Selected Scleractinians

The rate of diagenetic alteration of all scleractinians is independent of the terrace or reef cycle in which they occur and it is impossible to interpret the diagenetic history of the limestone mass from any given coral colony. This has already been demonstrated by Schroeder (1972), Scherer (1975) and has resulted in the petrogenetogram for scleractinians (Schroeder 1984). On the other hand, the spatial and, especially, the rate of diagenetic alteration for scleractinians depend

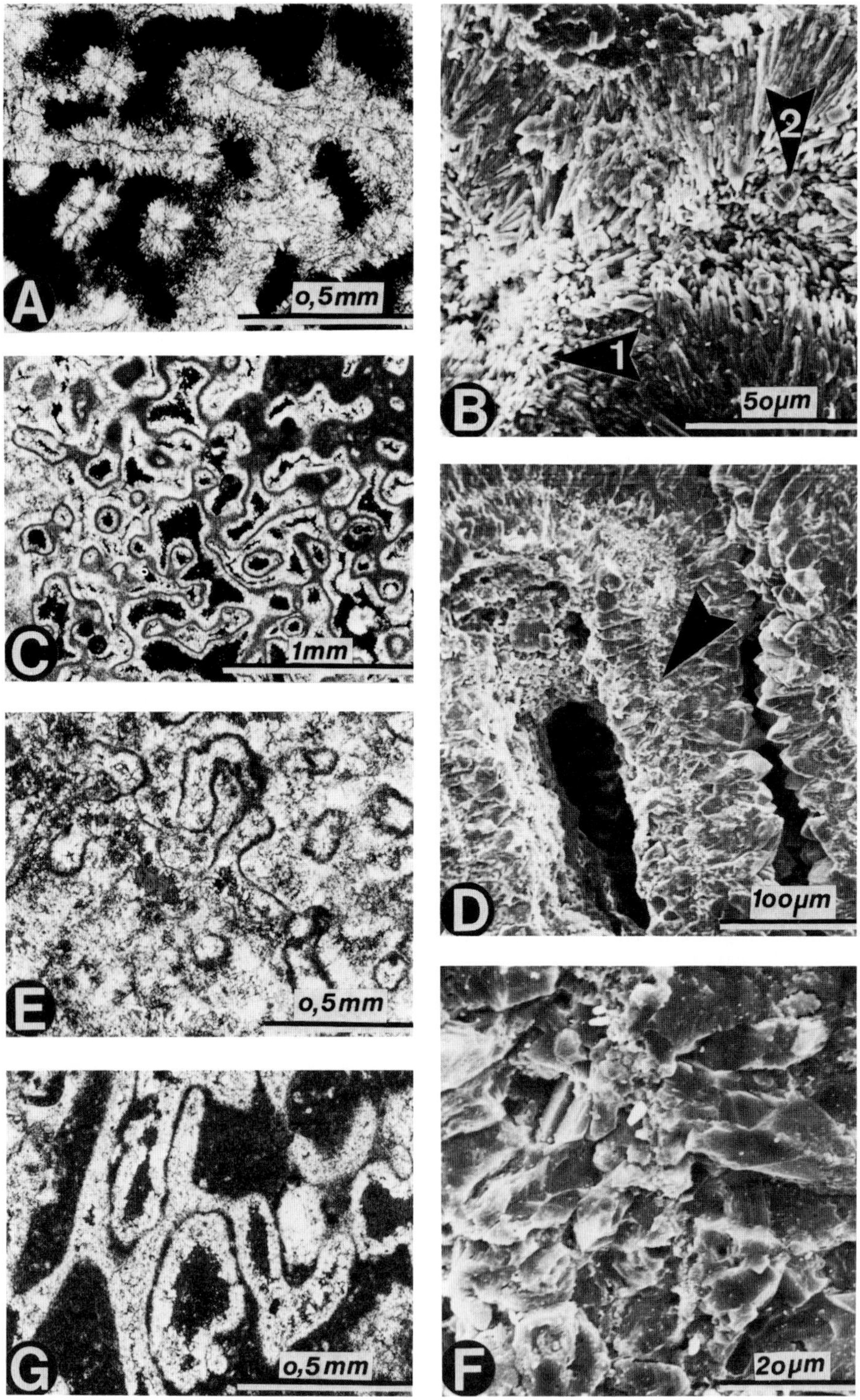

Fig. 4

on the primary microstructure and microarchitecture, as noted by Constantz (1984, and this Vol.) for *Acropora palmata* and *Acropora cervicornis*. Thus, taxonomic differences seem to influence the rate of diagenetic alteration.

4.1 Section I

The selected species *Stylophora pistillata* and *Favia laxa* which occur in reef 1 exhibit no alteration of the primary microstructure (Figs. 5A, 6A). Marine cementation by fibrous aragonite also is preserved. These needles grow either syntaxially on the trabecular structure of the coral (Hubbard 1972) or on micritic envelopes (Bathurst 1966). The influence of meteoric diagenesis is indicated only by rare vadose silts and micrites (Fig. 6A). The same features can be seen in the colonies of *Porites lutea* (Fig. 4A). However, meteoric leaching of the coral is common. This process starts in the center of the trabecular structure (James 1974) and results in open trabecular centers rimmed by a chalky layer of partly dissolved aragonite needles. In addition, newly formed blocky calcites may grow around the remaining aragonite needles (Fig. 4B), resulting in aragonite relics within fresh water calcites (Sandberg and Hudson 1983).

In addition to minor leaching, completely altered colonies of *Porites lutea* occur in reef 1, where vugs resulting from leached corals are partly filled with dog-tooth or blocky calcites (Fig. 4C). The primary arrangement of the septa is recorded only in areas where sediment infillings or micritic envelopes were present prior to dissolution (Fig. 4C, E). The thin micritic envelopes all consist of low Mg-calcite, confirmed by electron microprobe and X-ray diffraction and these form a thin substratum for cementation following meteoric leaching (Fig. 4D).

The diagenetic alteration of *Stylophora pistillata* and *Favia laxa* is expressed initially in reef 2. Both corals are characterized by preferentially dissolved trabecular centers (Figs. 5B, 6C), but most of the primary microstructure is unaltered. The interseptal cavities contain dog-tooth and blocky cements (Fig. 8B, C). Primary aragonite cements may occur locally. The thick walls and septa in *Stylophora pistillata* are replaced mainly by neomorphic calcite (Fig. 5F) which grows from the center of the trabecular structure. This common feature (James 1974, Sorauf 1980, Dullo 1984) was not recognized in all specimens of *Porites lutea*. *Favia laxa* also exhibits neomorphic crystals within the trabecular structure (Fig. 6B). However, dissolution with subsequent calcite precipitation

Fig. 4A–G. Diagenesis of *Porites luteae*. **A** Unaltered microstructure on which marine aragonite cements have grown. Section I, reef 1, thin section, crossed nicols. **B** Leached trabecular centers, which are rimmed by a chalky layer (*1*). Newly formed calcite occurs in the leached centers (*2*). Section I, reef 1, SEM. **C** Completely altered microstructure with sediment infillings, micritic envelopes, primary and secondary porosity. The cement is blocky and dog-tooth cement. Section I, reef 1, thin section, crossed nicols. **D** The thin fragile micritic linings (*arrow*) act as substratum for the subsequent meteoric cementation. Section I, reef 1, SEM. **E** Poorly preserved micritic linings and almost complete cementation with little secondary porosity. Section I, reef 1, thin section, crossed nicols. **F** Small relics of micritic lining within a well-cemented colony. Section II, terrace 3, SEM. **G** Microstructure completely altered by dissolution and subsequent cementation. Leaching has also affected the sediment infilling. Section I, reef 3, thin section, crossed nicols

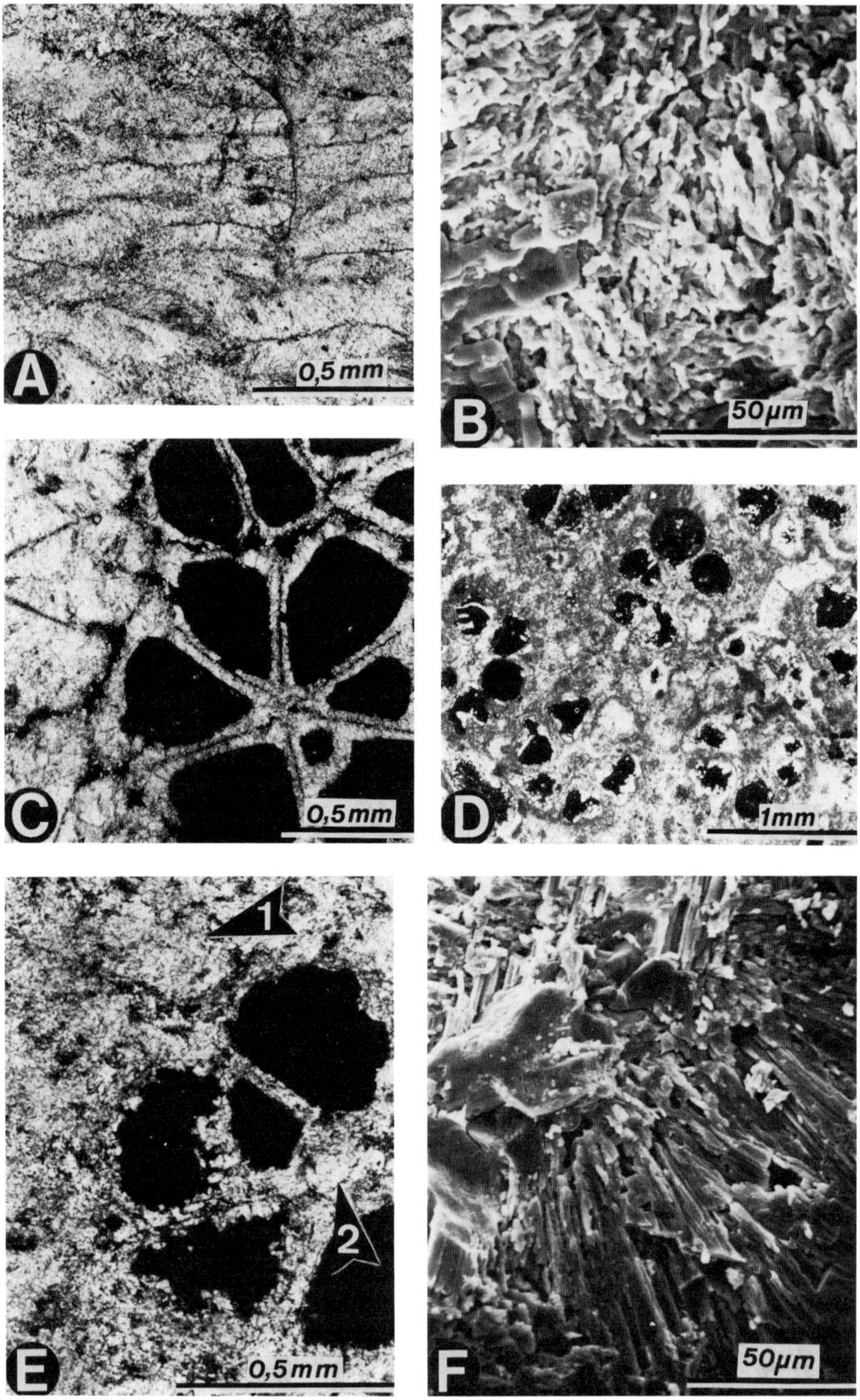

Fig. 5

prevails (Fig. 6C). The crystals never cut across the primary outline of the skeleton, indicating a vadose origin according to Pingitore (1976). Reef 2 is characterized by the complete alteration of *Porites lutea,* but primary porosity is retained in some parts (Fig. 4D). In contrast, the other two species locally exhibit unaltered microstructure (Figs. 5C, 6B).

Porites lutea in reefs 3 and 4 show the same diagenetic features as in reef 2 (Fig. 4G). The preservation of *Stylophora pistillata* and *Favia laxa* is similar. Both species are characterized by locally preserved primary microstructure within the thick walls (Fig. 5E). However, the septa of *Stylophora pistillata* are all replaced by neomorphic calcites or cements subsequent to dissolution. In contrast, the septa of *Favia laxa* exhibit relics of primary microstructure, while the dissepiments occur as ghost structures, even in reef 4 (Fig. 6D). The main difference between reefs 3 and 4 is the distinctly higher secondary porosity, which may attain 35% in reef 4.

4.2 Section II

Porites lutea is already replaced by calcite, following dissolution in terrace one. Neomorphic crystals also are lacking. *Stylophora pistillata* and *Favia laxa* are characterized both by leaching with subsequent precipitation of calcite, and by neomorphism. Traces of primary microstructure occur only in *Favia laxa.* Due to the lack of deeper erosional cuts in the seaward part of terrace one, it was impossible to collect good samples in the area of the onlap on terrace two. The corals represent an advanced stage of diagenetic alteration, comparable to that of reef 4 in Section I.

Terrace two is characterized by the lack of preserved microstructures in the species investigated. Only ghost structures of septa occur because of micritization by endolithic organisms or micritic cements (Schroeder 1984). These micritic linings must have been transformed into low Mg-calcite by incongruent leaching (Bathurst 1971) prior to dissolution of the aragonitic skeleton. Deformations and fractures of these micritic lines demonstrate that they had not been affected by dissolution (Fig. 6E, F).

The diagenetic features of terrace three are characterized by a relatively high degree of cementation of both, primary and secondary voids. This leads to the disappearance of distinct scleractinian structures. Even in those parts of the coral colony which are characterized by micritic linings, it is almost impossible to demonstrate replacement of primary septa or cementation of primary and secondary pores (Fig. 4F).

Fig. 5A–F. Diagenesis of *Stylophora pistillata.* **A** Unaltered microstructure with preserved trabecular centers. Section I, reef 1, thin section crossed nicols. **B** Leached trabecular centers with few newly formed calcites. Section I, reef 2, SEM. **C** Leached trabecular centers. Section I, reef 2, thin section, crossed nicols. **D** Neomorphic calcite occurs within the thick walls and septa. Section I, reef 2, thin section, crossed nicols. **E** Relics of primary microstructure occur (*1*) only within the thick walls, while the septa are altered by neomorphism (*2*), dissolution, and cementation. Section I, reef 3, thin section crossed nicols. **F** Neomorphic calcite in the center of the trabecula. Section I, reef 2, SEM

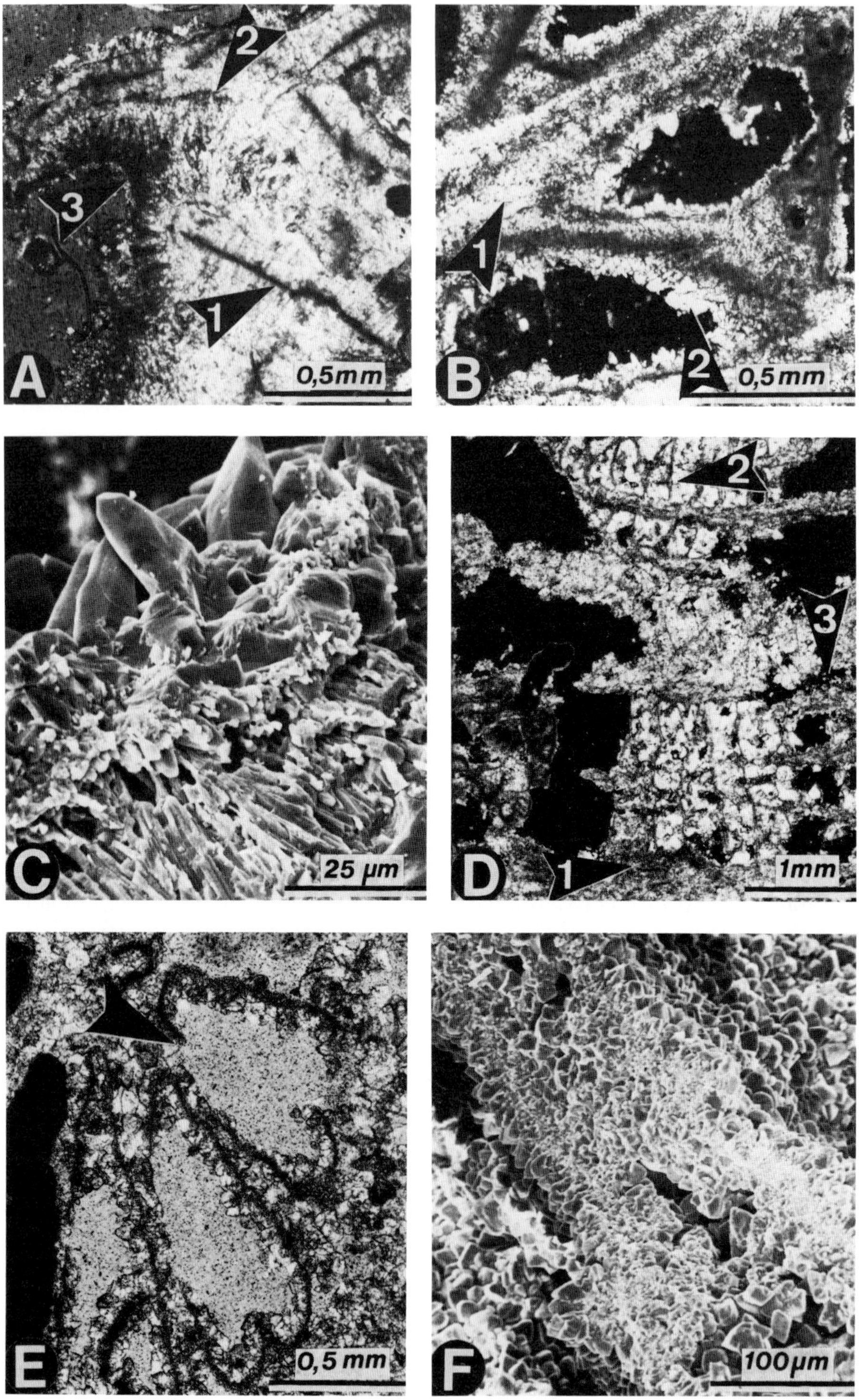

Fig. 6

5 Discussion and Conclusions

The four reef cycles of section I are characterized by a progressive destruction of the fossils. The differences in fossil preservation can be related to small-scale sea level changes which have occurred between the formation of the reef cycles. Hence there have been repeated stages of emergence during which the reef was subjected to the influence of meteoric diagenesis. For this reason the fossil record in reef 4 is poor and the secondary porosity high with respect to reef 1 (Fig. 3). The overall diagenetic development of section I corresponds to that of terrace one in section II. However, secondary porosity decreases because of increasing cementation in terrace two where dissolution is always followed by cementation. This development attains a climax in terrace three (Fig. 3). However, such a bulk diagenetic alteration of one reef or a terrace is not necessarily reflected in the diagenetic alteration of specific scleractinian species.

Lateral diagenetic variation occurs over very short distances. Even within a thin section unaltered microstructures coexist with neomorphic or blocky calcites formed after dissolution. These differences depend partly on the early diagenetic history of the coral colony. As pointed out by Hubbard (1972), Gvirtzman and Friedman (1977), Macintyre (1984), and others, marine diagenesis may begin directly below living coral tissue, where marine cementation may help to stabilize the coral. On the other hand, bioerosion by boring organisms results in destabilization of the skeleton by the formation of a pore system which subsequently acts as pathway for leaching meteoric waters. Thus it is almost impossible to systematically relate diagenetic evolution only to external factors.

The internal properties of the coral, such as microstructure and microarchitecture, are a better key for understanding the rate of diagenetic alterations. *Porites lutea* is altered completely before the other two species. Even between *Stylophora pistillata* and *Favia laxa* a different rate of diagenetic alterations exists. In reef 4 of section I *Favia laxa* retains traces of primary microstructure, while *Stylophora pistillata* is transformed totally into calcite (Fig. 7).

Stylophora pistillata belongs to the family Pocilloporidae, which is part of the suborder Astrocoeniina. According to Wells (1956), this suborder is characterized by an isolated and dotted arrangement of the trabecular centers within the septa. In the family Poritidae, a Fungiina, even the septa consists of small, loosely arranged sticks with isolated trabecular centers. In contrast, the

Fig. 6A–F. Diagenesis of *Favia laxa*. **A** Unaltered microstructure with the trabecular center (*1*). Marine aragonite cements (*2*) occur as syntaxial extensions of the trabecula aragonite needles. Meteoric diagenesis is expressed by vadose silts and micrites (*3*). Section I, reef 1, thin section, nicols 50°. **B** The trabecular centers are partly replaced by neomorphic calcite (*1*). The cements in the primary pores are blocky (*2*) and dog tooth calcites. Section I, reef 2, thin section, crossed nicols. **C** Leached trabecular centers and dog-tooth cements growing into the interseptal pore space. Section I, reef 2, SEM. **D** Almost totally altered colony with rare relics of primary microstructures (*1*), ghost structures of the dissepiments (*2*), and secondary pores of variable size (*3*). Section I, reef 4, thin section, crossed nicols. **E** Ghost structures of the septa. The deformations and fractures (*arrow*) of the micritic linings demonstrate that they have not been involved in the leaching process. Section II, terrace 2, thin section, parallel nicols. **F** Tiny relics of micritic linings on which meteoric cements have grown. Section II, terrace 2, SEM

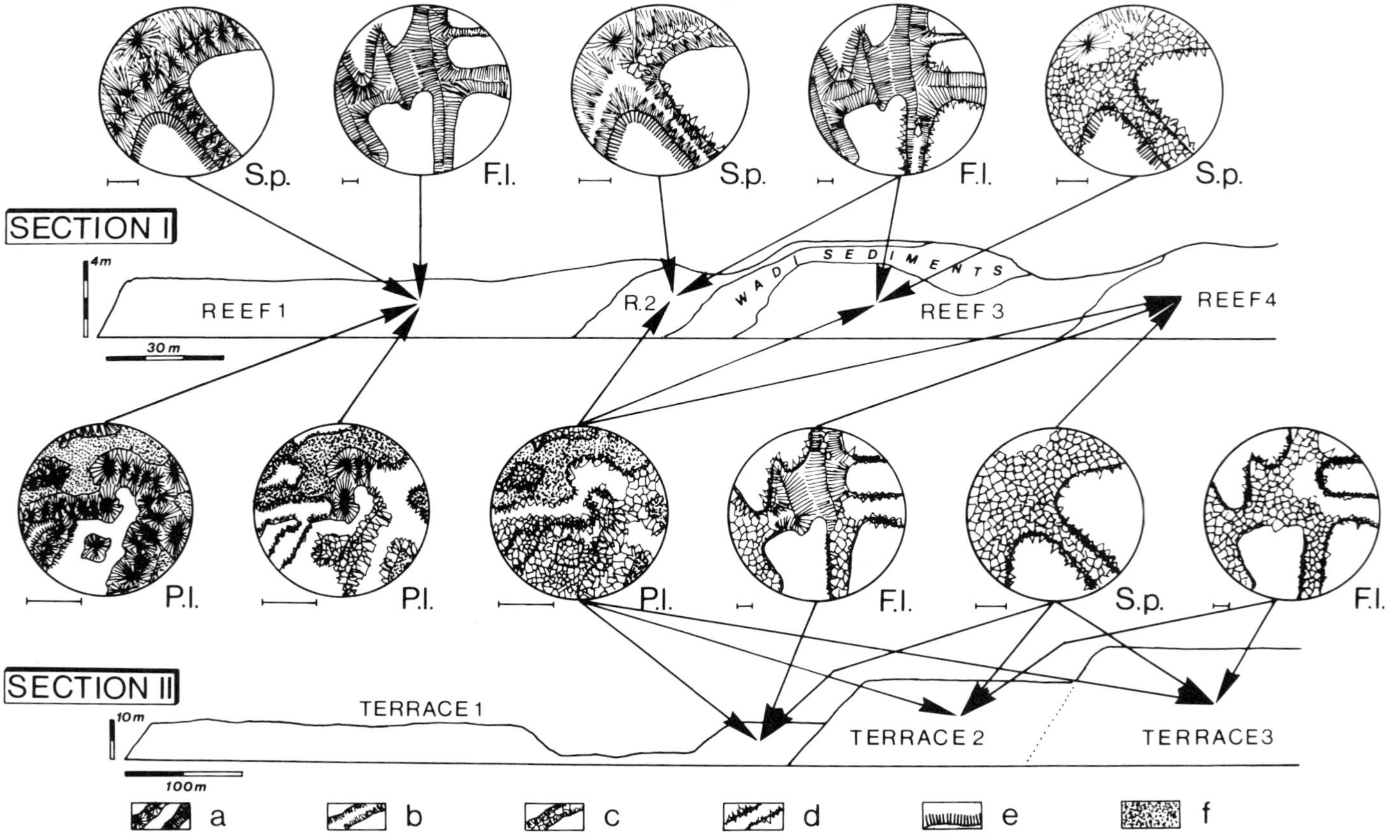

Fig. 7. Summary sketch showing the distribution of diagenetic fabrics in *Porites lutea* (*P.l.*), *Stylophora pistillata* (*S.p.*), and *Favia laxa* (*F.l.*) in section I and II. *a* unaltered microstructure; *b* leached trabecular centers; *c* neomorphic calcite within the trabecular center; *d* micritic linings with freshwater calcite cement growing on them; *e* marine cement; *f* sediment. Scale bar for each coral = 100 μm

Table 1. Differences in primary and diagenetic features between the three investigated species

	Dimension of aragonite needles (μm)	Arrangement of trabecular structure	Thickness of septa and calix walls (μm)	Width of leached trabecular centers (μm)	Dimension of fresh-water calcites replacing the primary microstructure	
					Formed after dissolution (elongated shape) (μm)	Neomorphic (blocky shape) (μm)
Stylophora pistillata	0.20 – 0.30	Isolated and dotted	50 – 130	10 – 20	10 – 50	10 – 50
Porites lutea	0.15 – 0.20	Isolated and dotted	30 – 80	10 – 50	5 – 50	–
Favia laxa	0.35 – 0.50	Linear	50 – 250	10 – 20	10 – 100	10 – 50

massive septa of the Faviidae are characterized by a linear arrangement of the trabecular centers. These differences in the microarchitecture are an important factor controlling the rate of diagenetic alterations. On the other hand, the size of the aragonite needles comprising the trabecular structure is also important. Based on large-scale SEM micrographs the different crystal sizes were measured: *Porites lutea:* 0.15 – 0.20 μm; *Stylophora pistillata:* 0.20 – 0.30 μm; *Favia laxa:* 0.35 – 0.50 μm. Because of similar crystal sizes in *Porites lutea* and *Stylophora pistillata,* a comparable behavior during meteoric diagenesis may be expected. However, the distinctly thicker wall of *Stylophora pistillata* tends to resist leaching. Thus, although primary porosity influences the rate of diagenetic alteration, it is of minor importance when comparing *Porites lutea* with *Favia laxa,* which show almost the same amount of total primary porosity.

The diagenetic behavior of these three species is comparable to that demonstrated for *Acropora palmata* by James (1974). However, there are minor differences between the species which are summarized in Table 1. The most important concern *Porites lutea* where neomorphic calcites do not occur and the diameter of the leached trabecular centers reaches a maximum dissolution being facilitated by the small size of the aragonite needles.

Different rates of diagenetic alterations occur not only on the taxonomic family level, but also exist on a species level, as it was noted by Constantz (1984, and this Vol.) for *Acropora palmata* and *Acropora cervicornis.* Further investigations may determine the taxonomic groups having comparable response to meteoric diagenesis, and thus establish an order of diagenetic susceptibility among scleractinians.

Acknowledgments. These investigations were carried out with the financial support of the Deutsche Forschungsgemeinschaft (Du 129-1) and the University of Petroleum and Minerals in Dhahran (Saudi Arabia). I wish to thank both organizations. I also thank all collegues from the Universities of Karlsruhe and Dhahran for their help in the field as well as Prof. Flügel and Prof. Bandel for their critical comments. Mrs. Neufert and Mrs. Sporn kindly prepared the photos.

References

Al Sayary SS, Dullo C, Hötzl H, Jado AR, Zötl JG (1983) The Quarternary along the Gulf of Aqaba. In: Zötl JG, Jado AR (eds) Quarternary period in Saudi Arabia. Springer, Wien New York, pp 27–47

Bathurst RGC (1966) Boring algae, micritic envelopes and lithification of molluscan biosparites. Geol J 5:15–32

Bathurst RGC (1971) Carbonate sediments and their diagenesis. Elsevier, Amsterdam, 620 p

Behairy AA (1983) Marine transgressions on the west coast of Saud Arabia (Red Sea) between Mid-Pleistocene and Present. Mar Geol 52:M25–M31

Berry L, Whiteman AJ, Bell SV (1966) Some radiocarbon dates and their geomorphological significance: Emerged reef complex of Sudan. Z Geomorphol 10:119–143

Constantz BR (1984) Functional comparison of the microarchitecture of *Acropora palmata* and *Acropora cervicornis*. Palaeontogr Am 54:548–551

Dullo WC (1983) Fossildiagenese im miozänen Leithakalk der Paratethys von Österreich: Ein Beispiel für Faunenverschiebungen durch Diageneseunterschiede. Facies 8:1–115

Dullo WC (1984) Progressive diagenetic sequence of aragonite structures: Pleistocene coral reefs and their modern counterparts on the eastern Red Sea coast, Saudi Arabia. Palaeontogr Am 54:254–260

Dunham RJ (1962) Classification of carbonate rocks according to depositional texture. Mem Am Assoc Petrol Geol 1:108–121

Flügel E (1982) Microfacies analysis of limestones. Springer, Berlin Heidelberg New York, 633 p

Gvirtzman G, Buchbinder B, Sneh A, Nir J, Friedman GM (1977) Morphology of Red Sea fringing reefs: a result of the erosion pattern of the last glacial low stand sea level and the following Holocene recolonization. Sec Symp coraux et récifs corallines fossils, Paris. Mem Bur Recher Geol Miner 89:480–491

Gvirtzman G, Friedman GM (1977) Sequence of progressive diagenesis in coral reefs. Am Assoc Petrol Geol Stud Geol 4:357–380

Hubbard JAEB (1972) Cavity formation in living scleractinian reef corals and fossil analogues. Geol Rundsch 61:551–564

James NP (1974) Diagenesis of scleractinian corals in the subaerial vadose environment. J Paleontol 48:785–799

Macintyre IG (1984) Preburial and shallow-subsurface alterations of modern scleractinian corals. Palaeontogr Am 54:229–244

Matthews RK (1973) Relative elevation of Late Pleistocene high sea level stands: Barbados uplift rates and their implications. Q Res 3:147–153

Matthews RK (1984) Late Oligocene-Pliocene transgressive-regressive cycles of sedimentation in northwestern Puerto Rico. In: Schlee JS (ed) Interregional unconformities and hydrocarbon accumulation. Am Assoc Petrol Geol Mem 36:89–107

Mergner H, Schuhmacher H (1974) Morphologie, Ökologie und Zonierung von Korallenriffen bei Aqaba (Golf von Aqaba, Rotes Meer). Helgol Meeresunters 26:238–358

Mesolella KJ (1969) The astronomical theory of climatic change: Barbados data. J Geol 77:250–274

Pingitore NE (1976) Vadose and phreatic diagenesis, process, products, and their recognition in corals. J Sediment Petrol 46:985–1006

Sandberg PA, Hudson JD (1983) Aragonite relic preservation in Jurassic calcite-replaced bivalves. Sedimentology 30:879–892

Scherer M (1975) Cementation and replacement of Pleistocene corals from the Bahamas and Florida: Diagenetic influence of nonmarine environments. N Jahrb Geol Palaeontol Abh 149:259–285

Schroeder JH (1972) Fabrics and sequences of submarine carbonate cements in Holocene Bermuda cup reefs. Geol Rundsch 61:708–730

Schroeder JH (1984) The petrogenetogram of corals: Spatial variations in diagenetic sequences. Palaeontogr Am 54:261–271

Sorauf JE (1980) Biomineralization, structure, and diagenesis of the coelenterate skeleton. Acta Palaeontol Pol 25:327–343

Wells JW (1956) Scleractinia. In: Moore R (ed) Treatise on invertebrate paleontology, part F, Coelenterata. Kansas Univ Press, Lawrence, pp 328–444

Internal Hydrology and Geochemistry of Coral Reefs and Atoll Islands: Key to Diagenetic Variations*

R. W. BUDDEMEIER[1] and J. A. OBERDORFER[2]

1 Introduction

1.1 Hydrology and Diagenesis

Diagenetic changes in coral reef carbonates continue to generate both interest and controversy. Considerable attention has been devoted to carbonate porosity (Choquette and Pray 1970, Enos and Sawatsky 1981), pore-water geochemistry and diagenetic environments (Berner 1980, Longman 1980), and general geochemistry (Curtis 1977), but the subject of formation-scale fluxes and material balances has not received the attention it deserves. The rates, distributions, and mechanisms of fluid transport through the reef structure are basic to the understanding of diagenetic processes. Although some mineralogic transitions may be able to occur in a closed system (e.g., aragonite to calcite), others, such as dolomitization, require the passage of many volumes of pore water through the system in order to remove calcium and supply magnesium. Thus, the hydrogeology of a formation – porosity, permeability, the forces available to drive fluids through the system, and the evolution of these parameters over time – will exert important controls over the rate, nature, and extent of diagenesis. At the same time, diagenetic changes may provide important clues to formation hydrogeology and its development.

Some idea of the interaction between hydrologic characteristics and diagenesis may be obtained by consideration of the proposed mechanisms for dolomitization. The hypersaline brine hypothesis (Friedman and Sanders 1967) requires a surface which is both sufficiently impermeable so that ponded seawater neither drains nor exchanges readily with adjacent ocean water, and sufficiently permeable so that the altered fluids can penetrate the carbonate formation. Dorag dolomitization (Badiozamani 1973) requires mixing of a freshwater body with marine water; in this case, the formation permeability must permit percolation of rainwater and tidal mixing with the seawater, but cannot be high enough to maintain nearly marine conditions nor low enough to inhibit mixing (presumably tidal) of the freshwater and the seawater. Saller (1984) has proposed that dolomitization may result from the flux of large volumes of deep ocean water (below the carbonate compensation depth) through high-permeability zones in

* Work performed under the auspices of the U.S. Department of Energy by Lawrence Livermore National Laboratory under Contract W-7405-Eng-48.

[1] Lawrence Livermore National Laboratory, Livermore, CA, USA
[2] Geology Department, San Jose State University, San Jose, CA, USA

Reef Diagenesis
Edited by J. H. Schroeder and B. H. Purser

atoll substructures. In the discussion which follows, we will relate modern reef hydrogeology and geochemistry to these models.

1.2 Background

There has been relatively little systematic investigation of reef hydrogeology. Drilling logs and core logs provide qualitative and anecdotal information related to formation permeability. Petrographic observations provide data on the porosity, mineralogy, and post-depositional alteration of selected samples. Studies of beachrock formation have included consideration of mixing and CO_2 system fluxes, but have seldom treated in a quantitative fashion the issues of permeability and water flux. Most hydrologic studies have been oriented toward island groundwater resource evaluation, and hence have provided little data on the dynamics of formation water movement or its relationship to structure, mineralogy, or geochemistry. Quantitative studies of flow dynamics, particularly in combination with geochemical measurements relevant to the characterization of diagenetic environments, are still in their infancy.

1.3 Aims and Limitations of this Presentation

The purpose of this contribution is to set forth the basic principles and concepts of hydrogeology as they relate to the movement of water through coral reef structures, and to illustrate them and their relationships to diagenetic environments by discussion of two coral reef systems for which the authors have obtained quantitative hydrologic and geochemical data.

The first system discussed is a shallow marine environment: Davies Reef, in the central Australian Great Barrier Reef. Measurements in test wells in the upper few meters (Holocene) of the reef give us estimates of the magnitude and heterogeneity of permeability of constructional origin, some insight into water transport rates, and some data on the importance of biogeochemical activity in the pore water.

The second system concerns Enewetak Atoll, where studies of island hydrology have provided some insight into the controls over freshwater distribution and mixing, and the effects of erosional permeability on the hydrology of the reef structure.

These case studies, although far from definitive, provide some idea of the range of chemical and physical environments currently available for early diagenesis, and of the importance of considering the interaction of water flow with the chemical environments. We hope that they will also serve to demonstrate the importance of hydrogeological considerations to diagenetic models, and to provide some concepts and data that may be of general utility.

2 General Hydrogeology

2.1 Determing Flow Rates and Directions

Knowledge of flow rates and directions permits calculation of pore water residence times and volumetric throughput of water within the reef. Determination of pore water velocities and flow directions can be made from data obtained in the field or in the laboratory, either by calculation or by direct measurement.

In order for flow to occur through geologic materials, they must be sufficiently porous and permeable. The larger, less tortuous, and more interconnected the pore openings, the easier it is for water to flow. In addition to a permeable medium, it is necessary to have differing levels of hydraulic head across the system. It is this head difference which provides the driving force for flow. The greater the head gradient, or the change in head per unit distance, the greater the driving force. Darcy's Law, which governs flow through porous media, describes these relationships:

$$v_p = K/n\,(dh/dl) \;,$$

where v_p is the average linear pore water velocity (L/T), K is the hydraulic conductivity (L/T), n is the effective porosity, h is the hydraulic head (L), l is the distance (L), and dh/dl is the hydraulic head gradient.

Hydraulic conductivity reflects the permeability of the porous medium, or the ease with which the water can flow through it. It can be measured in the laboratory on drill cores or on hand samples by measuring the flow rate and head gradient as water flows through the sample in a permeameter. It can also be measured in the field by pump tests or slug tests on a well where discharge rates and water level changes are monitored. While somewhat more complicated, the field determinations measure the value of a much larger volume of porous medium, and hence more closely approximate in-situ values. Reef deposits are generally considered to be quite permeable, with formation-scale hydraulic conductivity values of 10 to 2000 meters/day (m d^{-1}).

The effective porosity is that portion of the void spaces through which flow actively takes place. Carbonate fragments generally have very high intraparticle porosities in addition to interparticle pore space (Fig. 1 a); however, it is clear that most of the flow will take place between the fragments rather than though them (Choquette and Pray 1970). On a larger scale, large void systems (such as solution cavities, Fig. 1 b) may dominate the flow, and it would be their volume which would correspond to the effective porosity, rather than the total porosity of the material. Thus, the effective porosity is less than the true porosity; values in the range of 10 to 30% are typical for effective porosities, while total porosities are commonly 40 – 60% (Enos and Sawatsky 1981).

The hydraulic head is the energy per unit weight of water, that energy being imparted by gravity and pressure. Very simply, in reef systems it can be considered as a water level elevation either of ocean or lagoon water, or of water in a well. If water has the same head or elevation everywhere, there can be no flow through the system. However, where water level differences occur, flow will take place in the direction of lower head or lower water level. In reef systems,

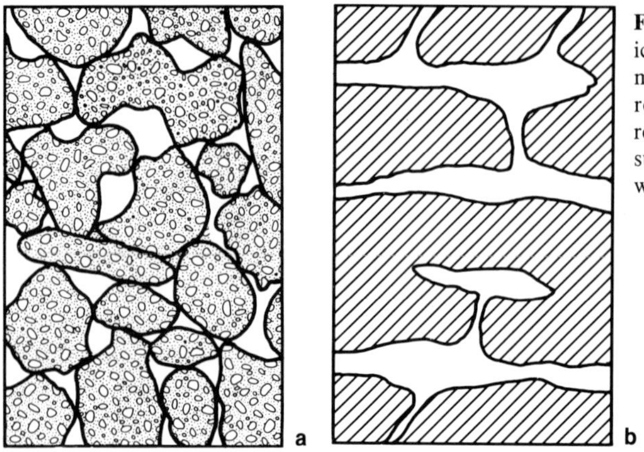

Fig. 1a, b. Hydrologically significant porosity. **a** Carbonate fragments with high interparticle porosity as well as intraparticle porosity. **b** Large void porosity resulting from solution effects on a well-cemented limestone

water level differences can be created by wind, tide, or current-driven water elevation changes across the reef flat. Even small and isolated reef structures can have such differences. Atolls with relatively enclosed lagoons can exhibit ponding within the lagoon, with average water level elevations relative to mean sea level of tens of centimeters (Atkinson et al. 1981). This occurs because water can enter the lagoon both by tidal flow through reef passes and by wind-driven cross-reef transport, but can exit only by flow through the reef passes on the falling tide. This ponding in the lagoon creates gradients oceanwards, and hence net flow through the reef structure in the direction of the reef front. In reef and atoll island situations, water level differences can be created by the accumulation of freshwater within the island, usually with a mounding of the water in the interior of the island (Davis and De Weist 1966).

Once the hydraulic conductivity, effective porosity, and the hydraulic gradient have been determined, it is possible to calculate the pore water velocity using Darcy's Law. Direction of flow can also be determined from the gradient, since flow will occur down gradient, i.e., in the direction of lower head. Experimentally, it is also possible to measure the pore water velocity directly by labeling that water with a tracer (dye, chemical, radioactive), and measuring the time required for the labeled water to travel a known distance. Dividing the distance the water must travel to traverse the reef by the pore water velocity gives the residence time of the pore water within the reef. Thus, the tracer technique can indicate both the direction and rate of flow; however, because of dispersion or spreading of the dye plume, multiple sampling points must be observed over a substantial period of time in order to measure net flow or deduce the exact direction of the maximum gradient.

2.2 Heterogeneity of Reef Porous Media

For calculational purposes, it is often convenient to assume that the porous medium is homogeneous. However, most reef deposits are extremely heterogeneous, both on large and small scales. The origin of this heterogeneity can be

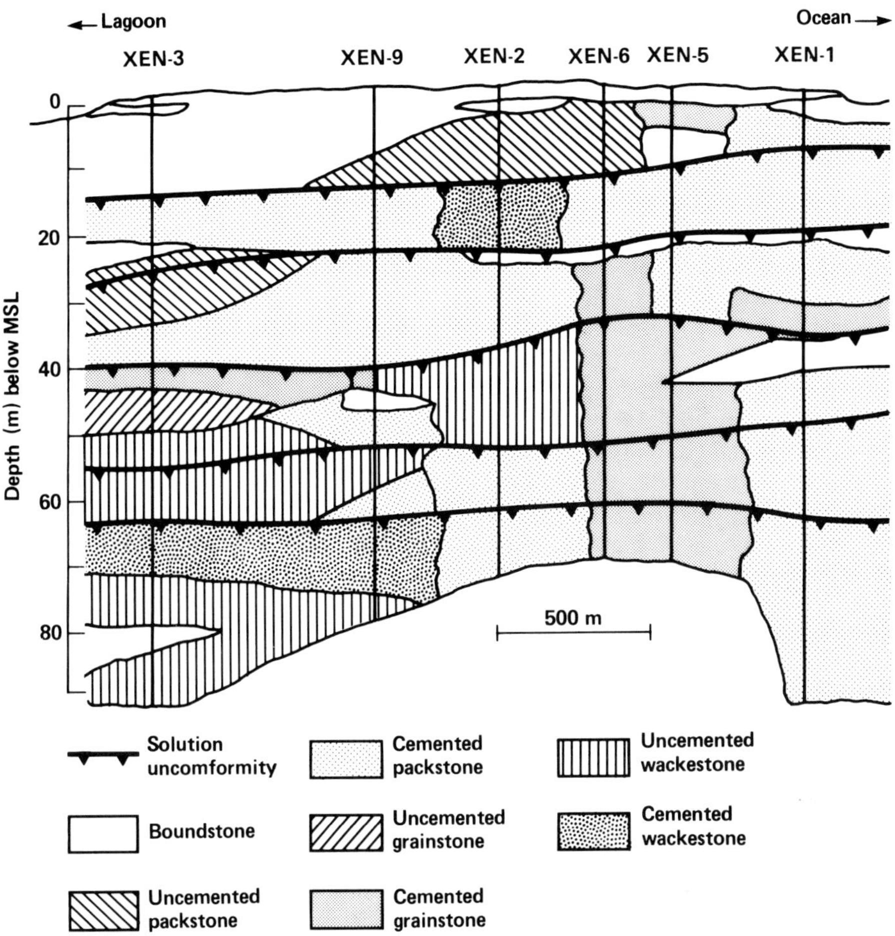

Fig. 2. Geologic cross-section of Enjebi Island, Enewetak Atoll, showing borehole locations, the variable lithologies, and the multiple solution unconformities. (After Ristvet et al. 1978)

understood by considering the structure of modern reefs, where short-range vertical and horizontal variations in texture, degree of lithification, and the construction of large voids are commonly observed. In addition, subaerial exposure during glacial low stands of the sea have created karst phenomena which now exist at depth within the reef structure. Such large voids have been documented repeatedly during drilling (Ladd and Schlanger 1960). The variation in lithology and the presence of multiple solution unconformities can be seen in the cross-section from Enjebi Island, Enewetak Atoll (Fig. 2).

Dense coral heads, reef rubble, and cemented and/or eroded carbonate rock all have very different porosites and permeabilities. A distinction can be made between open and restricted porosity. Restricted porosity includes isolated pores, where water exchange occurs only at a very slow rate and may be limited by diffusion. There is little or no advective water movement through the restricted porosity; water and solute exchange with the mobile pore water occurs primarily by diffusion, with rates controlled by micropore diameters and lengths (tortuosity). Open porosity refers to the larger voids (megapores, mesopores, fractures, and solution cavities) which control the permeability, are generally well interconnected, and where water movement can occur relatively rapidly.

Similarly, there is also considerable variability in the deposits' hydraulic conductivity. Individual coral heads, algal boundstones, and fine grained lagoonal sediments could all be expected to have some of the lowest hydraulic conductivities (0.001 to 1 m d^{-1}). Coarse sands and reef rubble have high hydraulic conductivities (10 to 1000 m d^{-1}). The major voids, if they are interconnected, would have the greatest hydraulic conductivities (greater than 2000 m d^{-1}). Here flow would be much closer to pipe or fracture flow than to flow through porous media. Enos and Sawatsky (1981) have determined laboratory porosities and permeabilities for a variety of carbonate sediments.

The spatial variability in hydraulic conductivity and temporal variations in hydraulic gradient will create considerable variability in pore water velocities even over relatively short distances within the reef. The bulk of the flow will be through the zones of high hydraulic conductivity, where pore water residence times will be relatively short. Flow velocities as high as 1 to 100 m d^{-1} may occur. Zones of low hydraulic conductivity will have slow pore water velocities (on the order of millimeters to centimeters per year), and hence much longer residence times. Water in restricted pores will exchange with the open-pore water only at an extremely slow rate.

2.3 Hydrogeology of Submerged Reefs

Pore waters in submerged reefs have essentially seawater salinity, though their chemistry may be quite distinct from seawater. Movement of water through submerged reefs on a large scale is induced by ocean water level differences across the reef, the result of wind, tide, or ocean current. In addition to flow caused by water level differences, flow in the shallow subsurface of the reef may be caused by the pumping action of waves passing over the reef flat. This is probably only significant in parts of the reef which are close to the surface or to large flow channels open to the ocean, as the wave energy is attenuated rapidly with depth in the water column and with depth into the porous medium. Such short-range (centimeters to meters) transport will not be dealt with here, although it may play an important role in near-surface submarine cementation.

2.4 Hydrogeology of Reef and Atoll Islands

For reef and atoll islands, the system is complicated by the fact that islands of sufficient size will hold rainwater in the subsurface, creating a freshwater lens. This lens, because of the lower density of freshwater, floats on the seawater (Fig. 3). It is described by the Ghyben-Herzberg relationship based on density differences which states that for every meter of freshwater head above sea level,

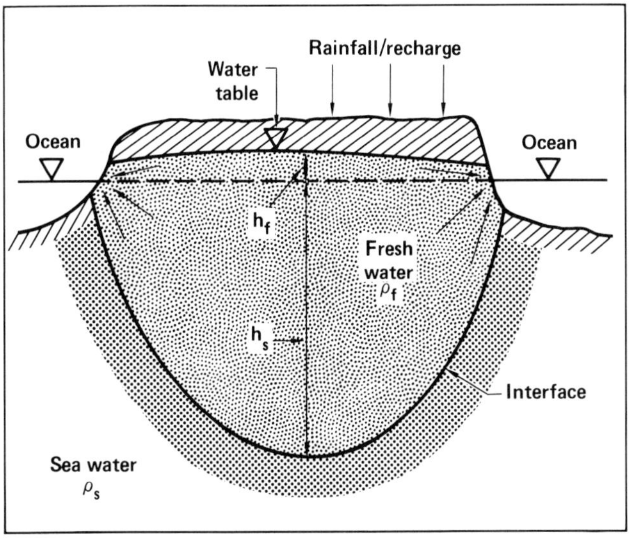

Fig. 3. Ghyben-Herzberg conceptual model of a freshwater lens in an island consisting of homogeneous porous material. The densities of seawater and freshwater are ϱ_s and ϱ_f, respectively, the freshwater head above mean sea level = h_f, and the depth of freshwater below MSL = h_s = $h_f (\varrho_f/\varrho_s - 1)$

there should be a corresponding 40 m of freshwater below sea level. The density contrast is a function of salinity and temperature, and may therefore vary from this 1 : 40 generalization. This conceptual model applies reasonably well to large volcanic islands such as Hawaii, but its unqualified application to reef islands is questionable. Smaller islands may exhibit reduced near-surface salinities from rainfall infiltration without having a permanent freshwater lens.

Islands with well-developed freshwater lenses generally do not exhibit a sharp interface between fresh and salt water, but rather have a zone of mixing, or transition zone, between the two. The thickness of that transition zone is usually proportional to the degree of perturbation of the lens by such processes as well pumpage and tidal water level fluctuations.

3 Davies Reef

3.1 Hydrogeology

Davies Reef is a submerged reef in the central portion of the Australian Great Barrier Reef. The reef flat is covered with water depths of 0 to 3 meters depending upon tidal stage. Figure 4a presents permeameter determinations of hydraulic conductivity performed on core and hand samples from the reef flat. The upper 25 m of the reef consist of Holocene deposits, both consolidated and unconsolidated. The unconsolidated deposits range from fine sand to coarse coral gravel with hydraulic conductivities in the range of 10 to 2000 m d^{-1}. The

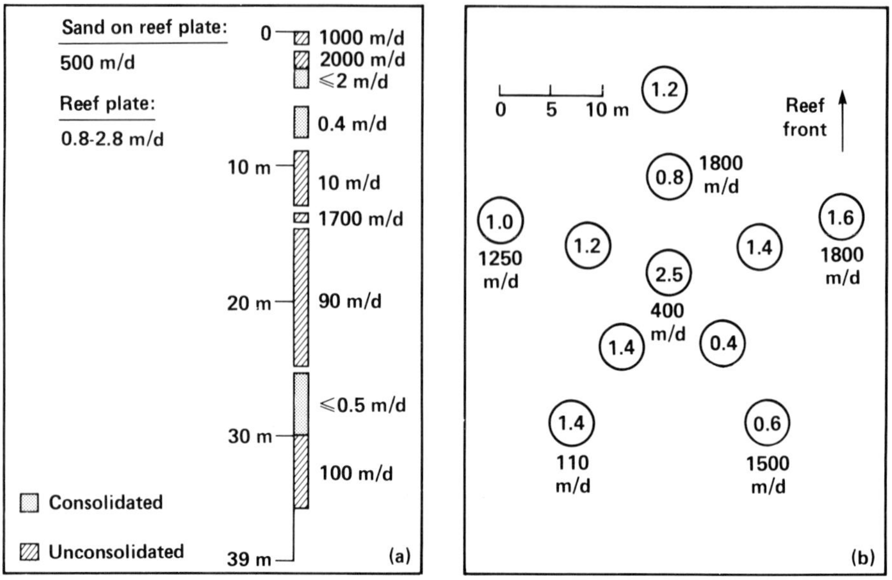

Fig. 4a, b. Vertical and horizontal variability in permeability in Davies Reef. **a** Vertical heterogeneity of near-surface reef deposits, based on laboratory permeametry of samples from AIMS Davies Reef Core No 1. Hydraulic conductivities are in m d^{-1}. **b** Horizontal distribution of hydraulic conductivity in an array of shallow test wells. *Numerical values inside circles* are well depths (m); hydraulic conductivities (m d^{-1}) were determined by field slug tests. (See Oberdorfer and Buddemeier 1986 for more detailed discussion)

consolidated deposits have hydraulic conductivities in the range of 0.1 to 2 m d^{-1}. Major vertical heterogeneities having hydraulic conductivities varying over four orders of magnitude are present. However, this vertical profile does not reflect the possible presence of large voids which cannot be sampled by coring or measured by permeametry. The underlying Pleistocene sediments show evidence of subaerial exposure and extensive cementation, but measured hydraulic conductivities are similar to those of the Holocene sediments. Lateral variability can be observed in Fig. 4b, which presents hydraulic conductivities determined by slug tests on shallow wells drilled into the reef flat. Values vary by over an order of magnitude over a 10-m distance.

The algal boundstone has a hydraulic conductivity which is smaller by several orders of magnitude than that of the underlying sediment. This means that it probably acts as a confining layer which restricts flow through the surface of the reef flat, and which could create pressurized conditions within the reef. This boundstone appeared to be continuous over much of the reef flat surface. However, it is a leaky confining layer, at least locally. Such effects have been reported by Ayers et al. (1984). This leakiness became clear when dye was observed to leak through the rock after injection into a test well; fissures or small holes permitted water to flow much more swiftly through limited areas of the boundstone surface than through the adjacent intact rock.

Head differences were measured using submersible tide gauges on the reef front and in the lagoon. Water level differences can be analyzed only in relative

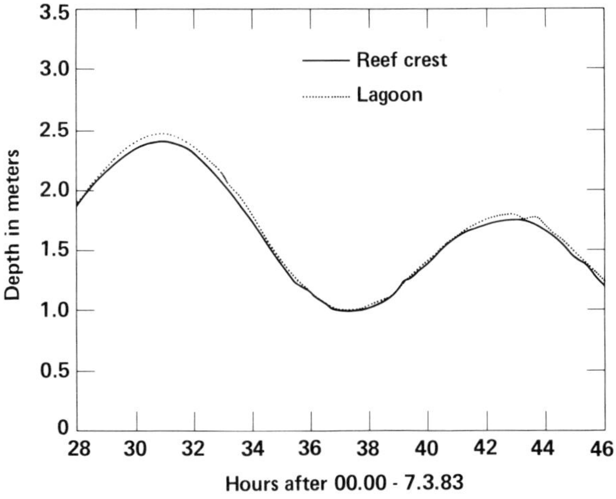

Fig. 5. Tide gauge measurements indicating water level differences across the reef flat of Davies Reef. Water elevations have been arbitrarily plotted to coincide at low tide in order to demonstrate the existence of net hydraulic gradients across the reef

terms, since it was not possible to survey the tide gauge elevations. There is no unique, correct way to watch the two tidal curves; Figure 5 presents one possible matching. Because the two curves do not match for the entire tidal cycle, for some period of time the water on one side of the reef flat clearly was higher by a number of centimeters than water on the opposite side.

Based on data collected, estimates of pore water velocity using Darcy's Law can be made. Assumptions for this calculation are:

1. Internal flow is dominated by high permeability zones ($K = 1000 \text{ m d}^{-1}$);
2. These high permeability zones (that portion of the open porosity where high-rate flow is taking place) comprise approximately 5% of the total reef volume ($n = 0.05$); and
3. An average head difference of 5 cm exists across the 300 m width of the reef ($dh/dl = 1.67 \times 10^{-4}$).

Substituting these values into the Darcy equation, we obtain an estimate for the pore velocity of $v_p = 3.3 \text{ m d}^{-1}$.

The probable range of pore water velocity values would be plus or minus one order of magnitude around this estimate. Using the value calculated above as a unidirectional pore velocity, a pore water residence time (or travel time through the 300-m-wide reef) of approximately 0.25 year can be estimated. Actual residence times could be significantly longer if the hydraulic gradient (and therefore flow) changes direction over time.

Similar calculations based upon the predominance of vertical flow can be made. Such transport is supported by some of the dye tracer data and is suspected to originate from a highly permeable Pleistocene aquifer (see, for example, Buddemeier 1981, Herman et al. 1986). The distance over which the head difference exists becomes <300, but $\geqslant 25$ m, and the travel distance then becomes the 25-m thickness of the Holocene aquifer. The maximum vertical pore water velocity calculated is 10 m d^{-1}, and the minimum residence time is 2.5 days.

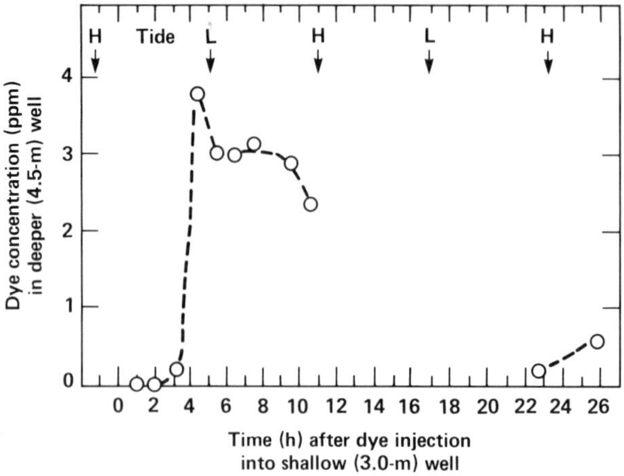

Fig. 6. Vertical transport as shown by the time history of dye concentration in a 4.5 m Davies Reef well after tracer injection into an adjacent 3.0 m well

Table 1. Pore velocities (V) and residence times (τ) for Davies Reef, determined by various experimental methods

	Horizontal		Vertical	
	V (m d^{-1})	τ (days)	V (m d^{-1})	τ (days)
Calculation (Darcy's Law)	3.2	90	10	3
Dye tracer	>0.2	<1500	>1.4 to <5	>6 to <20
	400	<1	10	3
			20 to 50	<1 to 2
O_2 Renewal	–	–	10	<17

Dye tracer studies were made in which direct calculation of pore water velocity was based on the travel time and distance from injection point to observation point. Both vertical and horizontal tracer tests were made. It was often impossible to detect the tracer at the observation points, probably reflecting the heterogeneity of the system, the convoluted flow paths, the difficulty of encountering a flow path with an observation well, and uncertainty concerning the actual gradient. Figure 6 presents results from one of the vertical dye tracer experiments. There is some suggestion that there is upward movement on the rising tide and downward movement on the falling tide, although more data need to be collected to substantiate that conclusion.

Pore water velocities and residence times are summarized in Table 1. While there is considerable variability in these results, they indicate that pore water residence times in the shallow subsurface are most likely on the order of a few weeks to a few months. These values reflect bulk water movement in the open-pore, high permeability regions; residence times for water entrapped within the less permeable sediments, coral fragments, or dead end pores will be much greater.

The relatively short residence times imply that large volumes of water flow through the reef. This, coupled with geochemical data (discussed below) which shows that both solution and precipitation occur, means that relatively large portions of the reef structure can be affected by these diagenetic processes. Porosities, hydraulic conductivities, and flow paths will change accordingly.

While frequently difficult, it is possible to make direct observations of hydrologic processes in a submerged coral reef. Future investigations need to be directed towards detailed characterizations of flow velocities and directions.

3.2 Pore-Water Geochemistry

Extensive sampling of the Davies Reef wells followed by prompt analysis of the pore waters permits us to deduce the conditions within the uppermost few meters of the reef structure. Table 2 summarizes the data for representative samples on pH, alkalinity, total CO_2, salinity, and the saturation index for the major carbonate minerals. These were calculated using the equilibrium geochemical code EQ3 (Wolery 1983), measured pH, alkalinity and nutrient values, and chemical concentrations derived from the measured salinities and standard seawater composition tables. EQ3 calculates saturation index ($SI = \log q/K_{sp}$, where q = ion activity product and K_{sp} is the corresponding saturation product constant) values for aragonite, calcite, and dolomite; high magnesium calcite (18% Mg) values were calculated from the EQ3 aragonite value and ratios of the Ksp(Mg-calcite)/Ksp(aragonite) published by Morse et al. (1985) and Bernstein and Morse (1985). EQ3 calculates values for both ordered and disordered dolomite; we have tabulated the disordered dolomite values on the assumption that these are more relevant for the process of initial dolomite formation.

The chemistry of the system apparently is dominated by the oxidation of organic matter within the sediment matrix. Total nitrogen and phosphorus concentrations are one to two orders of magnitude higher than those in the overlying seawater, pH is lower, and total CO_2 is elevated relative to surface seawater in all samples. The system is basically anoxic, with measurable sulfide concentrations in virtually all samples. However, traces of dissolved oxygen were also measurable, indicating the coexistence of nonequilibrated microenvironments. Dissolved oxygen presumably is advected into the system through the larger, interconnected, and rapidly flushed open pores, while bacteria in the more restricted pores maintain generally anaerobic conditions.

Although most alkalinities are less than the seawater alkalinity, indicating that precipitation has occurred, a few of the samples show elevated alkalinities, suggesting net solution. The data indicate that essentially all samples are very close to equilibrium with Mg-calcite, slightly supersaturated with respect to calcite and aragonite, and substantially supersaturated with respect to dolomite. These results are very similar to those obtained by Morse et al. (1985) in their study of carbonate sediments in the Bahamas. It appears, therefore, that Mg-calcite is the phase controlling the CO_2-system equilibrium. The alkalinity shifts relative to seawater indicate that initial precipitation may be followed by gradual dissolution in response to the CO_2 added by the oxidation of organics.

The effect of this early, biologically stimulated diagenesis must be the net replacement of Mg-calcite by less soluble carbonates; on the basis of the data

Table 2. Pore water chemistry in Davies Reef

Well	Depth	pH	Alkalinity meq l^{-1}	CO_2 mol l^{-1}	Salinity ‰	Mg-Calcite log q/K	Aragonite log q/K	Calcite log q/K	Dolomite (Dis) log q/K
A-1	0.8	7.817	2.283	2.145	35.346	0.12	0.32	0.48	1.30
A-2	1.2	7.883	2.212	2.051	35.353	0.16	0.36	0.53	1.39
A-3	1.2	7.683	2.212	2.127	35.368	−0.03	0.17	0.34	1.00
A-4	1.0	7.698	2.311	2.219	35.360	0.00	0.20	0.37	1.07
A-5	1.4	7.666	2.320	2.239	35.373	−0.02	0.18	0.34	1.01
A-6	1.4	7.692	2.165	2.078	35.400	−0.03	0.17	0.34	1.00
A-7	0.4	7.757	2.362	2.247	35.356	0.07	0.27	0.43	1.20
A-8	0.6	7.718	2.547	2.441	35.405	0.07	0.27	0.43	1.19
A-9	1.4	7.698	2.223	2.134	35.385	−0.02	0.18	0.35	1.02
A-10	1.6	7.650	2.429	2.354	35.343	−0.03	0.17	0.34	1.00
BD	4.5	7.701	2.178	2.089	35.394	−0.02	0.18	0.34	1.01
BS	3.0	7.736	2.280	2.175	35.395	0.03	0.23	0.40	1.12
FD	5.3	7.743	2.319	2.212	35.310	0.04	0.24	0.40	1.13
FS	2.5	7.758	2.272	2.160	35.290	0.04	0.24	0.41	1.14
MD	1.9	7.750	2.110	2.298	35.369	0.01	0.21	0.38	1.08
MS	0.5	7.761	2.197	2.053	35.293	0.04	0.24	0.41	1.14
Seawater	0	8.229	2.346	1.988	35.330	0.52	0.72	0.88	2.10

available, we cannot determine whether there is a net gain or loss of solid carbonates within the reef structure. We can, however, place some reasonable limits on the relative amounts of both precipitation and dissolution with the system. If we assume that the high permeability, high water flux zones which dominate the observed pore velocities represent about 5% of the total reef volume (about 20% of the effective porosity and 10% of the total porosity for the reef structure as a whole), then the shortest estimated residence times (approximately 3 days) correspond to the turnover of 50 l of porewater m^{-3} reef per 0.01 yr, or about 5 m^3 of water m^{-3} reef yr^{-1}. Based on maximum alkalinity depletions and average total CO_2 increases, an order of magnitude estimate of both precipitation and solution values is 20 g calcium carbonate m^{-3} pore water. The annual flux would thus be 100 g m^{-3} reef volume yr^{-1}. At 50% total porosity, m^3 of reef would contain about 1300 kg of solid; these estimated solution/precipitation rates would therefore represent about 8% of the total substrate per 1000 years. This estimate, although crude, represents a reasonable upper limit. A few percent per 1000 years seems to be the best estimate of present $CaCO_3$ dissolution and/or precipitation rates in the upper few meters of Davies Reef.

Two further comments may be based on these observations. First, we note that the conceptual model suggests that the concurrent precipitation and dissolution may not occur in the same locations; the anoxic environments of the stagnant micropores may be the primary dissolution sites, while precipitation may preferentially occur where the initially supersaturated ocean water first interacts with the matrix material – at shallow depths below the reef surface and along the high porosity, high permeability flow channels that dominate the internal flow regime. This suggests the possibility of a gradual equilibration of permeabilities as small pores are enlarged and precipitation occurs in more porous regions. Second, we note that microbial activity within the reef creates relatively large volumes of water which is at, or below, saturation equilibrium with respect to some of the carbonate minerals. In the case of the Dorag model, use of such water in place of the ocean water end-member would significantly extend the salinity range over which dolomitization might be expected to occur. In the case of Saller's (1984) marine dolomitization, it suggests that one need not necessarily go down to the carbonate compensation depth to find marine waters theoretically suitable for dolomitization.

4 Enewetak Atoll

4.1 Hydrogeology

Island groundwater studies conducted at Enewetak during the 1970's have produced data which lead to a model of reef and island hydrogeology considerably different from the traditional Ghyben-Herzberg lens concept (Buddemeier and Holladay 1977, Buddemeier 1981, Wheatcraft and Buddemeier 1981, Herman and Wheatcraft 1984, Herman et al. 1986, Oberdorfer and Buddemeier 1986). In summary, the hydrogeology of the Quaternary deposits may be described as follows.

Islands consist largely of unconsolidated sandy deposits with hydraulic conductivities of the general magnitude of 70 m d^{-1} (Wheatcraft and Buddemeier 1981). Underlying this material at a depth of 10–15 m is the Holocene-Pleistocene contact, below which Pleistocene materials exhibit varying degrees of lithification and a series of solution unconformities indicating subaerial erosion during emergence (see Fig. 2). Tidal response observations (Buddemeier and Holladay 1977, Wheatcraft and Buddemeier 1981) supported by computer simulations (Herman and Wheatcraft 1984, Herman et al. 1986) indicate that the hydraulic conductivity of the Pleistocene aquifer must be of the order of magnitude of thousands of m d^{-1}, presumably as a result of both constructional permeability, as was seen on Davies Reef, and erosional permeability associated with karst. Consolidated surficial deposits (reef rock and beach rock) have hydraulic conductivities of a few m d^{-1} (Wheatcraft and Buddemeier 1981), but these appear to have little effect on large-scale island hydrology beyond the possible reduction of wave and tide related mixing at the island margins.

The occurrence of high permeability strata has important consequences, and is not limited to the shallow Pleistocene formations. Ladd and Schlanger (1960) and Swartz (1958) have reported tidal fluctuations in two deep wells at Enewetak which indicate substantial oceanic water movement through the carbonate rocks at depths greater than 600 m. Saller (1984) has invoked this in his marine dolomitization model, and Samaden et al. (1985) have developed geothermal models which indicate hydraulic conductivities on the order of 10 m d^{-1} over depth intervals of hundreds of meters – presumably the result of averaging the effects of alternating layers of higher and lower permeability.

The existence of a high permeability aquifer at shallow depths exerts two very important influences on the island hydrology. First, the aquifer transmits the ocean tidal signal beneath the island with high efficiency, providing a mechanism for vertical mixing of the fresh water lens with the underlying seawater. Tidal response data from Kwajelein Atoll wells (Fig. 7) show that the immediacy and strength of the tidal signal in wells is more closely related to depth (and hence proximity to the Pleistocene contact) than to distance from shoreline as is generally thought. This results in the formation of an extensive transition zone of brackish water and a freshwater lens which is much smaller than that predicted from assumption of a homogeneous island and the traditional Ghyben-Herzberg lens model. Figure 8 shows the broad transition zone beneath Kwajelein Island (Hunt and Peterson 1980). Second, this aquifer provides a pathway for through-reef flow in response to differences in lagoon and ocean mean water levels. In the case of Enewetak, elevation of lagoon water levels due to wave-driven flow across the reef probably causes oceanward flow through the reef structure, entraining fresh water which is mixed downward into the transition zone (Buddemeier 1981, Atkinson et al. 1981). Under different conditions, such as a more nearly enclosed lagoon and high evaporation rates, lagoonward flow of ocean water through the reef could result (Rougerie 1985).

Overall, the groundwater at Enewetak is characterized by a relatively shallow and variable body of freshwater (typically a few meters or less) underlain by an extensive (tens of meters) brackish transition zone which varies in salinity and extent in response to fresh water recharge. The subsurface, as at Davies Reef, is

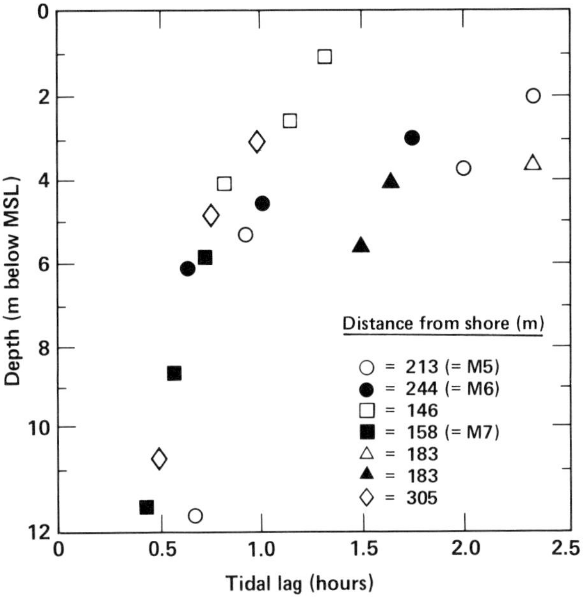

Fig. 7. Well tidal lags and their relationship to well depth and distance from shoreline for Kwajelein Island. Well nests M-5, M-6, and M-7 may be seen in cross-section in Fig. 8

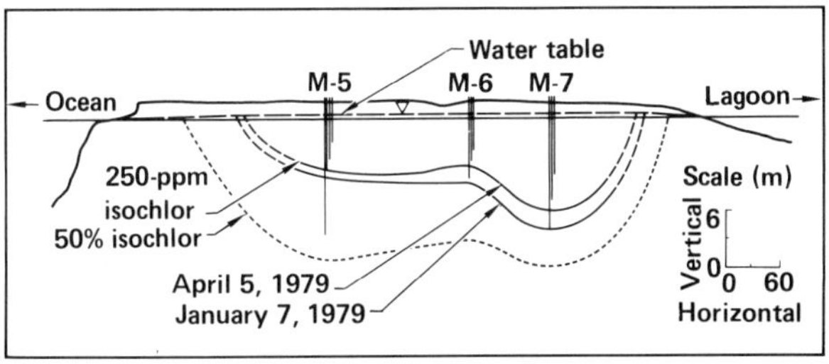

Fig. 8. Field observations of extended transition zone in Kwajelein Island groundwater. (After Hunt and Peterson 1980); see Fig. 7 for depths and tidal responses of nested wells M-5, M-6, and M-7

inhomogeneous with zones of low permeability. However, net water residence times are dominated by the effects of very high permeability strata and channels which permit through-reef transport. Estimates of overall groundwater residence times based on fresh water budgets are on the order of several years, which imply Pleistocene aquifer pore velocities of a few tenths of a meter/day (Buddemeier 1981). This is consistent with the lower end of the range of pore velocities estimated from observations on the smaller and shallower Davies Reef system.

4.2 Groundwater Geochemistry

The distributions and relationships of groundwater salinities and carbonate saturation states for various island groundwater conditions have been developed from available chemical data (Buddemeier, unpublished data). All results are based on field pH and prompt alkalinity determinations and subsequent laboratory measurements of major and minor ions and nutrients. The EQ3 computer code was used to calculate saturation indices. Salinity profiles were calculated from downhole conductivity probes. Samples selected for comparison include seawater and island well water from several classes of wells: those exhibiting consistent fresh water layers at the surface, those showing very little freshwater influence over their entire depth, and those with reduced surface salinity and an extensive transition zone. It is important to note that the deeper wells were perforated over the entire cased length, so that tidal mixing in the well itself does not necessarily represent the mixing in the formation itself (Buddemeier and Holladay 1977). However, the processes are presumably representative of the chemical relationships, even though imperfectly coupled with formation pore-water conditions.

Table 3 presents the data for selected Enewetak ocean and groundwater samples. Samples, grouped by source and depth, are neither sequential nor synoptic. Although some of the well profiles were sampled on the same day, the total time span for these analyses is approximately 2 years; the data thus give some idea of the system variability as well as of trends and patterns.

Table 3. Ocean and groundwater chemistry, Enewetak Atoll

Source/ well	Depth (m)	[Cl^-] (ppt)	pH	Alk (meq l^{-1})	Aragonite log q/K	Calcite log q/K	Dolomite(Dis) log q/K
Reef	0	19.6	8.57	2.32	1.00	1.15	2.62
Lagoon	0	19.5	8.35	2.30	0.79	0.95	2.23
Reef	0	19.4	8.20	2.30	0.64	0.81	1.99
Reef	0	19.5	8.25	2.30	0.69	0.85	2.07
XRU-1	0	19.2	7.63	2.34	0.08	0.24	0.85
XRU-1	4	19.2	7.92	2.30	0.40	0.56	1.47
XRU-1	15	19.0	7.81	2.70	0.34	0.51	1.35
XRU-1	15	19.5	7.90	2.25	0.35	0.52	1.38
XRU-1	40	19.3	8.10	2.26	0.55	0.71	1.77
BDF-1	0	0.09	7.30	7.48	0.19	0.35	0.33
BDF-2	0	0.11	7.65	6.47	0.77	0.89	0.91
BDF-3	0	1.01	7.25	6.94	0.15	0.32	0.26
AJA-1	2.5	0.59	7.30	5.25	0.11	0.27	0.10
AJA-1	3	0.46	7.10	4.80	−0.08	0.08	0.48
XEN-1	0	0.45	7.42	4.70	−0.33	−0.16	0.09
XEN-1	0	0.91	7.50	3.80	0.12	0.29	0.78
XEN-1	0	0.37	7.32	4.90	0.04	0.20	0.39
XEN-1	2.3	0.15	7.25	4.90	−0.06	0.10	0.04
XEN-1	3	0.28	7.32	5.30	−0.29	−0.13	0.03
XEN-1	7	0.89	7.61	3.00	0.13	0.29	0.81
XEN-1	10	1.28	7.43	3.27	0.22	0.39	1.06
XEN-1	15	1.64	7.66	2.70	0.18	0.34	1.04
XEN-1	20	1.71	7.52	2.35	0.09	0.26	0.83
XEN-1	20	1.78	7.49	2.40	−0.03	0.13	0.59
XEN-1	40	1.84	7.27	2.30	−0.11	0.06	0.42
XEN-1	40	1.82	7.60	2.32	−0.20	−0.03	0.25
XEN-1	40	1.88	7.43	2.30	−0.11	0.06	0.45
XEN-1	80	1.92	7.39	1.88	0.07	0.24	0.60
XEN-1	80	1.92	7.55	1.95	−0.06	0.11	0.55

Because of temperature variations and metabolic activity on the reef flat and in the near-shore lagoon, the chemistry of these samples probably show more variation than is to be expected from open ocean or lagoon water. Well XRU-1 is located near the shoreline of Runit Island, a narrow motu lacking a significant freshwater lens. All samples from this well are at least 95% seawater based on chlorinity, and have alkalinities similar to seawater. The pH values are reduced, however, presumably because of atmospheric CO_2 incorporated into the rainfall and/or microbially produced CO_2 added to the freshwater in the soil zone or to the seawater as it passes through the sediment interface (see the Davies Reef case above).

The samples from the BDF wells (open pits) and AJA-1 were included to demonstrate conditions near the water table surface where very low salinities occur. Below about 5% seawater, the water chemistry is controlled primarily by dissolution of the carbonate materials of the island surface, and chlorinity is not related to ionic strength, Ca or Mg concentrations, etc.

Well XEN-1 is the best example of a deep well with an extensive transition zone and a significant freshwater inventory. Figure 9a plots the salinity profiles obtained from downhole probes at three different times. In this borehole, solution unconformities were encountered at depths of approximately 8, 20, 35, 50, and 63 m below MSL (see Fig. 2); at least three of the persistent inflections in the salinity profile correspond reasonably well to depths of these unconformities. Both the down hole data and the chemical analyses (Table 3) indicate that measurable amounts of fresh water are mixed to depths of as much as 80 m; although it is possible that some of this mixing may be an artifact of the open borehole, the fact that the major features of the salinity structure are independent of tide stage and season suggests at least qualitative correspondence with formation salinities.

Figure 9b presents a plot of the saturation index values from Table 3 on the same vertical scale as Fig. 9a. The dashed lines at ± 0.2 represent the probable maximum uncertainty in saturation index values derived from the EQ3 code (T. Wolery, personal communication). At shallow depths, conditions are quite variable, but below a few meters depth, a consistent pattern emerges. Saturation values are highest at a depth of about 10 m; we note from the discussion of Davies Reef data (see above) that this water would be approximately saturated with respect to high magnesium calcite. At depths of 20 and 80 m, the water is approximately in equilibrium with aragonite, while at 40 m, calcite appears to be the equilibrium species. Disordered (and also ordered) dolomite is supersaturated throughout the profile.

These data indicate that, on a thermodynamic basis, one or more of the more soluble minerals should be dissolving throughout the profile, and that precipitation of the less soluble forms is possible. The data do not permit any conclusions concerning the rate or extent of such reactions.

Several further observations on the groundwater chemistry may be presented to complete the discussion. Nutrient levels were somewhat elevated in nearly all groundwater samples; nitrogen was roughly an order of magnitude more concentrated than in seawater, but phosphate showed little, if any, increase. There was no consistent relationship between salinity and nutrient levels, and we

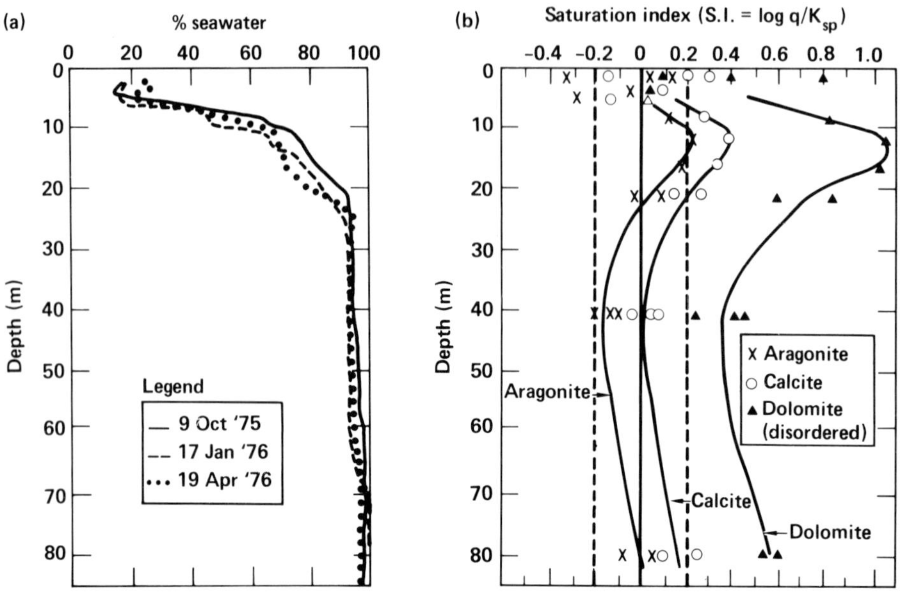

Fig. 9a, b. Salinity and carbonate saturation state profiles in well XEN-1, Enjebi Island, Enewetak Atoll (see Fig. 2). **a** Vertical salinity distribution (expressed as percent of seawater) on three different dates. **b** Vertical distribution of calculated saturation indexes (S.I. $= q/K_{sp}$, where q = ion activity product and K_{sp} = equilibrium solubility product) for three different carbonate minerals. Data are based on samples collected at three different times. Minerals are at equilibrium when S.I. = 0, and supersaturated at positive values. *Vertical lines* at ± 0.2 represent maximum limits of uncertainty in the calculated values

cannot determine what proportion of the excess nutrient loading originated from freshwater leaching of the soil zone and what may have come from nutrient regeneration in the marine sediments as noted at Davies Reef.

Downhole oxygen probes (to a maximum depth of 15 m) showed that dissolved oxygen levels dropped to a few tenths of a ppm below the upper few meters. Sulfide was not detected in the wells drilled with bentonite drilling mud, but because of the likelihood of artificial vertical mixing in the borehole, this does not prove conclusively that the formation porewater was not anoxic. In several wells drilled with an organic drilling fluid (of which AJA-1 is the only one reported here), sulfide-bearing water was found consistenly at the bottom of the well over the course of the study (>4 years).

The Enewetak data indicate the existence of a substantial through-reef water flux in zones of high permeability, vertical mixing of freshwater to depths of tens of meters below the islands, and the potential for biogenic effects on the chemistry of porewater within the reef and island substructure. As was the case in the shallow sediments of Davies Reef, the chemical conditions required for carbonate mineral dissolution and precipitation were found throughout the system, and the estimated water fluxes were sufficiently high to suggest that diagenetic changes could be occurring at a geologically significant rate.

5 Discussion and Summary

We have offered two specific examples, Davies Reef and Enewetak Atoll, to demonstrate the major points we wish to make: that coral reefs can be highly (albeit variably) permeable with substantial fluxes of water through them, and that the microbial activity and its chemical effects on water characteristics are ubiquitous inside reefs and islands. We do not suggest that these examples are representative of all reefs and atolls. Indeed, there is reason to believe that the internal water circulation in these examples may be more dynamic than the average; drilling on the more seaward Australian barrier reefs has shown greater lithification in the subsurface than was found at Davies Reef (P. Davies and D. Hopley, personal communication), and Enewetak Atoll has an unusually large number of lagoon pinnacle reefs developed on Pleistocene outcrops (E. Shinn, personal communication), which may provide direct hydraulic connections with the karst aquifer.

Nonetheless, these reefs and islands are by no means unusual in their environment, structure, and history, and we emphasize that the conditions described may be expected to occur commonly throughout coral reef systems. Although there are certainly reef sediments which are lithified or finegrained such that water movement and biological activity are negligible, we have chosen to stress the observations of high water movement and biological activity because these conditions dominate the large-scale hydrology and geochemistry of the systems. Furthermore, they are contrary to the intuitive view that the inside of a reef is a quiescent environment potentially understandable in terms of strictly chemical equilibria and kinetics.

In emphasizing the importance of hydrologic and biogenic effects, we also wish to stress that these effects are experimentally accessible. The concepts and methods of hydrology can be adapted to reef studies, and when combined with chemical observations (as described above) and microbial and mineralogical measurements may, in the future, provide new dimensions to the understanding of coral reef diageneses.

Disclaimer. This document was prepared as an account of work sponsored by an agency of the United States Government. Neither the United States Government nor the University of California nor any of their employees, makes any warranty, express or implied, or assumes any legal liability or responsibility for the accuracy, completeness, or usefulness of any information, apparatus, product, or process disclosed, or represents that its use would not infringe privately owned rights. Reference herein to any specific commercial products, process, or service by trade name, trademark, manufacturer, or otherwise, does not necessarily constitute or imply its endorsement, recommendation, or favoring by the United States Government or the University of California. The views and opinions of authors expressed herein do not necessarily state or reflect those of the United States Government or the University of California, and shall not be used for advertising or product endorsement purposes.

Acknowledgments. We gratefully acknowledge the support of portions of this work by the Australian Institute of Marine Sciences (AIMS). The editors of this volume and many other colleagues have patiently worked with us to help transport relevant ideas and concepts across disciplinary boundaries. Although much remains to be done, we are grateful for the interest and assistance we have received, and for the opportunity to make some contribution to an effort of synthesis.

References

Atkinson M, Smith SV, Stroup ED (1981) Circulation of Enewetak Atoll Lagoon. Limnol Oceanogr 26:1074–1083

Ayers JF, Vacker HL, Clayshulte JN, Stout D, Stebrinsky R (1984) Hydrogeology of Deke Island, Pingelap Atoll, Eastern Caroline Islands. Technical Report No 52, Water and Energy Research Institute of the Western Pacific, Univ Guam

Badiozamani K (1973) The Dorag Dolomitization Model – application to the Middle Ordovician of Wisconsin. J Sediment Petrol, 43, 4:965–984

Berner RA (1980) Early Diagenesis: A Theoretical Approach. Princeton Univ Press, Princeton

Bernstein LD, Morse JW (1985) The steady-state calcium carbonate ion activity product of recent shallow water carbonate sediments in seawater: a comparison of field observations and laboratory experiments. Mar Chem 15:311–326

Buddemeier RW (1981) The geohydrology of Enewetak Atoll Islands and Reefs. Proc 4th Int Coral Reef Symp 1:339–345

Buddemeier RW, Holladay G (1977) Atoll hydrology: island groundwater characteristics and their relationship to diagenesis. In: Proc 3rd Int Coral Reef Symp 2:167–173

Choquette PW, Pray LC (1970) Geologic Nomenclature and Classification of Porosity in Sedimentary Carbonates. Am Assoc Petrol Geol Bull 54:207–250

Curtis CD (1977) Sedimentary geochemistry: environments and processes dominated by involvement of an aqueous phase. Philos Trans R Soc Lond A 286:353–372

Davis SN, De Wiest RJM (1966) Hydrogeology. Wiley, New York

Enos P, Sawatsky LH (1981) Pore networks in holocene carbonate sediments. J Sediment Petrol 51:961–985

Freeze RA, Cherry JA (1979) Groundwater. Prentice-Hall, Englewood Cliffs, New Jersey

Friedman GM, Sanders JE (1967) Origin and occurrence of Dolostones. In: Chilingar GV, Bissell HJ, Fairbridge RW (eds) Carbonate rocks. Elsevier, Amsterdam, p 267–348

Herman ME, Wheatcraft SW (1984) Groundwater dynamics investigation of Enjebi Island, Enewetak Atoll: an interpretive computer model simulation. In: Proc 5th Int Conf Finite Elements in Water Resources. Burlington, Vermont, pp 133–142

Herman ME, Buddemeier RW, Wheatcraft SW (1986) A layered aquifer model of Atoll Island hydrology: validation of a computer simulation. J Hydrol (Amst) (in press)

Ladd HS, Schlanger SO (1960) Drilling operations on Enewetak Atoll. USGS Prof paper 260-Y, US Government Printing Office, Washington, DC

Longman MW (1980) Carbonate diagenetic textures from nearsurface diagenetic environments. Am Assoc Petrol Geol Bull 64:461–487

Morse JW, Zullig JJ, Bernstein LD, Millero FJ, Chappin GR (1985) Chemistry of calcium carbonate-rich shallow water sediments in the Bahamas. Am J Sci 285:147–185

Oberdorfer JA, Buddemeier RW (1985) Coral Reef hydrogeology. Proc 5th Int Coral Reef Congr (in press)

Oberdorfer JA, Buddemeier RW (1986) Coral Reef hydrology: field studies of water movement within a Barrier Reef, Coral Reefs (in press)

Ristvet BL, Tremba EL, Couch RF, Fetzer JA, Goter ER, Watter DR, Wendland VP (1978) Geologic and geophysical investigations of the Enewetak Nuclear Craters. US Air Force Weapons Laboratory AFWL-TR-77-242. Kirtland Air Force Base, NM

Rougerie F (1985) Main characteristics of the Takapoto Lagoon and internal functioning. Proc 5th Int Coral Reef Congr 1:363–367

Saller AH (1984) Petrologic and geochemical constraints on the origin of subsurface Dolomite, Enewetak Atoll: an example of dolomitization by normal seawater. Geology (Boulder) 12:217–220

Samaden G, Dallot P, Rocke R (1985) Enewetak Atoll: a geothermal system in natural conditions heat and mass transfer. Ions transport between the Ocean and the Lagoon. In: Euromech 194, 2–5 July 1985, Nancy (France)

Swartz JH (1958) Geothermal measurement on Enewetak and Bikini Atolls, Bikini and Nearby Atolls. US Geol Surv Profess Pap 260-U, pp 711–739

Walter LM, Morse JW (1984) Magnesian calcite stabilities: a re-evaluation. Geochim Cosmochim Acta 48:1059 – 1069
Wheatcraft SW, Buddemeier RW (1981) Atoll Island hydrology. Groundwater 19(3):311 – 320
Wolery TJ (1983) EQ3NR: a computer program for aqueous speciation – solubility calculations: user's guide and documentation. Univ California UCRL-53414, Livermore, CA

Diagenesis of a Miocene Reef-Platform: Jebel Abu Shaar, Gulf of Suez, Egypt

D. M. Aissaoui[1], M. Coniglio[2], N. P. James[2], and B. H. Purser[1]

1 Introduction

Abu Shaar El Quibli is one of many Miocene carbonate complexes exposed along the west side of the Gulf of Suez (Fig. 1). These platforms are developed on basement horsts and, when buried beneath evaporites, may be hydrocarbon reservoirs. The main structural features of the Gulf of Suez area consist of a series of tilted blocks related to the Neogene Rifting of the Red Sea and Gulf of Suez (Burollet et al. 1985). Two dominant fault trends, clysmic (NW-SE) and Aquaba (NE-SW), have cut crystalline basement and pre-Neogene cover forming a structural mosaic, most blocks being tilted antithetically, i.e., away from the axis of the Red Sea rift. Their existence during the Miocene has profoundly influenced Neogene sedimentation and diagenesis.

The Abu Shaar complex is a thin Miocene carbonate platform developed on the southern end of the Esh Mellaha range, an elongate fault-block, and is bordered to the east by Miocene evaporites. The carbonates now form a nearly flat plateau situated some 120 m above the surrounding coastal plain. The platform has steep, clearly defined east and south-facing margins, but grades gently westward into a sand gravel plain formed by alluvium from the Red Sea hills. The platform is undeformed and the present topography is close to the original depositional morphology.

Abu Shaar was recognized as a Miocene structure by Gregory (1906), Hume (1916) and Madgwick et al. (1920), based mainly on corals. Heybroek (1965) suggested that development of reefs on the Abu Shaar platform preceded evaporite sedimentation. Rouchy et al. (1983) confirmed that the evaporites lie unconformably against the base of the Miocene carbonate slope and thus are clearly younger. Recent studies (Cofer et al. 1984, Haddad et al. 1984) have synthesized stratigraphy, sedimentation, and paleogeography of the complex. Haddad et al. (1984) were the first to recognize that *all* Abu Shaar carbonates are dolomites and stressed the relationships between distensional tectonics and mixed siliciclastic-carbonate sedimentation.

[1] Laboratoire de Pétrologie Sédimentaire et Paléontologie, Bâtiment 504, Université de Paris-Sud, 91405 Orsay Cedex, France

[2] Department of Earth Sciences, Memorial University of Newfoundland, St. John's, Newfoundland, A1B 3X5, Canada

Reef Diagenesis
Edited by J. H. Schroeder and B. H. Purser

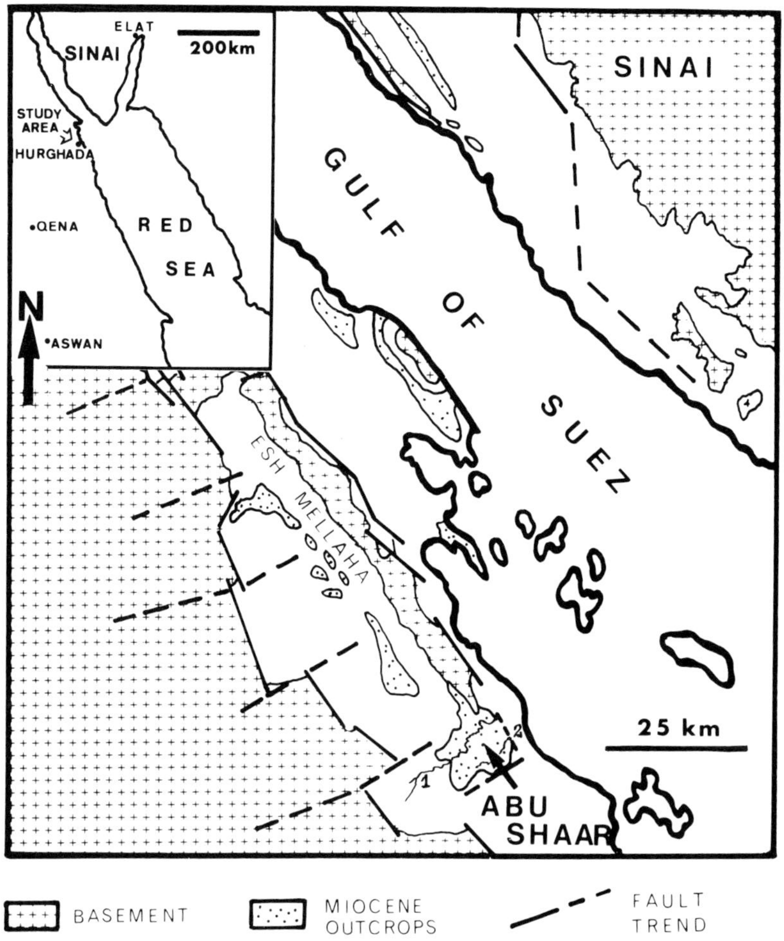

Fig. 1. Localization of the study area and schematic structural map of Abu Shaar where 1 is wadi Bali'a and 2, wadi Kharaza

2 Depositional Facies

Rapid vertical and lateral facies changes are related to pre-existing topography. Three principal facies (Fig. 2) are distinguished in these Miocene strata (Haddad et al. 1984).

2.1 Platform Interior Facies

These consist of alternating white chalky dolomites and quartz sands overlain by tidal-flat facies with stromatolitic domes. The terrigenous clastics are most abundant to the NW, reflecting proximity to the crystalline basement of the Esh

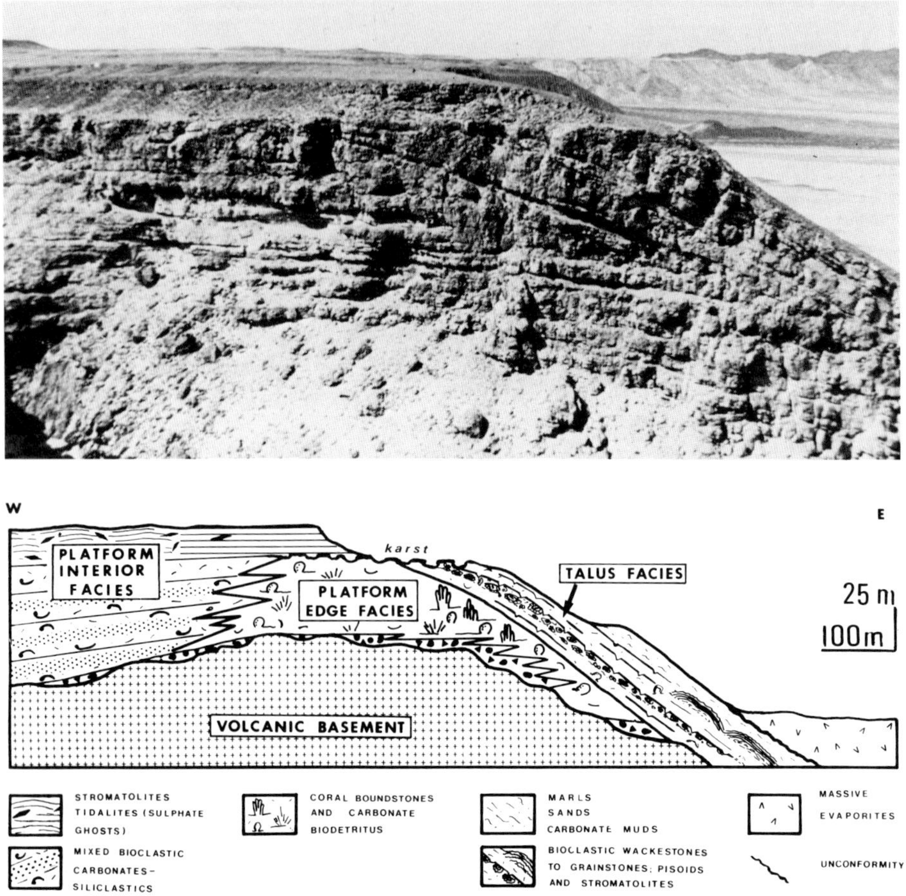

Fig. 2. Photo and schematic cross-section of the eastern periphery of Abu Shaar Plateau showing relationships between the principal facies

Mellaha Range, but decrease in proportion toward the southwest where a relatively pure carbonate succession prevails.

2.2 Platform Edge Facies

Along the eastern periphery of Abu Shaar plateau, these include lenticular reef bodies with coral floatstones, rudstones and local framestones, pavements of algal boundstones, and grainstones dominated by varying amounts of foraminifera, molluscs and peloids. Coral-rich facies may attain 20 m in thickness and are massively bedded at the platform edge. Toward the platform interior, bedding thickness diminishes as the number of corals decreases. Platform edge facies are limited to the eastern side of the complex, reflecting the more open marine conditions prevailing to the east (basin center).

2.3 The Talus Facies

This consists of grey colored, well bedded dolomites containing mollusc, coralline algal wackestones to grainstones, coral boundstones and spectacular, 5 m wide stromatolitic domes. It also includes 1 to 4 m thick, lenticular beds composed of cm-sized pisoids located near the top of the talus slope. The talus strata are restricted to the east-facing slope of the plateau, their present dip 40° to 60° being essentially depositional in origin.

3 Diagenesis

The Miocene carbonates at Abu Shaar exhibit numerous and varied diagenetic modifications including marine and meteoric cementation, karstic dissolution, and dolomitization. Dolomitization, which is regional, has affected most of the Miocene carbonates along the Egyptian Red Sea coast. At Abu Shaar, the replacement dolomite has retained textural details to the extent that these Miocene carbonates could readily be mistaken for limestone. Due to this extraordinary preservation, many of the original characteristics of the pre-dolomitization fabrics are visible, and can be interpreted.

3.1 Cement Petrography

There are two cement mineralogies, dolomite and minor calcite. Most of the dolomitized cements recognized at Abu Shaar are synsedimentary, as indicated by their associated internal marine sediments. Several types of cement are recognized. The calcite cement is local and post-dates all other cements.

3.1.1 Fibrous Cements

Fibrous cements occur as single or multiple layers around grains, within dissolution cavities or in synsedimentary fractures. The fibrous structure, outlined by dark inclusions, is clearly visible both under the light microscope and SEM in spite of total dolomitization. However, there are examples where these cements develop a secondary spar fabric consisting of a mosaic of equant dolomite crystals, their original fibrous morphology being visible as relics. Four types of fibrous cement are distinguished (Fig. 3):

Type 1 fibrous cement (Fig. 4A, B) is restricted to the talus facies, occurring within and around pisoids. Individual crystals average 10 μm in width and 300 μm in length. They constitute multiple rinds, whose total thickness may attain 5 mm, filling primary pores. The terminations of type 1 crystals are commonly irregular, suggesting erosion or corrosion (Fig. 6D). Some fringes are overlain by internal marine sediments, confirming their marine phreatic origin.

Type 2 fibrous cements (Fig. 4C, D) are found especially in coral reefs directly overlying the volcanic basement in the northern part of the Abu Shaar

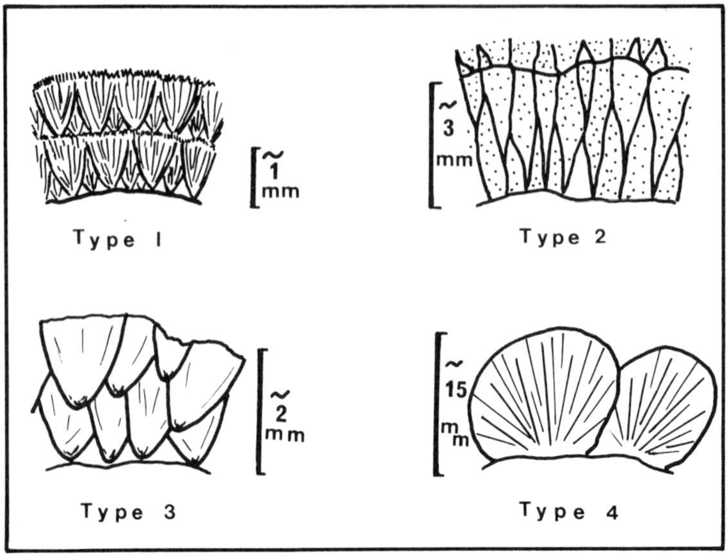

Fig. 3. Sketch of the four types of dolomitized fibrous cement

plateau. Cement fringes may reach 20 mm in thickness and typically are a milky-white colour in slab samples. In large secondary cavities created by the dissolution of the coral heads, some up to 50 cm across, type 2 cement occurs as cerebroid concretions which line the cavities (Fig. 4D). In thin section, the fibrous structure outlined by inclusions is seen to be relict; the cement now consists of a mosaic of equant crystals ranging from 15 to 40 μm in size. Under crossed-polarized light, type 2 cement is composed of parallel, elongate bands (Fig. 3) in which the crystals have a common extinction. The fibrous precursor of type 2 cements suggests a marine origin.

Type 3 fibrous cement (Fig. 5A) is found in the stromatolitic domes of the talus facies, and it is also associated with type 1 cement in the pisoid beds. In thin section, type 3 cement is composed of nested, cone-shaped zones of crystals which exhibit undulose extinction. In some samples, nested cones grade laterally to slightly individualized crystals averaging 10 μm in width and 400 μm in length. This transition is often outlined by an increase of dark inclusions, only thin boundaries between the radiating crystals remaining limpid. Type 3 cement resembles the coconut-meat calcite described by Folk and Asseretto (1976). These cements are interpreted to be marine, possibly originally Mg-calcite, based on analogy with similar cements described in other studies (Meyers and Lohmann 1977, James and Klappa 1983, Sandberg 1985).

Fig. 4A – D. Miocene cements (now dolomite). **A** Type 1 fibrous cement layer around dissolved coral (thin section). Talus facies. **B** Typical undulose extinction of type 1 cement (thin section). Talus facies. **C** Type 2 fibrous cement (white layer) lining inter-coralline cavity shown in **D**. The cement overlies dissolved coral substratum (polished surface). Platform interior facies. **D** Outcrop of type 2 fibrous cement; note the irregular, bulbous form of the cement lining this cavity formed within dissolved coral from the platform interior facies (*arrow* indicates coin as scale)

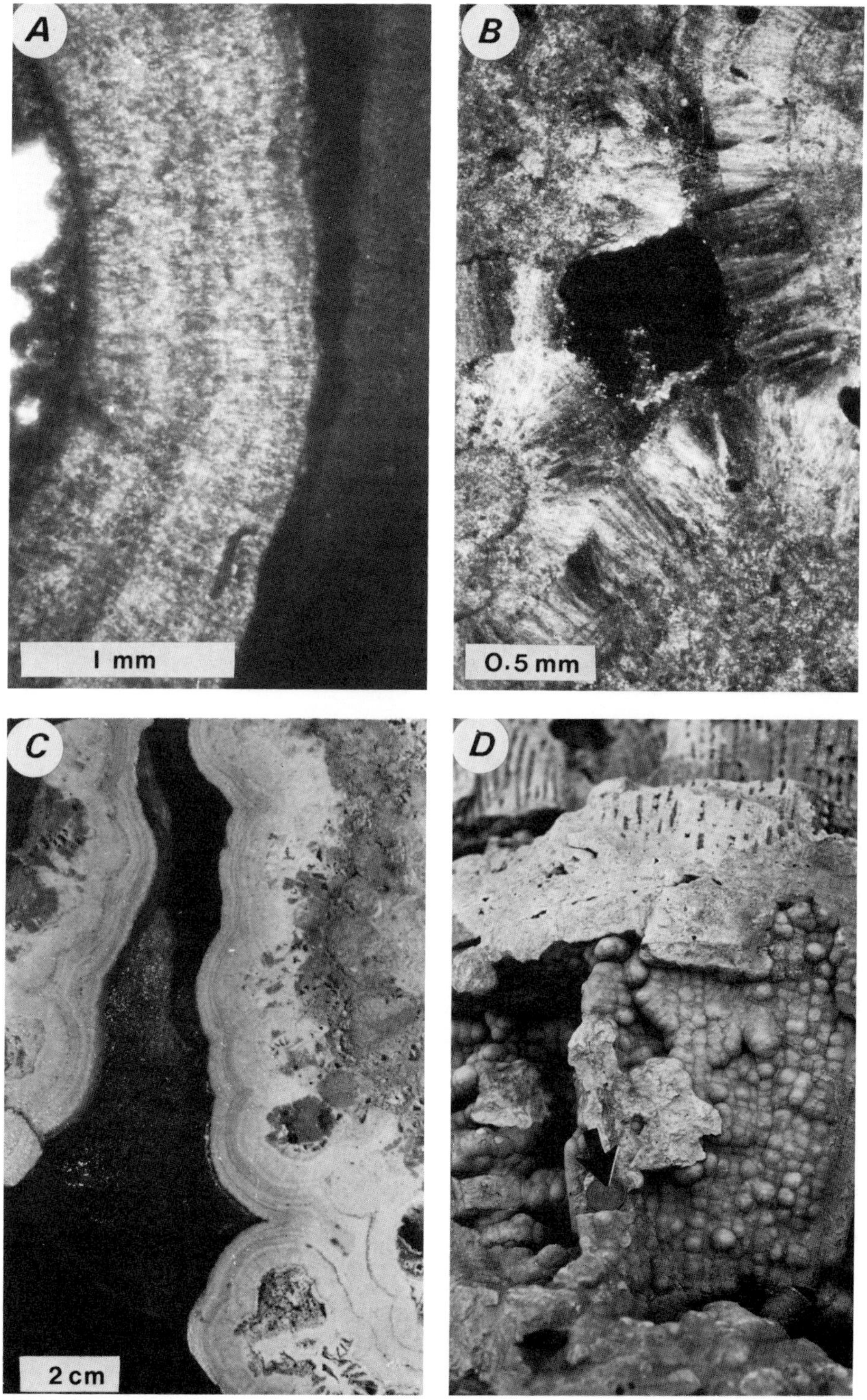

Fig. 4

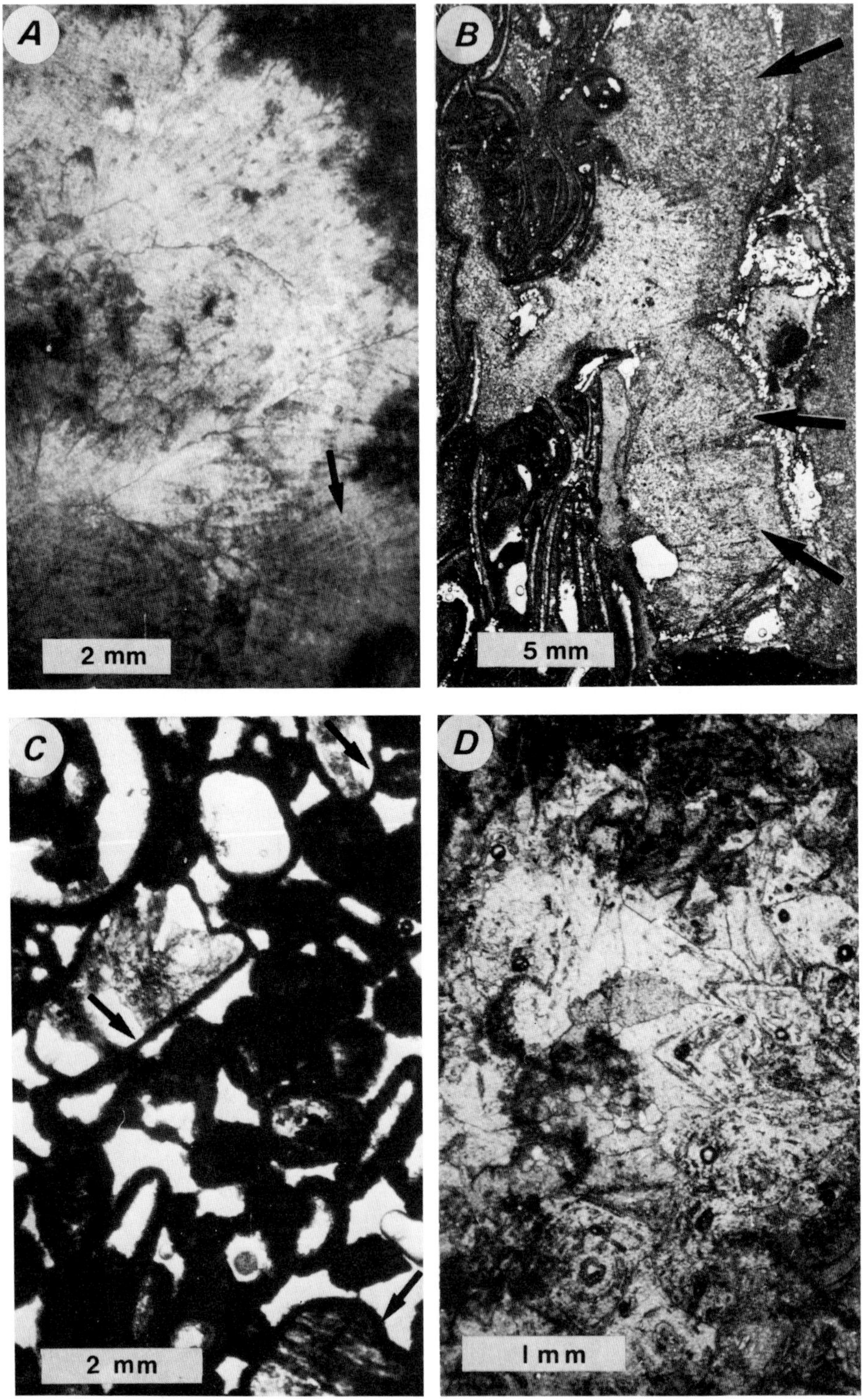

Fig. 5

Type 4 fibrous cement (Fig. 5B) occurs as rare isolated or coalescing mammelons ranging from 2 to 15 mm in diameter precipitated within primary cavities within some beds of the talus facies, notably within the pisoid beds. These cements consist of large fans of radiating crystals and are similar to modern botryoidal cements reported from Holocene and Pleistocene reefs (James and Ginsburg 1979, Aïssaoui 1985). Mammelons typically exhibit fibrous relics, suggesting that these cements were originally composed of tightly packed aragonite crystals (Aïssaoui 1985). Botryoidal cements from Abu Shaar are often post-dated by marine internal sediment.

3.1.2 Micrite Cement

These are the most ubiquitous early cements at Abu Shaar and occur both within the talus and platform interior facies, especially within the mixed carbonate-siliciclastic sediments (Fig. 5C). The abundance of bored hard-grounds exhibiting mud-supported texture in the talus facies suggest that the micritic cement is quantitatively important within the Miocene carbonates. Numerous bored hard-grounds at the top of these muddy substrates confirm that the micritic cements are synsedimentary. In grainy sediments, these cements appear as dark layers surrounding particles. The micritic cement rinds average 100 μm in thickness and are clearly not micrite envelopes (Bathurst 1966) because they also occur around quartz grains (Fig. 5C).

Much of the micritic cement is the result of subtidal cementation. However, there are some examples where micritic cements are clearly related to subaerial exposure, now in the form of dolocretes on the main exposure surface at Bir Abu Shaar and wadi Kharaza. There are also Miocene soils with plant roots, notably within the platform interior facies. Here, the associated micritic cements are certainly nonmarine (Bernouilli and Wagner 1971).

3.1.3 Spar Cement

Sparitic cementation is very minor, despite repeated Miocene emersion of the Abu Shaar platform. Only one type of spar cement, underlying Miocene karst, has been recognized. As all other previously described cements, this spar is now dolomite. It is found within dissolution cavities, especially dissolved bioclasts, occurring as a mosaic of limpid, zoned crystals ranging from 10 μm to 900 μm in size (Fig. 5D). Rare dolospar cements (Fig. 5D) at Abu Sharr may have originally been either calcite or dolomite. It is thus probable that spar cements

Fig. 5A – D. Miocene cements (now dolomite; thin sections). **A** Type 3 fibrous cement formed by nested cone-shaped zones; note the inclusions which underline the fibrous disposition (*arrow*). Stomatolitic domes from the talus facies. **B** Type 4 fibrous cement occurring as coalescing mamelons (*arrows*) similar to botryoidal cements. Talus facies. **C** Micritic cement within mixed carbonate-siliciclastic sediment; note the regular cement fringe around non-carbonate grains (*arrows*). Platform interior facies. **D** Dolosparitic cement with zoned crystals. Platform edge facies were originally calcite wich has later been replaced by dolomite with preservation of the typical calcite crystal morphology.

3.2 Pisoid Beds from the Talus Facies

Miocene cements are concentrated along the eastern margin of Abu Shaar, where they occur mainly within the talus facies. Included in the talus facies are pisoid beds (Rouchy et al. 1983; Cofer et al. 1984) which are remarkable in two respects: (1) their restriction to this facies; and (2) their composition of encrusting coralline algal and type 1 fibrous cement.

3.2.1 Description

The main features of the pisoid beds are the following:

- they are enclosed conformably within bioclastic mudstones and wackestones, rarely grainstones, and extend laterally for at least 2 km along the talus slope;
- the pisoid beds are asymmetrically lenticular, with thickness varying from 1 to 4 m, the maximum always being furthest upslope;
- the diameter of the pisoids tends to increase from the lower to the upper part of the beds;
- except for scattered small, trapped bioclasts and intrapisoid micrites, these beds are composed entirely of pisoids;
- there is commonly a bored hard-ground developed on these pisoid beds.

Both single and compound pisoids occur at Abu Shaar, their size varying from 2 to 100 mm in diameter, but averaging 20 mm (Fig. 6). Pisoids are subspherical, ovoid or lumpy in shape, and are generally complete, although broken ones are also present. Unlike some coated grains reported from other areas (Dunham 1969, Esteban and Pray 1983), no irregular pisoids exhibiting cerebroid shapes have been noted from Abu Shaar. In thin section, many pisoids have a dissolved nucleus, typically arcuate in shape. In other pisoids, the nuclei are preserved and are often broken *Halimeda* plates. Pellets and grapestones also form the nucleus of some pisoids.

The cortex of the pisoids has a complex but invariable micro-structure. The first cortical layer is usually a discontinuous, single or multiple layer of coralline algae (Fig. 6B) still showing conceptacles in some cases (Bosence, personal communication). This is followed by multiple layers of fibrous cement (Fig. 6C). These layers may be separated by thin, smooth, or irregular microcrystalline films more or less continuously developed around the pisoids. The fibrous crys-

Fig. 6A – D. Pisoids from the Talus facies (now dolomite; thin sections). **A** General view in thin section showing the variety of shape and size of pisoids. **B** Detail of the pisoid cortex showing the encrusting algae (*arrow*). **C** Pisoid cortex showing alternating layers of fibrous cement (clear lamina) and microcrystalline film (dark lamina) probably of organic origin, **D** Broken cement fringe (*arrow*) of pisoid indicating erosion and/or reworking of some grains

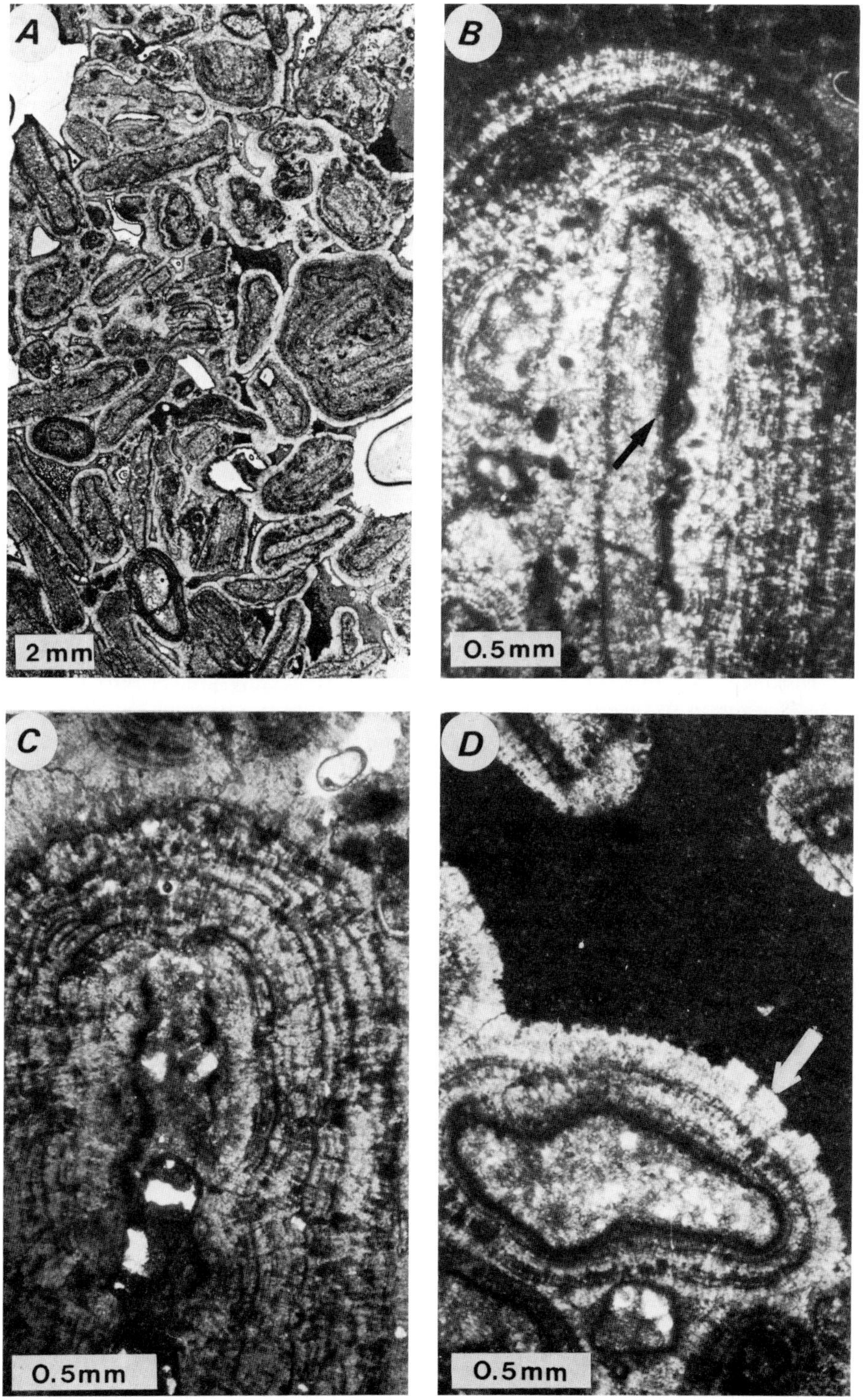

Fig. 6

tals are 10 to 20 μm wide and average 2 mm in length, have undulose extinction, and resemble crystals of type 1 fibrous cement, particularly when the outermost cortical layer extends between grains as an intergranular cement. The cortex may include abundant small grains such as peloids, foraminifera, and other bioclasts.

3.2.2 Interpretation

The coated grains in the talus facies at Abu Shaar have been described as oncoids by previous workers, who suggested that their origin was due mainly to encrusting blue-green algae (Rouchy et al. 1983, Cofer et al. 1984). However, coating by the coralline algae is clear in many instances. On the other hand, the presence of microcrystalline films within the cortex of Abu Shaar pisoids resembles the micro-structure of cyanoids, grains coated by cyanobacteria (Riding 1983). In all cases, however, radial fibrous crystals, which dominate the cortex, are present between the layers of either cyanobacteria or red algae. The origin of these grains may be similar to that suggested by Mazzulo and Cys (1983) for unusual algal-crystalline carbonate coated grains from the Capitan reef of New Mexico which these authors interpret as an alternation of red algae and fibro-radial calcite replacements of marine, acicular aragonite.

These mm-cm-sized pisoids occur as distinct beds only on the talus; they have never been observed within the horizontal beds on the platform. Although the presence of broken pisoids and other detritus suggests some reworking (Fig. 6D), the fibrous, crystalline nature of the cortex which locally grades into fibrous cement common to several grains clearly indicates that these grains developed in place, at least during their later stage of formation. The absence of vadose sedimentary or diagenetic criteria and their localization within typical open-marine facies, suggests that they have formed in situ under shallow subtidal conditions as coarse, dense gravel lenses on what was clearly an inclined sedimentary surface.

3.3 Internal Sediments

Both marine and nonmarine internal sediments are common within the Miocene carbonates of Abu Shaar, where they fill cavities ranging from few mm to more than 30 cm in diameter. Their color, which differs from that of their host rock, facilitates their recognition in outcrop (Fig. 7A, D). Like the cements which they often overlie, internal sediments are totally dolomitized.

Fig. 7A–D. Internal sediments and dissolution. **A** Nonmarine internal sediments on polished section: reddish internal sediment containing rock minute fragments (*1*) overlain by green internal sediment composed of crystal debris (*2*); note that these two generations of internal sediments are separated by type 2 fibrous cement (*arrow*). Talus facies. **B** Thin section showing marine internal sediment (*dark*) overlain by nonmarine internal sediment (*arrows*). Pisoids from the talus facies (*top right*). **C** Selective dissolution of components; note that coralline-algae have resisted dissolution (thin section). Platform edge facies. **D** Outcrop view of generalized selective dissolution of components, some cavities being filled with geopetal internal sediment (*arrows*). Talus facies. The outcrop shown is approximately 2 m high

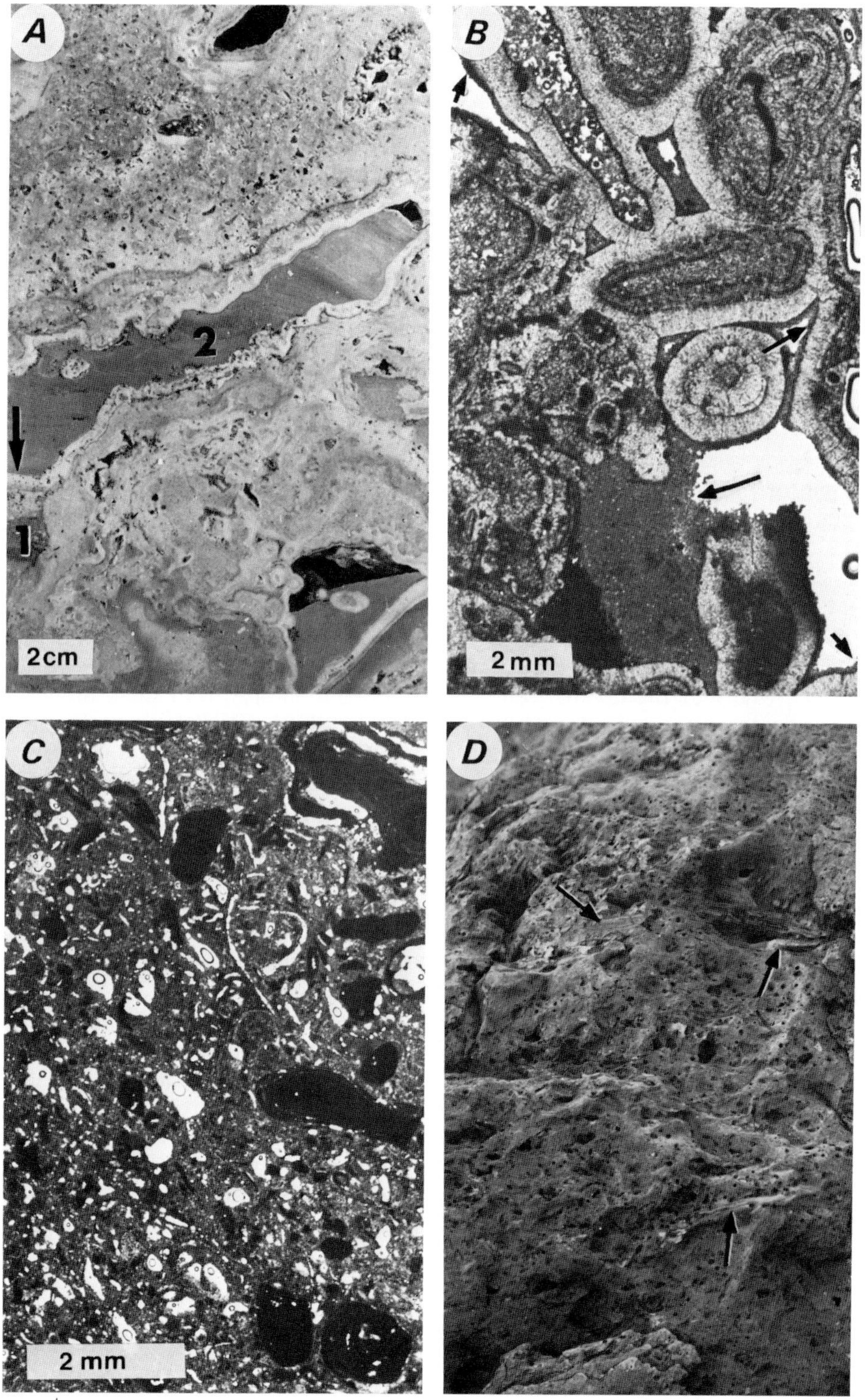

Fig. 7

Due to their geopetal disposition and horizontal top, marine internal sediments are critical evidence for the sedimentary origin of the steeply dipping talus beds (Haddad et al. 1984). Their marine origin is deduced from the enclosed fine bioclastic debris (Aïssaoui and Purser 1983).

Two types of nonmarine internal sediments are recognized: (1) internal sediment associated with karst; and (2) vadose internal sediment. The first (Fig. 7 A) consists of red-colored, iron-rich fillings whose distribution is closely related to Miocene karst in the eastern margin of the platform. Geopetally deposited, this internal sediment is slightly argillaceous, especially where the sediment includes angular fragments of volcanic rocks.

Vadose internal sediment is common within the talus beds, especially north of wadi Bali'a, where its green color is typical. It is composed of equigranular crystals less than 10 µm in size, and is characterized by the total absence of both bioclastic and terrigenous debris. Although geopetal in large cavities, this internal sediment has a centrifugal arrangement in millimetre-size pores (Fig. 7 B) suggesting a supratidal origin (Aïssaoui and Purser 1983). The near-surface origin of this sediment is confirmed by the presence of mud-cracks within this internal sediment when occurring within large cavities.

3.4 Dissolution

The Miocene carbonates at Abu Shaar have been strongly affected by meteoric waters. Many of the organic components, ranging in size from small mollusc shells to large coral heads, are now totally dissolved (Fig. 4 D). As expected during meteoric alteration, dissolution is mineralogically selective (James and Choquette 1984). Original Mg-calcite bioclasts, especially coralline algae and foraminifera are largely preserved (Fig. 7 C), their dissolution being usually minor. Sediments with a preponderance of Mg-calcite components therefore have mostly intergranular porosity. On the other hand, aragonitic particles were totally dissolved and are now represented by voids, dolomite spar or internal sediment-filled cavities. Sediments with a high abundance of aragonite grains are now the most porous dolomites, characterized by interparticle, moldic and vuggy porosity.

The Abu Shaar carbonates have been affected by several phases of dissolution. Coral bioherms at wadi Bali'a, the only major valley which crosses Abu Shaar plateau, have been dissolved and the remaining cavities lined with type 2 fibrous marine cement (Fig. 4 A). This dissolution probably occurred during emergence of the reef. It could be related, however, to a well-developed Miocene karst (Fig. 2) whose cavities locally exceed 0.5 m in size. Evidence for a karstic origin is suggested by the local development of Miocene soils (dolocretes) at Wadi Kharaza and by iron-oxide mineralization along the western edge of Abu Shaar plateau.

3.5 Dolomitization

Detailed analysis of the dolomites is the subject of a separate report (Coniglio, James, and Aïssaoui, in preparation), but the main observations are presented here for completeness.

3.5.1 Dolomite Petrography

In spite of the near-perfect preservation of the original sedimentary and early diagenetic fabrics, petrographic variability of the dolomite crystals is minor. In addition, the mineralogy of the Abu Shaar dolomites is relatively constant, with $MgCO_3$ content varying between 49 and 50.5 mol % whatever the position of the sample in the Miocene section. Three dolomite petrotypes are recognized.

Type 1 (Fig. 8 A) includes all dolomites exhibiting well-preserved original textures, both under plane and cross-polarized light. Fibrous cements commonly have their undulose extinction preserved and, under high magnification, the distinction of the individual dolomite crystals is vague. Under SEM, spherical components, such as peloids, retain their rounded form, but are composed of small, interpenetrating dolomite crystals. Well-developed rhombohedra are rare, their size being less than 10 μm. These crystals may exhibit curved rather than planar rhombic faces.

Type 2 includes dolomite in which primary or early diagenetic fabrics are preserved as ghosts. This dolomite petrotype occurs as finely crystalline, equigranular mosaic in which the distinction between grains such peloids, coralline algae and interparticle matrix is greatly diminished. The cores of certain large dolomite crystals are dissolved (Fig. 8 B).

Type 3: Both type 1 and 2 may be cemented locally by small amounts of late coarse, anhedral dolospar (Fig. 8 C, D). This later constitutes type 3 and often exhibits etched crystal faces and blunt edges presumably resulting from Quaternary dissolution (Fig. 8 D).

The distribution of type 1 is restricted to the east-facing margin sediments, principally within the platform edge and talus facies. Type 2 is more widely distributed and occurs throughout the Miocene section. Type 3 is closely associated with the main exposure surface at the top of the edge and talus facies.

3.5.2 Stable Isotopes

Two types of sample were analyzed: (1) bulk rock samples consisting of dolomicrites and dolomicrosparites; and (2) selective sampling of cements and internal sediments.

$\delta^{18}O$ values vary considerably (Fig. 9), both in bulk and selected samples (from +1.2 to −8.5‰ PDB), suggesting variations either in the nature of the dolomitizing fluids or in the variable contribution of inherited oxygen isotopes. However, in spite of this wide scatter, *bulk* samples (Fig. 9 A) generally have a strong predominance of lighter isotopes, suggesting dolomitization from essentially meteoric waters (Hudson 1977). Thermal fractionation is unlikely to

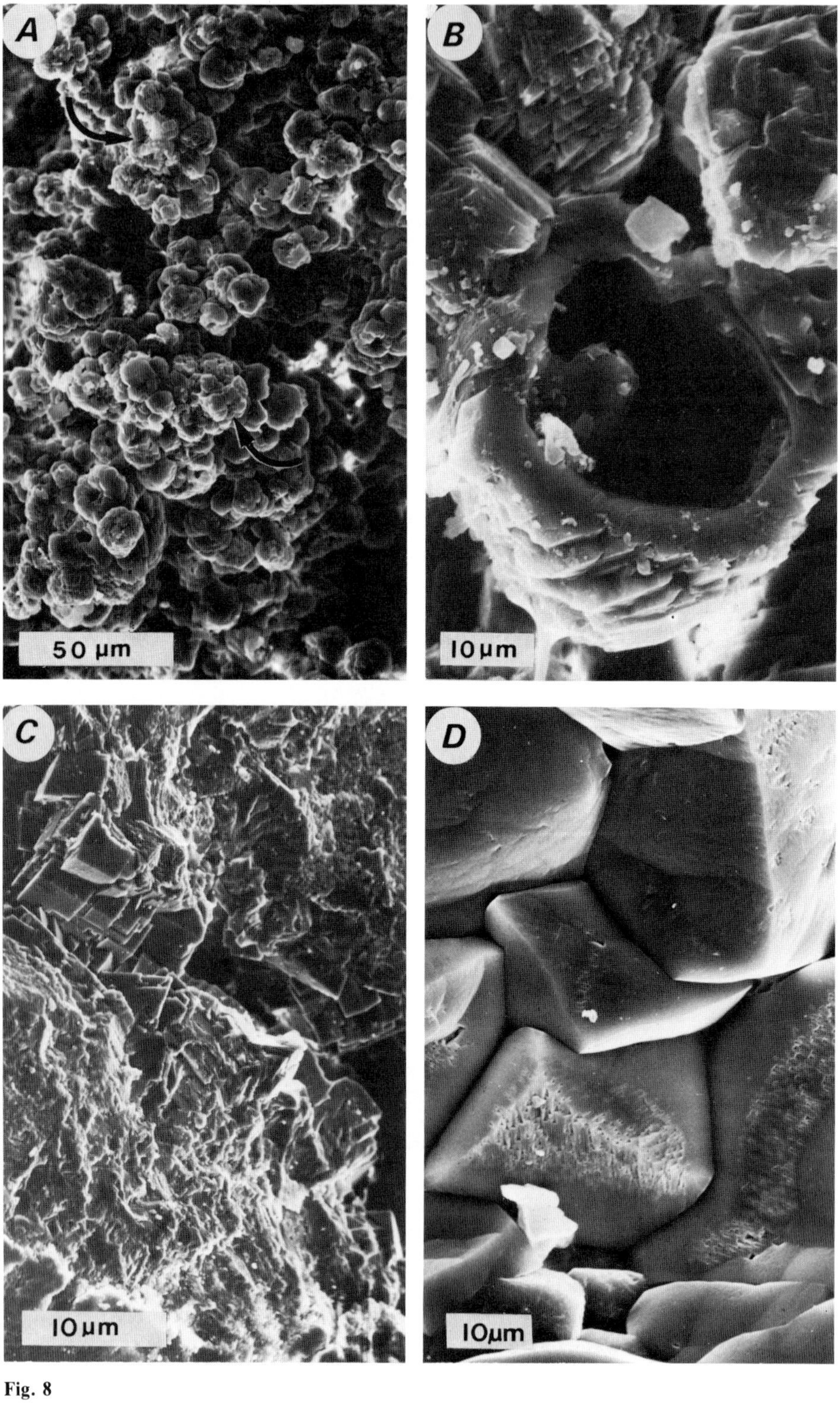

Fig. 8

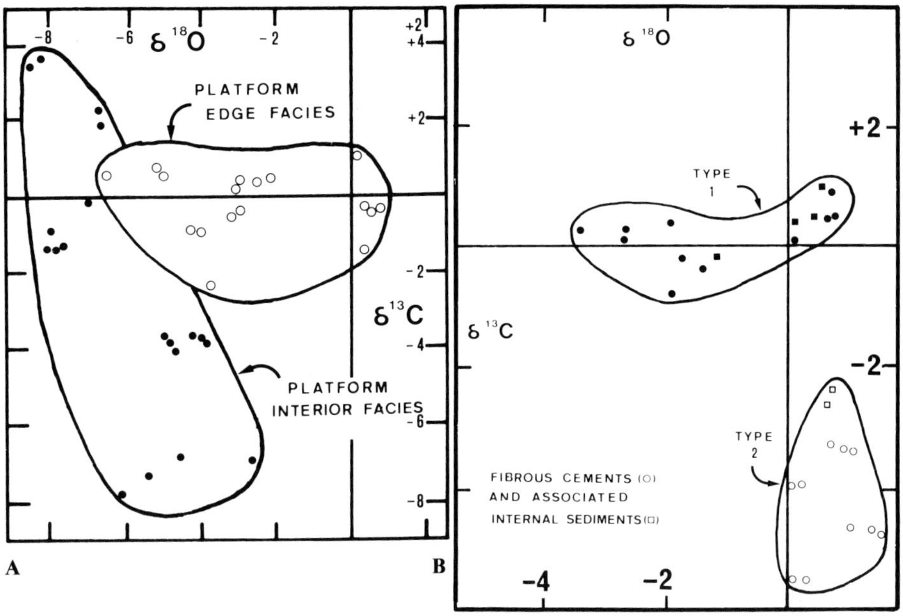

Fig. 9 A, B. Stable isotope diagrams of **A** bulk-rock samples essentially dolomicrites and dolomicrosparites; **B** selective samples of cements and internal sediments

have caused these lighter negative $\delta^{18}O$ values (−8.5‰ PDB) which could require precipitation temperatures of 70 °C. Such elevated temperatures are not consistent with the known burial history of these strata in this region.

Certain selectively sampled dolomites show a clear grouping of their isotopic signatures. This is true for the dolomitized fibrous marine cements which exhibit distinct oxygen and, especially, carbon isotope qualities (Fig. 9B); the dolomitized type 2 fibrous marine cements are characterized by grouped, relatively light $\delta^{13}C$ which may attain −6‰ PDB. The original (pre-dolomite) type 1 and 2 cements almost certainly were of marine origin and thus had comparable isotopic values. Therefore, present differences must result mainly from subsequent diagenesis, notably dolomitization, or, possibly, from the presence of light organic carbon (oil) within the intrarhombohedral micro-porosity, as suggested by odor and oil stains.

The distinctly different groupings of type 1 and 2 dolomitized marine cements may possibly be related to the contrasting settings of these dolomitized cements. While type 1 cements pisolite beds which are now dense dolomite, type 2 cement lines large (up to 1 m) karst cavities which have retained their cavernous character during and subsequent to dolomitization. This karst system, lined with

Fig. 8 A – D. SEM view of Abu Shaar dolomites. **A** Petrotype 1: spherical components (*arrows*) are still recognizable, but are composed of interpenetrating subhedral rhombohedra. Platform interior facies. **B** Petrotype 2 showing crystal with dissolved core or grain. Platform interior facies. **C** Petrotype 3: dolosparitic cement in secondary cavity. Platform edge facies. **D** Detail of dolosparitic cement showing pitted surfaces which are dissolution or crystal growth effects. Platform edge facies thick crusts of marine cement, may have constituted localized conduits for dolomitizing fluids which affected these cements. In other words, the distinct difference in isotopic properties (Fig. 9B) between these two groups of dolomite expresses differences in the nature of the dolomitizing fluids, the lighter $\delta^{13}C$ signatures of the dolomitized type 2 (karst-lining) marine cements indicating localized flow of surface waters considerably enriched in organic CO_2.

The overall isotopic qualities of the dolomitized platform reefs, as already noted, suggests dolomitization from variable, but essentially nonmarine waters. In that these carbonates are partially buried in Miocene evaporites, this is perhaps surprising. However, it is recalled that there exist several intra-Miocene, pre-evaporite, post-reefal, karst surfaces. Furthermore, the thick, essentially terrigenous Plio-Quaternary sequence, which expresses regional uplift of the peripheral parts of the Red Sea rift, includes well-developed continental facies, implying that these Miocene carbonates may have been affected by the lateral influx of meteoric waters.

4 Sedimentary and Diagenetic Evolution

4.1 Sea Floor Diagenesis

The sea floor was occupied mainly by corals, coralline algae, molluscs, and stromatolites. Although a substantial part of the sea floor was hardened by binding and encrusting organisms, intensive submarine cementation was also responsible for consolidation of these sediments (Fig. 10). This is best exemplified by the pisoids in the talus facies which consists of algal crusts alternating with fibrous, marine cements.

Common fibrous and micritic cements, and rare botryoid cements are the main elements contributing to the synsedimentary lithification of the Miocene sea floor, especially on the seaward margin of the Abu Shaar platform. The importance of submarine cementation is especially clear in the steeply dipping talus facies which, at the time of deposition, were inclined at 40° to 60°.

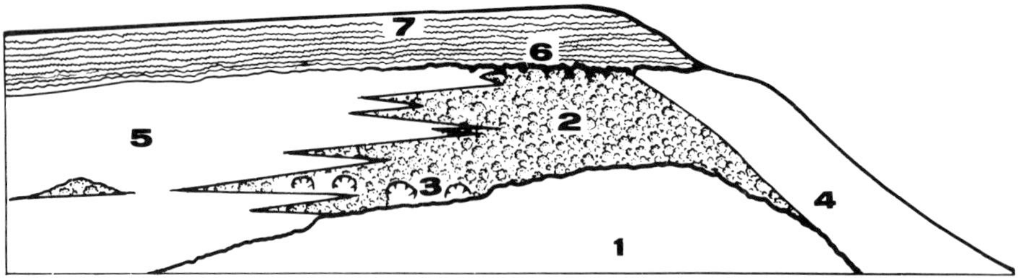

Fig. 10. Sketch showing main diagenetic units at Abu Shaar; all carbonates are dolomite. *1* Volcanic basement (fault-block); *2* massive reef core, generally porous; *3* dissolved massive corals; caverns lined with fibrous, type 2 dolomitized marine cement (see Fig. 4D); *4* highly cemented (pre-dolomite) talus including pisoid bed; cements are mainly marine (see Fig. 6); *5* highly porous, leached, platform carbonates; *6* karst surface well developed along E margin of the platform; *7* fine, dense, algal laminites

4.2 Meteoric Diagenesis

A local but important phase of subaerial exposure occurred during the Miocene at Abu Shaar, prior to deposition of the uppermost stromatolitic unit of the platform interior facies. This karst surface was limited to the eastern edge, where there was a more or less continuous string of islands. Diagenetic effects included dissolution, development of calcretes (soils), and filling of dissolution cavities by coarse internal nonmarine sediments. Somewhat lower in the sequence (Fig. 10), corals are also dissolved and their molds lined by type 2 marine cement (Fig. 4D). This dissolution, obviously Miocene in age, may also be related to the karst surface.

4.3 Dolomitization

Most carbonate examined at Abu Shaar is fine, fabric-preserving dolomite whose isotopic values, although highly variable, are dominated by meteoric signatures. As such, dolomitization has not been conditioned by the evaporitic environment which followed these marine carbonates. That this dolomitization is not necessarily the result of local emergence of the Abu Shaar platform is amply demonstrated by the regional distribution of Miocene dolomites; undolomitized limestones are generally an exception along the Egyptian shores of the Red Sea (BHP personal observation). Although multiple local effects undoubtedly exist, dolomitization of the Miocene series appears to have been conditioned essentially by eustatic or regional evolution of the Red Sea basin. In particular, the considerable increase in post-Miocene clastics including frequent nonmarine facies, clearly implies peripheral uplift and influx of meteoric aquifers potentially capable of dolomitizing the pre-existing Miocene carbonate substratum.

5 Diagenetic Model and Hydrocarbon Potential

There is a strong contrast in the degree of synsedimentary cementation between the talus and platform edge facies located along the eastern margin of the platform, and the remainder of the Miocene platform sediments. The former are well cemented while the latter are poorly lithified or friable. The weak lithification is not due only to dissolution or other alteration and includes all sediment types regardless of depositional texture (mud- or grain-supported) or mineralogy (pure carbonate or mixed siliciclastic-carbonate). The Abu Shaar strata have not been buried deeply, as suggested by the total absence of compaction. In other words, the variation in synsedimentary lithification is inherited and remarkably consistent with those noted from Mururoa (Aïssaoui et al., this Vol.).

The petrophysical characteristics of the Abu Shaar complex have not been measured but hand-specimen and thin-section estimations demonstrate clearly the reservoir potential of this Miocene platform. The most interesting targets, in terms of reservoir potential, are the platform interior facies (Haddad et al. 1984).

Because these deposits are poorly lithified and consist mainly of coarse mixed terrigeneous-carbonate sediments, they have primary, intergranular porosities which commonly attain 40%. Dolomitization and dissolution of the carbonate fraction have significantly increased the original permeability. This amelioration also includes mud-supported sediments.

Platforms comparable to Abu Shaar are common along the Egyptian Red Sea coast and similar subsurface structures are interesting hydrocarbon reservoirs in the Gulf of Suez. The Abu Shaar complex thus serves as a depositional and diagenetic model for these exploration targets.

Acknowledgments. The authors extend their thanks to the Marathon Oil Company, in particular to J. L. Wray, J. D. Denton, and C. R. Cofer for help in the field and for isotope data. Thanks are also offered to our colleague J. C. Fontes of the University of Paris Sud-Orsay, for useful discussion concerning the interpretation of the isotope data.

References

Aïssaoui DM, Purser BH (1983) Nature and origin of internal sediments in Jurassic limestones of Burgundy (France) and Frioud (Algeria). Sedimentology 30:273 – 283

Aïssaoui DM (1985) Botryoidal aragonite and its diagenesis. Sedimentology 32:345 – 361

Bathurst RGC (1966) Boring algae, micrite envelopes and lithification of molluscan biosparites. J Geol 5:89 – 109

Bernouilli D, Wagner CW (1971) Subaerial diagenesis and fossil caliche deposits in the calcare Massiccio Formation (Lower Jurassic, Central Appenines, Italy). N Jahrb Geol Paläontol Abh 138:135 – 149

Burollet PF, Montenat C, Ott D'Estevou P, Purser BH, Jarrige JJ, Orszag-Sperber F, Philobbos E, Plaziat JC, Prat P, Richert JP, Roussel N, Thiriet JP (1985) Principal sedimentary and tectonic events during the Neogene evolution of the Gulf of Suez and the NW edge of the Red Sea (Egypt) (in press)

Cofer CR, Lee KD, Wray JL (1984) Miocene carbonate microfacies Esh El Mellaha Range, Gulf of Suez. Econ Geol Petrol Cong, 7th Expl Seminar, Cairo 1984

Dunham RJ (1969) Vadose pisolite in the Capitan Reef (Permian), New Mexico and Texas. In: Friedman GM (ed) Depositional environments in carbonate rocks. Soc Econ Paleontol Miner Spec Publ 14:182 – 191

Esteban M, Pray LC (1983) Pisoids and pisolite facies (Permian), Guadelupe Mountains, New Mexico and West Texas. In: Peryt TM (ed) Coated grains. Springer, Berlin Heidelberg New York, pp 503 – 537

Folk RL, Assereto R (1976) Length-slow and length-fast calcite and calcitized aragonite in a Holocene speleothem, Carlsbad Caverns, New Mexico. J Sediment Petrol 46:486 – 496

Gregory JW (1906) Fossil coral from Egypt. Geol Mag 3:50 – 58

Haddad EA, Aïssaoui DM, Soliman MA (1984) Mixed siliciclastic carbonate-siliciclastic sedimentation on a Miocene fault-block, Gulf of Suez, Egypt. Sediment Geol 37:185 – 202

Heybroek F (1965) The Red Sea evaporite basin. In Salt basins around Africa, London Inst of Petroleum, pp 14 – 40

Hudson JD (1977) Stable isotopes and limestone lithification. J Geol Soc Lond 133:637 – 660

Hume WF (1916) Report on the oilfield region of Egypt with a geological map of the region from surveys by Dr John Ball. Egypt Surv Dep Cairo, 103 p

James NP, Choquette PW (1984) Diagenesis 9. Limestones – The meteoric diagenetic environments. Geosc Canada 11:161 – 194

James NP, Ginsburg RN (1979) The seaward margin of Belize barrier and atoll reefs. Int Assoc Sedimentol Spec Publ 3:1 – 191

James NP, Klappa CF (1983) Petrogenesis of early Cambrian reef limestones, Labrador, Canada. J Sediment Petrol 53:1051 – 1096

James NP, Ginsburg RN, Marszalek DS, Choquette PW (1976) Facies and fabric specificity of early subsea cements in shallow Belize (British Honduras) reefs. J Sediment Petrol 46:523 – 544

Madgwick TG, Moon FW, Saded H (1920) Preliminary geological report on the Abu Shaar El Quibli (Black Hill) district. Petrol Res Bull 6, Gov Press Cairo, 11 p

Mazzulo SJ, Cys JM (1983) Unusual algal-crystalline carbonate coated grains from the Capitan Reef (Permian, Guadalupian, New Mexico, USA). In: Peryt TM (ed) Coated grains. Springer, Berlin Heidelberg New York, pp 599 – 608

Meyers WJ, Lohmann KC (1978) Microdomite-rich syntaxial cements: proposed meteoric-marine mixing zone phreatic cements from Mississippian limestones, New Mexico. J Sediment Petrol 48:475 – 488

Riding R (1983) Cyanoliths (Cyanoids): Oncoids formed by calcified cyanophytes. In: Peryt TM (ed) Coated grains. Springer, Berlin Heidelberg New York, pp 276 – 284

Rouchy JM, Bernet-Rollande MC, Maurin AF, Monty C (1983) Signification sédimentologique et paléogéographique des divers types de carbonates bioconstruits associés aux évaporites du Miocène moyen du Gebel Esh Mellaha (Egypte). Compte Rend Acad Sci 296:457 – 462

Sandberg P (1985) Aragonite cements and their occurrence in ancient limestones. In: Schneidermann N, Harris P (eds) Carbonate cements. Soc Econ Paleontol Miner Spec Publ 36:33 – 57

Diagenetic Diversity in Paleocene Coral Knobs from the Bir Abu El-Husein Area, S Egypt

J. H. Schroeder[1]

1 Introduction

During stratigraphic fieldwork in S. Egypt, coral knobs within a specific carbonate bed were found to have undergone a complex diagenesis. In a preliminary report (Schroeder 1984) they served as a demonstration of the multigeneration diagenetic history and its spatial variation, which were illustrated in a petrogenetogram. The purpose of this paper is to delineate more thoroughly the diagenesis of these corals, to describe and illustrate various diagenetic products, to show their sequences and their spatial distributions, and to discuss respective diagenetic environments and factors that determine variations. This is of interest to the carbonate sedimentologist/petrologist, and, in addition, contributes to the knowledge of the post-depositional history of underlying and overlying terrigenous rocks: Although they provide some diagenetic information of general importance (used below), they do not seem to yield much for a detailed analysis.

2 Location, Stratigraphy, Materials, Methods

The area of study is located about 45 km E of Bir Abu El-Husein (Fig. 1). The coral-bearing layer is 80 cm thick and occurs in the lower portion of the E-W-trending escarpment. The best outcrops surround an outlier in front of the escarpment, a conical hill (Fig. 2A) which rises from the limestone layer studied. Because it is overlain by slightly calcareous shales, the layer forms a marked step which was traced 8 km towards the east and 4 km towards the west before it was lost due to faulting.

The regional stratigraphic analysis of the present project was outlined by Hendriks et al. (1984) and studied in detail by Luger (1985); these authors provide ample references to previous work. The profile measured at the conical hill is presented in Fig. 1C. Luger (1985) considers the limestone layer to be part of the upper transgressive sequence of the Kurkur Formation, and places it in the lowermost part of the Upper Paleocene, based essentially on the occurrence of *Planorotalites pseudomenardii* at the base of the Garra Formation about 3 m above the limestone layer.

[1] Institut für Geologie und Paläontologie, Technische Universität Berlin, Hardenbergstr. 42, 1000 Berlin 12, FRGermany

Reef Diagenesis
Edited by J. H. Schroeder and B. H. Purser

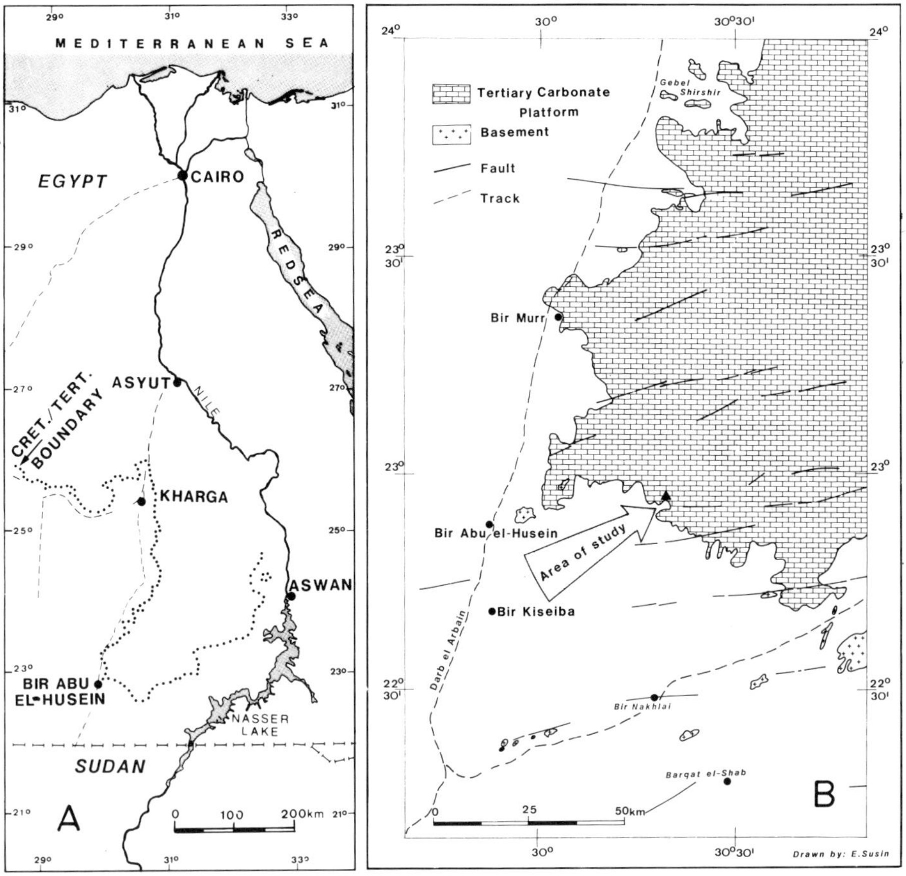

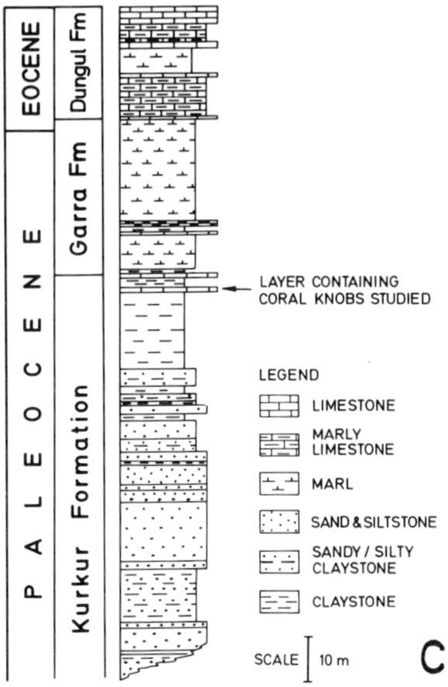

Fig. 1A – C. Area of study: location and stratigraphy. **A** General map. **B** Specific map. **C** Measured section at the main outcrop of this study, shown in Fig. 2A. (After Luger 1985)

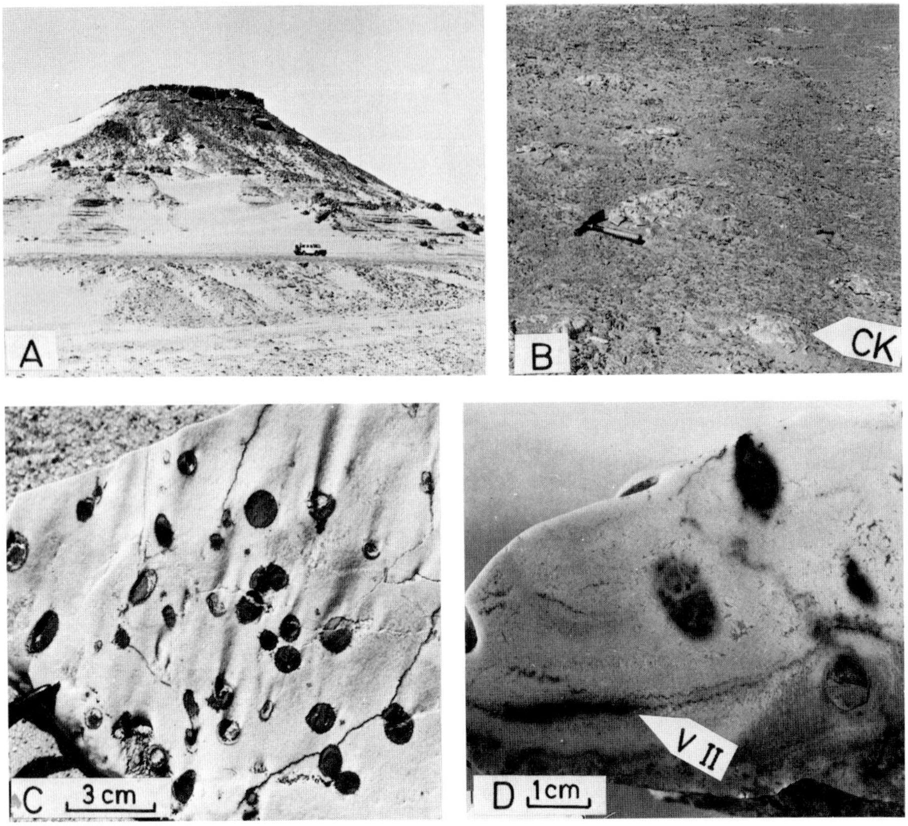

Fig. 2A – D. Outcrops and hand specimens. **A** Outcrop with conical hill showing the upper part of the section presented in Fig. 1C; vehicle is parked on limestone layer studied. **B** Upper surface of limestone layer with coral knobs (*CK*) forming small mounds at the surface. **C** Specimen of *A. hexaphylla* I: Horizontal surface with pelecypod borings truncated by sandblasting; most fillings of holes have been impregnated and stained by iron minerals. **D** Polished vertical section of *A. hexaphylla* I specimen; note second generation vug (*V II*) oriented subparallel to growth bands and pelecypod borings which penetrate the specimen and are filled in various ways

The limestone layer is a biosparite containing skeletal detritus of pelecypods, gastropods, echinoids, foraminifers, and corals, suggesting a shallow marine environment. Ninety percent of the layer are carbonates; the insoluble residue contains kaolinite, illite and smectite as ascertained by X-ray diffraction and glauconite pellets as observed in thin section.

The coral knobs are found in the upper third of the layer. They are disc-shaped and up to 135 cm, but commonly 40 – 80 cm in diameter. The maximal height measured was 18 cm. However, exposed knobs are truncated by sandblasting from the top downwards, and some knobs were dissolved from the bottom upwards (see below) and therefore some may have been higher than measured. Horizontally, the knobs are irregularly scattered (Fig. 2B); in a contiguous horizontal outcrop of 100 × 20 m 28 knobs were counted.

As to taxonomy and relative abundance, most corals collected were identified as *Actinastraea* ex gr. *hexaphylla* (Quenstedt). Within this species two distinct morphotypes, designated below I and II, have been recognized, which are characterized by the diameter of their corallites of 1 and 3 mm, respectively. In addition, occasional colonies of *Actinastraea* ex gr. *subdecaphylla* (Oppenheim), *Astrarea* aff. *media* (Sowerby), *Procladocora* cf. tenuis (Reuss) and only one of *Koilomorpha* aff. *asperrima* (Reuss) were noted. The paleontological implications of these occurrences are being investigated. The coral colonies are disc- to dome-shaped, except for the branching *Procladocora.* Large coral knobs consist of several colonies of one or more species, while small ones may consist of a single colony only. There is no evidence of extensive hard substrate, thus the corals probably grew initially on large pieces of skeletal detritus. This interpretation is supported by some studies of Höfling (in preparation), who found Cretaceous *Actinastraeidae* and *Procladocora tenuis* on unequivocal soft substrates. The morphology of the coral knobs, the low species diversity, the probable lack of extensive suitable substrate, and the relatively high proportion of insoluble residue in the host rock suggest that the environment was far from ideal for coral growth.

These coral knobs, although among the least impressive fossil reefs on record, had the potential to build complex structures. They provided a primary framework as substrate for a number of secondary framebuilders such as encrusting coralline red algae, serpulid worms, and bryozoa; they were bored by

Table 1. Methods of study

Method/instrument	Sample preparation	Parameter analyzed
Petrographic microscope plain light (PON) or crossed nicols (PXN)	Thin section (some stained with alzarin red S and potassium ferricyanide)	Min. composition, morphology, identity and textures of components and cements
Cathodoluminescence microscope (CL)	Polished thin section	Composition, distribution and growth patterns of crystals
Scanning electron microscope a) electron image (SEM) b) electron diffraction X-ray analysis (EDAX)	Fractured or polished/ etched blocks	a) Morphology and texture of components and cements b) Qualitative chem. composition of components and cements
Electron microprobe a) energy dispersive X-ray analysis b) secondary electron image	Polished thin section	a) Quantitative chem. composition of cements and components b) Surface features
X-ray diffractometer	Powdered sample	Bulk mineralogy
X-ray diffractometer	Powdered sample dissolved: Residue	Mineralogy of insoluble residue
Carbonate bomb	Powdered sample	Carbonate content

Note: Abbreviations used in figure captions.

pelecypods, sponges, worms and endolithic algae (or possibly fungi); some intraskeletal pores and borings were filled by sediment, and early submarine cements are suggested by relics and circumstantial evidence. Thus, all processes interacting in Recent reefs (Schroeder and Zankl 1974) were active, although not very successful: they were neither able to build larger structures, nor did they continue for a long time, as no knobs are found below or above.

Thirty coral colonies were sampled and studied using the methods listed in Table 1. The present report is focused on the coral knobs: the diagenesis of the sediment below, between, and above them is only of marginal concern here.

3 Diagenetic Processes and Products

3.1 Bioerosion

The coral knobs were attacked in the marine environment by various borers. This process probably started during growth and continued afterwards. Most conspicuous are the holes of boring pelecypods, up to 1.2 cm in diameter and up to 3 cm in length (Fig. 2C, D). These pelecypods clearly preferred *A. hexaphylla* I, possibly because of its small corallite diameter and the resulting denser skeleton: While specimens of this morphotype are riddled with holes (Figs. 2C, 8B), only one single pelecypod hole was found in all *A. hexaphylla* II samples. As in recent reefs, subsequent borings truncate earlier ones. Sponge borings are irregular holes of millimetre size, whereas worms (?) made winding borings of a millimetre in diameter and up to several centimeters in length. Tubes several micrometres in diameter, ascribed to algae, have rarely been preserved.

Borings constitute additional pores which cut through existing skeletal structures. The sponge and worm holes were generally filled by sediment and some were later stained by iron minerals. The pelecypod holes were fully or partially (geopetally) filled by sediments and/or coated or filled by various cements; in addition they contain the shell of their maker; these components may then be altered in many possible ways, like the host coral skeleton itself. Where the pelecypods are volumetrically important, as in *A. hexaphylla* I, the alteration of the contents of the borings should not be ignored; however, in the interest of brevity and clarity, the resulting complex diagenetic pathways are not pursued here. As indicated below, these holes also influence the alteration of the surrounding coral material at various stages of its diagenetic history.

3.2 Internal Sedimentation

Both borings and intraskeletal pores of corals were filled with sediments. In some fillings, the skeletal detritus is clearly recognized and indicates early deposition; other fillings are micrite and this may be a sedimentary or a diagenetic product. In some pores, glauconite pellets were found within the internal sediment, while in other fillings Edax analysis indicated the presence of silica and aluminum, suggesting terrigeneous components, which are considered a major source of the 1 – 3% insoluble residue found in bulk samples of coral.

3.3 Neomorphism

A considerable portion of the coral was neomorphically altered into low Mg Fe Calcite (Figs. 3, 9: Neomorphic Fe Calcite I) exhibiting familiar features: a mosaic of equant crystals with boundaries which are frequently curved, indented or even serrate, with relics of skeletal structure, earlier cements or sediment fills. (Characteristics of these and the following minerals are summarized in Table 2.)

The original skeletal structure is clearly distinguished from the original filling of the intraskeletal pores in cathodo-luminescence; therefore minor compositional differences, particularly with respect to Fe and Mn contents, are present. As to the fill, generally neither relics nor CL suggest particular precursor cements or sediment fillings, although there are rare exceptions where CL suggests an isopachous lining. The absence of fabric patterns indicated by the relics suggest that the intraskeletal pores were filled either by finely crystalline cements, such as micrites, or by fine-grained internal sediments. As internal sediments are preserved in portions of many samples, a precipitative original filling is more likely (presursor cement I in Fig. 9). Fe calcite neomorphism also affected cements in secondary pores, in which case relics indicate specific cement fabrics (Figs. 5, 8C; precursor cements II and III in Fig. 9). In some places this neomorphism was fabric-selective: It did not affect intraskeletal pores filled with sediment (Figs. 3B, 5A), sponge borings filled by sediments, or pelecypod borings. (Within the pelecypod borings, however, components such as the pelecypod shell or some cement may be neomorphically altered.)

The neomorphic portions of the samples are not uniform but exhibit some variety. Firstly the size of the neomorphic crystals varies; in all *Actinastraea* specimens distinct bands of fine and coarse crystals are seen in vertical section; these are interpreted as outlining the original growth bands of the corals, also called density bands, which have been used for dating Recent corals (Buddemeier and Kinzie 1976, and others).

The relation between crystal boundaries and relics of the original skeletal structure also varies. In some samples, crystal boundaries cut across relics, while in others they coincide with the relics; in many samples, however, some boundaries cross, others coincide with the relics (Fig. 3A). Within a given thin section all types of mutual relations are found and no pattern of distribution was recognized.

In *A. hexaphylla* I, two distinct results of neomorphism were observed next to each other: one portion was rich in relics and the original skeletal structure readily discernible in thin section. In the other portion there were considerably fewer relics, but in hand specimen the original skeletal structure also is seen (Fig. 3B, C). Whereas in CL the first has the mottled appearance characteristic of recrystallized material, the latter portion shows radiating or zoned crystals (Figs. 3C, 9: Neomorphic Fe calcite II). The first type occurs as a halo around pelecypod borings and along fractures, the second type in the portions in between

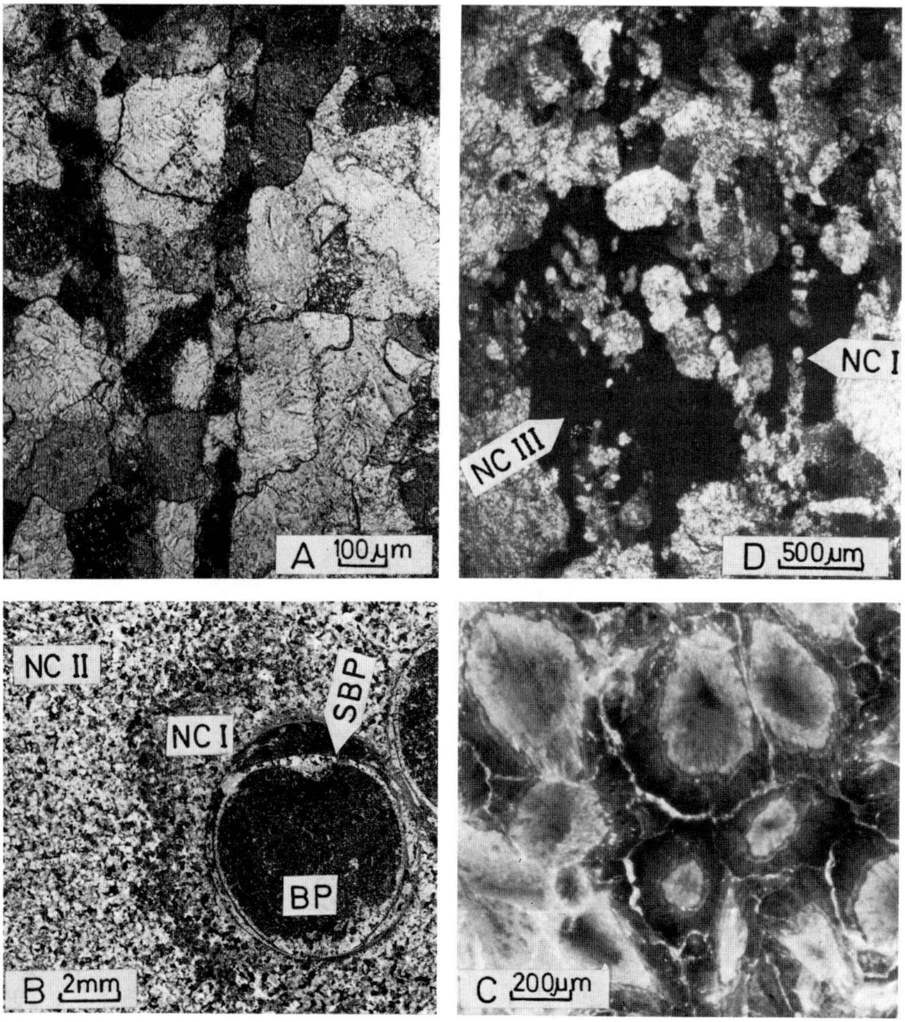

Fig. 3A–D. Fe calcite neomorphism. **A** Vertical sections of *A. hexaphylla* I: Some crystal boundaries coincide with relics outlining original skeletal structures, others are independent; PXN. **B** Horizontal section of *A. hexaphylla* I showing pelecypod borings (*BP*) with its altered shell (*SBP*) surrounded by neomorphic Fe calcite with many relics (*NC I*) and neomorphic Fe calcite with few relics (*NC II*); note distinct boundary between the different neomorphic portions; for details see Fig. 3C; cf. also Fig. 8B; PXN. **C** *A. hexaphylla* I: Zoned crystals of relic-poor neomorphic crystals: cf. Fig. 3C (*NC II*) and Fig. 8B; CL. **D** Two generations of neomorphic crystals: earlier smaller crystals (*NC I*) float in larger ones (*NC III*; *black*); PXN

(Figs. 3B, 8B). This difference may reflect earlier variations in diagenesis/preservation, for example more intensive cementation around borings and fractures versus less intensive cementation away from them. Alternatively, it may result from variations in the rate of the neomorphic process: around the boring and next to the fractures, circulation of solutions along neomorphic fronts was presumbly faster, leaving more relics, while in the portions in between, the

Table 2. Characteristics of neomorphic, replacive and cement minerals encountered. A: Fabrics and dimensions. B: Average composition of carbonate minerals based on microprobe spot analyses

A	Fig.	Fabrics/habits	Dimensions
Neomorphic Fe calcite			
I: Coarse	3A, B, D	Equant	200 – 500 μm
Fine			50 – 150 μm
II:	3B, C		200 – 400 μm
III:	3D		– 2 mm
Precursor Cement II	5, 8C	Bladed aggregates of bladed subcrystals	100 μm – 5 cm
III			
Fe Calcite Cement I		Equant to bladed	
Pyrite		Equant,	5 – 15 μm
Ankerite Cement	4, 6	Equant saddle-shaped rhombohedra	50 μm – 1 mm
Replacive Ankerite I	4A, B, 5A		
Calcian Dolomite	4D, 6	Crust with plain surfaces	15 – 20 μm thick
Replacive Ankerite II	4A – E, 6B	Equant	1 μm
Fe Calcite Cement II	7A, B		
crust		Composed of rhombohedral crystals	– 20 μm thick
scattered crystals		equant rhombohedra	– 4 mm
Gypsum Cement	7C	Needles	– 60 μm

B	Mol % Carbonate			
	Ca	Fe	Mg	Mn
Fe Calcite				
Neomorphic I, II, III	96.4	2.0	1.6	Trace
Precursor Cement II, III				
Cement I				
Ankerite				
Cement	57.7	15.0	26.4	0.9
Replacive I				
Replacive II	52.1	15.4	31.4	1.1
Ca Dolomite				
Cement	56.3	–	42.5	1.2
Fe Calcite				
Cement II	97.7	1.4	–	0.9

process was slower. Thus, the coexistence of different neomorphic fabrics is interpreted to be caused by the microhydrology of the coral, either before or during neomorphism.

In some portions of *A. hexaphylla* I specimens, there is evidence of two phases of neormophism. A first phase produced a mosaic with crystal boundaries following skeletal relics, while in a second phase remnants of this mosaic were preserved within large cross-cutting crystals (Figs. 3D, 9: Neomorphic Fe calcite III). Patches affected by two phases of neomorphism are surrounded by portions affected only by one.

All these varieties of neomorphism are expressed only in fabrics; neither staining nor microprobe analyses revealed significant differences in compositon.

3.4 Replacement

3.4.1 Replacive Ankerite I

At a later diagenetic stage, the neomorphic Fe calcite is locally replaced by coarsely crystalline ankerite (Figs. 4A, 5A); this process is restricted to portions of the samples where later solutions were able to circulate, i.e., either near borings or near secondary pores. Like dolomite, this replacing ankerite is virtually idiomorphic (Fig. 4B). However, the spatial relationship with the precursor material as well as the relics indicate its replacive origin. Crystal boundaries of this ankerite are straight to curved, but definitely not serrate.

3.4.2 Replacive Ankerite II

Still later in diagenesis, the replacive coarsely crytalline ankerite (I) as well as the cogenetic ankerite cement (see below) are replaced partially by a micro-crystalline ankerite (II) with admixed Fe oxides/hydroxides which were identified as hematite and goethite by X-ray diffraction (Fig. 4A – E). These iron minerals impart a dark brownish red color (rusty to black in thin section) to the respective portions. Alteration frequently remained incomplete and therefore is seen to have progressed centripetally from the outside of a crystal inward, and laterally from cleavage plains (Fig. 4D). It also moved from the front between neomorphic Fe calcite and replacive coarsely crystalline ankerite into the latter (Fig. 4A, B). As far as the ankerite cement is concerned, this replacement affected crystals in one part of the pore, but not in the other. Although this process is comparable with the dedolomitization described by Frank (1981), in this case the hematite and goethite crystals are submicroscopic.

Fig. 4A – E. Replacement by ankerite. **A** Replacive ankerite front (*RAF*) moving into coral skeleton previously neomorphically altered to Fe calcite (*NC I*) at bottom and upper right hand corner. In the upper center there is ankerite cement (*AC*) covered by Ca dolomite cement (*DC*) and gypsum cement (*GC*) in the remaining pore space. Note that part of ankerite cement and coarse replaced ankerite (*RA I*) are replaced by micro-crystalline ankerite (*black*); PON. **B** Detail of ankerite replacement front (*RAF*) in *A. hexaphylla* I. Replacive ankerite I (*RA I*) grew from left to right into neomorphic Fe calcite (*NC I*) and formed idiomorphic crystals. Later, micro-crystalline ankerite with admixed Fe oxides/hydroxides (*RA II; black*) progressed from the outside of these crystals inward, more generally, from former replacement front into the coarse replacive ankerite; PXN. **C** Micro-crystalline ankerite with Fe oxides/hydroxides (*RA II*) replacing ankerite cement (*AC*) underneath a layer of Ca dolomite cement (*DC*); vug II in *A. hexaphylla* I; SEM; etched surface. **D** Centripetal replacement of ankerite cement (*AC*) by microcrystalline ankerite with Fe oxides/hydroxides (*RA II*; *black*). Alteration progressed along cleavage planes inward; crystal faces and Ca dolomite cement (*DC*) overlying ankerite cement, but preceding its replacement, were not affected. Remaining porespace was later filled by gypsum cement (*GC*). Vug II in *A. hexaphylla* I; PXN. **E** Detail of Fig. 4C showing micro-crystalline ankerite (*RA II*); even at this small scale the Fe oxides/hydroxides cannot be discerned. SEM, etched surface

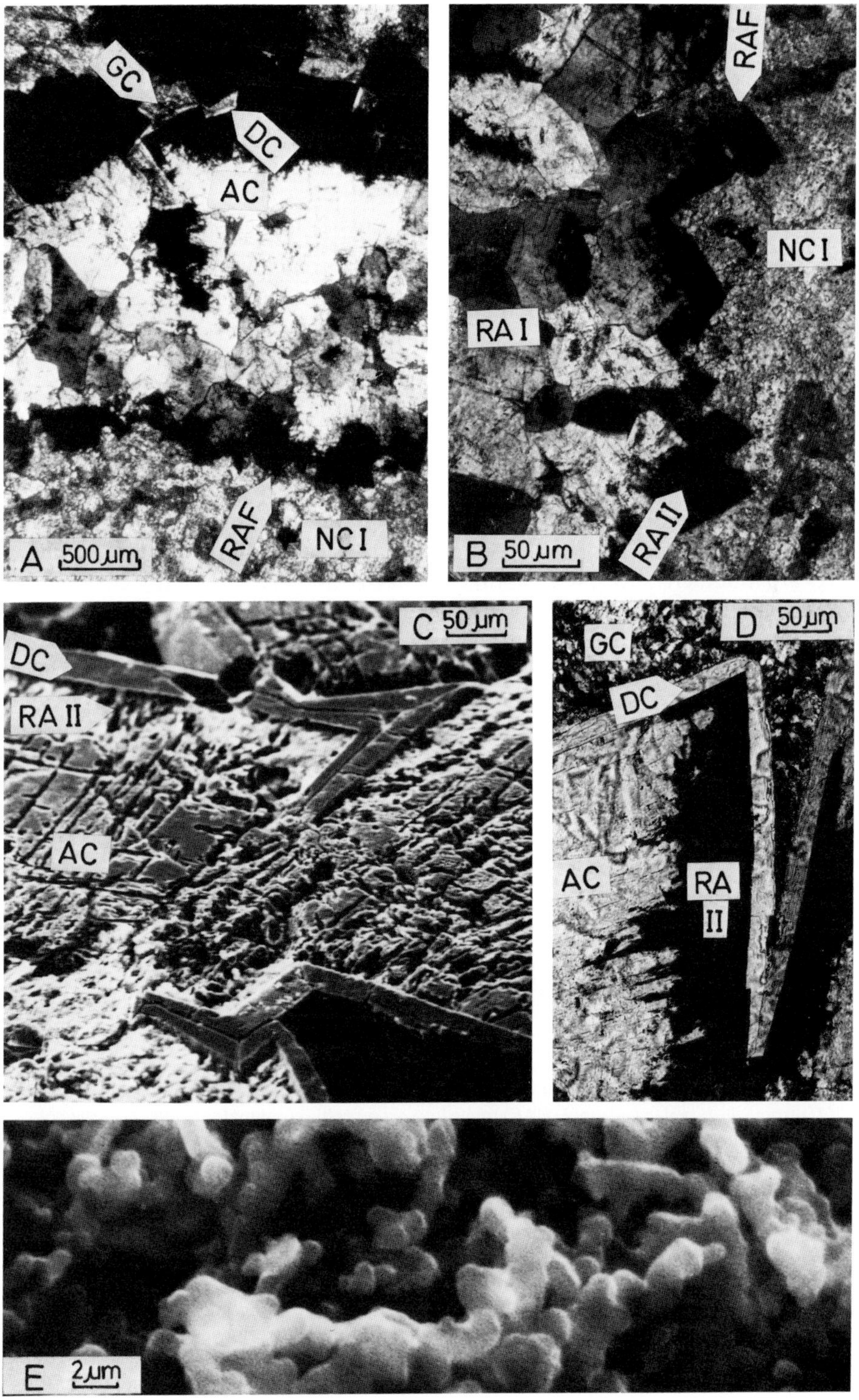
GC
DC
AC
RAF
NCI
A 500 µm
RAF
NCI
RA I
RA II
B 50 µm
C 50 µm
DC
RA II
AC
D 50 µm
GC
DC
AC
RA
II
E 2 µm

Fig. 4

3.5 Dissolution

Within the corals the non-fabric-selective vugs (terminology from Choquette and Pray 1970) are the only dissolution pores. However, within the pelecypod borings, fabric-selective dissolution pores occur where the shell was dissolved and molds either remained open or, more frequently, were filled in a variety of ways.

There are two generations of vugs which can be distinguished by their mutual spatial relations and by their respective fillings. The second truncates the first (Fig. 5 A).

3.5.1 Vug I

The first generation of vugs is highly irregular in outline and apparently extends in every direction. The outline is scaloped to amoeboid and in section, many peninsulas and islands of neomorphic material are left at, or near, the margin (Fig. 5 A). Like neomorphism, dissolution does not cut across pelecypod borings, but leaves a thin frame of skeletal material, now neomorphic, around it. These vugs are millimeters to a few centimeters in size and are often fully enclosed by skeletal material. However, pores of this generation also developed between the coral and its substrate (Fig. 5 B). As indicated by variations in thicknesses and by the irregular lower outline of the coral, as well as by the truncation of borings, these vugs certainly developed at the expense of the original coral; it is possible that they also developed at the expense of the substrate. As a result of dissolution from below, remnants of the coral colonies of this species were reduced from several centimeters in height to millimeter-thin layers. In some larger colonies of *A. hexaphylla* II, early vertical fractures (see Sect. 3.7; Fig. 8 A) permitted micrite to enter the coral skeleton laterally and to form thin linings which later escaped dissolution. Together with the upper surface as a roof, the fracture linings formed a centimeter-sized boxwork of downward-open secondary pores.

Fig. 5 A – C. Vugs in corals. **A** Vertical section of *A. hexaphylla* I showing first generation vug (*V I*) with precursor cements (*PC II/III*) and Fe calcite cement (*FC I*) truncated by second generation vug (*V II*) from which replacive fronts (*RAF*) move into neomorphic coral material (*NC I*) at the bottom and into precursor cements of vug I at the top. Within vug II there is ankerite cement (*AC*) partly neomorphically altered by microcrystalline ankerite and (not visible because too small) Ca dolomite cement (see Fig. 4 C, D); both vug I and II originally contained gypsum cement, lost during preparation of the thin section. Note that the pelecypod boring (*top center*) contains the same cement sequence in its upper part as vug I (*PC II/PC III/FC I*); its lower part underwent a different diagenetic development involving ankerites. Direct enlargement of thin section, plain light. **B** Specimen of *A. hexaphylla* II, where the first-generation vug was formed by dissolution from the bottom up. At the top there is a thin layer of neomorphically altered coral with some internal sediment fillings; the bulk is composed of neomorphically altered precursor cements and Fe-calcite cement (*FC I*); PXN. **C** Crystal aggregate of neomorphic Fe calcite in *A. hexaphylla* II showing individual components with respective extinction and serrate boundaries of aggregates (*BA*); PXN

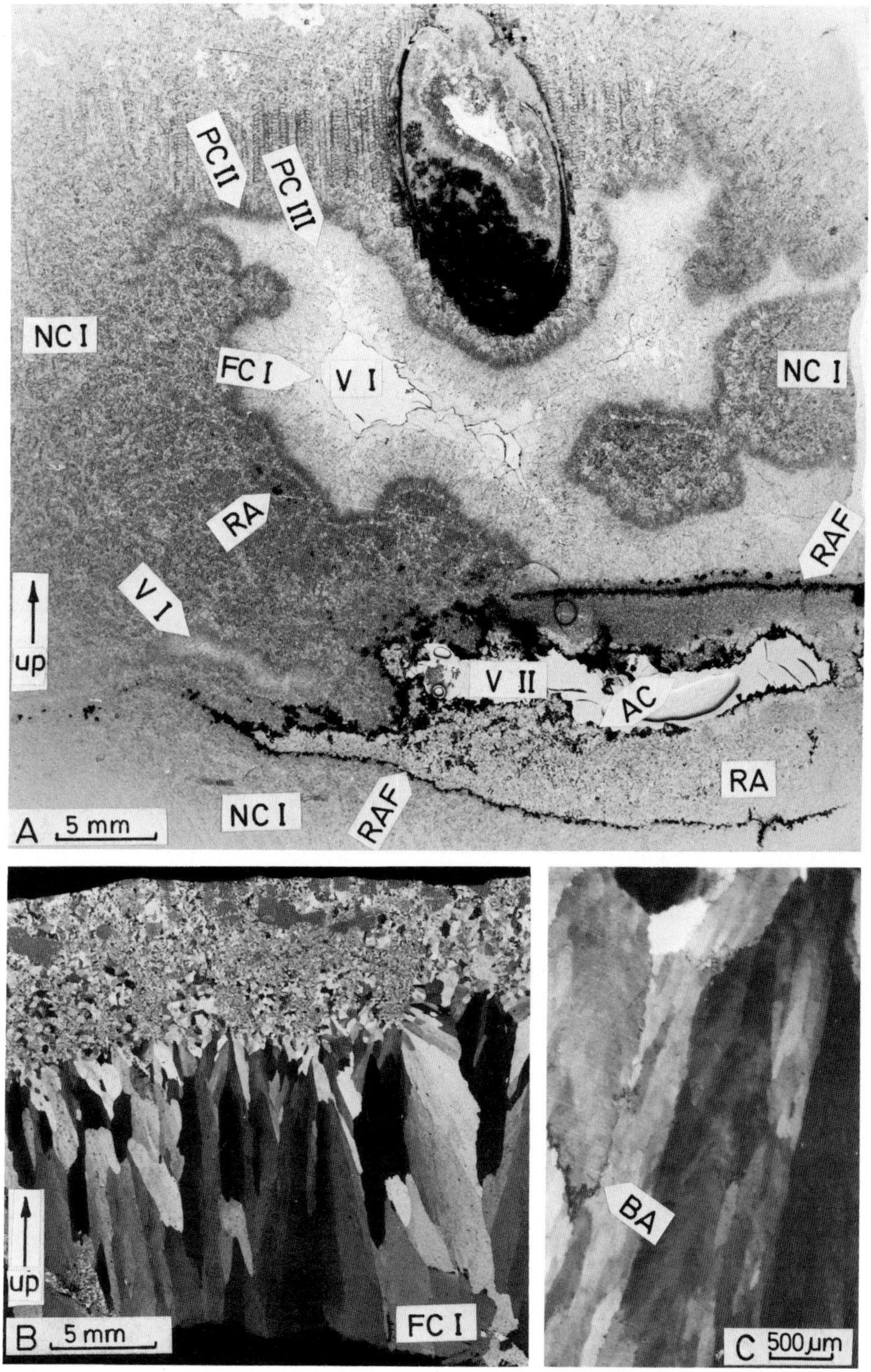
PC II
PC III
NC I
FC I
V I
NC I
RA
RAF
V I
up
V II
AC
RA
A 5 mm
NC I
RAF
up
B 5 mm
FC I
BA
C 500 µm

Fig. 5

3.5.2 Vug II

The second vug generation is found exclusively in *A. hexaphylla* I (Figs. 2D, 5A). It truncates neomorphic coral material or first generation vugs. These vugs, which extend mainly perpendicular to the growth direction of the coral, rarely exceed 1 cm in height, may be several centimeters wide, and tend to have an irregular discoid geometry. Often the original outline of this vug is not readily discernable as the replacive ankerite advanced outwards, while the ankerite cement was precipitated inwards; relics, to some extent, delineate the original outline.

3.6 Cementation

At various stages of diagenetic history, different crystals precipitated within the pore space available. Their respective characteristics are summarized in Table 2.

3.6.1 Precursor Cement I

As mentioned above, there is little preservation of very early cements. Cementation generally follows immediately below the living tissue in Recent corals (Hubbard 1972). In these Paleocene corals, relics, mostly lacking fabric pattern, suggest that a precursor cement, possibly marine, had filled intraskeletal pores. Information concerning composition and/or fabric was destroyed by a later neomorphism.

3.6.2 Precursor Cements II/III

The following generation was also altered by Fe calcite neomorphism, but in this case the relics mimic fanshaped arrangements of fibrous crystals (Figs. 5A, 8C). Similar fabrics occur as aragonitic cements in Recent reefs and have been described as spherulitic cements (Schroeder 1972a), and botryoidal cements (Ginsburg and James 1976). Their replacement was studied by Mazullo (1980) and several fossil occurrences are known. However, linings of fanshaped aggregates of bladed to fibrous Mg calcite crystals have been reported as encrustations of algal filaments in Recent reefs (Schroeder 1972b); two and three successive linings may occur. Both fabrics strongly suggest that this precursor cement was marine.

The subsequent cement generation (Fig. 5A), also neomorphically altered, shows a relic pattern very similar to the preceding one, although the fibers appear to have been somewhat coarser and possibly not as densely packed. These two generations often cannot be clearly distinguished, and a common origin is proposed.

3.6.3 Pyrite Cement

Minute idiomorphic pyrite crystals were found within the neomorphic portions of the corals; as both Fe calcite and pyrite require reducing conditions for their formation, they are considered contemperaneous.

3.6.4 Fe Calcite Cement I

The equant Fe calcite spar with an average content of 1.6 mol % $MgCO_3$ is the first cement preserved as such (Fig. 5A). Where space permitted, it formed the terminations of the coarse-bladed crystal aggregates which replaced precursor cements II and III (Fig. 5A, B, C). These aggregates are now composed of numerous crystals, each of which extinguishes separately in polarized light (Fig. 5C), while the spar cement shows unit extinction. The boundaries between crystal aggregates are serrate, those between cement crystals straight to curved. In CL, the neomorphic portion shows mottled luminescence, while the cement shows a clear luminescence with growth zonation reflecting minor compositional variations. The cement contains no relics. Where the cement crystals show terminations, these are obtuse rhombohedra. Where these vugs dissolved on the underside of the coral, neomorphic and Fe-calcite cement crystals formed a downward array of coarse crystals with a pavement of obtuse rhombohedra towards the bottom (Figs. 5B, 8C). In the boxwork, these crystals grew from the roof downwards and from the walls sideways until they mutually abutted (Fig. 8A). As to composition, neither staining nor microprobe spot analyses revealed significant differences between neomorphic Fe calcite and the cement.

3.6.5 Ankerite Cement

In the second vug generation, the first cement is a manganese-bearing ankerite (Figs. 5A, 6A, B). It is rarely found in residual pores of first generation vugs, and only where these are truncated by a second generation vug. The coarse ankerite crystals exhibit the well-known saddle shape reported by Radke and Mathis (1980) in dolomites. These authors also reviewed previous work. The saddle dolomite described by them contains 1 – 33.5 mol % $FeCO_3$, and thus includes the minerals described in this section. In order to facilitate the distinction of dolomites, ferroan dolomites and ankerites, the nomenclature of Hey (1962) is used, according to which ankerites contain more than 10 mol % $FeCO_3$. In addition to habit and composition, they are characterized by curved cleavage planes (Fig. 6B) and sweeping extinction. As mentioned above, these crystals may be partly or fully replaced by microcrystalline ankerite with goethite and hematite (Figs. 4, 6B). In this process, however, the characteristic shape is preserved.

Fig. 6 A, B. Filling of second generation vug in *A. hexaphylla* I. **A** Saddle shaped crystals of ankerite cement coated by layer Ca dolomite cement; SEM. **B** Ankerite cement (*AC*) coated by Ca dolomite cement (*DC*), and partially replaced by microcrystalline ankerite (*RA II*; *black*); remaining pore filled by gypsum cement; PXN

3.6.6 Ca Dolomite Cement

The next cement, again found only in second generation vugs, is a thin layer of isopachous Ca dolomite (Figs. 4C, D, 6B) which lines ankerite surfaces. On the basis of fabric which is determined by its substrate, this cement would be considered an additional zone of the coarsely crystalline ankerite cement. However, composition, luminescence, and the fact that it is not affected by microcrystalline ankerite replacement, distinguish it very clearly from its substrate (Table 2). Both lower and upper boundaries are planar and distinct. As observed in etched thin section and confirmed by series of Edax spot analyses and probe transects (Ca, Mg, and Mn contents), as well as CL, this dolomite shows distinct zonation interpreted as reflecting slight fluctuation of pore water chemistry (see Schroeder 1985, for discussion).

3.6.7 Fe Calcite Cement II

An Fe calcite cement containing no magnesium follows (Fig. 7A, B). It occurs as clusters of larger individual rhombohedra or as a crust of interlocking crystals showing many rhombohedral faces (Fig. 7B). These clusters or crusts are irregularly distributed on the dolomite-coated ankerite crystals.

3.6.8 Gypsum Cement

Any remaining space of both first and secondary generation vugs may be partly or completely filled by gypsum cement (Figs. 6B, 7A, C). Fibrous crystals form irregular clusters or loose spherulites. In SEM, crystal shape is diagnostic of gypsum, but in X-ray diffraction, small quantities of anhydrite were also identified, suggesting an alteration.

3.6.9 Microspherulites

Microspherulites are unevenly scattered on Ca dolomite, Fe calcite, and gypsum cements (Fig. 7D). SEM-Edax analysis indicated the presence of Ca and Fe. Although the crystal pattern suggests some organic origin, specialists consulted on various microorganisms could not help with the identification.

3.7 Fracturing

In addition to dissolution, fracturing produced secondary pores in these coral knobs. The process was effective at various stages throughout diagenetic history, and fracture generations are distinguished by the material broken, by the filling, whether it is sediment or cement, or by their effects on adjacent material.

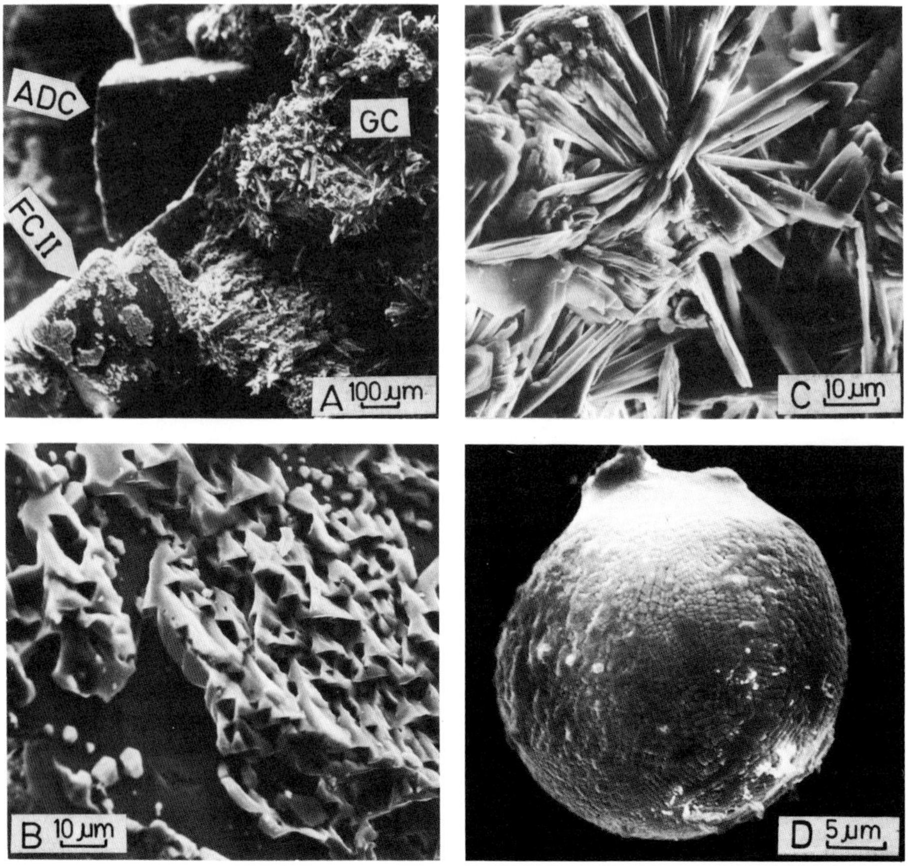

Fig. 7 A – D. Details of fillings in second generation vug in *A. hexaphylla* I. **A** General view showing ankerite/Ca dolomite cement crystals (*ADC*), patches of Fe calcite cement crusts (*FC II*), and clusters of gypsum cement needles (*GC*), SEM. **B** Fe calcite cement II forming grid-like crust composed of rhombohedral crystals; SEM. **C** Spherulite of gypsum cement; SEM. **D** Undetermined microspherulite found on Ca dolomite cement, Fe calcite cement, and on gypsum cement; note pattern on surface; SEM

Fig. 8 A – D. Fractures. **A** Horizontal section of *A. hexaphylla* II colony showing first generation fractures (*F I*) which permitted micrite to enter adjacent intraskeletal pores and to form linings which served as substrate for precursor cements and for the neomorphic calcite replacing them; fracture IV (*F IV*) later broke obliquely across Fe calcite and was filled with ankerite cement (*AC*) (which was replaced) and by gypsum cement; close-up photograph. **B** Top view of sandblasted specimen of *A. hexaphylla* I with two fracture generations: Fracture III with a halo of neomorphic Fe calcite rich in relics (*NC I*) which also surrounds pelecypod borings; in between the distinctly darker neomorphic Fe calcite poor in relics (*NC II*; see Fig. 3 B, C). Fracture IV (*F IV*) cuts obliquely across halos and is partly filled by ankerite cement; close-up photograph. **C** Vertical section of *A. hexaphylla* II specimen (cf. Fig. 5 B, C). Fracture IV broke through Fe calcite, in part neomorphic as indicated by fanshaped aggregates of relics from fibrous crystals, in part cement. This fracture was filled by ankerite cement which was mostly replaced by microcrystalline ankerite; the fracture opened again and was filled by another generation of Fe calcite cement: PXN. **D** Detail of **C** showing ankerite cement (*AC*) and later Fe calcite cement (*FC II*); PXN

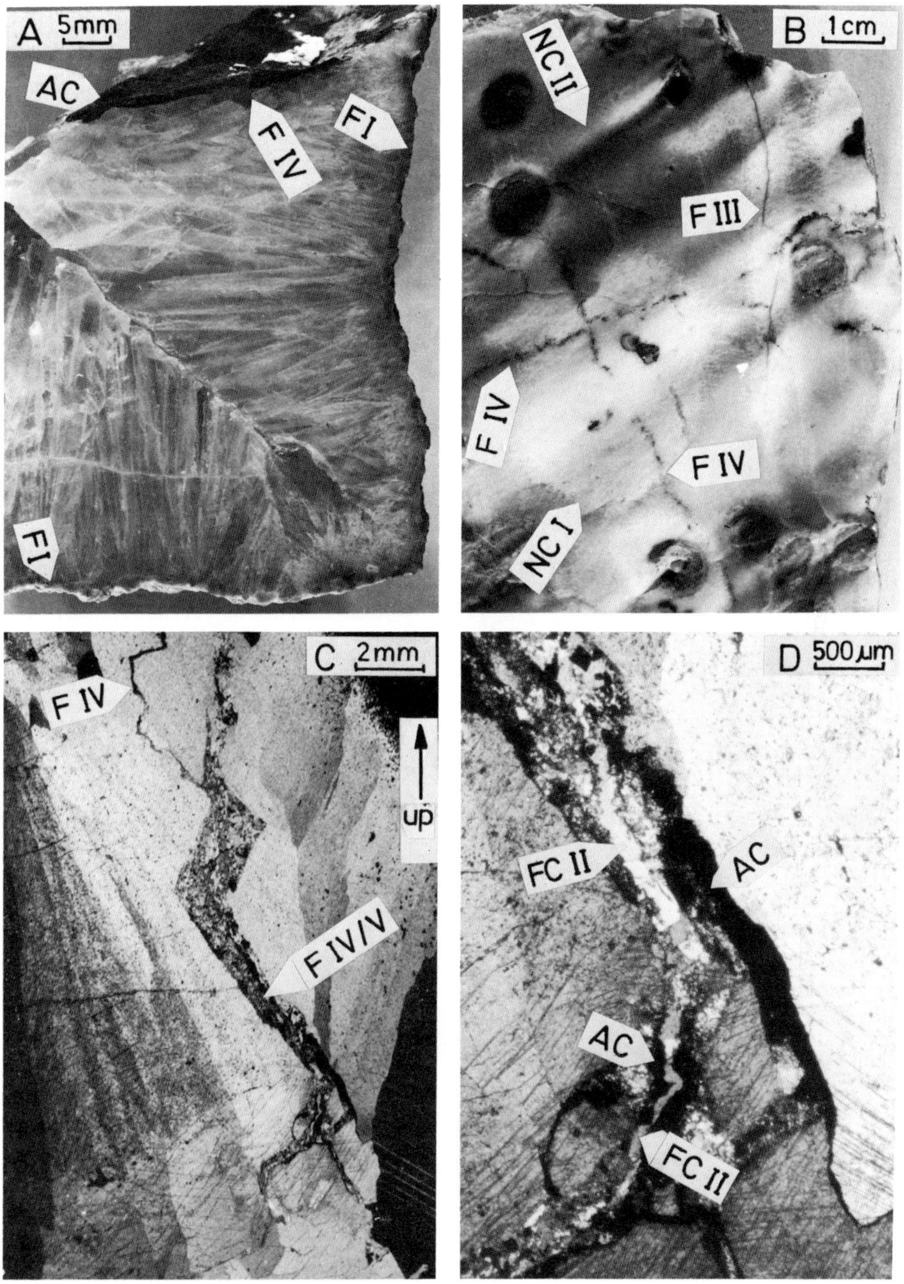
A 5mm
AC
F IV
F I
F I
B 1cm
NC II
F III
F IV
F IV
NC I
C 2mm
F IV
up
F IV/V
D 500µm
FC II
AC
AC
FC II

Fig. 8

The earliest fractures (Figs. 8A, 9: Fracture I) are synsedimentary, and these provided space for micritic material which entered the coral skeleton and formed the linings of the fracture boxwork during the first phase of dissolution.

Another early fracture generation (Fig. 9: Fracture II) contains relics of a precursor cement, probably of the second or third generation (see above), as suggested by the relics which mark a boundary line/surface at which cement crystals, growing towards each other from opposite fracture walls, abutted. Like its surroundings, the fracture filling was affected by Fe calcite neomorphism, whose crystals cut across the walls of the fracture.

Also predating neomorphism are fractures lined by halos of relic-rich neomorphic material (Figs. 8B, 9: Fracture III). Although there is no filling, the fractures must have served as conduits for solutions.

Postdating neomorphism are fractures which cut neomorphic skeletons, precursor cements and Fe calcite cement (Figs. 8A, B, C, 9: Fracture IV). Because some terminate within second-generation vugs, they probably formed after them. They were filled with coarse crystalline ankerite cement which was largely altered to micro-crystalline ankerite with hematite/goethite (see above). Occasionally, a fracture was observed to have re-opened (Figs. 8C, D, 9: Fracture V), and the resulting pores filled with Fe calcite cement. In some samples still another fracture generation (Fracture VI) is found, which was filled with gypsum cement.

Some of the later fractures cut across earlier generations (Fig. 8B). However, the earlier fractures did not receive cement precipitated in the later ones.

4 Sequences and Spatial Distribution of Diagenetic Products

These diagenetic products record a sequence of processes relating to a series of different diagenetic environments. However, all diagenetic products are never found in one sequence. At a given point, some are mutually exclusive, as for example, neomorphism and dissolution of a skeleton. Others are distributed irregularly due to a blend of systematic reasons and randomness. As a result, adjacent samples, or even adjacent portions of a given sample, exhibit quite different, but incomplete sequences. The different diagenetic pathways observed are presented in a petrogenetogram (Fig. 9), a method of illustration introduced by Schroeder (1984).

Like stratigraphic correlation, petrogenetic correlation involves a degree of interpretation. An example is the relation between vug I and the neomorphic Fe calcites. In view of the identical composition of the neomorphic coral surrounding vug I and the precursor cements II and III as well as the Fe calcite cement I within vug I, I consider the first vug generation to be the result of an earlier diagenetic phase and all Fe calcites mentioned above as products of a subsequent phase. Although one could invoke several diagenetic phases to produce these Fe calcites, there is no evidence to justify such complex pathways.

Another problem is the position of Ca dolomite cement in the diagenetic sequence. As its fabric is entirely determined by the ankerite cement substrate,

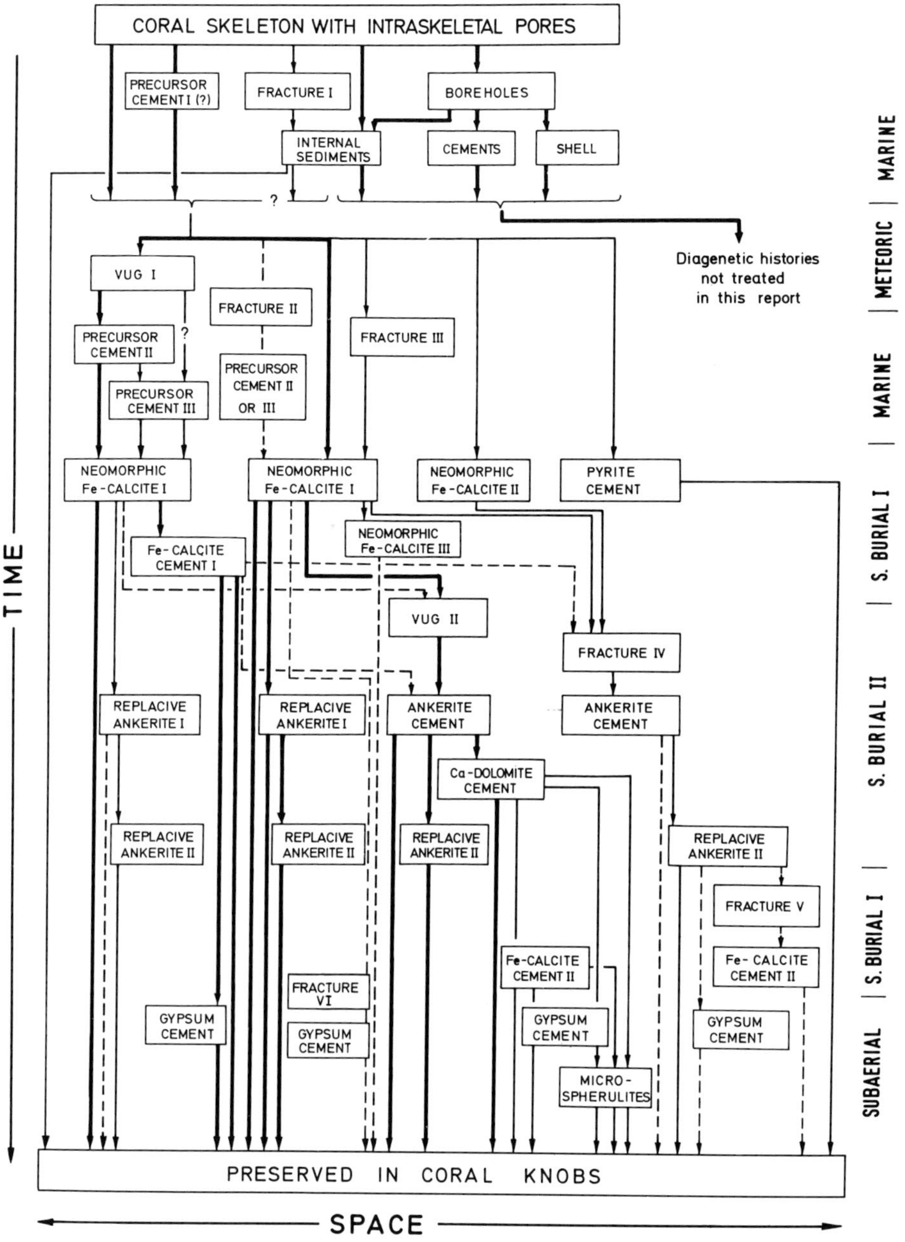

Fig. 9. Petrogenetogram for the corals studied showing diagenetic pathways observed. Sequences are indicated in vertical direction, progressing *from top to bottom*; in *horizontal direction* the distribution in space is illustrated. Along the *right hand margin* the diagenetic environments as discussed in Section 5 are indicated. *Connecting lines* indicate relative abundance based on visual estimates: *bold lines* abundant; *thin lines* common; *dashed lines* rare; *S. burial* shallow burial

even where that cement was later replaced by micro-crystalline ankerite, the Ca dolomite precipitation must have preceded the selective replacement of its substrate. Supporting evidence includes the preservation of the initial shape of the coarse ankerite crystals: had replacement taken place adjacent to open pores, some deformation would have been likely.

It is essential to recognize the diagenetic diversity of reefs in order to understand the development of their porosity. However, to appreciate the effect of the respective pathways, some quantification of the relative importance of a given pathway is required. As a first step in this direction, categories of abundancy based on visual estimates are introduced indicating whether a given sequence is abundant, common or rare (Fig. 9). Of the 24 sequences distinguished in these corals, nine are abundant, nine common, and six are rare. Systematic observation, for example using a grid or a point counter, would certainly provide more reliable information.

When examining the relative abundance of various diagenetic products in various specimens, one of the basic controls for the variations observed appeared to be skeletal morphology. The two morphotypes *A. hexaphylla* I and II, exhibit significant differences (Table 3), in spite of the fact that they belong to the same species.

5 Diagenetic Environments

Growth of corals, bioerosion, internal sedimentation and probably internal cementation (precursor cement I) took place in the marine environment. Extensive dissolution (vug I) therefore requires a major change, probably to an active meteoric environment. Although the evidence supplied by the vugs is compelling, there is no erosional unconformity, no hard-ground or any similar indication to support this interpretation. Such evidence is needed if emergence is to be postulated. However, meteoric phreatic conditions may well have been effective under a thin cover of younger sediments. This depends on the structural and sea-level histories of the particular area, which have not yet been worked out.

Precursor cements II and III, whose fabrics suggest a marine origin, imply a return to marine conditions.

The subsequent diagenetic phase is dominated by Fe calcite formed either by neomorphism or cementation. It is generally agreed that reducing conditions are required, and that meteoric waters are involved (Scholle and Halley 1985). While earlier reports suggested formation in the mixing zone (Scherer 1977), most later studies suggest a certain degree of burial and temperatures of tens or about 100 °C (e.g., Moritz 1983, Grover and Read 1983). Boles et al. (1985) indicate a minimal temperature of less than 25 ° – 30 °C for ferroan calcite formation, as determined by vitrinite reflectance.

The ankerite-dominated phase could then be placed in a similar shallow burial environment. Evidence includes the saddle-shaped habit, which, according to Radke and Mathis (1980), is found in the range of 60° – 150 °C. Further support can be drawn from the diagenesis of clastic rocks in other areas (e.g., Boles and Franks 1979), where ankerite has been observed at temperatures exceeding about 100 °C. This result, transferred to the area of study, conflicts

Table 3. Observed relative abundance of diagenetic products in morphotypes I and II of *Actinastraea* ex gr. *hexaphylla*

	Morphotype I	Morphotype II
Neomorphic Fe Calcite I	+ + +	+ +
Neomorphic Fe Calcite II	+ +	+
Neomorphic Fe Calcite III	+	+ ?
Replacive Ankerite I	+ + +	–
Replacive Ankerite II	+ +	+ +
Vug I enclosed	+ +	+
Vug I basal	+	+ + +
Vug II	+ +	–
Precursor Cement I; isopachous	–	+
Precursor Cement I; micro-crystalline	+ +	+ +
Precursor Cement II	+ + +	+ + +
Precursor Cement III	+ +	+ +
Fe Calcite Cement I	+ + +	+ + +
Pyrite Cement	+ +	+ +
Ankerite Cement	+ +	+
Ca Dolomite Cement	+ +	–
Fe Calcite Cement II	+ +	+
Gypsum cement	+ +	+
Fracture I (Sediment filled)	–	+ +
Fracture II (Precursor cement fill)	+	–
Fracture III (Without fill)	+ +	–
Fracture IV (Ankerite cement fill)	+ +	+ +
Fracture V (Fe Calcite cement II fill)	+	+

Visual estimate; + + + abundant; + + common; + rare; – not observed.

with the interpretation of the composition of clay minerals in the insoluble residues of the limestone layer studied, and in the surrounding shales. In addition to some kaolinite and illite they contain a high proportion of poorly crystallized smectite (Hendriks, in preparation) which would gradually be converted to illite at temperatures between 50 ° and 100 °C (Boles and Franks 1979, Hoffman and Hower 1979). The temperature in this case probably did not rise far beyond 50 °C. Considering a geothermal gradient of 2 ° – 3 °C/100 m, this temperature would still signify burial depth of at least 1000 m, which clearly exceeds the depths implied by the stratigraphic record. However, the thermal gradient of this area may deviate from the normal; this is likely in view of the proximity of the Red Sea Graben where the present thermal gradient is 4 °/100 m (Wong and Degens 1980).

In order to appreciate thermal influence on diagenesis, the entire post-Paleocene thermal history, i.e., for the last 60 million years, needs to be considered: Meneisy and Kreuzer (1974) have dated volcanic rocks from the region to be 41 and 33 million years old. Schandelmeier and Darbyshire (1984) studying rocks west of the region report a peak of both intrusive and extrusive magmatic activity 45 million years ago, and small scale volcanic activity in Late Tertiary and Quaternary; Franz et al. (in press) documented volcanic activity from Northern Sudan at 59 and 26 – 20 million years B.P. Thus there are numerous indications for possible changes in the thermal regime which may compensate for the minor extent of vertical movement of the coral knobs.

Initial conversion of smectite to illite may well have provided the Mg and Fe required for the formation of ankerite (for review see McHargue and Price 1982). This process depends on temperature, and either increased burial or increased geothermal gradient due to magmatic processes may have triggered it, thus leading to ankerite formation.

The Ca dolomite cement and subsequent replacive ankerite II are ascribed to changes in pore water chemistry. Similar changes have been observed by Katz (1971). The replacive microcrystalline ankerite with its hematite and goethite indicates intermittent oxidizing conditions. The second-phase Fe calcite cementation may signal some uplift/cooling, i.e., somewhat shallower burial.

The gypsum cement is attributed to present arid surface conditions, as it is almost ubiquitous throughout the area.

6 Discussion

6.1 Neomorphism and Replacement

Neomorphic fabric is well known (for review see Folk 1965). The above observations should draw attention to the complexity of this process and stimulate further observations and, possibly, experimental work. The observations suggest that in certain cases the original skeleton functions to varying degrees as a template. However, the factors determining distribution and intensity remain elusive. The fact that certain patterns are genetically related to apparent solution conduits suggests that permeabilities may be important.

This question relates to the more general problem evoked by Sandberg (1984, and earlier papers), who stresses the close spatial proximity of dissolution and neomorphism; the factor(s) determining one or the other remain unknown.

In this study a close relationship between cementation and replacement is evident. Around second generation vugs, the coarsely crystalline ankerite moved outward from the original pore as a replacive front, and grew inward as a cement. Indeed, large individual crystals were observed to be replacing at one end and precipitated as cement at the other. Replacive and cement ankerite are distinguished by the presence and absence of relics, as well as by spatial relations to pre-existing material and open/subsequently cemented pore, respectively. On the basis of this close spatial relationship I have interpreted coarse replacive ankerite (I) and ankerite cement as products of the same diagenetic stage (Fig. 9); this interpretation may be subject to debate. However, it may lead to the general consideration of interrelationships of the major diagenetic processes of dissolution, neomorphism, replacement, and cementation in a given diagenetic environment at a given diagenetic stage.

6.2 Factors Determining Spatial Distribution in Reef Diagenesis

On a large scale, facies, sea level history, and respective morphological positions are known to cause spatial diagenetic variations (Scherer 1975, Schneider 1977,

Walls et al. 1979, Grover and Read 1983). These factors were excluded from this study by taking corals from one stratigraphic horizon, one facies, and equivalent morphological positions.

On a smaller scale, different pathways in primary and secondary pores, respectively, were found by Bebout et al. (1979), while James and Klappa (1983) distinguished three groups of allochems with different diagenetic patterns, and Mattes and Mountjoy (1980, Fig. 19) presented a series of diagenetic sequences which affected various bioclasts. On a somewhat smaller scale, the variation of diagenesis in adjacent pores of identical nature has led others (Schroeder 1972a, Lighty 1985) to suggest microenvironmental differences as being responsible. On a similar scale, James (1974) observed that different structural units within a given coral were altered in different ways, and James et al. (1976) found that the pore sizes of the coral structure influenced diagenesis, a view basically confirmed by Gvirtzmann and Friedman (1977). On the ultra-structural level, Constantz (1986) and Dullo (1986) have shown that taxon-specific characteristics such as dimensions, surface area and packing of skeletal elements play a decisive role in determining diagenetic behavior.

By studying corals only, most of which even belong to one species, taxonomically determined primary mineralogic and structural differences were partly excluded. However, sufficient factors remained to provide considerable diagenetic diversity which can be explained to some extent by differences in corallite diameter (small versus large leading to dense versus open and relatively massive versus relatively fragile skeletons). These skeletal properties may also explain the preference of boring organisms, which certainly contributes to the diversity shown. The striking diagenetic differences observed even between morphotypes of a given species suggests that comparison between different species, let alone higher taxonomic levels, may be fortuitous.

Although in the case studied there is no evidence that morphotypes were determined environmentally, the differences demonstrated in their diagenesis add a new dimension to other investigations: Numerous observational studies (Grauss et al. 1977, Fricke and Schuhmacher 1983, many others) and the experimental study of Foster (1979) have shown that skeletal morphology in a given species is modified in response to environmental factors such as light intensity and wave energy. Thus these factors may effectively program coral diagenesis.

As exemplified by the relation of borings and fractures to types of neomorphism, the results of Buddemeier and Oberdorfer (1986) can be applied on a very small scale. Micro-hydrology may provide additional explanations for diagenetic variations observed, especially when this parameter is altered by every diagenetic process at every diagenetic stage: evolving micro-hydrology is considered an important determinant of diagenetic diversity.

6.3 Cement Stratigraphy

Cement stratigraphy as developed by Meyers (1974) has been an important step in tracing cements and porosites. However, the spatial diversity of sequences and the fact that one is dealing with a number of incomplete sequences certainly merits caution in the use of this tool.

As shown above, cement generations may be missing from a given pore. In other cases, cement generations are precipitated in secondary pore space leached below or between older cements (Schroeder 1979). Furthermore, it has been observed that a cement generation may be enveloped by another cement which is both a substrate and a cover, and both precedes and suceeds it (Schroeder 1985). Such deviations from simple stratigraphic principles should be considered when applying this method.

6.4 Porosity Assessment

By illustrating the variety of diagenetic pathways, this report will hopefully contribute to the understanding of porosity evolution in corals and thus in reefs. For the practical purpose of porosity assessment, however, taking diversity into account may be more detrimental than beneficial, and the petrogenetic analysis must be followed by a synthesis, and simplification. A synthesis can be achieved by partly excluding rare pathways and those with minimal volumetric effects, and possibly by using suitable species or biogenic fragments as indicators.

7 Summary

The Paleocene coral knobs studied passed through a sequence of diagenetic environments ranging from marine to meteoric and back to marine, on to shallow burial at various depths and then back to the surface.

In the course of this development, various portions of corals were affected at various times by neomorphism, replacement, dissolution, cementation and/or fracturing. As a result their diagenetic history comprises a large number of parallel, variously incomplete sequences progressing in spatial proximity, many of which may be recognized in a given coral colony or a single thin section. This study shows that skeletal morphological differences within a given species may program such a variation, and that microhydrology, evolving throughout diagenesis and thus itself highly variable, is another important determinant of diagenetic diversity.

Acknowledgments. Fieldwork was carried out as part of the Special Research Project "Geoscientific problems of arid regions"; I would like to express my sincere appreciation to the chief scientist, E. Klitzsch, to the project's national counterpart, the General Petroleum Company of Egypt, and to its sponsor, the German Research Foundation (DFG), which also financed some of the laboratory equipment used.

H. Kallenbach, F. Hendriks, and P. Luger worked in the field with me; the following helped in the laboratory: U. Zinkernagel (CL microscope), D. Ackermand, F. Galbert, and T. Lund (microprobe), F. Hendriks, H. Lange, W. Vorwerk, and P. Wallenburg (X-ray diffractometer), W. Reimann (SEM/Edax), B. Gabriel (thin section preparation). R. Höfling identified the corals. I was assisted in preparing the manuscript by F. Lyons (linguistic advice, typing), B. Dunker and E. Susin (drafting), B. Kleeberg, and U. Schuldt (photography). Many colleagues at the Universität Kiel and the Technische Universität Berlin discussed various aspects of the projects; G. Franz, F. Hendriks, and B. Purser reviewed the manuscript. I thank all of the above for their time and efforts, interest and cooperation.

References

Bebout DG, Schatzinger RA, Loucks RG (1979) Porosity distribution in the Stuart City Trend, Lower Cretaceous, South Texas. Am Assoc Petrol Geol Course Note Ser 11:B3 – B25

Boles JR, Franks SG (1979) Clay diagenesis in Wilcox sandstones of southwest Texas: Implications of smectite diagenesis on sandstone cementation. J Sediment Petrol 49:55 – 70

Boles JR, Landis CA, Dale P (1985) The Moeraki boulders – anatomy of some septarian concretions. J Sediment Petrol 55:398 – 406

Buddemeier RW, Kinzie RA (1976) Coral growth. Oceanogr Mar Biol Annu Rev 14:183 – 225

Buddemeier RW, Oberdorfer JA (1986) Internal hydrology and geochemistry of coral reefs and atoll islands: key to diagenetic variations. This volume

Choquette PW, Pray LC (1970) Geologic nomenclature and classification of porosity in sedimentary carbonates. Am Assoc Petrol Geol Bull 52:207 – 250

Constantz BR (1986) The primary surface area of Quaternary Caribbean corals and variations in their susceptibility to diagenesis. This volume

Dullo WC (1986) Variations in diagenetic sequences: an example from Pleistocene coral reefs, Red Sea, Saudi Arabia. This volume

Folk RL (1965) Some aspects of recrystallization in ancient limestones. Soc Econ Paleontol Miner Spec Publ 13:14 – 48

Foster AB (1979) Phenotypic plasticity in the reef corals *Montastrea annularis* (Ellis & Solander) and *Siderastrea siderea* (Ellis & Sollander). J Exp mar Biolo Ecol 39:25 – 54

Frank JR (1981) Dedolomitization in the Taum Sauk limestone (Upper Cambrian Southeast Missouri). J Sediment Petrol 51:7 – 18

Franz G, Puchelt M, Pasteels P (in press) Petrology, geochemistry and age relations of Triassic and Tertiary volcanic rocks from SW Egypt and NW Sudan. J Afr Earth Sci

Fricke HW, Schuhmacher H (1983) The depth limits of Red Sea stony corals: An ecophysiological problem. PSZN Mar Ecol 4:163 – 194

Ginsburg RN, James NP (1976) Submarine botryoidal aragonite in Holocene reef limestones, Belize. Geology (Boulder) 4:431 – 436

Grover G, Read JF (1983) Paleoaquifer and deep burial related cements defined by regional cathodoluminescent patterns, Middle Ordovician carbonates, Virginia. Am Assoc Petrol Geol Bull 67:1275 – 1303

Grauss R, Chamberlain JA, Boker AM (1977) Structural modification of corals in relation to waves and currents. Am Assoc Petrol Geol Stud Geol 4:135 – 153

Gvirtzmann G, Friedman GM (1977) Sequence of progressive diagenesis in coral reefs. Am Assoc Petrol Geol Stud Geol 4:357 – 380

Hendriks F, Luger P, Kallenbach H, Schroeder JH (1984) Statigraphical and sedimentological framework of the Kharga-Sin El-Kaddab Stretch (Western and Southern Part of the Upper Nile Basin), Western Desert, Egypt. Berliner Geowiss Abh A50:117 – 151

Hey MH (1962) An index of mineral species and varieties arranged chemically, 2nd edn. British Museum London 728

Hoffman J, Hower J (1979) Clay mineral assemblages as low grade metamorphic geothermometers: Application to the thrust-faulted disturbed belt of Montana, USA. Soc Econ Paleontol Miner Spec Publ 26:55 – 79

Hubbard JAEB (1972) Cavity formation in living scleractinian reef corals and fossil analogues. Geol Rundsch 61:551 – 564

James NP (1974) Diagenesis of scleractinian corals in the subaerial vadose environment. J Paleontol 48:785 – 799

James NP, Klappa CF (1983) Petrogenesis of Early Cambrian reef limestones, Laborador, Canada. J Sediment Petrol 53:1051 – 1096

James NP, Ginsburg RN, Marszalek DS, Choquette PW (1976) Facies and fabric specificity of early subsea cements in shallow Belize (British Honduras) reef. J Sediment Petrol 46:523 – 544

Katz A (1971) Zoned dolomite crystals. J Geol 79:38 – 51
Lighthy RG (1985) Preservation of internal reef porosity and diagenetic sealing of submerged Early Holocene barrier reef, Southeast Florida shelf. Soc Econ Paleontol Miner Spec Publ 36:123 – 151
Luger P (1985) Stratigraphie der marinen Oberkreide und des Alttertiärs im südwestlichen Obernil-Becken (SW-Ägypten) unter besonderer Berücksichtigung der Mikropaläontologie, Palökologie und Paläogeographie. Berliner Geowiss Abh A63:1 – 151
Mattes BW, Mountjoy EW (1980) Burial dolomitization of the Upper Devonian Miette buildup, Jasper National Park, Alberta. Soc Econ Paleontol Miner Spec Publ 28:259 – 297
Mazullo SJ (1980) Calcite pseudospar replacive of marine acicular aragonite, and implications for aragonite cement diagenesis. J Sediment Petrol 50:409 – 422
McHargue TR, Price RC (1982) Dolomite from clay in argillaceous shale-associated marine carbonates. J Sediment Petrol 52:873 – 886
Meneisy MY, Kreuzer H (1974) Potassium-argon ages of Egyptian basaltic rocks. Geol Jahrb D9:21 – 31
Meyers WJ (1974) Carbonate cement stratigraphy of the Lake Valley Formation (Mississippian) Sacramento Mountains, New Mexico. J Sediment Petrol 44:837 – 861
Moritz W (1983) Fazies und Diagenese des Briloner Karbonatkomplexes anhand einiger ausgewählter Bohrungen. PhD Thesis Univ Braunschweig, pp 1 – 181
Radke BM, Mathis RL (1980) On the formation and occurrence of saddle dolomite. J Sediment Petrol 50:1149 – 1168
Sandberg PA (1984) Recognition criteria for calcitized skeletal and non-skeletal aragonites. Paleontogr Am 54:272 – 281
Schandelmeier H, Darbyshire F (1984) Metamorphic and magmatic events in the Uweinat-Bir Saf Saf Uplift (Western Desert/Egypt). Geol. Rundsch 73:819 – 833
Scherer M (1975) Cementation and replacement of Pleistocene corals from the Bahamas and Florida: diagenetic influence of non-marine environments. N Jahrb Geol Paläontol Abh 149:259 – 285
Scherer M (1977) Preservation, alteration and multiple cementation of aragonitic skeletons from the Cassian Beds (U. Triassic, southern Alps): Petrographic and geochemical evidence. N Jahrb Geol Paläontol Abh 154:74 – 124
Schneider W (1977) Diagenese devonischer Karbonatkomplexe Mitteleuropas. Geol Jahrb D21: 1 – 109
Scholle PA, Halley RB (1985) Burial diagenesis: Out of sight, out of mind. Soc Econ Paleontol Miner Spec Publ 36:309 – 334
Schroeder JH (1972a) Fabrics and sequences of submarine carbonate cements in Holocene Bermuda cup reefs. Geol Rundsch 61:708 – 730
Schroeder JH (1972b) Calcified filaments of an endolithic alga in Recent Bermuda reefs. N Jahrb Geol Paläontol Monatsh 1972:16 – 33
Schroeder JH (1979) Carbonate diagenesis in Quaternary beachrock of Uyombo, Kenya: Sequences of processes and coexistence of heterogenic products. Geol Rundsch 68:894 – 919
Schroeder JH (1984) The petrogenetogram of corals: spatial variation in diagenetic sequences. Palaeontogr Am 54:261 – 271
Schroeder JH (1985) Sparry calcite cement in Miocene corals from Khor Eit, NE Sudan. Proc 5th Int Coral Reef Congr 3:283 – 288
Schroeder JH, Zankl H (1974) Dynamic reef formation: A sedimentological concept based on studies of recent Bermuda and Bahama reefs. Proc 2nd Int Coral Reef Symp 2:413 – 428
Walls RA, Mountjoy EW, Fritz P (1979) Isotopic composition and diagenetic history of carbonate cements in Devonian Golden Spike reefs, Alberta, Canada. Geol Soc Am Bull 90:963 – 982
Wong HK, Degens ET (1980) The Red Sea and Gulf of Aden – a geological and geophysical review. Proc Symp Coastal and Marine Environment of the Red Sea, Gulf of Aden and Tropical Western Indian Ocean 3:213 – 266

II Mesozoic Reefs

Diagenesis of Mid-Cretaceous Rudist Reefs, Valles Platform, Mexico

Paul Enos[1]

1 Introduction

The Taninul ("reef") facies of the El Abra Formation in the Sierra de El Abra and the adjacent Sierra Madre Oriental (Fig. 1) is renowned for exposures of mid-Cretaceous rudist assemblages. These outcrops have long served as the most accessible depositional model (Kellum 1930, Muir 1936, Bonet 1952, Griffith et al. 1969), and, recently, as a diagenetic model (Aguayo 1978), for the spectacular oil-producing trend of the Golden Lane. Early submarine cement of rudist boundstones and grainstones is extensive (Shinn et al. 1974, Aguayo 1978), and Minero (in press) documents significant early meteoric diagnesis as well in the backreef facies. This study deals with the diagenesis of the platform-margin rocks, generally referred to as the Taninul facies or "reef", at the eastern margin of the Valles platform (Fig. 1).

2 Setting

During the mid-Cretaceous (Albian-Cenomanian), a series of shallow-water platforms developed in eastern Mexico, surrounded by vast expanses of intracratonic basins (Carrillo 1971, Enos 1974, 1983) that isolated them from any source of siliciclastic detritus. The platforms developed relief of about 1000 m and slopes as steep as 20° to 45° (Enos 1977, 1980, Suter 1984). This morphology was a consequence of high rates of relative sea-level rise that generally precluded progradation of the platform margins. One exception to this geographic stability is indicated by significant propagation in both basin-margin and platform-margin facies west of Xilitla, San Luis Potosí (Fig. 1, LC, Enos 1982, Minero et al. 1983, p. 90).

Mid-Cretaceous basinal and platform rocks comprise most of the Sierra Madre Oriental, a classical fold-and-thrust belt (Suter 1984). An exposure of Taninul facies dominated by rudist rubble near the eastern front of the Sierra Madre was used in this study (Fig. 1, LC). Further east the mid-Cretaceous interval is generally buried beneath Upper Cretaceous (Turonian-Maastrichtian) basinal carbonates and the Tertiary clastic wedge of the Gulf coastal plain, but platform-margin reefs and upper slope deposits are exhumed in the open anticline of the Sierra de El Abra (Fig. 1, TQ and CA, resp.).

[1] Department of Geology, University of Kansas, Lawrence, KS 66045, USA

Reef Diagenesis
Edited by J. H. Schroeder and B. H. Purser

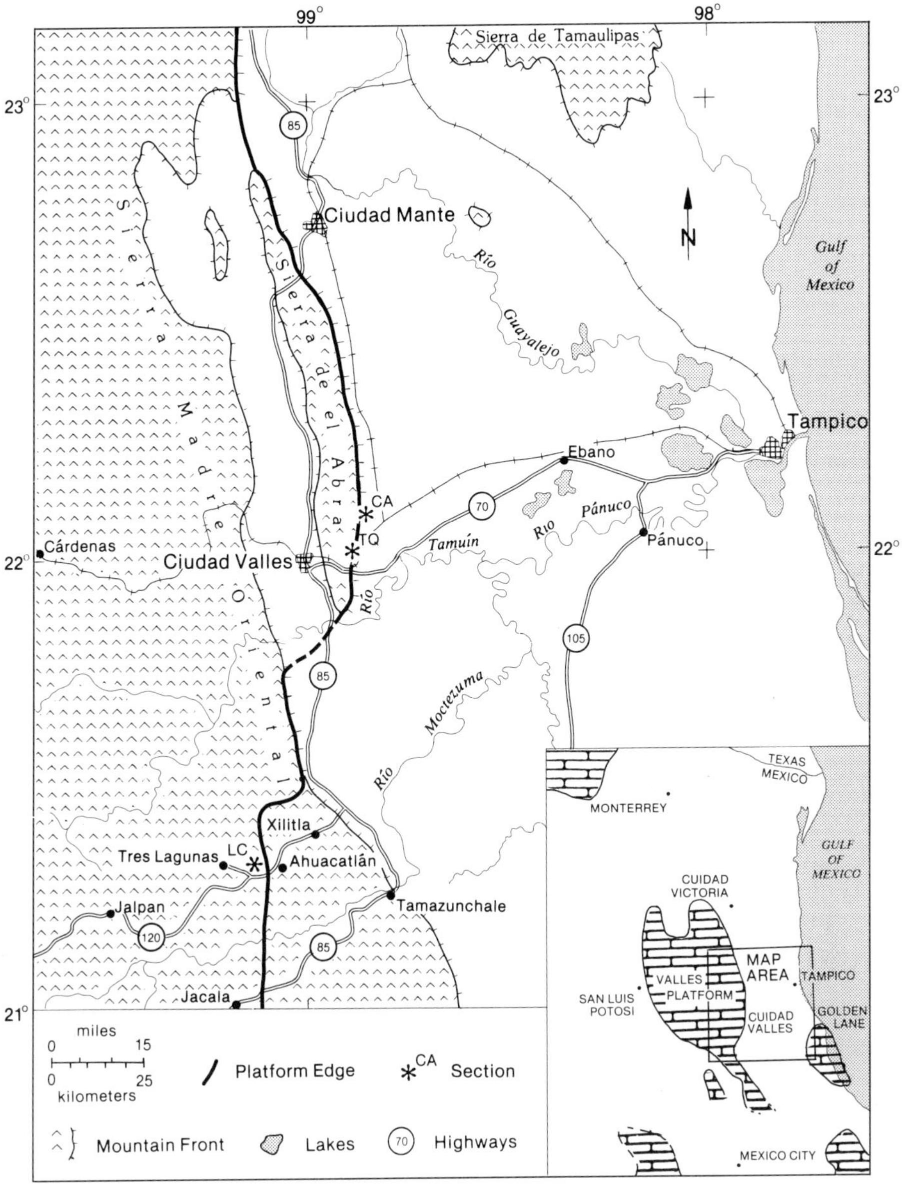

Fig. 1. Location of measured sections at eastern margin of Valles platform. Sections are: *CA* Cementos Anahuac quarry; *TQ* Taninul quarry; *LC* Laguna Colorada. Regional map (*inset*) shows location of mid-Cretaceous platforms in brick pattern. (After Enos 1974)

The shallow-water platforms varied in width from 10 km (El Doctor platform, Wilson et al. 1955, Ward 1979) to over 200 km (Valles platform, Carrillo 1971). Regardless of size, the platform interior is predominantly thick-bedded, peloid-foraminifer wackestones (Ward 1979, Minero, in press). Within about 3 km of the platform edges, the peloid-foraminiferal rocks are interspersed with fenestral limestones, cryptalgal laminites, and lenticular requienid (rudist) limestones, (Griffith et al. 1969, Aguayo 1978, Minero, in press). Cyclic patterns within these lithofacies indicate repeated shoaling and emergence of the platform margin in the shelter of reefs or islands. About a kilometer from the slope break at the platform edge, locally cross-bedded bioclastic lime grainstone lenses interbedded with oncolitic packstone reflect stronger current and storm influence (Minero, in press). Development of pronounced subaerial disconformities at the tops of shoaling-up cycles and attendant meteoric diagenetic effects show that the platform margin was periodically exposed as islands (Ward 1979, Minero, in press).

The slope break at the platform margin is marked by lenses of caprinid (rudist) boundstone interspersed with, and ultimately overlain by, skeletal and intraclastic lime grainstone. The caprinid lenses measure 30 to 480 cm thick and 1 to 30 m or more in lateral dimension; they are thicker and more abundant immediately adjacent to the slope break. The Taninul facies is less than a kilometer wide; skeletal debris dominates toward the platform (Aguayo 1978, Minero, in press).

Three exposures provided the material for this study (Fig. 1). Two are of rudist assemblages near the slope break at the platform margin (TQ, LC) and the third (CA) is an upper slope deposit of rudist debris. Taninul quarry (TQ, Fig. 2A), exposes the upper 45 m of the El Abra Formation that may attain 1800 m in thickness (Aguayo 1978, p. 84). The top 21 m consists of skeletal debris with accretion beds that dip about 6° seaward (Fig. 2A, Minero et al. 1983, p. 24). The lower 24 m is boundstone interspersed with skeletal debris (Table 1). Caprinid rudists dominate all lithologies including the boundstones, but small boundstone lenses are formed by radiolitids and arborescent corals. Secondary porosity is distributed throughout the section as molds and vugs in bioclastic intervals and large molds in boundstones (Fig. 2B). The lower part of the section is cut by numerous disconformities (Figs. 2B, 3) marked by lithologic boundaries that truncate grains, by sharp local relief (Minero et al. 1983, p. 29), or encrustations. None of the disconformities could be traced as far as 100 m before being obscured by changes in character of the bounding lithologies.

The boundstone intervals, and perhaps some caprinid-rich portions of the bioclastic rocks, are the reef or at least a platform-edge organic community. The community lived in small patches surrounded by their own rubble. The patches were a few meters or tens of meters in lateral extent and covered about a third of

Fig. 2A – D. Outcrops of Taninul facies. **A** Taninul quarry. Exhumed slope break of Valles platform is at left. Bedding and geopetal fabrics give a structural dip of 5 °E (*left*); accretion bedding *above the arrow* shows an additional dip of 6 °E (Minero et al. 1983, p. 24). Height of exposure at the slope break is 28 m. TQ section is to the *right* of this view, 200 to 260 m from the slope break. **B** Lens of caprinid boundstone (*top*) overlying bioclastic grainstone, separated by planar disconformity at hammer head (*center*). Pores in boundstone are molds of caprinids. Hammer is 40 cm long. Taninul quarry (TQ, Fig. 1); disconformity is at 3.2 m (Fig. 3). Note: Location of samples throughout are by section initials and elevation. **C** Disconformity in skeletal grainstones that truncates grains (*center*). Some of large caprinids below disconformity may be in growth position. Coarse-fine layering above the disconformity involves some cut-and-fill and, locally, truncation of grains. LC 15 m. **D** Dike filled with alternations of cement crusts and skeletal sediment. Laguna Colorada

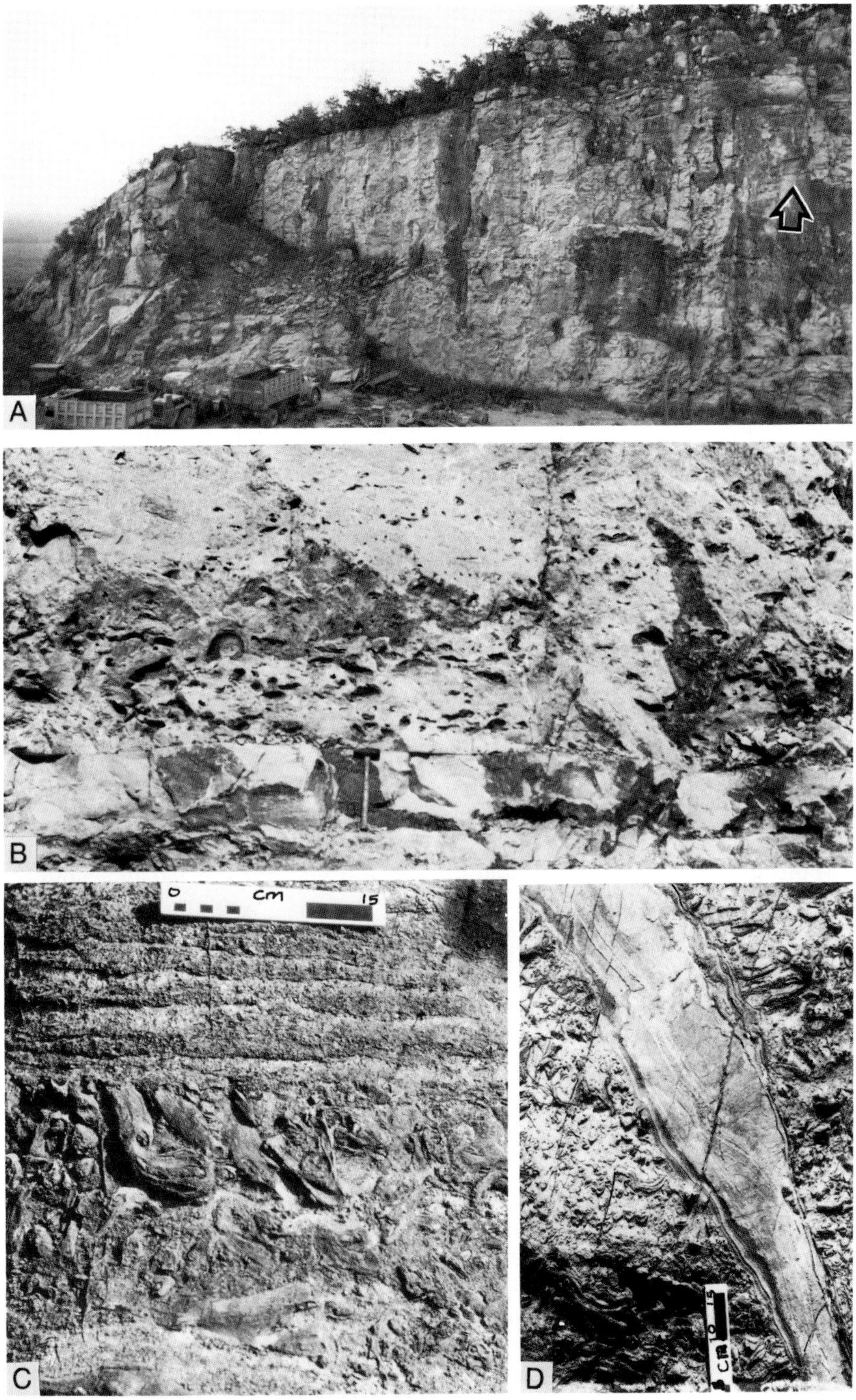

Fig. 2

Table 1. Lithologies of measured sections

Section	TQ (Base)	TQ (Top)	LC (Base)	CA
Thickness (m)	24	21	105	44
Lithologies (percent)				
Boundstone	35	–	–	–
Grainstone	40	55	19	19
Packstone	25	45	79	59
Wackestone	–	–	2	12

the area, judging from their vertical distribution. Encrusting organisms were probably more effective within the rubble rather than among living organisms. The bioclastic upper half of the El Abra section at Taninul quarry (Fig. 3) is interpreted as platform-margin sand shoals that prograded over the reefs and eventually reached the platform edge.

The section at Laguna Colorada in the Sierra Madre Oriental (LC, Fig. 1) is at the leading margin of a major thrust fault that nearly coincides with the platform edge (Suter 1984). No boundstone intervals were noted (Table 1), although small clusters of radiolitids, some overturned, are widely scattered (Minero et al. 1982, p. 91, 96). Alternating coarse- and fine-grained layers with cut-and-fill structures are prominent near the base of the section (Fig. 2C). Sedimentary dikes (Fig. 2D), disconformities that truncate grains (Fig. 2C), and pholad bivalve borings indicate early cementation. Present visible porosity is nil.

The Laguna Colorada rocks probably formed somewhat further from the slope break than those at Taninul quarry. They contain no demonstrably in situ caprinids, but small radiolitid clusters are more common. Although mud is scarce, packstones are prevalent at Laguna Colorado (Table 1). This sequence was previously interpreted as a sand flat (Minero et al. 1983, p. 94) analogous to that described by Harris (1979) from the Joulters Cays area of the Great Bahama Bank. However, in view of the coarse rubbly nature of the deposits, numerous truncation surface in parts of the sequence, extensive cementation, and intermittent early exposure (see Sect. 5), it appears more analogous to the cemented, planated reef flats of Pacific reefs. Encrusting and binding organisms seem anomalously rare, however.

At Cementos Anahuac quarry, the upper surface of the El Abra Limestone has a primary basinward dip of 43°, as reconstructed from geopetal fabrics (Enos 1980). The upper contact, presumably the ultimate depositional surface of the El Abra, formed in a minimum water depth of 270 m below the slope break. The section is entirely bioclastic limestone (Table 1) with no recognizable disconformities in the section. Several sediment-filled dikes cut the section. Visible porosity is 1 to 12%, mostly in skeletal molds, but with local intraparticle pores. It seems improbable that many of the rudists and green algae that make up the Cementos Anahuac section derived from organisms living at the inferred depositional depth of 270 m. They probably were shed from reefs at the top of the slope.

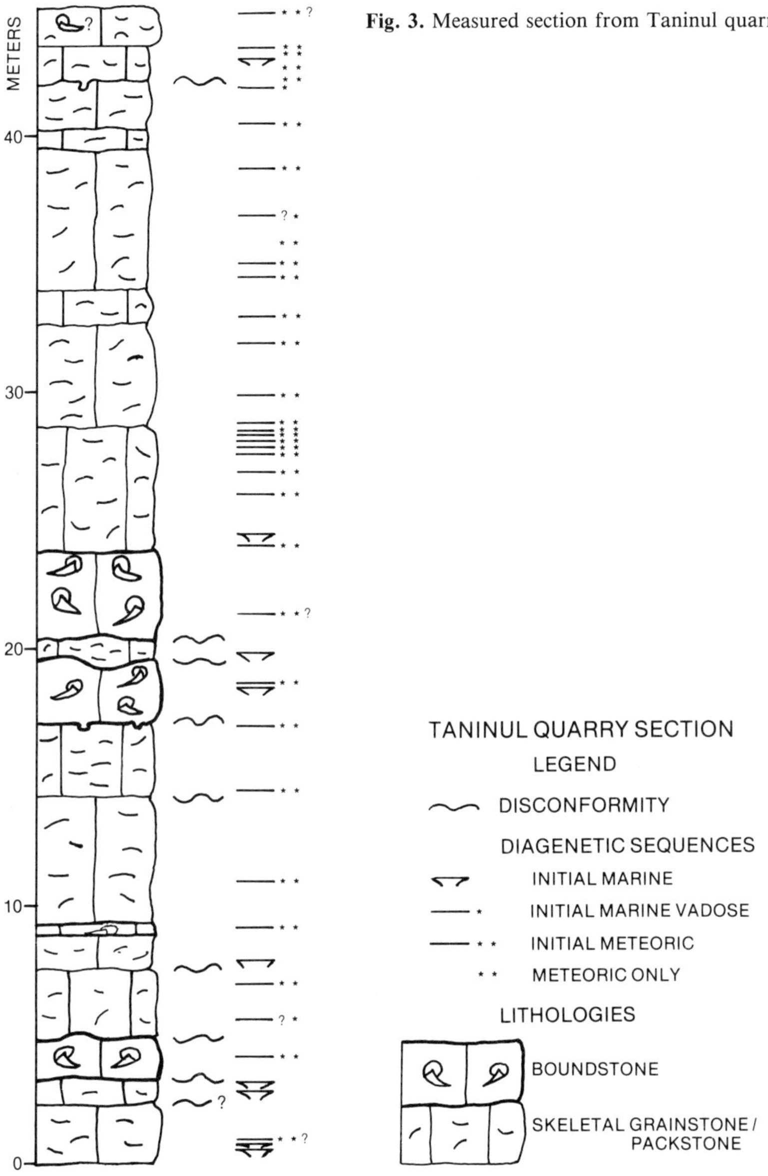

Fig. 3. Measured section from Taninul quarry (TQ, Fig. 1)

3 Diagenetic Fabrics

Differences in diagenetic fabric among sections are more in degree and sequence than in kind. It is therefore convenient to characterize the full range of diagenetic effects, then to consider their origins, to identify sequences of diagenetic fabrics, and, finally, to consider the significance of the sequences. Diagenetic effects include porosity reduction by crystalline or muddy calcite and porosity enlarge-

Table 2. Diagenetic fabrics of El Abra Formation, Taninul facies

Fabric	Characteristics	Space occupied[a]		Interpretation
		Primary	Secondary	
Pore reduction				*Internal sediment*
Microfossil mud	Marine microfossils, interlayered[b]	BP, WP, SH	MO, VUG	Marine internal sediment
Skeletal, peloidal sand	Marine fossils	BO, WP, SH	MO, VUG	Marine internal sediment
Crystal silt[c]	Uniform size, loose packing	BP	VUG	Vadose internal sediment
				Cement
Isopachous mud	Sparse fossils, fragments, interlayered[b]	BP, WP	MO, VUG	Submarine cryptocrystalline cement
Patchy mud	Grain contact, meniscus, pendant, tubes	BP, WP		Meteoric vadose cement
Pendant crystal crust	Pendant, grain contact, undulose extinction	BP, WP, SH		Marine vadose (intertidal) cement. Possibly meteoric
Isopachous crystal crust	Radiaxial fibrous mosaic[d] cloudy w/microdolomite[e]	BP, WP, SH	MO, VUG	Submarine cement, Mg calcite
Equant spar	Linear crystal boundaries, clear, uniform extinction	BP, WP, SH	MO, VUG, FR	Meteoric phreatic cement, shallow
Pore enlargement				*Dissolution*
Microvug	Tubular shapes	Matrix, muddy cement		Meteoric dissolution, vadose zone?
Mold	Replaces metastable skeletons	Caprinids, gastropods, algae		Meteoric dissolution
Vug	Enlarged molds	Not selective		Meteoric dissolution
				Other
Boring	Disconformity surfaces	Not selective		Postdates initial cement
Fracture	Locally circum-granular	Not selective		Near surface

[a] Choquette and Pray (1970). Porosity types: BP interparticle, WP intraparticle, SH shelter, MO mold, VUG vug, FR Fracture.
[b] Interlayered with isopachous crystal crust.
[c] Dunham (1969).
[d] Kendall and Tucker (1983).
[e] Lohmann and Meyers (1977).

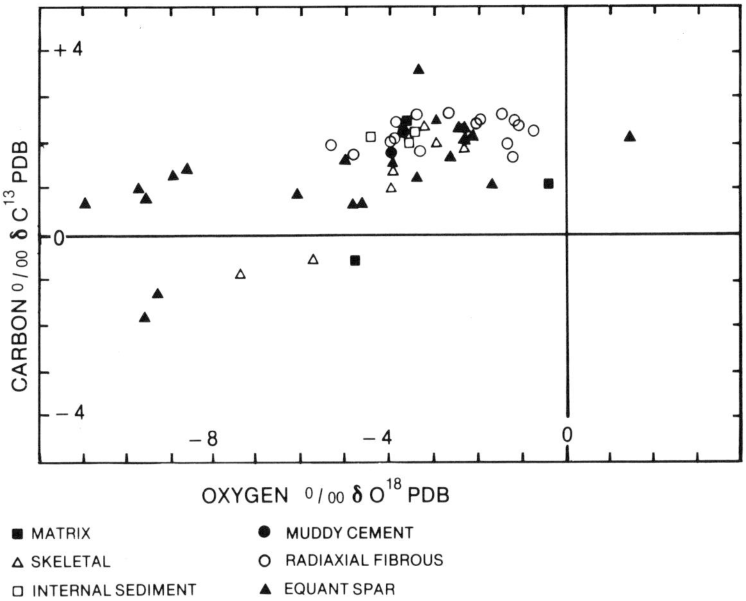

Fig. 4. Stable isotopic analyses from Taninul facies

ment (Table 2). Isotopic and minor-element composition of various diagenetic phases are given in Figure 4 and Tables 3 and 4.

3.1 Muddy Fabrics

In coarse-grained, relatively matrix-free rocks such as the platform-margin facies of the El Abra Formation, mud should present few problems. Ironically, the muddy fabrics are quite varied and difficult to categorize and interpret. Some end members can be readily recognized, but even the distinction between cement and internal sediment (cf. Choquette and Pray 1970) may involve considerable interpretation.

3.1.1 Microfossil Mud

In the simplest case, lime mud containing unbroken benthic microfossils and peloids overlies crusts of cement in primary or secondary pores. Most of it is otherwise indistinguishable from matrix mud or perched sediment in primary pores. However, some muds in vugs contain planktonic calcispheres or globigerinids (Fig. 5A) that have not been observed in the matrix of the Taninul facies, although Aguayo (1978, p. 87) has reported them from back-reef sections.

Microfossil mud is marine sediment that has filtered into open pores from the overlying depositional interface. It is distinguished from infiltrated primary sedi-

Table 3. Isotopic analyses El Abra platform – margin limestones (Taninul facies)

Component	All samples[a]					CA˙		TQ		LC		Shinn et al. (1974)[b]				
	$\delta^{18}O$	SD	$\delta^{13}C$	SD	N	$\delta^{18}O$	N	$\delta^{18}O$	N	$\delta^{18}O$	N	$\delta^{18}O$	SD	$\delta^{13}C$	SD	N
Skeletal, neomorphic	−3.4	±0.4	1.8	±0.6	3							−1.32	±0.69	2.67	±0.57	11
	(−2.9 /	−4.0)	(1.0 /	2.3)								(−2.2 /	−0.3)	(1.2 /	3.3)	
calcitic	−4.5	±2.1	0.8	±1.2	3											
	(−7.4 /	−2.3)	(−0.8 /	1.9)												
Matrix	−2.0		1.8		2							−1.6	±0.8	2.5	±0.1	8
	(−3.6 /	−0.4)	(1.1	/ 2.4)								(−2.5 /	0.0)	(2.3 /	2.7)	
Microfossil mud	−3.8	±0.4	2.2	±0.1	3											
	(−4.4 /	−3.4)	(2.0	/ 2.2)												
Muddy cement	−4.1	±0.6	1.2	±1.5	3											
	(−4.8 /	−3.7)	(−0.5 /	2.3)												
Isopachous crust	−2.67	±1.44	2.24	±0.30	15	−3.7	4	−2.96	7	−1.1	4	−1.63	±0.44	2.08	±0.52	12
	(−5.3 /	−0.7)	(1.7	/ 2.7)								(−2.3 /	−0.9)	(1.4 /	2.8)	
Equant spar	−5.17	±3.29	1.28	±1.21	23	−8.0	5	−6.24	9	−2.51	9	−2.50	±1.44	1.59	±1.01	12
	(−11.0 /	+1.5)	(−1.8 /	3.7)								(−6.8 /	−1.3)	(1.4 /	2.3)	

[a] In PPT (‰) relative to PDB standard. Range of values given in parentheses below mean.
[b] All samples from Laguna Colorada (LC).

Table 4. Minor element concentrations[a]

Component	Mg	Sr	# Analyses
Isopachous crust	0.96 ± 0.37	0.06 ± 0.04	23
Equant spar	0.87 ± 0.43	0.07 ± 0.03	18
Isopachous mud	0.73 ± 0.26	0.06 ± 0.06	4
Skeletal, neomorphic	0.95 ± 0.06	0.10 ± 0.07	4
Peloids	0.53	0.05	2

[a] Electron microprobe analyses, in mole percent of carbonate fraction. Mn, Fe, Ba, and Na were consistently below detection limits (500 ppm).

mentary matrix only where it overlies cement or occupies secondary pores. The origin of mud with planktonic microfossils is less obvious. Aguayo (1978, p. 78) reported a planktonic microfauna of Campanian age from solution cavities in the uppermost surface of the Taninul facies at Cementos Anahuac. The occurrences noted here are from as much as 20 m beneath the upper surface at Taninul quarry and are Cenomanian in age (Roy Waite, personal communication) or roughly contemporaneous with deposition. A pore network in communication with open marine water at the platform edge or supporting a cryptic fauna (Ginsburg et al. 1971b) might explain this occurrence. In one example where the disconformity surface could be recognized, mud infiltration was limited to less than a centimeter (Fig. 5C). However, the depth of infiltration undoubtedly depends on the size and tortuosity of the pore network and on wave and storm energy. Internal sediment in some secondary pores consists of skeletal and peloidal sand (Fig. 5B), indicating both larger pore connections and higher energy.

3.1.2 Isopachous Mud

Isopachous mud coats pore surfaces uniformly in defiance of gravity (Fig. 6). It also forms prominent bands interlayered with multiple isopachous layers of cloudy crystals (Figs. 5A, 6C). Such bands are difficult to distinguish from micritized rims on the crystals, but where terminations of crystals project into the mud, it was clearly an addition, not a modification (Fig. 6C). The mud may be vaguely layered, peloidal, or clotted with microrelief on the surface, unlikely fabrics in micrite rinds. Although isopachous mud lining secondary pores was not noted, it overlies cements within secondary pores. Skeletal material including highly micritized encrusting organisms, but more commonly bioclastic debris, is locally incorporated in the mud.

Isopachous mud is interpreted as cryptocrystalline cement deposited from marine pore waters (Land and Goreau 1970, Ginsburg et al. 1971a, Macintyre 1977, 1985, James and Ginsburg 1979, p. 113). The uniform coating of pore walls regardless of orientation indicates a cement that precipitated in water-saturated pores. A marine origin is suggested by the presence of a few encrusting organisms and bioclastic debris and by the intimate interlayering with crystalline cements interpreted as marine (see below). The isotopic composition (−3.7‰

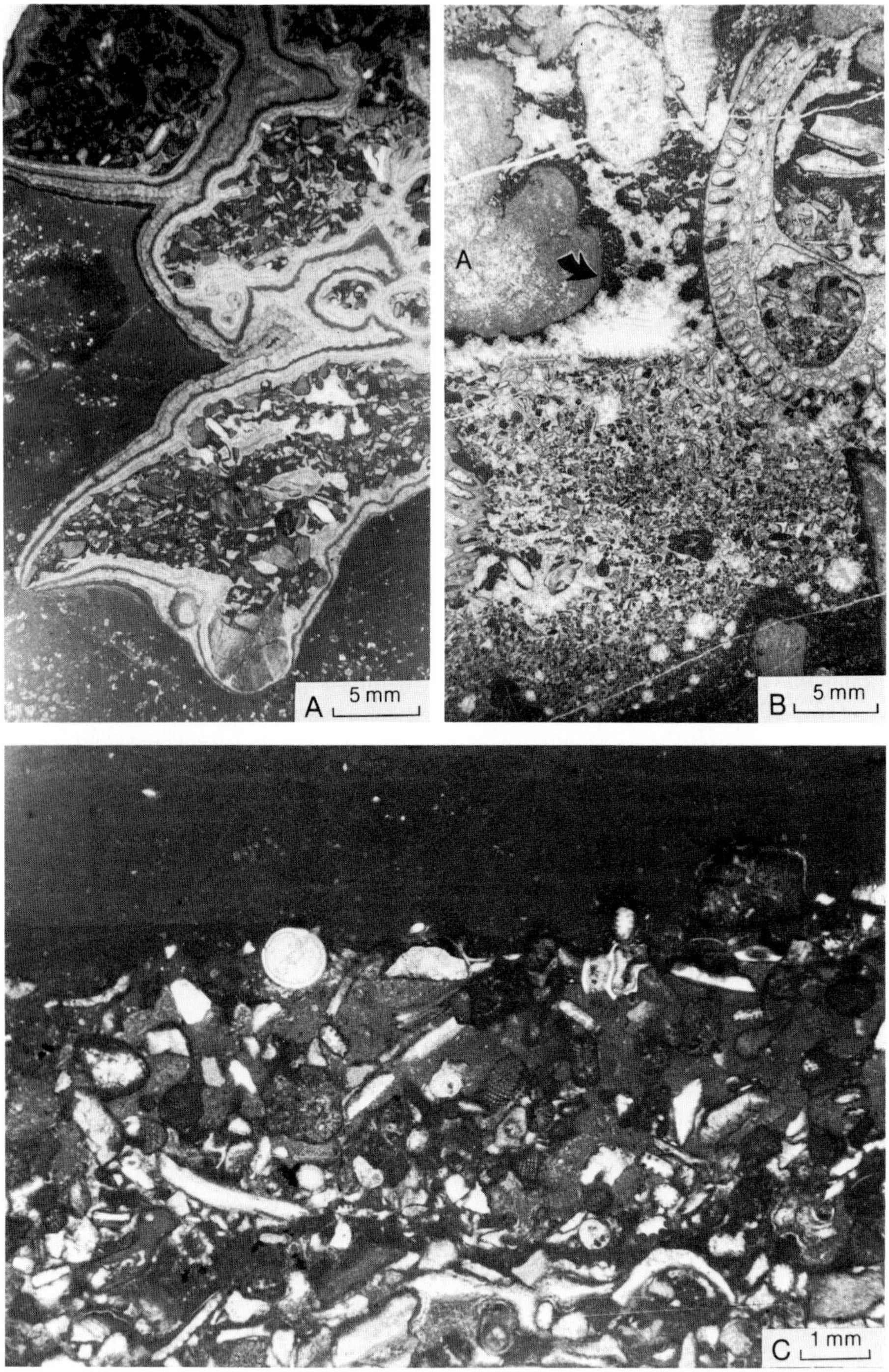

Fig. 5

$\delta^{18}O$, 2.3‰ $\delta^{13}C$, one sample) is very similar to that of other marine components (Table 3). Incorporation of detrital fragments and encrusting organisms, peloidal texture, and a vague scalloped layering, features of isopachous mud in the Taninul facies, are reported from modern examples interpreted as submarine cement (Macintyre 1977, 1985).

3.1.3 Patchy Mud

Patchy mud, for want of a better name, has a distinctly nonisopachous, erratic, distribution. Characteristic fabrics (Fig. 7 A) are grain-contact concentration, pendant lumps, narrow bridges between grains, and curved "meniscus" surfaces. Muddy cellular or reticulate structures in protected pores are difficult to distinguish from micritic walls of unidentified cellular encrusters (Fig. 8 A, B). Elongate tubes and circum-granular cracks that extend through the mud and around the grains are associated with the patchy mud (Fig. 8 C). Patchy mud very rarely contains fossils and is only vaguely clotted. It has been identified only in primary pores as the initial pore-reducing phase or overlying thin pendant crusts of cement (Fig. 7 A, B).

Distribution of the patchy mud indicates formation in pores containing both air and water (Dunham 1971, Müller 1971). Whether the water was marine or meteoric is best determined chemically, but no satisfactory analyses have been obtained. Minero (in press) showed that comparable cements in the adjacent El Abra backreef facies are downward extensions of caliche crusts formed on paleokarst. The patchy mud has many features of caliche (Fig. 8 C, D). Tube-like structures that extend through the interparticle mud and reticulate structures within protected pores indicate subaerial exposure and possible soil formation. The tubes may have formed by penetration of roots through the mud or by precipitation around roots as rhizoliths (Klappa 1980, Esteban and Klappa 1983, p. 25). The irregular size and shape of some suggests micropipes formed by selected dissolution of mud. The reticulate structures resemble alveolar texture (Esteban and Klappa 1983, p. 26), also formed by plant roots. Circum-granular cracks are considered indicative of caliche (Esteban and Klappa 1983, p. 30). Some patchy mud could be sediment plastered against the pore boundary by air-

Fig. 5 A – C. Dissolution and disconformity. **A** Large vug filled by microfossil mud and packstone, following reduction by multiple layers of isopachous mud and crystalline cement. Microfossils are planktonic, including *Hedbergella washitensis*, a Cenomanian guide fossil. Initial cementation of the rock was by patchy mud (cryptocrystalline cement). The initial isopachous crusts filled remnant interparticle as well as secondary porosity. Crusts thicken in concavities and thin over projections in the etched rock. TQ 26.6 m. **B** Vug formed by preferential dissolution of matrix. Irregular matrix adheres to bioclasts (*arrow*); micropipe is at upper right. Vug was reduced by laminated peloidal mud (base) and bioclastic grainstone containing ragged intraclasts of the matrix. Fill is by isopachous bladed crust and equant spar. Codiacian alga was also etched to form partial mold (*A*). LC 5.0 m. **C** Disconformity marked by microrelief and a dark crust that extends downward as point-contact cement, interpreted as caliche. Sediment similar to overlying mudstone, but more loosely packed, fills pores for a half centimeter beneath the surface. Pores at bottom of photomicrograph are only partly filled by mud; those a few millimeters lower are free of mud. Pendant cement is the initial pore-reducing material at that level. LC 21 m

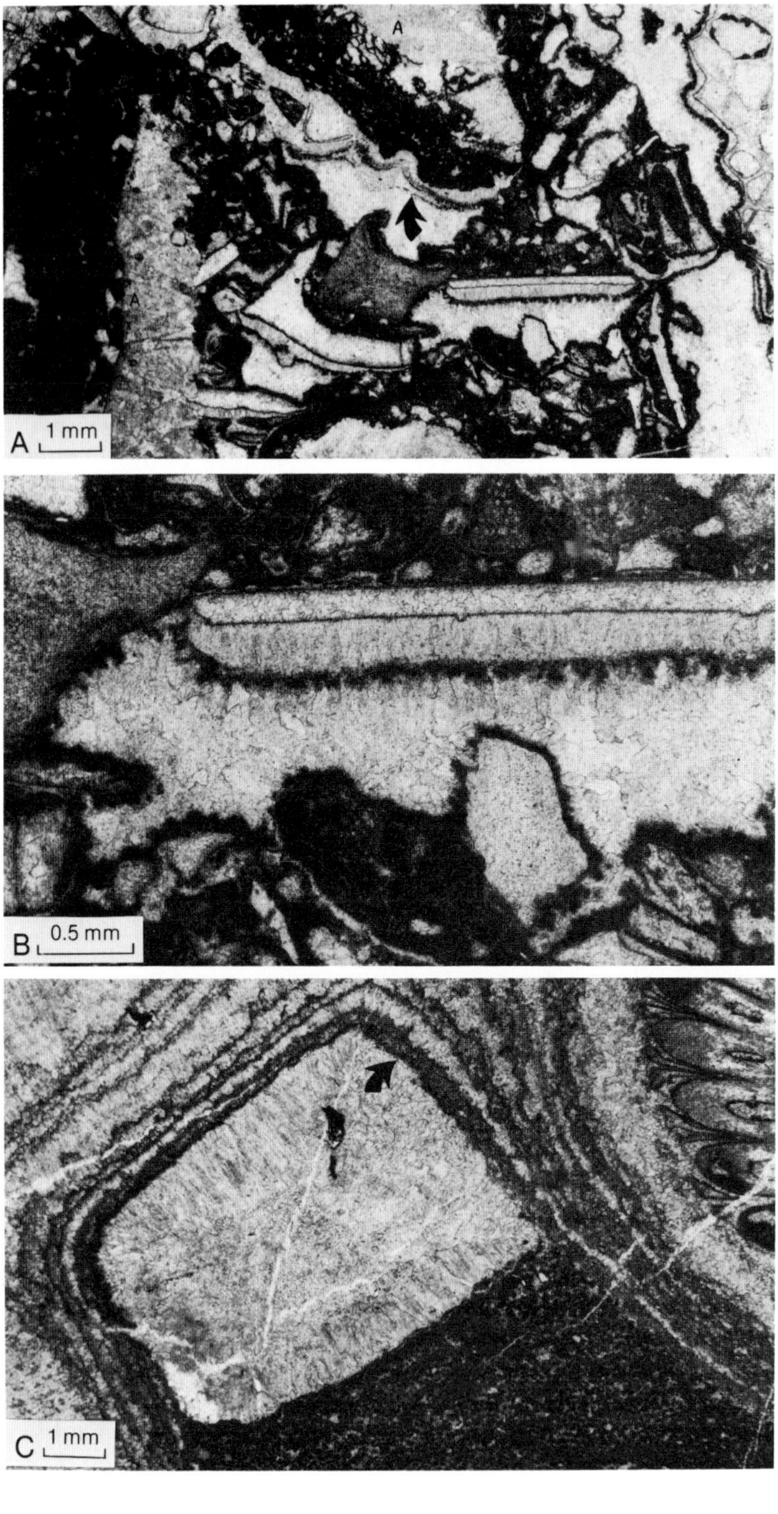

Fig. 6

water interfacial tension (Perkins 1977, Enos and Freeman 1978, p. 430). It would be difficult to distinguish cement from sediment, but both reflect vadose conditions.

3.1.4 Crystal Silt

Light-gray, unfossiliferous mud with uniformly silt-sized crystals occurs rarely as floors and perched or meniscus-shaped accumulations in intergranular pores and microvugs (Fig. 8E). These characteristics distinguish it from the cements and marine sediments described above; they are those ascribed by Dunham (1969) to vadose internal sediment.

3.1.5 Other Muddy Fabrics

The diagenetic mud fabrics described above are the end members; many intermediate examples with nondiagnostic or seemingly conflicting characteristics occur. Differential dissolution (see Sect. 3.3) has locally removed sedimentary matrix, particles, and diagenetic mud to produce a fragile patchwork of mud and bioclasts. Mud or cement may later fill the patchwork (Fig. 5A). Alternatively, erosion results in intraclasts of microfossil mud (Fig. 5B), bioclasts coated with microfossil or isopachous mud, and clasts with uneven coats of poorly laminated, porous, unfossiliferous mud (Fig. 8E) interpreted as caliche (patchy mud). Petroleum residues that are prevalent in Taninul quarry have selectively stained mud-coated intraclasts (Fig. 8E), presumably porous ones. This obscures the texture of the coatings, but helps differentiate them from the matrix. In a few cases, however, the oil stain extends into the matrix, blurring the distinction, or locally reversing the relationship.

3.2 Crystalline Cement Fabrics

Three distinct crystalline calcite fabrics are recognized, with some gradations: pendant crusts, isopachous crusts, and equant apar.

Fig. 6A – C. Early cements. **A** Skeletal packstone consisting of heavily encrusted large bioclasts (*A*) and perched infiltrated sediment. Initial lithification was by cloudy pendant cement, locally forming multiple crusts (*arrow*), followed by peloidal isopachous mud and cloudy cement that is isopachous in large pores. LC 1.8 m. **B** Detail of **A**. Pendant fibrous crust in shelter pore beneath platy skeletal fragment is succeeded by peloidal "isopachous" mud and cloudy bladed cement. Grains above and below shelter also show thin pendant crusts followed by pore-filling mud, probably that which developed as an isopachous layer in the larger pores. **C** Isopachous mud (*arrow*) sandwiched between fibrous cement crusts within solution-enlarged interparticle pore. Peloidal sediment with inclined layering overlies isopachous mud at the base of the pore. Muddy bands separating earlier cement crusts may be partly or entirely micritized rinds, but note the crystal terminations beneath the last layer in contrast to the smooth exterior (*arrow*). LC 49.2 m

Fig. 7

3.2.1 Pendant Crusts

Thin, pendant crusts of calcite are widespread in Laguna Colorada and portions of Taninul quarry (Figs. 6A, B, 7B). Although volumetrically insignificant, they commonly constitute the only evidence of a distinctive diagenetic environment. The fibrous or bladed crystals (Folk 1965) are clear to slightly cloudy with sweeping extinction indicating convergent optic axes. The thickness of the crusts varies considerably, even within single pores. The greastest thickness, up to 200 μm, is on the undersides of grains, as gravitational or pendant fabric (Purser 1969, Freeman 1971, Müller 1971, Enos and Freeman 1978). This cement has been seen only in primary pores (Table 2) and is invariably the first pore-reducing phase. It is probably more widespread than is recognized because, where it is not delineated by a micritic rind or a muddy layer, it is typically overlain in optical continuity by crystalline isopachous cement and cannot be differentiated. The pendant cements are nonluminescing, and iron content is below detection limits of potassium-ferricyanide stain. The crusts are too thin and discontinuous to be separated for isotopic analysis.

Pendant crusts are interpreted as vadose cements based on the point-contact, meniscus, and pendant fabrics that reflect the distribution of water in the vadose zone. Whether the water was marine or meteoric is more difficult to ascertain. Circumstantial evidence of marine origin is its occurrence as the initial diagenetic phase in marine sediment and succession, typically, by marine cements in optical continuity. Undulose extinction with convergent optic axes, generally considered indicative of neomorphic stabilization of a metastable fibrous precursor (Kendall and Tucker 1973), has recently been demonstrated in a primary marine cement (Kendall 1985). Rare pendant crusts in backreef intervals adjacent to Laguna Colorada have unit extinction and have been interpreted as meteoric phreatic products (Minero, in preparation). A marine vadose (intertidal) origin is favored for the pendant cements from the platform margin, but the evidence is equivocal.

3.2.2 Isopachous Crusts

The most voluminous cement in the Taninul facies is coarsely crystalline fibrous-to-bladed calcite of fairly uniform thickness (isopachous) where pores are incompletely filled (Figs. 6C, 8A). In large pores many successive isopachous crusts, separated by micritic rinds, isopachous muddy cement, or internal sediment,

Fig. 7A – D. Disconformity and related cements, TQ 39.9 m. **A** Patchy mud lining interparticle pores, concentrated at grain contacts, curved (meniscus?) boundaries (*center*), and "bridges" between grains (*bottom*). Very thin pendant crust locally underlies patchy mud (*arrows*); bladed crust overlies it. **B** Pendant cement showing point-contact concentration and curved meniscus boundaries (*arrows*). Very sparse patchy mud and cloudy bladed crust succeeded pendant cement. **C** Thin section from sample in **D**, showing boring and infiltration by overlying peloidal sediment. Sediment has also infiltrated skeletal molds (*bottom*) and interparticle pores rimmed by pendant cement, patchy mud, and bladed isopachous crust (**A, B**). Skeletal molds, top of boring, and fractures are filled by equant spar. **D** Disconformity cutting bioclastic limestone, bored by clams (*arrows*) and larger organisms. Marine effects are overprinted on earlier intertidal and meteoric diagenesis (**A, B**). Slab

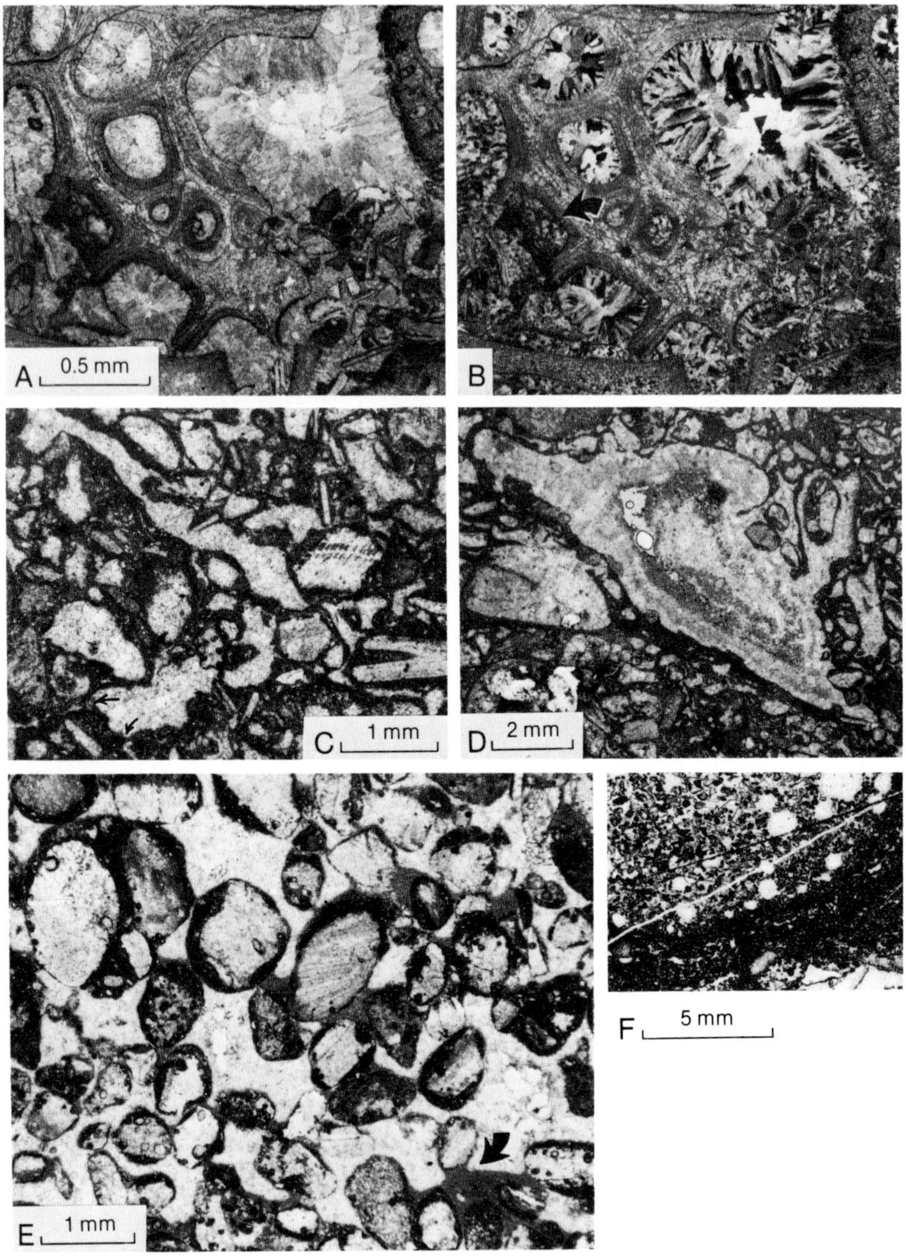

Fig. 8 A – F. Cements, dissolution, and internal sediments. **A** Bladed isopachous cement crust in skeletal packstone. Clear equant spar fills pore at *upper right.* LC 50.7 m. Plane-polarized light. **B** Cross-polarized light, showing undulose extinction. Cellular micrite within broken caprinid pallial canal (*arrow*) is probably cryptic encrusting skeleton. **C** Tube-shaped pores within patchy mud. Mud surrounding the pore extending from upper left is vaguely layered. Numerous small, discontinuous cracks cut the mud (*arrows*); locally these are circum-granular. The patchy, porous, unfossiliferous mud is interpreted as cement formed in a vadose environment, essentially a caliche. Dissolution is also prominent in this sample (**D**). TQ 26.1 m. **D** Vugs formed by dissolution of skeletal grains and

form flamboyant mosaics (Figs. 5A, 6C). Crystal boundaries are wavy and terminations are low-angle rhombohedra except where they are truncated by boring or dissolution or grade into an overlying equant spar. Extinction is undulose and indicative of convergent optic axes in the subcrystals (Fig. 8B). Cleavages traces are concave outward. These are the attributes of radiaxial fibrous mosaic (Bathhurst 1971, p. 426), although they can be seen only in large pores with favorable cuts. Elsewhere, the cements are identified by continuity into the large pores, on the basis of undulose extinction, and on the extreme cloudiness of the crystals. The persistent cloudiness results from numerous inclusions, many of which are microdolomite rhombs (Lohman and Meyers 1977), identified by staining with Clayton yellow and etching polished surfaces. Iron concentrations are too low to detect with stains or the electron microprobe. Cathodoluminescence is absent to very dull red indicating low Mn concentrations, as iron is also low. Despite the microdolomite inclusions, Mg contents are only about 1 mole percent (Table 3), which is not significantly higher than in clear cements. Isopachous cement produced the initial reduction in many primary pores (Table 2), but elsewhere it succeded pendant or cryptocrystalline cements (patchy or isopachous mud). It is not uncommon in molds and vugs, typically as the initial cement phase.

Isopachous crusts are the most diagnostic cements in the Taninul facies. Radiaxial fibrous mosaic, prevalent in the isopachous crusts, is generally considered to be practically synonymous with neomorphically altered marine cement because of sedimentologic evidence (cf. Purser 1969, Davies 1977), modern analogs (cf. Shinn 1969), geochemical signatures (cf. Davies 1977, Lohmann and Meyers 1977) and petrographic arguments (Kendall and Tucker 1973). Recently, however, Kendall (1985) has shown that radiaxial fibrous mosaic can be a primary fabric developed in calcite cement. Although the marine origin of radiaxial fibrous mosaic was not questioned, this fabric should be used only in conjunction with other evidence. The evidence of marine origin in the El Abra rocks includes interlayering with marine internal sediment and isotopic compositions comparable with other marine components (Table 3). Microdolomite inclusions in the crusts suggest original magnesium calcite, a characteristic marine mineralogy, although the present low magnesium content (Table 4) is a puzzle.

3.2.3 Equant Spar

The final pore-reducing cement is invariably transparent or white, equant spar that is relatively free of inclusions (Figs. 5B, 7C, 8A). Crystal boundaries are

possibly, matrix. Initial lithification was by patchy mud (caliche; not readily distinguishable from muddy matrix, if any) with tubular and meniscus fabrics that supported peninsular grains and very fragile boundaries of molds and vugs after dissolution. Vugs are filled by bladed isopachous crusts. Holes are flaws in thin section. TQ 26.1 m. **E** Crystal silt perched in intergranular pores. Note curved boundaries (*arrow*). Coatings on grains are interpreted as reworked caliche; the dark color is in part oil stain. Overlying cement is bladed isopachous crust. TQ 29.8 m. **F** Detail of Fig. 5B, showing round pores within grainstone fill of a vug. These are interpreted as keystone vugs (Dunham 1970), suggesting that the large vug was filled in the intertidal zone amid crashing waves. LC 5 m linear or curving compromise traces; terminations are obtuse rhombohedra where pore filling is incomplete. Extinction is sharp. Equant spar overlies isopachous crusts with markedly contrasting morphology, clarity, and extinction, yet the spar is typically syntaxial on the isopachous cement, and the contact is gradational over a few micrometers. Chemically, equant spar is surprisingly similar to the isopachous crusts and other components of the rock (Tables 3 and 4), although oxygen isotopic composition is generally depleted in ^{18}O and is more variable (Fig. 4). It does not luminesce except for bright-orange banding in the latest portions of a few crystals. Preliminary results do not indicate any relationship between luminescence and the extremes of isotopic composition. Equant spar is the final cement in primary pores, where space was left by earlier cements. Only one generation can be recognized petrographically; the few examples of late luminescent overgrowths and the wide range of oxygen isotopic values (12.5‰, Table 3), are the only indications of inhomogeneity.

Compromise crystal boundaries and euhedral terminations indicate precipitation in water-saturated pores. Relatively heavy oxygen isotopes exclude elevated temperatures and suggest only slightly depleted meteoric waters such as might occur in coastal areas. Spars depleted in ^{18}O may represent later additions under burial or more continental conditions. Low concentrations of minor elements (Table 4) are also consistent with low temperatures and meteoric waters. The limited magnesium and strontium probably derived from reactions of the water with marine components of the rock.

Cement in fractures has the same characteristics, including lack of luminescence, and merges with the equant cement in pores cut by the fractures. Most of the fractures are extensional, poorly oriented, and predate stylolites, even in the structurally deformed Laguna Colorada area. This suggests that fractures in the Taninul facies originated as early joints, like those in Pleistocene limestones of Andros Island, Bahamas (Newell and Rigby 1957, p. 28 – 30), and were filled with shallow phreatic cement.

3.3 Porosity Enlargement

The formation of molds and vugs enhanced porosity at various times in the diagenetic sequence at Taninul and Cementos Anahuac quarries. Considerable secondary porosity still exists. In contrast, secondary porosity was never volumetrically important at Laguna Colorada. Solution enlargement of joints in the Sierra de El Abra has produced small caves, apparently related to the present cycle of erosion and therefore not considered here.

Molds are sporadically distributed throughout the Taninul quarry and Cementos Anahuac sections. Molds formed primarily at the expense of caprinid shells (Fig. 2B) the largest, most numerous, and, with original aragonitic composition, among the most soluble of skeletal materials. The longitudinal pallial canals of the caprinids were shelters favorable for cryptic encrusters, micritization, and precipitation of submarine cement, increasing chances of preservation as the surrounding shell was leached away. Other originally aragonitic shells commonly represented by partial or complete molds are gastropods, corals, codiacian algae (Fig. 5 B) and the stromatoporoid, *Parkeria*.

Vugs formed by enlargement of molds (Fig. 8 D), as well as selective dissolution of mud. Dissolution of mud, both sedimentary and diagenetic, produced highly irregular pore margins with overhanging projections of mud (Fig. 5 B), solution pipes (Fig. 8 C), and grains supported by peninsular projections of mud (Fig. 5 A, C, 8 D). Microvugs were volumetrically important only in a few zones, but they record subtle dissolution episodes and locally influenced subsequent diagenesis.

All dissolution in the platform-margin facies is interpreted as the result of near-surface meteoric diagenesis. Carbonate dissolution on the shallow sea floor where skeletons and marine cements were forming in profusion is highly improbable. Even the Cementos Anahuac section, deposited in part at depths of 270 m, was the site of extensive submarine cementation. Discounting shallow-marine dissolution, although obvious, is important because much of the early secondary porosity was subsequently reduced by marine diagenesis. Dissolution in meteoric lenses, initially established in small backreef islands (Minero et al. 1983, Minero, in press) but extending toward the reefs and the lagoon during lower sea-level stands, is more plausible than wholesale changes in seawater chemistry, especially in view of the numerous discontinuities within the shallow-water sections (Fig. 3, Shinn et al. 1974, Aguayo 1978, Minero et al. 1983, Minero, in press). Erosion of the fragile microvug intervals produced intraclasts of mud and of mud-coated grains from the irrregular mud projections and the peninsular grains. The intraclasts were probably eroded and rounded during transgression of the disconformity surfaces.

Virtually all samples of the Taninul facies show some evidence of dissolution. In addition to the early episodes of dissolution overprinted by marine diagenesis, a later period of dissolution preceded equant-spar emplacement. This more pervasive dissolution may have resulted from near-surface meteoric water or from burial diagenesis. Petrographic evidence argues against burial; no crushed grains or sutured contacts were observed. Extensive early cementation may have produced a rock sufficiently rigid to undergo dissolution at depth without compaction, but where fractures extend without any signs of adjustment across molds that were open at the time of fracturing (Fig. 7 C), near-surface conditions appear more likely. This is consistent with the shallow meteoric origin proposed for the subsequent equant spar.

4 Diagenetic Sequences

The diagenetic fabrics can be grouped into three distinctive sequences that reflect differences in the detailed diagenetic histories of the rocks involved. Each sequence is characterized by the environment that imparted the initial diagenetic fabric to the sediment, as seems appropriate in rocks where early diagenesis was paramount. The sequences recognized are: (1) initially marine phreatic, (2) initially marine vadose, and (3) initially meteoric. The different diagenetic se-

Table 5. Diagenetic sequences[a]

	TQ	LC	CA
Initially marine phreatic	23	23	75
Initially marine vadose	4	46	–
Initially meteoric	64	31	25
Meteoric only	9	–	–

[a] Percentage of samples from each section (Fig. 1) that belongs to a given sequence.

quences are interlayered in each section (Fig. 3), but in quite different proportions (Table 5).

4.1 Marine Phreatic

This sequence is the simplest in that it begins with marine sediments that were affected first by marine diagenesis and subsequently by meteoric events reflecting emergence. Cloudy isopachous cement, typically radiaxial fibrous mosaic, may fill the finer primary porosity. In larger pores the initial cement may be overlain by meteoric equant spar or by additional isopachous crusts, microfossil mud, and isopachous muddy cement of marine origin. These sequences record as many as ten episodes (Fig. 6C) that cannot be correlated within adjacent samples (Fig. 8A) separated by about a meter of section. The cements presumably formed very near the sediment/water interface, judging from modern examples (cf. James and Choquette 1983), and were stimulated or interrupted by surface conditions such as periods of nondeposition or storms. The cementation of a given layer was probably completed before deposition of the layer a meter above.

The marine cements, although generally isopachous within a field of view, do not in fact extend uniformly through all pores in a single sample. Inner whorls of gastropod cavities or the inner rows of caprinid pallial canals locally contain equant spar but little or no marine cement. Apparently these most sheltered pores were closed to marine cement, although not to later meteoric cement, in exception to Alexandersson's (1972) principle that sheltered pores are the preferred locus for growth of marine cement.

4.2 Marine Vadose

The initial diagenetic episode in this sequence was precipitation of a thin crust of pendant cement interpreted as a marine precipitate during periodic emergence. In most cases the succeeding phase was isopachous marine cement. Sediment was very lightly cemented at sea level and subsequently extensively cemented in a submarine phreatic environment. Elsewhere exposure diagenesis was more extensive and included dissolution (Fig. 7C).

4.3 Meteoric

Initial meteoric effects may be dissolution, cementation, or internal sedimentation. Dissolution of skeletal grains formed partial molds that subsequently received marine cement in most cases. Dissolution of matrix in fine-grained intervals produced microscopic pipes, irregular intergranular pore boundaries, and peninsular grains. Initial meteoric cementation was limited to patchy mud with point-contact and meniscus distribution, indicating a vadose environment. Vadose silt was encountered locally in small primary or secondary pores. Initial meteoric diagenesis was followed by marine cementation in virtually all samples from the entire Taninul facies (Table 5).

5 Discussion

The volumetrically most important cements in the Taninul facies, the platform margin of the El Abra Formation, are isopachous crusts of submarine origin. However, diagenetic effects related to exposure and subsequent addition of the dominant submarine cement are pervasive. Later meteoric dissolution produced major secondary porosity in TQ and CA sections, but only minor effects in LC. Equant spar further reduced remnant primary and secondary porosity.

In the shallow platform-edge section at Taninul quarry, sea level fluctuated near the depositional interface to produce varied and irregularly alternating diagenetic sequences (Fig. 3). In two-thirds of the samples, the initial diagenetic phase involved meteoric waters. None of the many disconformities in this section (Fig. 3) preserves unequivocal evidence of subaerial origin at the outcrop, such as the paleokarst and laminated crusts developed in the back-reef area (Minero, in press). Petrographic evidence of initial subaerial exposure of many of them, however, suggests that some of the back-reef discontinuities probably extended to the slope break. At this slightly lower elevation, relative to the back-reef rubble islands, the duration of exposure should have been briefer and the expression of the surfaces weaker. Exposure to higher energy conditions at the platform edge is another factor in the muted record of subaerial exposure; marine erosion may have removed the vestiges of exposure that did not extend below the surface. Intraclasts coated with caliche and fossiliferous mud are evidence of this erosion. The marine erosional disconformities superposed on the subaerial exposure surfaces are marked by encrustation and boring (Figs. 7C, D).

Secondary porosity developed during early exposure was reduced by marine cement or internal sediment as marine waters transgressed the exposure surface. Dissolution that preceded submarine diagenesis must have been repeated episodes, because interspersed intervals of the section lack evidence of early dissolution. Moreover, marine cementation and internal sedimentation probably extended less than a meter below the sea floor (Fig. 5C). Repeated periods of exposure of sufficient duration and magnitude to produce some dissolution are indicated. Freshwater penetration into the reefs necessitates at least a slight drop in relative sea level.

Where meteoric diagenesis, necessarily post mortem, was the first effect, subsequent submarine cementation could not have created the primary wave-resistant structure during reef growth, as has been proposed in the Permian Capitan reef, for example (cf. Yurewicz 1977, Babcock 1977). Except for two thin lenses of rounded, cement-filled bioclasts, no examples of reworked marine-cemented clasts were encountered. On the other hand, transported clusters of radiolitid rudists are not rare, indicating that organic binding may have been more important.

Later, more pervasive dissolution produced extensive molds and some vugs that locally cut marine cement. Some of these vugs contain marine internal sediment, indicating yet another marine incursion, but most were reduced by equant spar. Pore reduction by equant cement is comparable volumetrically to that by submarine cement in TQ and CA sections, although significant secondary porosity and minor primary porosity remain. A complicated pattern of oil stains and tar residues records subsequent oil emplacement in TQ.

The diagenetic history at Laguna Colorada, although dominated by submarine cementation, typically started with minor cementation in an intertidal setting (Table 5). The early pendant cement may actually be more extensive than is apparent; where overlain by radiaxial fibrous cement without an intervening muddy fabric, the two are syntaxial and the early pendant crust cannot be recognized. The pendant cement indicates that the coarse bioclastic debris that constitutes most of the LC section was deposited in a regime with frequent exposure, perhaps a reef flat (see Sect. 2). Keystone vugs (Dunham 1970) noted at two levels (Fig. 8F) indicate periodic exposure and strong surf action. Some disconformities in the LC section apparently formed during the early period of exposure (Fig. 5C), but others cut the isopachous crusts of marine cement, indicating that they formed later. Definitive criteria for either submarine or subaerial processes are generally lacking.

The Cementos Anahuac section shows no disconformities nor initial marine vadose sequences, which is not surprising in deposits of the forereef slope 270 m below the shelf break. However, it does contain at least five stratigraphic intervals with dissolution followed by submarine cementation. How fresh water penetrated to this level repeatedly during El Abra deposition remains unsolved. The drops in sea level that exposed the reefs seem unlikely to have approached this magnitude, as do the consequent freshwater lenses. This section was subsequently extensively flushed by meteoric water to produce more secondary porosity, much of which remains despite reduction by equant spar.

Later dissolution, post-dating marine diagenesis, affected all sections to a greater (TQ, CA) or lesser (LC) degree, prior to, or overlapping, precipitation of equant spar. This more pervasive dissolution is probably the result of a freshwater lens that extended throughout the presently exposed intervals. There is evidence of prolonged exposure at the end of El Abra deposition, generally considered the end of the Cenomanian (cf. Coogan 1973), but possibly late Turonian (Aguayo 1978, p. 87). Cavernous porosity in the Golden Lane was probably developed at this time. Cavernous porosity of this age is unknown from the Valles platform, but the Aqua Nueva Formation (Turonian) is absent between the El Abra and San Felipe Formations (Santonian-Coniacian) at the eastern edge of the Sierra de El Abra and locally within the Sierra Madre Oriental, suggesting a major period of exposure. Aguayo's (1978, p. 87) discovery of Campanian microfossils in vugs at the top of the Taninul indicates even more prolonged exposure. This is the most likely time for the development of pervasive solution porosity at the top of the Taninul facies and emplacement of equant spar.

The oxygen isotopic composition of the equant spar is appreciably lighter in the TQ and CA sections that at LC (Table 3); burial and tectonism in the Sierra Madre produced no late isotopic reequilibration. If anything, the TQ and CA sections must have been buried more deeply during cementation or were exposed to meteoric waters more depleted in ^{18}O, perhaps through development of a larger freshwater lens. Minero has shown that burial diagenesis and deformation preceded late spar emplacement in the back-reef area near LC (Minero et al. 1983, Minero, in preparation), but there is no evidence of burial effects or a late cement phase in the Taninul facies. Virtually the entire diagenetic alteration of the Taninul facies apparently occurred near the surface and produced a massive, well-lithified rock that underwent shallow burial and orogeny without further modification.

6 Summary

The Taninul facies consists of rudist reefs and skeletal detritus that bordered mid-Cretaceous shallow-water platforms in eastern Mexico. The upper Taninul facies, exposed at the eastern margin of the Valles platform, was extensively cemented by isopachous crusts of fibrous submarine cement. However, subaerial cementation or dissolution commonly preceded marine cementation. Leaching of skeletal components and, locally, matrix during later exposure resulted in extensive secondary porosity. Equant spar, interpreted as near-surface meteoric cement, followed marine cementation, filling most of the remaining primary pores, and reducing secondary porosity. At this stage of the rock was sufficiently stable and rigid to withstand shallow burial and mountain building during the Late Cretaceous and early Tertiary with no detectable diagenetic alteration.

Three distinctive diagenetic sequences reflect differences in the earliest diagenetic history of the rocks: (1) initial marine phreatic (submarine cementation), (2) initial marine vadose (intertidal cementation), and (3) initial meteoric (dissolution, vadose cementation). Interlayering of these sequences in shallow platform-margin sections shows that sea level fluctuated about the depositional interface causing many periods of exposure. Even the slope facies shows some intervals with initial meteoric diagenesis. Submarine cementation did not generally contribute to forming wave-resistant reefs, because it was commonly emplaced only after initial exposure, thus after the demise of the reef.

Acknowledgments. Thanks to Johannes H. Schroeder and Bruce H. Purser for the opportunity to write this paper and for copious editing. Charles Minero's participation from field work to editing was invaluable. Dennis Presbindowski and Sam Savin provided isotopic analyses. Teresa Wilson and Betty Nemecheck did the typing. National Science Foundation (EAR 8026844) and Kansas General Research Fund (3053-20-0038) supported the research.

References

Aguayo C, JE (1978) Sedimentary environments and diagenesis of a Cretaceous reef complex, eastern Mexico. An Centro Cienc Mr Limnol, Univ Nac Auton Mex 5:83 – 140
Alexandersson T (1972) Intergranular growth of marine aragonite and Mg-calcite: Evidence of precipitation from super-saturated sea water. J Sediment Petrol 42:441 – 460
Babcock JA (1977) Calcareous algae, organic boundstones, and the genesis of the upper Capitan Limestone (Permian, Guadalupian) Guadalupe Mountains, West Texas and New Mexico. In: Upper Guadalupian facies, Permian reef complex, Guadalupe Montains, New Mexico and West Texas. Soc Econ Paleontol Miner Permian Basin Sect Publ 77-16:3 – 44
Bathurst RGC (1971) Carbonate sediments and their diagenesis. Elsevier, Amsterdam, p 620
Bonet F (1952) La facies Urgoniana del Cretácico Medio en la region de Tampico. Bol Asoc Mex Geol Pet 4:153 – 262
Carrillo B, J (1971) La plataforma Valles-San Luis Potosí. Bol Asoc Mex Geol Pet 23:1 – 113
Choquette PW, Pray LC (1970) Geologic nomenclature and classification of porosity in sedimentary carbonates. Am Assoc Petrol Geol Bull 54:207 – 250
Coogan AH (1973) New rudists from the Albian and Cenomanian of Mexico and south Texas. Rev Inst Mex Pet 5:51 – 83
Davies GR (1977) Former magnesian calcite and aragonite submarine cements in Upper Paleozoic reefs of the Canadian Arctic: a summary. Geology (Boulder) 5:11 – 15
Duham RJ (1969) Early vadose silt in Townsend mound (reef) New Mexico. In: Friedman GM (ed) Depositional environments in carbonate rocks. Soc Econ Paleontol Miner Spec Publ 14:139 – 181
Dunham RJ (1970) Keystone vugs in carbonate beach deposits (abs). Am Assoc Petrol Geol Bull 54:845
Dunham RJ (1971) Meniscus cement. In: Bricker OP (ed) Carbonate cements. Johns Hopkins Univ Stud Geol 19:297 – 300
Enos Paul (1974) Reefs, platforms, and basins of middle Cretaceous in northeast Mexico. Am Assoc Petrol Geol Bull 58:800 – 809
Enos Paul (1977) Tamabra limestone of the Poza Rica trend, Cretaceous, Mexico. In: Cook HE, Enos Paul (eds) Deep-water carbonate environments. Soc Econ Paleontol Miner Spec Publ 25:273 – 314
Enos Paul (1980) Lower Cretaceous reefs and sea-level changes in the Atlantic and Gulf Coasts (abs). Geol Soc Am Abs 12:422
Enos Paul (1982) Basin-to-platform transition, middle Cretaceous, Mexico (abs). Am Assoc Petrol Geol Bull 66:567
Enos Paul (1983) Late Mesozoic paleogeography of Mexico. In: Reynolds MW, Dolly ED (eds) Mesozoic paleography of west-central United States. Soc Econ Paleontol Miner Rocky Mountain Sect Symp 2:133 – 157
Enos Paul, Freeman T (1978) Shallow-water limestones from the Blake Nose, Sites 390 and 392. In: Benson WE, Sheridan RE et al. Init Repts D S D P 44:413 – 461
Esteban M, Klappa CF (1983) Subaerial exposure. In: Scholle PA, Bebout DG, Moore CH (eds) Carbonate depositional environments. Am Assoc Petrol Geol Mem 33:1 – 54
Folk RL (1965) Some aspects of recrystallization in ancient limestones. In: Pray LC, Murray RC (eds) Dolomitization and limestone diagenesis. Soc Econ Paleontol Miner Spec Publ 13:14 – 48
Freeman T (1971) Morphology and composition of Ordovician vadose cement. Nature Phys Sci 233:133 – 134
Ginsburg RN, Marszalek DS, Schneidermann N (1971a) Ultrastructure of carbonate cements in a Holocene algal reef of Bermuda. J Sediment Petrol 41:472 – 482
Ginsburg RN, Schroeder JH, Shinn EA (1971b) Recent synsedimentary cementation in subtidal Bermuda reefs. In: Bricker OP (ed) Carbonate cements. Johns Hopkins Univ Stud Geol 19:54 – 59
Griffith LS, Pitcher MG, Rice GW (1969) Quantitative environmental analysis of a Lower Cretaceous reef complex. In: Friedman GM (ed) Depositional environments in carbonate rocks. Soc Econ Paleontol Miner Spec Publ 14:120 – 138
Harris PM (1979) Facies anatomy and diagenesis of a Bahamian ooid sand shoal. Sedimenta VII, Comparative Sed Lab, Univ Miami, Miami Beach, Florida
James NP, Choquette PW (1983) Limestones – the sea floor diagenetic environment. Geosci Can 10:162 – 179

James NP, Ginsburg RN (1979) The seaward margin of the Belize barrier and atoll reefs. Int Assoc Sediment Spec Publ 3:1 – 191

Kellum LB (1930) Similarity of surface geology in front range of Sierra Madre Oriental to subsurface in Mexican south fields. Am Assoc Petrol Geol Bull 14:73 – 91

Kendall AC (1985) Radiaxial fibrous calcite: a reappraisal. In: Schneidermann N, Harris PM (eds) Carbonate cements. Soc Econ Paleontol Miner Spec Publ 36:59 – 77

Kendall AC, Tucker ME (1973) Radiaxial fibrous calcite: a replacement after acicular carbonate. Sedimentology 20:365 – 389

Klappa CF (1980) Rhizoliths in terrestrial carbonates; classification, recognition, genesis and significance. Sedimentology 276:613 – 629

Land LS, Goreau TF (1970) Submarine lithification of Jamaican reefs. J Sediment Petrol 40:457 – 462

Lohmann KC, Meyers WJ (1977) Microdolomite inclusions in cloudy prismatic calcites: a proposed criterion for former high-magnesium calcites. J Sediment Petrol 47:1078 – 1088

Macintyre IG (1977) Distribution of submarine cements in a modern Caribbean fringing reef, Galeta Point, Panama. J Sediment Petrol 47:503 – 516

Macintyre IG (1985) The peloidal question. In: Schneidermann N, Harris PM (eds) Carbonate cements. Soc Econ Paleontol Miner Spec Publ 36:109 – 118

Minero CJ (in press) Sedimentation and diagenesis along an island-sheltered platform margin, El Abra Formation, Cretaceous of Mexico. In: James NP, Choquette PW (eds) Paleokarst. Soc Econ Paleontol Miner Spec Publ

Minero CJ, Enos Paul, Aguayo CJE (1983) Sedimentation and diagenesis of mid-Cretaceous platform margin, east-central Mexico, with accompanying field guide. Dallas Geol Soc, Dallas

Muir JM (1936) Geology of the Tampico Region, Mexico. Am Assoc Pet Geol Tulsa

Müller G (1971) "'Gravitational" cement: an indicator for the vadose zone of the subaerial diagenesis environment. In: Bricker OP (ed) Carbonate cements. Johns Hopkins Univ Stud Geol 19:183 – 192

Newell ND, Rigby JK (1957) Geological studies on the Great Bahama Bank. In: Le Blanc RJ, Breeding JG (eds) Regional aspects of carbonate deposition. Soc Econ Paleontol Miner Spec Publ 5:15 – 72

Perkins RD (1977) Depositional framework of Pleistocene rocks in south Florida. In: Enos Paul, Perkins RD, Quaternary sedimentation in south Florida. Geol Soc Am Mem 147:131 – 198

Purser BH (1969) Syn-sedimentary marine lithification of Middle Jurassic limestones in the Paris Basin. Sedimentology 12:205 – 230

Shinn EA (1969) Submarine lithification of Holocene carbonate sediment in the Persian Gulf. Sedimentology 12:109 – 144

Shinn EA, Bloxsom WE, Lloyd RM (1974) Recognition of submarine cements in Cretaceous reef limestones from Texas and Mexico (abs). Am Assoc Petrol Geol Abs 1:82 – 83

Suter M (1984) Cordilleran deformations along the eastern edge of the Valles-San Luis Potosícarbonate platform, Sierra Madre Oriental fold-thrust belt, east-central Mexico. Geol Soc Am Bull 95:1387 – 1397

Ward JA (1979) Stratigraphy, depositional environments and diagenesis of the El Doctor platform, Querétaro, Mexico. PhD Thesis, State Univ New York Binghamton

Wilson BW, Hernandez M, JP, Meave TE (1955) Un banco calizo del Cretácico en la parte oriental del Estado de Querétaro, México. Bol Soc Geol Mex 18:1 – 10

Yurewicz DA (1977) Origin of the massive facies of the lower and middle Capitan Limestone (Permian), Guadalupe Mountains, New Mexico and West Texas. In: Upper Guadalupian facies, Permian reef complex, Guadalupe Mountains, New Mexico and West Texas. Soc Econ Paleontol Miner Permian Basin Sect Publ 77-16:45 – 92

A Comparative Study of the Diagenesis in Diapir-Influenced Reef Atolls and a Fault Block Reef Platform in the Late Albian of the Vasco-Cantabrian Basin (Northern Spain)

J. Reitner[1]

1 Introduction

The Mesozoic Vasco-Cantabrian Basin is situated in the Spanish Basque-Lands of northern Spain, close to the French border (Fig. 1). To the west the basin is delineated by the Paleozoic massifs of the Cantabrian Mountains, to the east by the Pyrenees, and to the south by the Late Tertiary Ebro Basin. The tectonic history of the Vasco-Cantabrian ranges from the Triassic to the Tertiary but the major part of its fill (8000 m) is attributed to Late Mesozoic deposition. During the Late Albian due to rapid subsidence, salt diapirs (Keuper age) developed between basement fault blocks. Shallow marine carbonate reefs (Urgonian Facies) developed on these highs. Two types of reef can be distinguished: (1) larger fault block reef platforms on tilted basement segments and (2) smaller reef atolls on top of rising diapirs. The diapir reefs can be differentiated from the fault block reefs through sedimentological, paleobiological, diagenetic, and geochemical criteria.

The purpose of this paper is to present the different diagenetic phases in diapir reefs and fault block reef development. To achieve this, the respective paleobiological and sedimentological frameworks will be described first and the petrographic and geochemical aspects of the diagenetic sequences will be treated subsequently.

2 Previous Studies

The first detailed work on the "Keuper" diapirs of northern Spain by Lotze (1953) did not include facies analysis. Brinkmann et al. (1967) studied some of these diapirs in greater detail using facies and structure analysis. Von Stackelberg (1967) first suggested that the small-sized carbonate reefs of the Late Campanian Oro Limestone situated above the Murguía diapir may have been indirectly linked to the diapir. Other similar smaller carbonate reefs also exist near the Villasana de Mena diapir of the Vasco-Cantabrian Basin. These latter reefs are especially interesting since the presence of large quantities of reworked Keuper material indicates a collapse event in the Villasana de Mena diapir during the Late Albian (Vraconian) (Schroeder 1980, Schroeder and Willems 1983, Reitner

[1] Institut für Paleontologie, Freie Universität, Schwendener Straße 8, 1000 Berlin 33, FRGermany

Reef Diagenesis
Edited by J. H. Schroeder and B. H. Purser

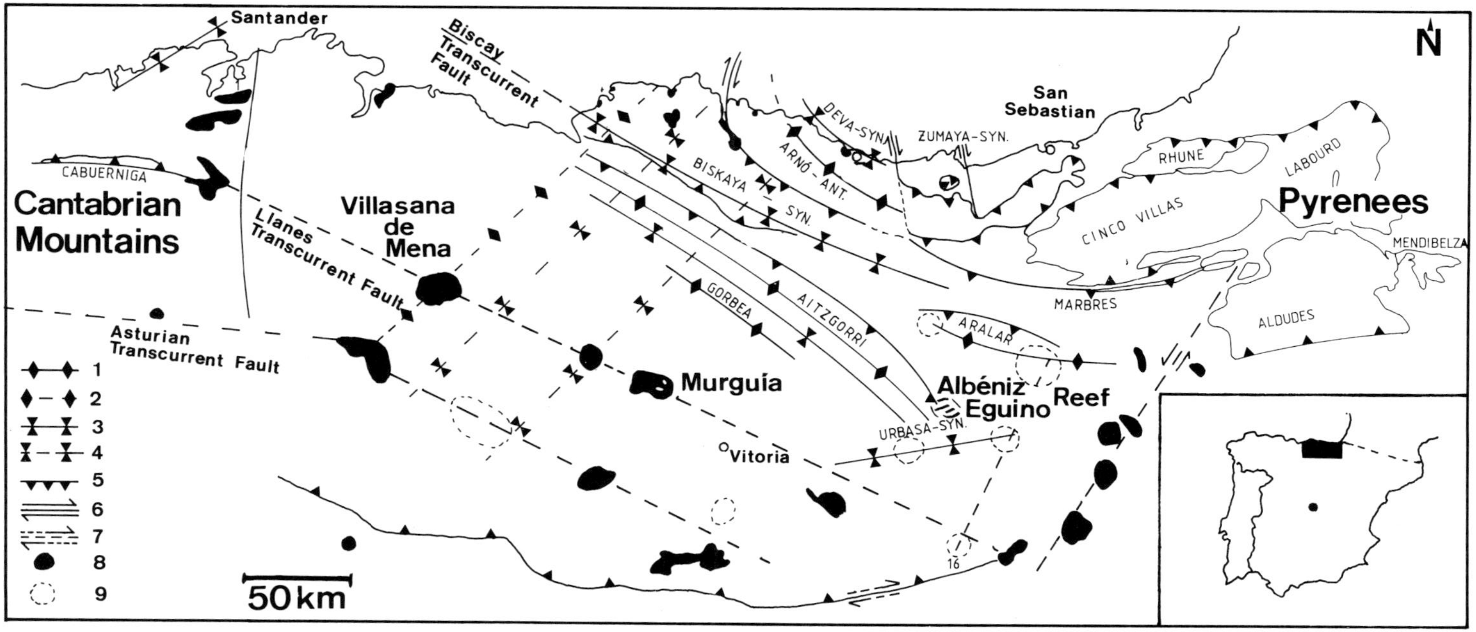

Fig. 1. Location map and geological structure of the Vasco-Cantabrian Basin in northern Spain (Reitner 1985). *1* Anticline; *2* assumed anticline; *3* syncline; *4* assumed syncline; *5* thrust fault; *6* strike slip fault; *7* assumed strike slip fault; *8* salt diapirs; *9* salt pillows

1980, 1982). According to Reitner (1985), this may also be the case in the Murguía diapir. This presentation deals with the carbonate reefs associated with these two diapirs.

3 Methods

The Villasana de Mena diapir and the Murguía diapir were mapped in detail. Sections across each diapir were measured and the lithological changes were established according to biologic, sedimentologic, and petrographic criteria. A similar procedure was followed in the fault block constituting the Albeníz-Eguino reef platform of the same area.

Diagenetic features were studied by petrographic microscope, scanning electron microscope (SEM) with energy dispersive X-ray analyzer, atomic absorption spectrometry (AAS), and X-ray fluorescence. Selected thin sections were stained with potassium ferricyanide in order to distinguish between Fe and Fe-free calcites.

4 Diapir Reef Atolls

4.1 The Late Albian Villasana de Mena Reef Atoll (Caniego Limestone)

4.1.1 Reef Structure

Due to outcrop conditions, the original reef structure cannot be determined with precision. The size of the reef, however, may correspond to the diameter of the diapir roof which may have measured 2 to 5 km. Sedimentation in the surrounding basins consisted of prodelta muds (Blank 1983). The ascending diapir formed a submarine dome which lifted the diapir roof out of the terrigenous influence into the photic zone and allowed the development of a reef. In an advanced stage, the diapir roof collapsed due to salt subrosion. In this process the reef carbonates were fractured and dissected by fissures and crevices through which diapiric material, such as Keuper pelites, hypersaline brines, and Keuper ophites, were extruded. A small diapir basin may have developed within the reef with a surrounding barrier of grainstone shoals and coralgal framework. With this atoll-like structure, water circulation between the open sea and the diapir lagoon was probably restricted.

4.1.2 Facies Belts of the Caniego Limestone (Fig. 2)

There are five distinctive facies belts within the Caniego Limestone, but only the key facies for diapir reef identification are summarized here. A more detailed profile, which includes 16 facies zones, is presented in Reitner (1985). The facies belts, ranging from the basin and slope to lagoonal, have sharp contacts.

The basin and slope facies belt is distinguished by reef debris and planktonic foraminifera while the reef core belt contains the coralgal framework. The reef flat belt is characterized by rudists and grainstone shoals while the lagoonal facies contains a hypersaline fauna (miliolid foraminifera and thin-shelled ostracods) and tempestites. The intertidal facies belt resembles the typical Loferite mudstone facies of the Late Triassic Dachstein Limestone of the Northern Calcareous Alps (Fischer 1964).

4.2 The Late Albian Murguía Diapir

Another example of an Albian diapir reef system is found 50 km east of the Villasana de Mena diapir. The Murguía diapir exhibits further facies and diagenetic peculiarities as a result of the influence of salt brines. The first major feature of the Murguía diapir is the ooid facies which was aragonitic. This is in sharp contrast to other Cretaceous ooid facies which were exclusively calcitic (Sandberg 1975, 1983). The Mg/Ca ratios in the Cretaceous open sea was 2:1, while in the area of the diapirs the Mg/Ca ratio may have reached 5:1 due to the Mg-rich salt brines. This may have promoted the production of aragonitic ooids in the diapir reef lagoon (Lippmann 1973) (Fig. 7e, f).

The second major feature of the Murguía diapir is the presence of sulfide ores in the fractures and fissures of the reef. The circulation of diapiric waters through the reef complex caused the formation of sphalerite, galena, pyrite, etc. (Lietz 1951) (Fig. 8d). The Vasco-Cantabrian Basin contains a large resource potential in sulfides and sulphates.

5 Fault Block Reef Platform

5.1 The Late Albian Albeníz-Eguino Reef Platform (Figs. 1 and 3)

This reef platform is situated about 70 km east of the Villasana de Mena diapir and is about 150 square km in area. The reef beds range from 200 to 500 m in thickness and consist of eight separate reef bodies, each separated by deep grabens filled by siliciclastic pelites and carbonate turbidites. The Albeníz-Eguino reef platform is situated on tilted fault blocks (the Aitzgorri high), this type of reef platform being a typical shelf-margin island platform. The sedimentary cycles comprising the platform carbonates are transgressive. They range in age from the uppermost Middle Albian to the lower Late Albian (*Rotalipora ticinensis* zone) (Reitner 1985).

Five facies belts are distinguished in this reef platform. The basin and slope facies belt contains common planktonic foraminifera and debris flows of reef material containing coralline sponges while the reef core facies belt is composed of a coralgal framework and the reef flat facies has caprinid rudists. The carbonate intertidal facies belt contains mudcracks and stromatolites. The hypersaline lagoon facies is characterized by miliolid foraminifera while the littoral intertidal facies consists of siliciclastics.

Facies belts

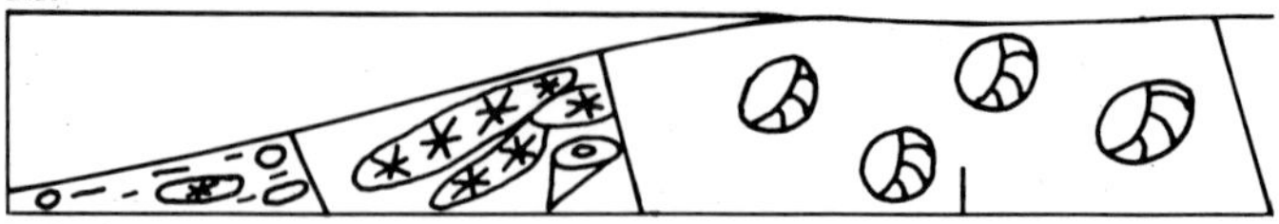

Facies belts	Basin/slope	Reef core	Reef flat: *Caprina/ Plagloptychus belt*	Reef flat: Grainstone shoal
Major Lithology Fauna Flora / **Sequences of Diagenesis**	Wackestones: Packstones; Planktonic foram; Planktonic crinoids; Cephalopods	Framestones; Floatstones; Microsolenid-corals; Coralline algae; Coralline sponges	Wacke-, Float-, Packstones; Caprinid rudists; Rhodoliths of coralline algae; Orbitolinid forams	Cross-bedded grainstones, rudstones; Trocholinid forams
Marine phreatic	Reworking of reef sediments; Stromatolitic hardgrounds with phosphorites; Partial replacement of the phosphates by pyrite; Partial reworking of the stromatolitic crusts		Micrite envelopes	Micrite envelopes, coated grains
Marine vadose				Beachrock; meniscus and dripstone cements
Fresh water vadose	Dissolution of aragonite	Dissolution of biogenics; Red micrite filling; Dog tooth II-dripstones	Aragonite dissolution; molds; Dog tooth II dripstones; Vadose slit; Reworking; lithoclasts; Micrite fill	
Marine phreatic			Radiaxial fibrous calcitic cement	
Marine vadose				
Fresh water phreatic	Neomorphic calcite; Bladed ferroan calcite	Red micrite filling; calcitic dog tooth III-cements; Equant calcite cement	Syntaxial calcitic overgrowth	Syntaxial calcitic overgrowth; Equant calcite
Fresh water vadose				
Fresh water phreatic				
Deep burial		Partial dolomitisation of cements		
	Stylolites Bitumen	Stylolites Bitumen	Stylolites Bitumen	Bitumen

Time ↓

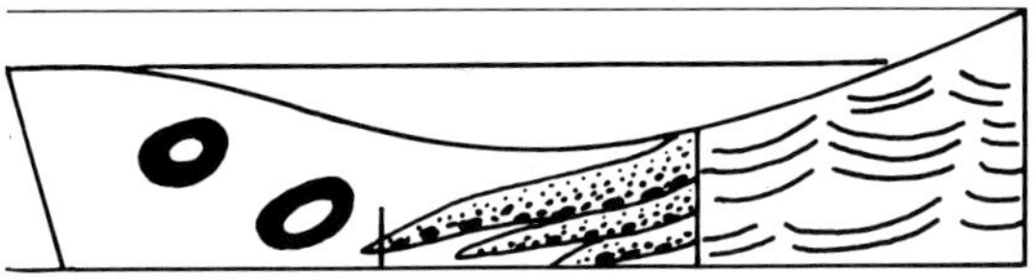

Lagoon: *Polyconites/ Toucasia belt*	Lagoon: Mudstone belt	Intertidal
Wackestones Floatstones Requienid rudists Carpotinid rudists Miliolid forams	Mudstones Wackestones Packstones Ostracods	Mudstones Wackestones Bindstones (Stomatolites) Ostracods
	Stromatactis Sulphates, proto-dolomites Replacement of sulfates by quartz Micrite fill	Storm layers
		Fibrous dripstone-cement
Dissolution of aragonite Dog tooth II-dripstone Vadose silts	Microkarst	Sheet cracks, tepees Intraformational breccia
Fibrous isopachus cement		
	Dog tooth I-dripstone in stromatactis	
Dog tooth III-calcite cement	Equant ferroan calcite	Equant ferroan calcite cement
Brecciation Rotation of components Fissures		
Micrite fill Dog tooth III-cement Equant ferroan calcite cement		
Stylolites Bitumen	Replacement of ferroan calcite by dolomite Calcedony Pyrite Gypsum cements Sulfur cements Stylolites Bitumen	Dolomite/ankerite Authigenic quartz cement Stylolites Bitumen

Fig. 2. Facies model and diagenetic environments of the Villasana de Mena diapir reef platform (Caniego Limestone)

Facies belts — **Tide**

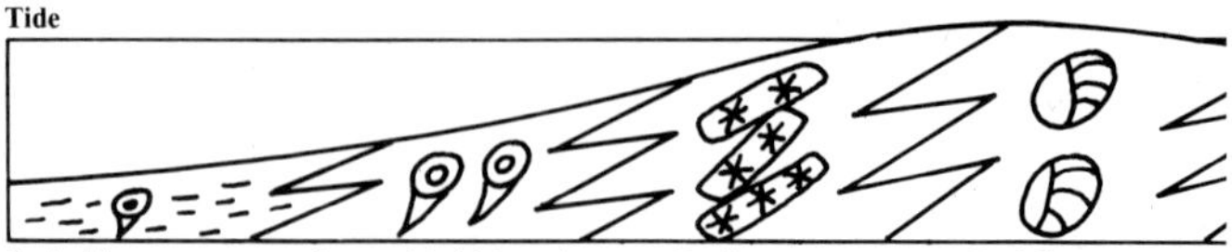

Facies belts	Basin/slope	Sponge Mud Mound	Reef core	Reef flat *Caprina belt*
Major Lithology Fauna Flora / **Sequences of Diagenesis**	Grey marls Turbidites Debris flows Planktonic forams Ammonites Sponges	Bindstones Packstones Siliceous sponges Coralline sponges Bluegreen algae	Framestones Floatstones Microsolenid-corals Coralline algae Coralline sponges	Floatstones Packstones Graintones Caprinid rudists Radiolitid rudists Articulate coralline algae Calcareous green algae Orbitolinid forams
Marine phreatic	Reworking of reef sediments Dissolution of the siliceous sponge scleres Fibrous cement	Stromatactis Sediment fill Short fibrous calcite cement Sediment fill Dissolution of the sponge scleres Radiaxial fibrous calcite cement Sediment fill	Micrite fill within the coralgal frame Short fibrous calcite cement Multiphased sediment fill	Reworking of large clasts Micrite envelopes coated grains Dissolution of sponge scleres Short fibrous calcite cement
Marine vadose				Brownish radiaxial fibrous meniscus cement
Fresh water vadose			Neomorphic calcite, dissolution of aragonite Molds Microkarst Dog tooth II-dripstone Sediment fill	
Fresh water phreatic	Neomorphic calcite high Mg-calcite to low Mg-calcite preserving the original crystal size Equant calcite	Neomorphic calcite Equant ferroan calcite	Dog tooth III-cement Equant calcite cement Syntaxial calcitic overgrowth	Neomorphic calcite micritisation of the Mg-calcite Equant ferroan calcitic cement Syntaxial calcitic overgrowth
Deep burial	Bitumen	Bitumen	Stylolites Bitumen	Stylolites Bitumen

Time →

Fig. 3. Facies model and diagenetic environmens of the Albeniz-Eguino fault block reef platform

Reef flat Grainstone belt	Intertidal mud-pank	Lagoon	Siliciclastic littoral
Grainstones Rudstones Trocholinid forams Coralline algae	Mudstones Wackestones Bindstones (Stromatolites) Ostracods	Well bedded wackestones, mudstones Miliolid and arenaceous forams	Well sorted medium and fine grained beach and intertidal sandstones
Reworking of clasts Micrite envelopes coated grains Short fibrous calcite cement	Short fibrous isopachous cement Shrinkage Brownish radiaxial fibrous calcitic cement	Glauconite in forams Reworking of clasts with micrite envelopes	Lithoclasts Micrite envelopes
	Dog tooth I-dripstone calcite cement		
	Dissolution of aragonite Molds		Dissolution of Biogenics Molds
Neomorphic calcite Dog tooth III-cement Equant ferroan calcite Syntaxial calcitic overgrowth	Dog tooth III-calcitic cement Equant ferroan calcite cement	Neomorphic calcite molds Dog tooth III-calcitic cements Chert replaced calcite Equant ferroan calcite cement	Dog tooth III-calcite cement Intergranular granular calcite cement Limonitic cement relacing the calcitic granular cement Syntaxial calcitic overgrowth
Stylolites Bitumen	Bitumen	Pyrite	Sparry dolomites replacing the granular cements

Table 1. Characteristics and interpretations of observed cements

Cement	Fig.	Composition	Habit/fabric	Dimensions	Interpretation	Reference
Bladed scalenohedral						
type I	4a	Neomorphic low-Mg calcite	Dripstone with relic structures of a fibrous cement	40 – 80 μm	Marine vadose aragonite cement	Purser (1969) Hanor (1978)
type II	4b	Low-Mg calcite	Dripstone without inclusions	50 – 60 μm	Freshwater vadose cement	Schneider (1977)
type III	4e, 7a, b	Low-Mg calcite	Isopachous crusts	80 – 330 μm	Freshwater phreatic cement	Pierson and Shinn (1985)
Radiaxial fibrous cement	4d, 5d, 6b	Neomorphic low-Mg calcite	Dripstone, meniscus, multiphased isopachous crusts	20 μm – 5 mm	Marine phreatic-marine vadose aragonite cements	Bechstaedt (1974)
Short fibrous calcite cement		Neomorphic low-Mg calcite (~2% $MgCo_3$, EDAX)	Isopachous crusts	60 – 100 μm	Marine phreatic high-Mg calcite cement	Longman (1980)
Micro-granular calcite cement		Neomorphic low-Mg calcite (~1.5% $MgCo_3$, EDAX)	Isopachous crusts	15 – 30 μm	Marine phreatic high-Mg calcite cement	Longman (1980)
Micritic cement		Neomorphic low-Mg calcite (~4 – 5% $MgCo_3$, EDAX)	Dripstone	8 – 10 μm	Marine vadose high-Mg calcite cement	
Fibrous calcite cement	5b	Neomorphic low-Mg calcite	Tannish colored isopachous crust	80 – 100 μm	Marine phreatic aragonite or high-Mg calcite cement	
Equant calcite cement	4c	Low-Mg calcite	Blocky crystals with enfacial junctions	100 μm	Freshwater vadose cement	Longman (1980)
Equant ferroan calcite cement	5d	Fe calcite	Blocky crystals with enfacial junctions	100 μm	Freshwater phreatic cement (shallow burial)	Oldershaw and Scoffin (1967)

Bladed cements		Low-Mg calcite	Bladed crystals	100 – 120 μm length 30 – 45 μm width	Meteoric cement	Longman (1980)
Syntaxial cements		Low-Mg calcite	Syntaxial overgrowth on echinoderm and ino-ceramid skeletal remains	200 – 500 μm	Freshwater phreatic cement	Neugebauer and Ruhrmann (1978)
Sulfate cements	7d	Gypsum, anhydrite	Limpid idiomorphic elongated crystals	50 – 800 μm	Burial cement	
Sulfur cements	8a, b, c	Elementary sulfur	Blocky crystals	100 – 530 μm	Burial cement	Feely and Kulp (1957) Ellison (1971)
Sparry dolomite	7c	Dolomite with Fe-rich zones (X-ray analysis)	Idiomorphic crystals replacing earlier cements	120 – 300 μm	Burial	Mattes and Mountjoy (1980)
Anhedral dolomite		Dolomite	Dolomicrite with brown coloured anhedral crystals	4 – 10 μm	Intertidal, early dia-genetic formation	Fischer (1964) Folk and Land (1975)
Ankerite cement		Ankerite (X-ray analysis)	Idiomorphic crystals	100 – 350 μm	Deep burial cement	Oldershaw and Scoffin (1967)
Early diagenetic authigenic quartz	8a, f	Quartz with inclusions of sulfates, dolomites, calcites, and Na-rich liquids (EDAX, X-ray fluorescence analysis)	Idiomorphic crystals commonly pseudomorphic after anhydrite	10 – 50 μm	Early diagenetic, prior to lithification	Grimm (1962)
Pyrite I		FeS_2	Framboidal aggregates	100 – 120 μm	Early diagenetic, linked with organic rich sediments presumably influenced by sulfur bacteria	Fabricius (1961)
Pyrite II		FeS_2	Idiomorphic crystals	80 – 210 μm	Burial, may be linked to the migration of hydrocarbons	

6 Cements

The cement types from both the diapir reefs and the fault block reef platform are listed in Table 1 and some are discussed below.

6.1 Bladed Scalenohedral ("Dog Tooth") Cements (Figs. 4a, b, e, 7a, b)

Low-Mg scalenohedral "dog tooth" cement is the most widespread cement in the diapir reef facies. In the fault block reef facies this cement is rare. These cements appear in cavities as dripstone cements or in grainstone fabrics as meniscus cements.

According to Schneider (1977) and Pierson and Shinn (1985), scalenohedral calcites are products of freshwater diagenesis. Dog tooth (I) cements exhibit relic structures of fibrous cement neomorphic crystals. This cement was probably originally formed in the marine vadose zone (Purser 1969, Hanor 1978). Dog tooth (II) cements do not show any neomorphic characteristics. It occurs as a dripstone cement and was probably formed in the freshwater vadose zone. Dog tooth (III) cements are interpreted to be a phreatic freshwater cement (Schneider 1977).

6.2 Radiaxial Fibrous Cements (Figs. 4d, 5d, 6b)

Two theories concern the precipitation of radiaxial fibrous cements. According to Bechstaedt (1974), this type of cement is formed under subtidal to supratidal conditions, whereas Kendall and Tucker (1973) suggest exclusively subtidal conditions. In the diapir reefs these radiaxial fibrous cements developed locally as a replacement of fibrous aragonitic dripstone, thus confirming the observations of Bechstaedt. That radiaxial fibrous cements developed as a replacement of aragonitic cements is clearly suggested by the presence of relic structures, notably prismatic and spherulitic crystal aggregates. Spherulites commonly occur in the paleo-intertidal facies, whereas prismatic structures are observed in subtidal sediments. It has also been demonstrated geochemically and microscopically by Assereto and Folk (1980) that the radiaxial fibrous cements of the Late Triassic intertidal carbonates of the southern Alps were originally aragonitic. Kendall (1985) recently investigated radiaxial fibrous cements in Devonian reefs of western Australia and suggests that these cements are primary low or high Mg calcite formed in the marine phreatic zone. These Devonian cements show no

Fig. 4. a Dog tooth (I) dripstone cement – calcitic cement from the grainstones of the Villasana de Mena diapir reef (VMDRP). **b** Solution cavity of the lagoonal facies of the VMDRP and dog tooth (II) dripstone cement. **c** Equant ferroan calcite cement with enfacial junctions (*1*) of a stromatactis fabric in the lagoonal facies belt of the VMDRP. Internal fine-grained calcareous sediment of the stromatactis fabric (*2*) (crossed nicols). *Arrow* indicates upper surface. **d** Radiaxial fibrous cement with relics of a spherulitic fabric. Intertidal belt of the VMDRP (crossed nicols). **e** dog tooth (III) cement in a large cavity of the reef flat belt of the VMDRP with *Caprina* (crossed nicols)

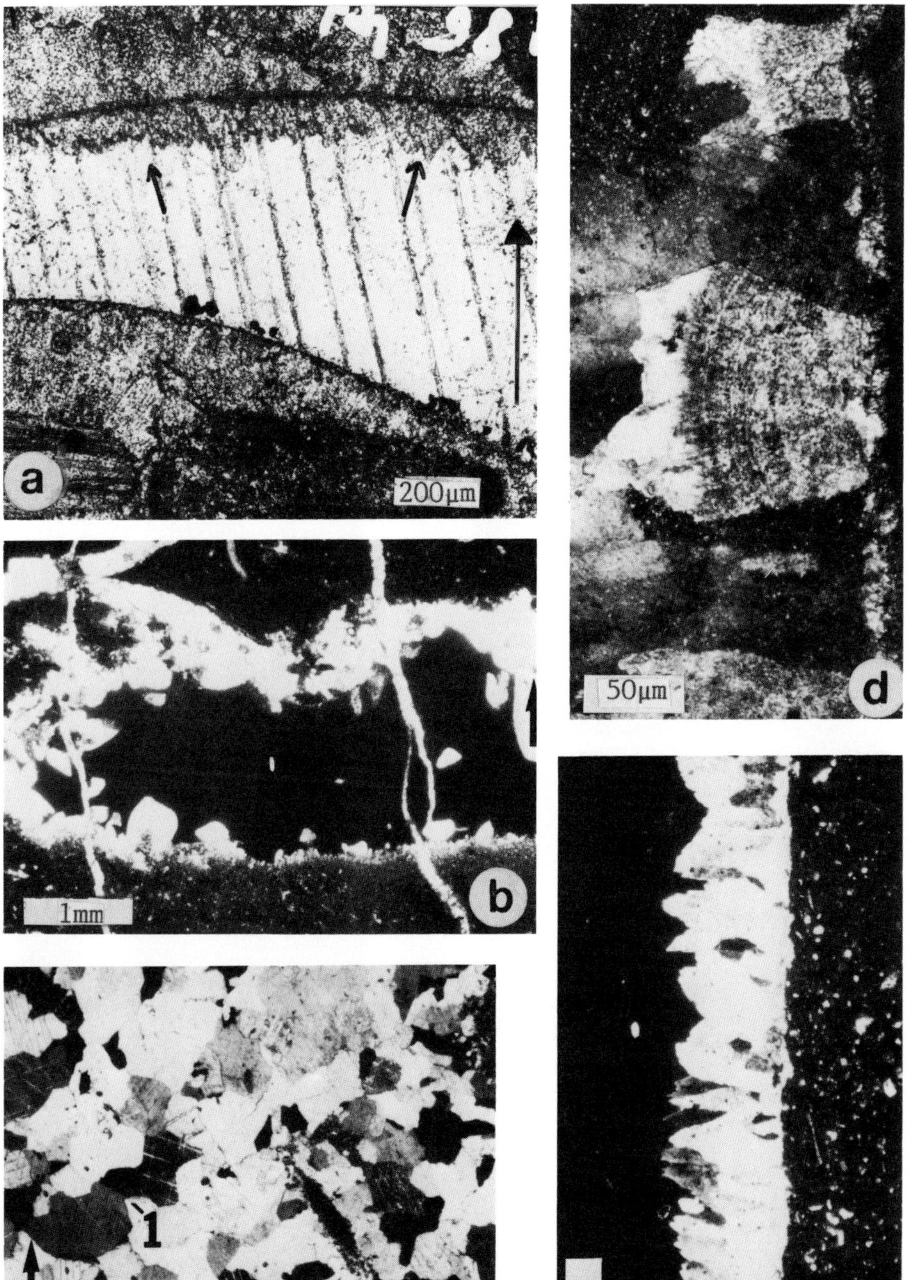

Fig. 4

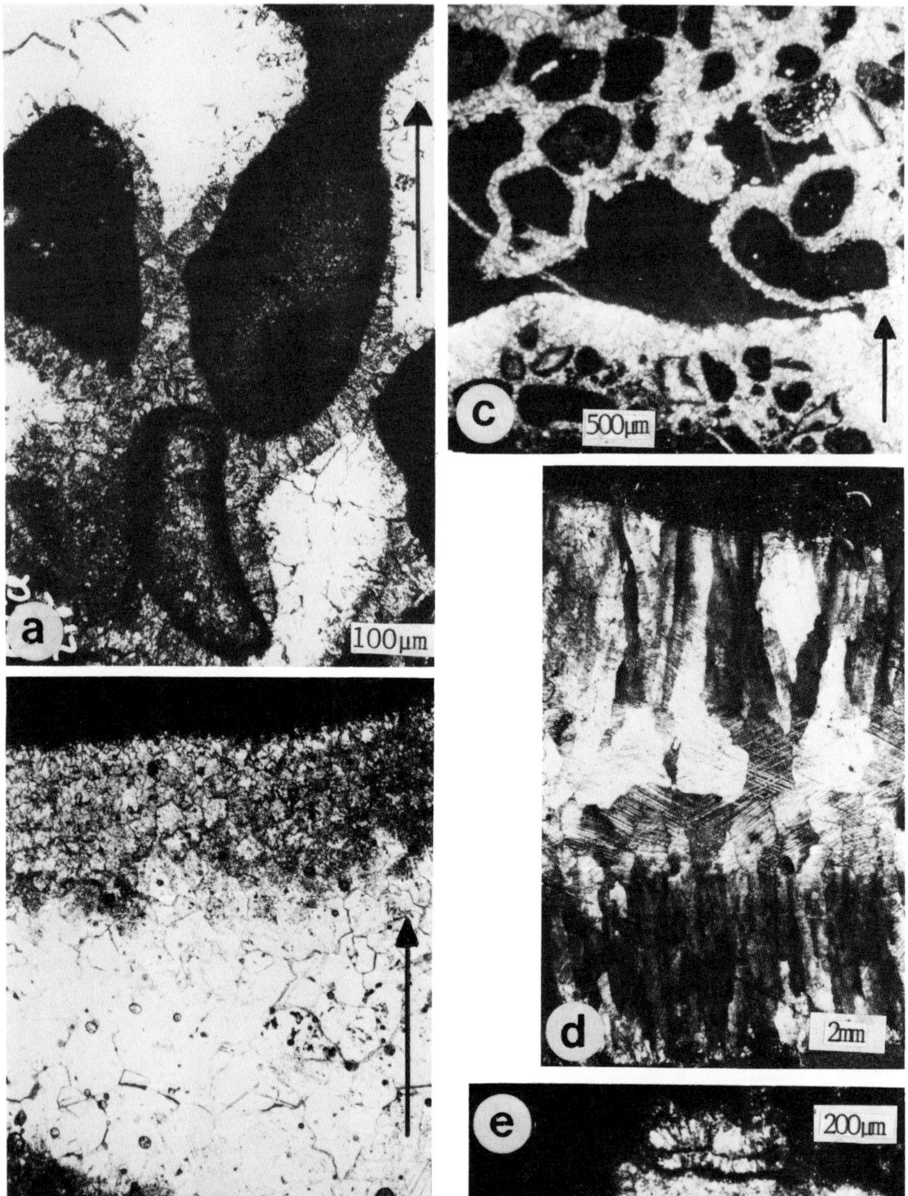

Fig. 5. a Meniscus cement of the reef flat belt of the VMDRP. **b** Microstalactitic and microstalagmitic dripstones from the reef core facies of the Albeníz-Eguino fault block reef platform (AEFBP). **c** Dog tooth (II) cements of the grainstones in the reef flat facies belt of the VMDRP. **d** Radiaxial fibrous cement (*1*) and equant cement (*2*) from a sheet crack of the intertidal facies in the VMDRP (crossed nicols); **e** Fibrous dripstone cement from the reef flat facies belt of the VMDRP. *Arrows* indicate upper surface

relic structures while the Spanish Albian radiaxial fibrous cements do contain fibrous relic structures, suggesting different precipitation conditions.

6.3 Sulfate Cements (Fig. 7d)

In the lagoonal facies belt of the diapir reefs, rare large cavities contain late diagenetic gypsum cements whose limpid idiomorphic crystals lack inclusions. There are two possible explanations for the formation of the gypsum cements. The sulphate may be derived from corroded pyrite which commonly occurs in the lagoon facies or it may come from the circulating diapir waters which may have been rich in dissolved sulphates.

6.4 Sulfur Cements (Fig. 8a, b, c)

These cements are found in large cavity fillings in the lagoonal facies of the diapir reefs. The elementary sulfur originates from the reduction of anhydrite to hydrogen sulfide and the oxidation of hydrogen sulfide to sulfur by means of sulfur bacteria in the cap rock of the diapir (Feely and Kulp 1957, Ellison 1971). Both the sulfur and gypsum cements formed under deep burial conditions.

7 Authigenic Quartz

7.1 Early Diagenetic Authigenic Quartz (Fig. 8e, f)

According to Grimm (1962), authigenic quartz forms in hypersaline conditions, an observation confirmed by Friedman and Shulka (1980). These authors describe quartz which is pseudomorphic after anhydrite and gypsum from lagoonal sediments. Bouroullec and Deloffre (1982) observe authigenic quartz pseudomorphic after anhydrite in Jurassic sabkha sediments. However, Richter (1972), Harder and Menschel (1967), and Lippmann and Schlenker (1970) demonstrate another process of authigenic quartz formation which is not related to hypersaline conditions.

The genetic relations between quartz precipitation and hypersaline conditions is evident in the Caniego Limestone of the Villasana de Mena salt diapir. However, pH is also an important factor in quartz precipitation (Grimm 1962, Friedman and Shulka 1980). Alkaline conditions ($>$pH 9) favor the precipitation of sulfates whereby quartz is dissolved, while under acidic conditions ($<$pH 5) quartz is precipitated. The formation of in-situ authigenic quartz in the Villasana de Mena diapir reef was probably influenced by both pH and hypersaline conditions. The hypersalinity was caused by the diapir waters, while the acidity probably was related to the presence of vadose freshwater. This type of quartz is not observed in the fault block reef platform.

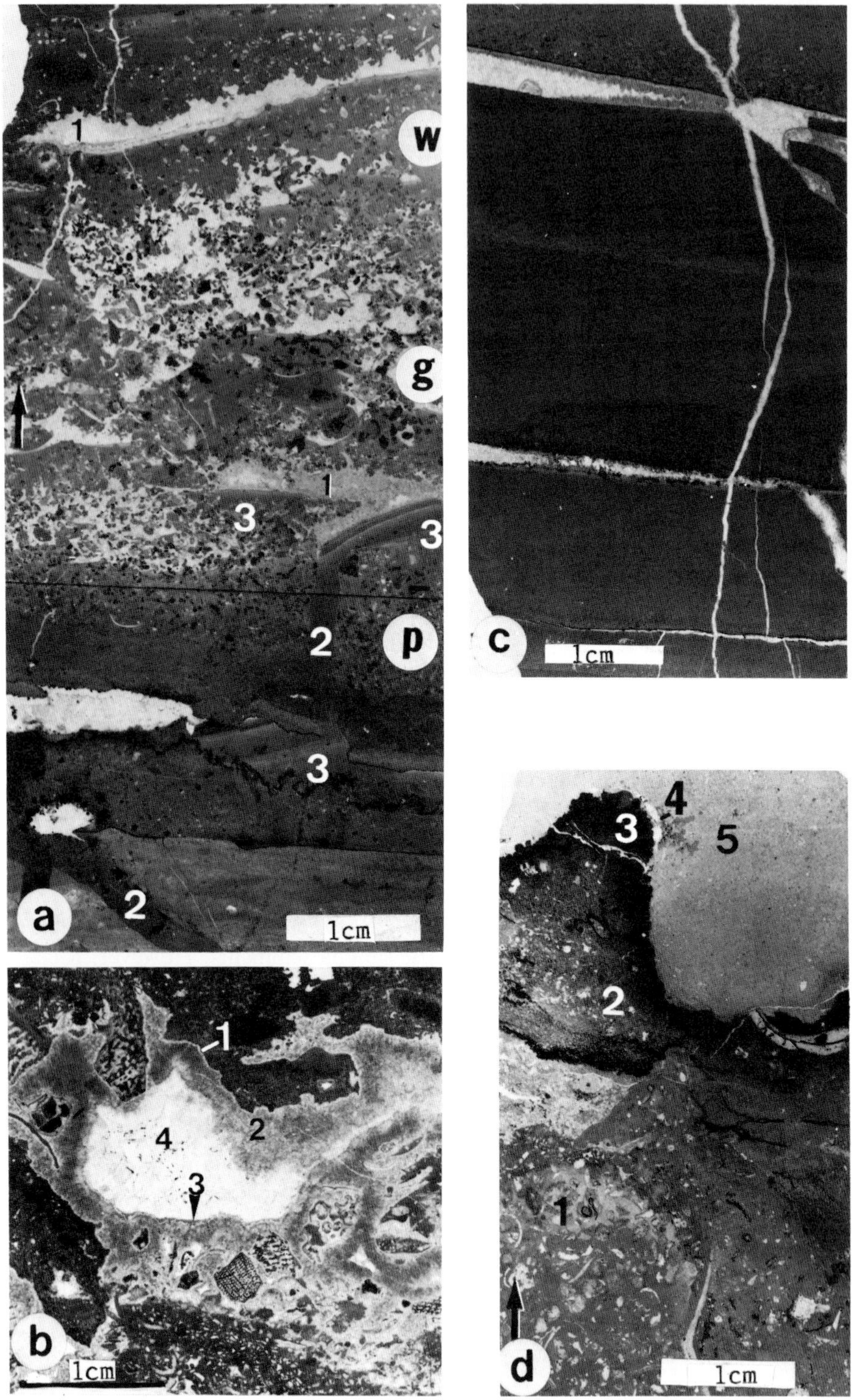

Fig. 6

7.2 Allochthonous Idiomorphic Quartz

In addition to the in-situ authigenic quartz, allochthonous idiomorphic quartz occurs in increased quantities at the base of the Caniego Limestone where a debris flow of diapir material (Keuper clays and ophites) is present (Schroeder 1980, Reitner 1982). This idiomorphic quartz is partly fractured and is identical to authigenic quartz from the Early Jurassic Carniolas Formation of this area.

8 Microstalagmitic Dripstone (Figs. 4b, 5b)

A new type of vadose cement which is reminiscent of typical cave dripstones is observed in both types of reef. According to Purser (1969), microstalactitic cement crusts (Fig. 5e) can be found in large open pores, especially in beachrocks. In some fissures and molds microstalagmitic cements are present. This type of microcave dripstone indicates a long period during which pores remained open. There is no preferred cement type within these dripstones, but dog tooth (II) and radiaxial fibrous cements are common. The cement crystal length varies between 100 and 280 μm.

9 Microkarst

In both diapir and fault block reefs 0.5 – 10 cm cavities with irregular shapes cut sedimentary and biogenic structures. Such cavities are characteristic of microkarst and are morphologically distinct from stromatactis fabrics, lacking the irregularly shaped roofs and flat bottom surfaces (see Sect. 10).

10 Stromatactis (Figs. 4b, 6a)

Stromatactis-like voids occur in carbonate-rich (more than 95%) mudstones, wackestones, and bindstones of both reef types. These elongate voids measure 1 – 5 cm in length with a vertical width 5 – 10 mm. The roof of these voids is irregular while the bottom is flat and partly filled with sediment. A halo of very

Fig. 6. a Tempestite in lagoonal facies of the VMDRP. The tempestite contains a basal packstone unit (*p*), a central grainstone unit (*g*), and an upper wackestone unit (*w*). *1* Stromatactis fabric; *2* synsedimentary fissures with multiple micrite fillings; *3* internal sediment of the stromatactis fabric; *3′* internal sediment of the stromatactis fabric rotated during a probable collapse event. **b** Large cavity in the reef core of the AEFBP. *1* Short fibrous marine cement; *2* brownish radiaxial fibrous cement; *3* vadose silt; *4* late diagenetic nonferroan calcite cement. **c** Intertidal mudstone with sheet and prism cracks filled by dripstone cements from the VMDRP. **d** Hard-ground of the basin and slope belt of the VMDRP. *1* Forereef debris (packstone); *2* phosphorite layer; *3* pyrite layer; *4* hard-ground surface with encrusting organisms; *5* pelagic mudstone. *Arrows* indicate upper surface

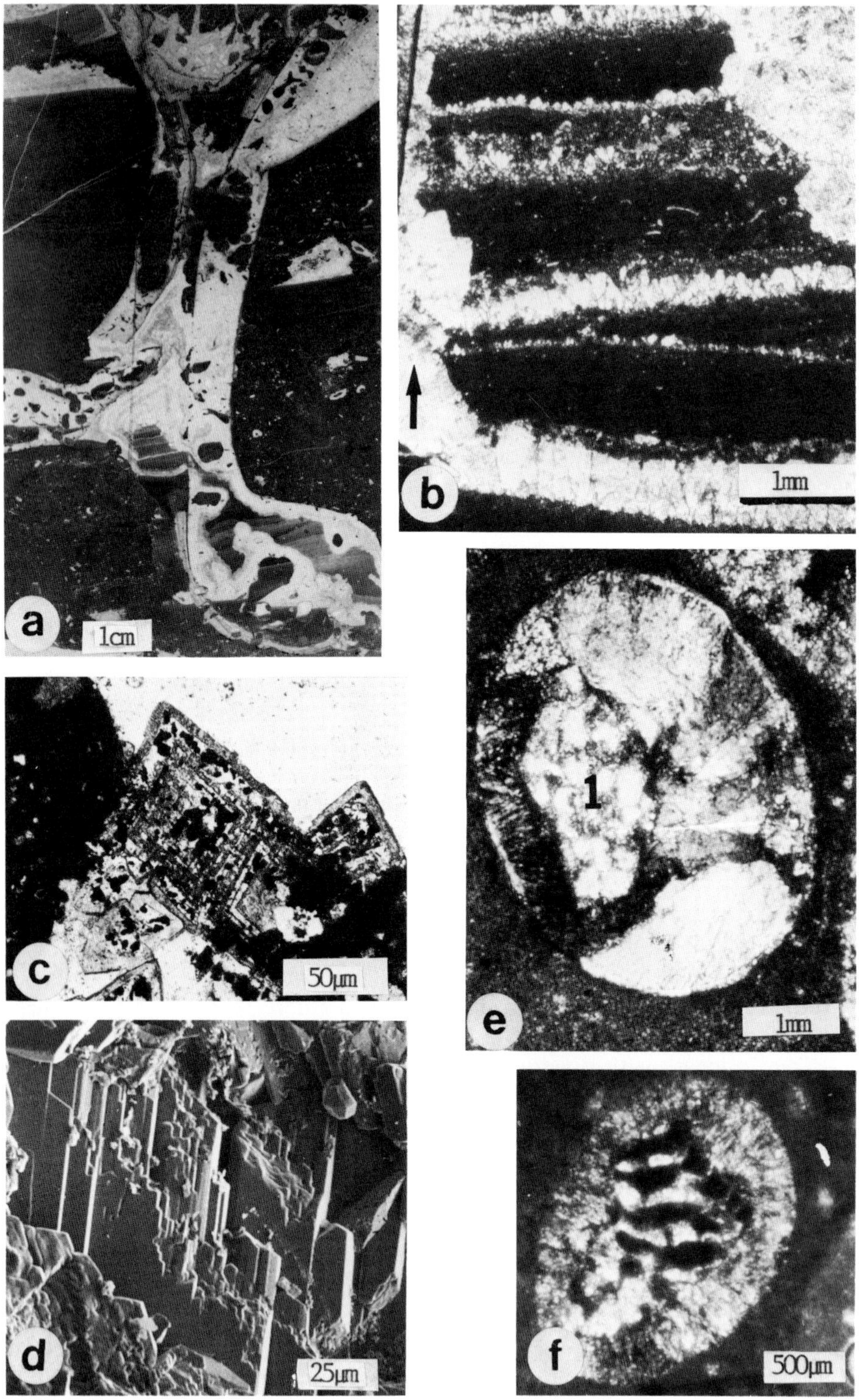

Fig. 7

dense micrite surrounds the stromatactis voids, which is in sharp contrast to the less dense micrite of the matrix rock. The grain size of both types of micrite is similar (0.5 – 1 μm, SEM).

These stromatactis voids presumably are the product of a very early compaction of the micritic sediment (Heckel 1972, Bechsteadt 1974). Tsien (1985), on the other hand, believes that stromatactis fabrics are biogenic in nature, as described from the type locality at St. Remi in Belgium. The origin and nature of stromatactis is unclear and Tsien gives an overview of the problem. The formation of stromatactic fabrics may well be due to a combination of both physical and biogenic processes. In the Albian reef complexes, no regularity of structures or ecological zonation is observed and the sizes of the voids are not compatible with the size of organisms such as sponges, corals, etc.

11 Interpretation of the Hard-Grounds and Stromatolitic Crusts of the Basin and Slope Belt of the Diapir Reefs (Fig. 6d)

The formation of hard-grounds and stromatolitic crusts occurred contemporaneously with sedimentation. The crusts are colonized with encrusting foraminifera, such as *Placopsilina* and *Coscinophragma*. Further evidence for synsedimentary formation is the presence of hard-ground clasts in the tempestites of the lagoon facies. The hard-grounds are cut by fine pyritized strings resembling fungi mycels, suggesting participation of fungi and bacteria in the formation of the stromatolitic crusts.

The presence of phosphatic crusts (apatite) may be explained by upwelling along the diapir slope. The same process also may be responsible for reduced sedimentation rates in this area favoring the establishment of reefs. This particular hydrodynamic situation was the consequence of a change in the tectonic patterns during the Late Albian (Vraconian) (Reitner 1985). Cool waters from the deeper parts of the Bay of Biscay may have found their way into the Vasco-Cantabrian Basin and caused upwelling around the tectonic highs and diapirs. This change in hydrodynamic situation during the Late Albian resulted in the disappearance of Albian coralgal reefs, which were partly replaced by crinoid/algae bioherms (Reitner 1985).

Fig. 7. a Diagenetically altered rudists which are now filled with multiple generations of micrite and dog tooth (III) cements. Lagoon facies with *Polyconites* in the VMDRP. **b** Details of **a** illustrating the laminar fill of alternating micrite and cement. *Arrow* indicates upper surface. **c** Zoned sparry dolomite of the intertidal facies belt of the VMDRP. Gray zones are Fe-rich. **d** SEM photo of gypsum cement from the lagoon facies of the VMDRP. **e** Ooid from the Murguía diapir reef platform (MDRP) showing neomorphic calcite (crossed nicols). *1* Bioclast. **f** Radiaxial fibrous ooid from the MDRP

Fig. 8

12 Comparison of the Two Reefs Types (Table 2)

In the reef atoll of the Villasana de Mena diapir there occur certain cement types which are not found or are rare in the Albeníz-Eguino fault block reef. These include scalenohedral cements, certain fibrous cements, sulfur cements, gypsum cements, dolomite, and ankerite. Conversely, marine cements are rare in the diapir reef and common in the fault block reef.

Scalenohedral low Mg-calcite cements are found in molds, microkarst cavities, and primary voids as first generation cement in the diapir reefs. All described types of scalenohedrals occur; dog tooth (II) and (III) are common while dog tooth (I) is rare. The abundant occurrence of dog tooth (II) as dripstone cements within the molds and microkarst cavities indicates freshwater vadose conditions. Dripstone formation in a marine vadose environment is observed in only a few cases (dog tooth I). Within the *Polyconites* facies of the diapir reef atoll scalenohedral cements also are present in large vugs as multiple isopachous cements (dog tooth III), where cements and sediment fills alternate with the latter consisting of graded packstone, mudstones with ostracods, and calcareous silt ("vadose silt"). According to Schneider (1977) and Pierson and Shinn (1985), all scalenohedral cements are formed under meteoric conditions. Closely linked with these freshwater cements is extreme aragonite dissolution, mold formation, and microkarst, which is present in all the facies of the diapir reef atolls. Scalenohedral cements are rare in the fault block reef; they are present as neomorphic cement only in the reef flat facies.

Early marine fibrous cements are very rarely observed in the diapir reef atolls but occur infrequently in the basin and slope and reef flat facies. Radiaxial fibrous cements commonly follow the scalenohedral cements as the next cement generation in the diapir reefs.

Certain burial cements of the diapir reef atolls appear to be closely related to the circulating diapir waters of the diapir body, these include the sulfur and gypsum cements, dolomite, and ankerite. Sulfur and gypsum cements are found in cavities in the lagoonal facies of the diapir reefs. Gypsum cement is found in the fault block reef associated only with corroded pyrite. The presence of sulfide ores in the Murguía diapir reef also tends to confirm the unique diagenetic influence of diapirs on reef carbonates since these phenomena are not observed in the fault block reef.

Dolomites and ankerites, the last cementation phase, destroy former cements and crystallize in tectonic fissures. The cement filling of the large fissures (width of several decimeters) are dark drusy and fibrous calcite and are presumably the youngest of the late cements, since the fissure systems show a radial pattern linked to the last diapir collapse in the Late Tertiary.

◄

Fig. 8. a Diagenetic sulfur cement in a large dissolution cavity of the lagoonal mudstone facies of the VMDRP (*1*) (SEM photo). **b** Calcium distribution pattern from the sulfur cement of **a**; scale as in **a**. **c** Sulfur distribution pattern from the sulfur cement of **a** (1000 counts s^{-1}; 20 kV); scale in **a**. **d** Ore paragenesis in a fissure of the MDRP (SEM photo). **e** Large early diagenetic authigenic quartz with Na-rich inclusions in the lagoonal facies of the VMDRP (crossed nicols). **f** Authigenic quartz from the lagoonal mudstone of the VMDRP (SEM photo)

Table 2. The differences between fault block reefs and diapir reefs

Fault block reefs	Diapir reefs
Reef Structure	
Island platforms with isolated reefs, ca. 15 × 20 km	Atolls and small pinnacle ca. 5 × 5 km
Organism diversity of selected groups	
Corals: 36 species	Corals: 5 species
Sponges: 11 species	Sponges: 8 species
Mg-calcite algae: 6 species	Mg-calcite algae: 4 species
Aragonite algae: 15 species	Aragonite algae: 2 species
Stromatolites: rare	Stromatolites: abundant
Mollucs: 18 species	Mollucs: 8 species
Microfacies (field observation, quantitative analysis of profiles)	
Micrite – sparite	Micrite – sparite
70 : 30	90 : 10
Mudstone rare	Mudstone with authigenic quartz common
Geochemistry	
Sr 300 – 1300 ppm	Sr 100 – 250 ppm
Mn 100 – 500 ppm	Mn 100 – 1600 ppm
Fe 100 – 20000 ppm	Fe 300 – 30000 ppm
Na 110 – 400 ppm	Na 60 – 3300 ppm
Mg 120 – 4000 ppm	Mg 140 – 5400 ppm
Diagenesis	
Early marine cements in all facies zones, except in the intertidal zones; Neomorphism of aragonite Calcite ooids	Early marine cements rare; Meteoric cements abundant; Early diagenetic authigenic quartz formed by hypersaline solutions; Late diagenetic gypsum and sulfur cements; Burial dolomite and ankerite; Multiphased sedimentation and cementation in cavities; Sulfide ores; Aragonitic ooids

The fault block reef exhibits a totally different cementation history. The abundance of marine cements, especially radiaxial fibrous and short fibrous cements, is in direct contrast to the diapir reefs. An intense freshwater diagenesis is not observed except in the intertidal and reef core facies of the Albeníz-Eguino fault block reef platform. The cementation sequence is controlled by subsidence of the fault block. It appears, therefore, that the principle diagenetic difference between the two types of reefs concerns the timing of the freshwater diagenesis which is related to differing fault block and diapir movements.

13 Discussion and Conclusions

The key difference between diapir reef atolls and fault block reef platform diagenesis appears to be the early introduction of meteoric water into the diapir reef complex. This is demonstrated by the presence of certain non-marine cements, dissolution of fossils, microkarst, extremely low Sr (100 – 250 ppm), and Mg values (500 – 1600 ppm) (results from bulk analyses of limestones in the diapirs). Also interesting is a comparison of high-Mg calcite sponges (*Acanthochaetetes*) between the two reef types. The sponges of the fault block reef contain more than twice the amount of Mg than the diapir reef sponges. Within the Villasana de Mena diapir reef geochemical anomalies were also noted for the trace elements Na, Fe, and Mn. The Na values of up to 3300 ppm, the Fe content of up to 30000 ppm, and the Mn values of up to 1500 ppm were measured in the soluble portion of the limestone from the reef facies. The anomalies are linked both with large amounts of reworked Keuper sediments from the diapir and with early diagenetic quartz caused by the extruding hypersaline brines associated with the synsedimentary collapse event of the diapir and the corroded Keuper ophites (continental tholeiites, Meschede 1985). It is evident that diapiric movement was active during reef development.

Fossil constituents also confirm both the diagenetic and the sedimentary differences between diapir and fault block reefs. In all facies within the diapir reef complex organism preservation is significantly poorer than in the fault block reef platforms. The reduction mainly affects organisms with aragonitic skeletons, the intense early meteoric diagenesis causing the selective dissolution of aragonite. Low animal diversity may also be due partly to unfavorable ecological conditions reflecting the hypersalinity in these diapir-influenced reefs.

All peculiarities of the diapir reef platform may therefore related to the movement of the diapir. Thus, meteoric diagenesis, the multiple sediment filling, and the cementation may have been conditioned by the repeated, but relatively brief, elevations of the diapir. These periodic movements brought the reef complex repeatedly into the vadose environment. Each emergence of the entire reef complex resulted in the multiple filling of the reef cavities and cementation of the sediments with a meteoric cement, as well as the death of the reef fauna. The mobility of the diapirs may have resulted from the complicated interplay of general subsidence of the Vasco-Cantabrian Basin, ascent of the diapir, sea level changes, and dissolution processes in the diapir subsequent to collapse events.

This oscillating vertical movement is not found in the Albeníz-Eguino fault block reef platform. The absense of certain cements, such as sulfur and sulfate cements, and of diapir-linked sulfur ores, as in the Murguía diapir reef, also demonstrates the lack of diapir influence in the burial conditions of the fault block reef. The geochemical data from the Albeníz-Eguino fault block reef platform exhibits high Sr values of 200 – 1500 ppm (mean values approx. 400 ppm), which are not observed in the diapir reef atolls.

Modern examples of diapir reef atolls can be found in the Gulf of Mexico as pinnacle reefs (West Flower Garden Bank). The faunal and floral diversity of these diapir pinnacle reefs are significantly less than in the fault block reef platforms in the Caribbean (Reitner 1982). However, the direct influence of diapiric brines and diapir material on the reef complex has not yet been observed. Ancient examples of diapir-influenced reefs outside the Vasco-Cantabrian Basin are known from the Albian of Tunisia, which exhibit typical collapse structures. Such diapir reefs are of considerable value in oil and gas exploration (Maurin 1980).

Acknowledgments. The author thanks Dr. T. Engeser and Dr. E. Gierlowski-Kordesch for assistance in translation and critical reading, and the Comisíon Nacional de Geología Madrid (Spain) for providing access to the field area (100/80). The field work was financed by the Deutsche Forschungsgemeinschaft (SFB 53, F20).

The studies were carried out in the Geological Institute of the University of Tübingen and in the Institute of Paleontology of the Free University of Berlin. Some SEM analysis were carried out at the Bundesanstalt für Materialprüfung (BAM) in Berlin.

References

Assereto R, Folk RL (1980) Diagenetic fabrics of aragonite, calcite, and dolomite in an ancient peritidal-spelenan environment: Triassic calcare rosso Lombardia, Italy. J Sediment Petrol 50:371 – 394

Bechstaedt T (1974) Sind Stromatactics und radiaxial-fibroeser Calcit Faziesindikatoren? N Jahrb Geol Palaeontol Monatsh 1974:643 – 663

Blank R (1983) Stratigraphie und Entwicklung der Oberkreide im Westteil des Diapir von Villasana de Mena (Prov. Burgos, Norspanien). Diplom Thesis, Eberhard-Karls Univ Tübingen

Bouroullec J, Deloffre R (1982) Les Paléosebkhas du Jurassique Terminal en Aquitaine (SW France). Bull Cent Rech Explor Prod Elf-Aquitaine 6:227 – 255

Brinkmann R, Loegters H, Pflug R, von Stackelberg U, Hempel PM, Kind HD (1967) Diapir-Tektonik und Stratigraphie im Vorland der spanischen Westpyrenaeen. Beih Geol Jahrb 66:1 – 183

Ellison SP (1971) Sulfur in Texas. Bur Econ Geol 2:1 – 48

Fabricius FH (1961) Die Strukturen der „Rogenpyrite" (Koessener Schichten, Raet) als Beitrag zum Problem der „Vererzten Bakterien". Geol Rundsch 51:311 – 319

Feely HW, Kulp KL (1957) Origin of Gulf Coast Salt Dome Sulphur Deposits. Am Assoc Petrol Geol Bull 41:1802 – 1853

Fischer AG (1964) The Lofer cyclothems of the Alpine Triassic. Bull Kansas Geol Surv 169:107 – 149

Folk RL, Land LS (1975) Mg/Ca ratio and salinity: two controls over crystallization of dolomite. Am Assoc Petrol Geol Bull 59:60 – 68

Friedman GM, Shulka V (1980) Significance of authigenic quartz euhedra after sulphates: example from the Lockport Formation (Middle Silurian) of New York. J Sediment Petrol 50:1299 – 1304

Grimm WD (1962) Idiomorphe Quarze als Leitmineralien für salinare Fazies. Erdöl Kohle Erdgas Petrochem 11:880 – 887

Hanor JS (1978) Precipitation of beachrock cements: mixing of marine and meteoric waters versus carbon dioxide degassing. J Sediment Petrol 48:489 – 501

Harder H, Flemhig W (1967) Bildung von Quarz aus verdünnten Lösungen bei niedrigen Temperaturen. Naturwissenschaften 54:140

Harder H, Menschel G (1967) Quarzbildung am Meeresboden. Naturwissenschaften 54:561

Heckel PH (1972) Possible inorganic origin for stromatactis in calcilutite mounds in the Tully Limestone, Devonian of New York. J Sediment Petrol 42:7 – 18

Kendall AC (1985) Radiaxial fibrous calcite: a reappraisal. In: Schneidermann N, Harris PM (eds) Carbonate cements. Soc Econ Paleontol Miner Spec Publ 36:59 – 77

Kendall AC, Tucker P (1973) Radiaxial fibrous calcite: a replacement after acicular carbonate. Sedimentology 20:365 – 389

Lietz J (1951) Sulfidische Klufterze im Deckgebirge des Salzstockes Reitbrook. Mitt Geol Staatsinst Hamburg 20:110 – 118

Lippmann F (1973) Sedimentary Carbonate Minerals. Springer, Berlin Heidelberg New York Tokyo

Lippmann F, Schlenker B (1970) Mineralogische Untersuchungen am Oberen Muschelkalk von Haigerloch (Hohenzollern). N Jahrb Miner Abh 113:68 – 90

Longman MW (1980) Carbonate diagenesis textures from nearshore diagenetic environments. Am Assoc Petrol Geol Bull 64:461 – 487

Lotze F (1953) Salzdiapirismus im nördlichen Spanien. Z Dtsch Geol Ges 105:814 – 822

Mattes BW, Mountjoy EW (1980) Burial dolomitisation of the Upper Devonian Miette Buildup, Jasper National Park, Alberta. Soc Econ Paleontol Miner Spec Publ 28:259 – 297

Maurin AF (1980) Halocinèse et sédimentologie – un thème de réflexion. Trav Lab Geol Hist Paleontol 11:18–19

Meschede M (1985) The geochemical character of volcanic rocks of the Basco-Cantabrian Basin, northeastern Spain. N Jahrb Geol Palaeontol Monatsh 1985:115–128

Neugebauer J, Ruhrmann G (1978) Experimentelle Karbonatzementation: Syntaxialer Calcit auf Echinodermen. N Jahrb Geol Palaeontol Monatsh 1978:545–555

Oldershaw AE, Scoffin TP (1967) The source of ferroan and non-ferroan calcite cements in the Halkin and Wendlock Limestones. Geol J 5:309–320

Pierson BJ, Shinn EA (1985) Cement distribution and carbonate mineral stabilization in Pleistocene limestones of Hogsty Reef, Bahamas. In: Schneidermann N, Harris PM (eds) Carbonate Cements. Soc Econ Paleontol Miner Spec Publ 36:153–168

Purser BH (1969) Synsedimentary marine lithification of Middle Jurassic limestones in the Paris Basin. Sedimentology 12:205–230

Ramirez del Pozo J (1971) Bioestratigrafia y microfacies del Jurásico y Creatácio del Norte de España. Mem Inst Geol Miner España 78:1–357

Reitner J (1980) Fazies, Bau und Stratigraphie der Riffkalk-Zone (Mittel-Oberalb) im Raum Araya-Alsasua (Prov. Alava/Navarra, Nordspanien) und ein Vergleich mit dem Caniego-Riffkalk (Oberalb) bei Villasana de Mena (Prov. Burgos, Nordspanien). Diplom Thesis, Eberhard-Karls Univ, Tübingen

Reitner J (1982) Die Entwicklung von Inselplattform und Diapir-Atollen im Alb des Basko-Kantabrikums (Nordspanien). N Jahrb Geol Palaeontol Abh 165:87–101

Reitner J (1985) Mikrofazielle, paloekologische und palaeontologische Analyse ausgewählter Vorkommen flachmariner Karbonate im Basko-Kantabrischen Strike-Slip Fault-Becken-System (Nordpanien) an der Wende von der Unterkreide zur Oberkreide. Dissertation, Eberhard-Karls Univ, Tübingen

Richter DK (1972) Authigenic quartz preserving skeletal material. Sedimentology 19:211–218

Sandberg PA (1975) New interpretation of Great Salt Lake ooids and of ancient non-skeletal carbonate mineralogy. Sedimentology 22:497–537

Sandberg PA (1983) An oscillating trend in Phanerozoic non-skeletal carbonate material. Nature 305:19–22

Schneider W (1977) Diagenese devonischer Karbonatkomplexe Mitteleuropas. Geol Jahrb 21(D):1–107

Schroeder R (1980) Le calcaire de Caniego: témoin d'une montée du diapir de Valle de Mena (Prov. Burgos, Espagne) dans l'Albian supérieur. Cuad Geol Iber 5:221–225

Schroeder R, Willems H (1983) Über einen submarinen Durchbruch des Diapirs von Villasana de Mena (Prov. Burgos, N-Spanien) an der Wende Unter-/Oberkreide. N Jahrb Geol Palaeontol Abh 166:65–85

Stackelberg von U (1967) Der Diapir von Murguia. Beih Geol Jahrb 66:63–94

Tsien MM (1985) Origin of stromatactis – a replacement of colonial microbial accretions. In: Toomey DF, Nitecki MH (eds) Paleoalgology. Springer, Berlin Heidelberg New York, pp 274–298

Micrite Diagenesis in Senonian Rudist Build-ups in Central Tunisia

A. M'RABET[1], M. H. NEGRA[1], B. H. PURSER[2], S. SASSI[1], and N. BEN AYED[3]

1 Introduction

Studies of reef diagenesis, intensified during the last two decades, have demonstrated the importance and the variability of diagenetic processes ranging from marine cementation to freshwater dissolution and neomorphism. Many examples, not all of which are reefs, are compiled in two volumes on *carbonate cements*, one edited by Bricker (1971), and the other by Schneidermann and Harris (1985). These diagenetic modifications have been recognized both in modern and ancient reefs generally composed of corals and algae. Noncoral reefs, including rudist build-ups, have received less attention, one notable exception being the contribution of Enos in this Volume.

Although sedimentary micrite is an important constituent in rudist reefs and, indeed, in most fossil reefs, it has received remarkably little attention. This is understandable. The coarser textures are often filled with spectacular, photogenic cements whose origins, although often debated, are relatively easy do deduce; their age, relative to the sediments they cement, is generally fairly clear, and the distinction between synsedimentary, shallow burial and deep burial origins is discussed, frequently with conviction.

In fine-grained sediments, obviously, the nature of the diagenetic phase is not readily defined and the distinction between sedimentary and diagenetic micrite presents many well-known problems. In this somewhat discouraging context a brief study has been made of the micrites comprising two Upper Cretaceous mounds in Central Tunisia.

The main objectives of this study were:

- to distinguish between the sedimentary and diagenetic fabrics within the micrites;
- to define the diagenetic effects responsible for the early lithification of these micrites;
- to interpret the environmental conditions under which this early diagenesis was active.

[1] Département des Sciences de la Terre, Faculté des Sciences de Tunis, Campus Universitaire, 1060-Tunis, Tunisia

[2] Laboratoire de Pétrologie Sédimentaires et Paléontologie, Bâtiment 504, Université Paris Sud-Orsay, 91405 Orsay Cedex, France

[3] Ecole Normale Supérieure de Bizerte, Bizerte, Tunisia

Reef Diagenesis
Edited by J. H. Schroeder and B. H. Purser

These objectives have been partly attained. However, as may be expected, the authors have been confronted with the habitual problems relating to micrite diagenesis.

The Upper Cretaceous rudist build-ups in Central Tunisia are relatively favorable sites for the study of micrite diagenesis. Not only are they composed mainly of fine sediment, but they occur under desert conditions with only very moderate weathering. More importantly perhaps, these bodies exhibit clear evidence of early lithification, notably in the form of thick conglomerates comprising lithified material derived from the adjacent rudist build-ups.

2 Sedimentary Characteristics of the Rudist Reefs

2.1 The Reef Complex at Jebel el Kebar

Several Senonian rudist reef complexes occur in Central Tunisia, one of which has been studied in considerable detail at Jebel el Kebar by Khessebi (1978) and by Negra (1984). Here, a series of lenticular bodies crop out along the SW, subvertical flank of the Kebar anticline (Fig. 1 b). This steep structural dip, combined with considerable morphological relief, enables three-dimensional study of the rudist build-ups. The reef bodies are developed on a disconformity situated at the top of an open marine shale, the Aleg Formation (Unit S_1, Fig. 1 c) which is marked by ferruginous crusts and iron-stained conglomerate.

The reef complex comprises a series of asymmetric rudist reef lenses laterally separated by conglomeratic deposits (Figs. 1 c, 2). Rudist lenses attain 60 m in thickness and their lateral extension varies from 150 m to about 1 km. Generally, the lower part of these massive lenses is composed of wackestones with scattered rudists, echinoid debris, pelletoids and some calcispheres. The middle and upper parts are constructed mainly of hermatypic corals and, especially, by various rudists including abundant Hippuritids (*Hippurites, Vaccinites*), Radiolitids (*Biradiolites, Pseudopolyconites* and *Pironea*) and some Caprinids (*Sabinia*). Rudists occur in vertical or slightly inclined growth position (Fig. 3 B) either joined or loosely packed, interrudist spaces being filled by fine-grained bioclastic sediments. In addition to these partly bioconstructed bodies, some relatively smaller lenses, less than 15 m in thickness, are totally built of rudists and associated corals.

The rudist reef bodies are separated by conglomeratic lenses up to 20 m thick and 100 m in lateral extent. The contacts between rudist reef and adjacent conglomerates are subvertical and sharp. They correspond to synsedimentary cliffs. These possibly tectonic fractures are sealed by conglomerates which comprise millimeter to meter sized clasts and blocs whose rudist debris and mud- to grain-supported textures are similar to those of the neighboring rudist reef bodies. These reef-derived clasts and blocks occur in a wackestone matrix containing calcispheres and planktonic foraminifera. These lateral facies relationships indicate that the rudist bodies formed relatively shallow, partly lithified mounds, whose related conglomerates have been trapped in adjacent, somewhat deeper depressions.

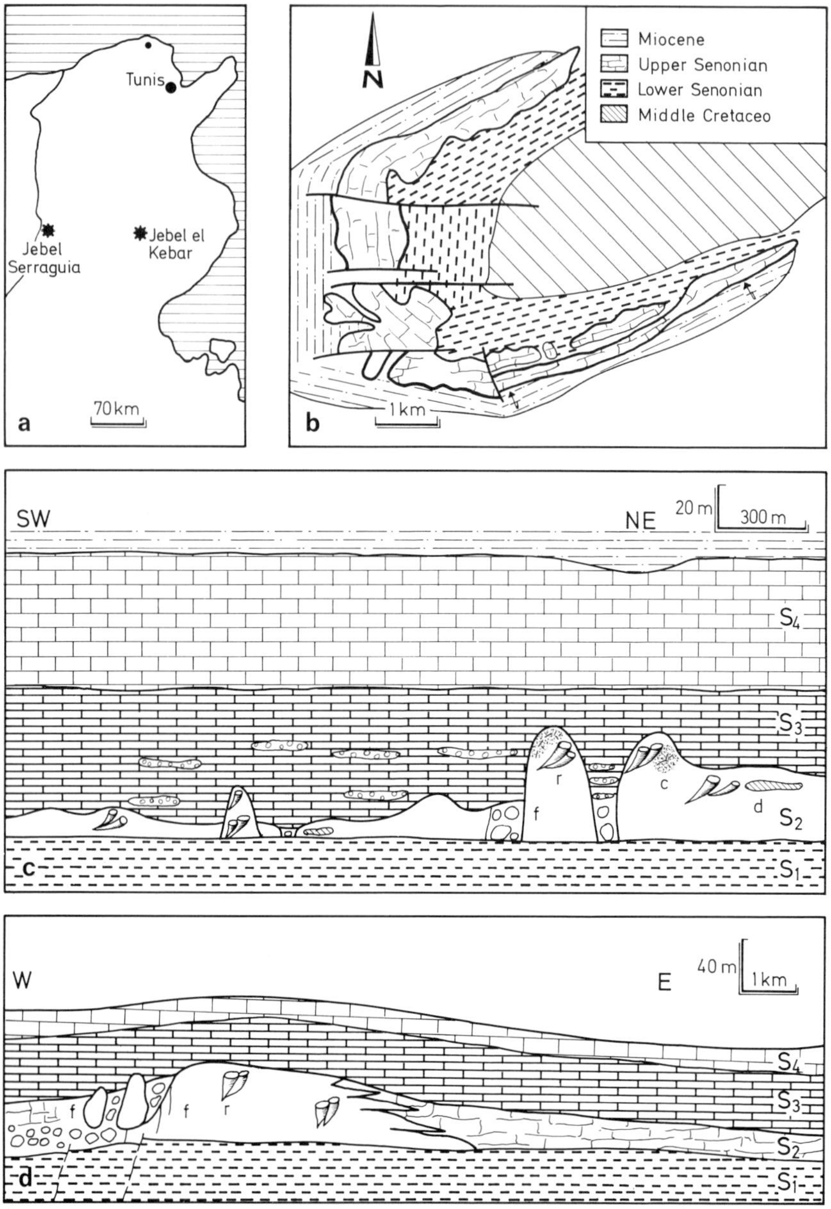

Fig. 1. **a** Location map of the two Senonian rudist reef complexes at Jebel el Kebar and Jebel Serraguia. **b** Simplified geological map of the south western part of Jebel el Kebar anticline especially showing the lateral extension of the Senonian carbonates. **c** and **d** Simplified diagrams of the Senonian rudist reef complexes (*S2*) and associated sequences (*S1, S3, S4*) at Jebel el Kebar (**c**) and Jebel Serraguia (**d**) respectively (*d* dolomite lenses; *r* rudistids; *c* corals; *f* faults)

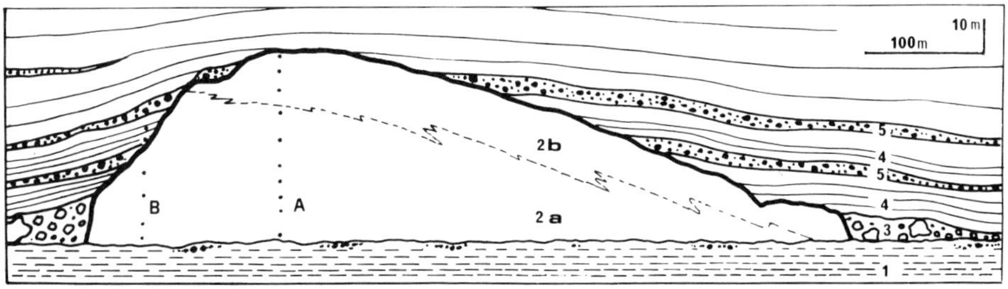

Fig. 2. Simplified section through a reef body at Jebel el Kebar showing the principal sedimentary units. *1* Aleg Formation with local conglomerates at top (unit S_I, Fig.1); *2* rudist mound; *2a* essentially micritic with scattered rudists; *2b* micrite with abundant rudists and scattered corals (unit S_2); *3* coarse conglomerate derived from reef (unit S_2); *4* fine-grained, slightly argillaceous mudstone with planktonic forams (unit S_3, Fig. 1). *5* Fine conglomerate (unit S_4). *A* and *B* are profiles; samples, indicated by points, have been studied under SEM and analyzed for Sr

Fig. 3. A General view at Jebel el Kebar showing a structurally overturned rudist reef lens (*1*) passing laterally into conglomeratic flank facies (*2*) and unconformably overlain by unit S_3 open marine mudstones (*3*). **B** Rudists (Hippuritids) in growth position within a reef body at Jebel el Kebar (Note pencil for scale)

The rudist reef complex and its associated conglomerates (unit S_2, Fig. 1 b) is overlain by 100 – 150 m of well-bedded mudstone (units S_3 and S_4, Fig. 1) containing planktonic foraminifera, thin lenses of conglomerate and occasional, relatively small (2 × 10 m) rudist-bearing lenses of massive carbonate. The Upper Cretaceous sequence is topped by an erosion surface overlain by Miocene clastics.

Fig. 4

2.2 The Rudist Build-up at Jebel Serraguia

A second Senonian reef body at Jebel Serraguia (Figs. 1 d, 4; Bernet-Rollande and Philip 1981, M'Rabet et al. 1985) has been studied, mainly for comparison with those at Jebel el Kebar. The reef is a single, asymmetric lense some 60 m thick the lower parts of which are a bioaccumulation of slightly reworked rudists and corals formed in shallow marine, but somewhat agitated environments.The upper part is clearly constructed, mostly by rudists. Towards the West, the rudist build-up exhibits a paleoslope locally overlain by conglomerates, and is affected by synsedimentary faults and fractures filled by reef-derived clasts. Here, the sloping rudist reef passes laterally into breccia and conglomeratic sediments containing clasts and blocks of millimeter to decameter size (Fig. 4). Certain megablocs are true olistoliths: the largest, measuring 40 m in size, is composed of beds tilted to 90° from the original horizontal position (Fig. 4 B). These reef-derived clasts and blocs are associated with a matrix rich in calcispheres, bryozoans, orbitoids and reworked lithoclasts.

As at Jebel el Kebar, this striking lensoid mass (Fig. 4 A) is overlain by well-bedded basinal lime muds (units S_3 and S_4, Fig. 1 d), with planctonic foraminifera.

3 Diagenesis of the Rudist Reef Complexes

As already noted, the essential part of this study concerns the micrites comprising the complex at Jebel el Kebar. However, the diagenesis of these Senonian lime muds must relate to the overall diagenetic evolution of the reef complex, the general aspects which are summarized as follows.

3.1 General Diagenetic Attributes of the Rudist Reefs and Related Sediments

3.1.1 Diagenesis of the Rudist Reefs

Rudists within the lenses are partly dissolved. Dissolution is frequently selective, the original aragonitic components of the skeleton having been replaced by void-filling, sparry calcite mosaics. The calcitic parts, however, are altered neomorphically to microspar, often with ghosts of the initial skeletal structure. When skeletal debris has been micritized, the envelopes are preserved selectively. This selective dissolution of the rudists could indicate the presence of meteoric waters relatively early (before aragonite inversion) in the diagenetic history of the reefs.

Fig. 4A–C. Reef body at Jebel Serraguia. **A** General view from Jebel Serraguia showing the lenticular rudist reef body (*1*) and associated fore-reef breccia (*2*) unconformably overlain by open marine mudstones of unit S_3. **B** General view showing a megablock (*1*) 40 m in height which is situated laterally to the reef body. It is overlain by well bedded open marine lime muds (*2*) of units S_3. **C** Detail of fore-reef conglomerates (scale is 5 cm)

The lower parts of many rudist mounds include well-defined lenses of brown dolomite each averaging 1 m in thickness and 20 m in length, the dolomite having selectively replaced particular facies of the rudist reefs, thus being probably a Senonian diagenetic phenomenon. The dolomites are characterized by relatively light $\delta^{18}O$ signatures which average -3.6 while the $\delta^{13}C$ averages $+0.2‰$ POB, suggesting dolomitization from mixed or, possibly, from non marine waters (J-C Fontes, personal communication). These dolomite lenses are generally silicified and dedolomitized, possible additional evidence of non marine diagenesis.

3.1.2 Diagenetic Characteristics of the Conglomerates

Because the abundant rudist-bearing conglomerate boulders and gravels clearly express an important early diagenesis of the rudist lenses, their diagenetic properties are of obvious interest. Many cobbles are composed of micrite, the diagenetic

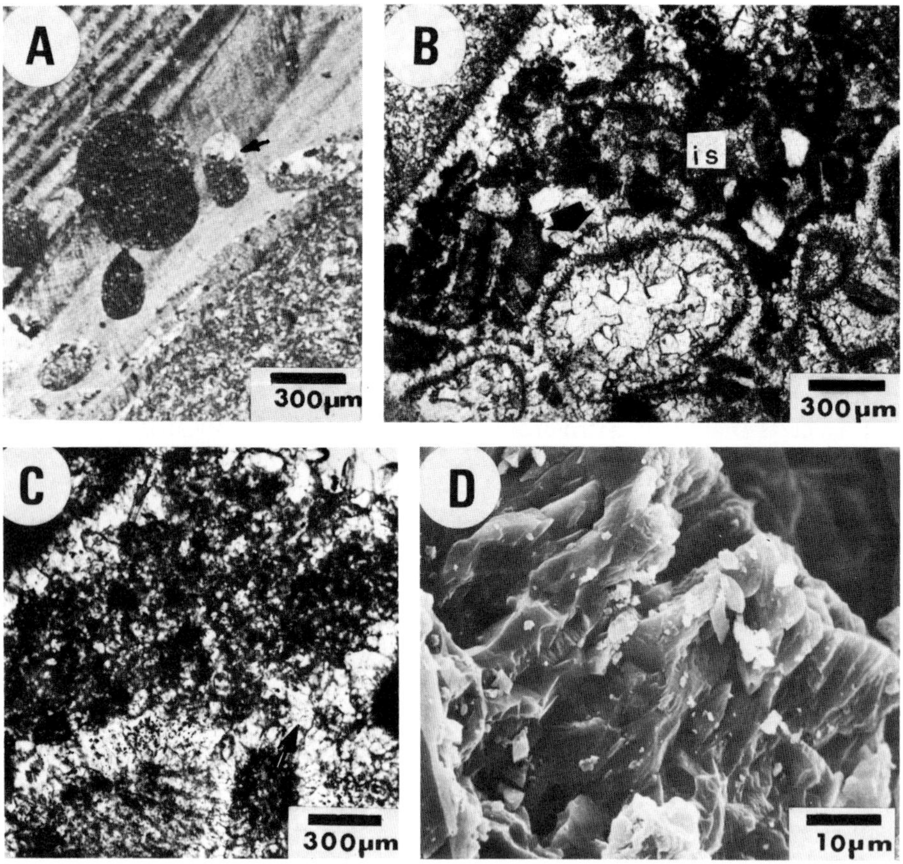

Fig. 5. A Rudist debris affected by perforations completely or partly (*arrow*) filled by micritic sediment. **B** Thin section from the rudist reef body and **C** a derived clast showing early blocky calcite cement (*arrow*) overlain by internal sediment. **D** SEM photo of the early cement

aspects of which are discussed in the following section. Gravels composed of bioclastic packstone and, rarely, grainstone are cemented by a well-developed microspar. Occasionally, component grains are coated by an isopachous crust of blocky calcite crystals (Fig. 5) overlain by internal sediment.

Rarely, pebbles reworked from the reef show traces of early dissolution and possible pedogenesis; one sample included fossil molds filled with skeletal debris, while another exhibited solution channels filled with brown micrite. The conglomeratic elements therefore show occasional traces of meteoric diagenesis which must have affected the source area, i.e., the rudist reef. Although very clearly defined in the field (Fig. 3 A), the tops of the rudist mounds do not show traces of caliche crust, karst, fissuration or any other features commonly associated with emergent surfaces. In addition, boulders obviously reworked from the adjacent reef, only rarely exhibit solution cavities filled by ferruginous or other material of possible pedological origin. Therefore, the authors, while proposing an early meteoric diagenesis of the rudist reef complex at Jebel el Kebar, nevertheless stress that evidence for nonmarine diagenesis is far from conclusive.

3.2 Diagenesis of the Micrite and the Problem of Microsparitization

It is recalled that micrite, generally with silt-sized bioclastic debris, foraminifera, and coccoliths, is the major constituent, both of the lensoid rudist reefs, the conglomeratic detritus which surrounds these reefs, and the slightly argillaceous lime muds enclosing the conglomerates. It is clear that those micrites comprising both the rudist mounds and the fine-grained conglomeratic elements must include an early diagenetic phase, the identification of which constitutes one of the essential aims of this research. Its identification is based on a simple comparative study of these micrites.

3.2.1 The Micrites and Microsparites Within the Rudist Reefs

Under SEM these are seen to consist of several microfabrics:

- Equant grains of micrite (Fig. 6A) measuring 1 – 5 μm in diameter are the main component. Because they are fairly well sorted and generally have a regular morphology, they are unlikely to be of bioclastic origin. However, in view of the homogeneous nature of certain microdebris such as coccolith fragments, this cannot be entirely excluded. These subovoid microparticles generally exhibit overpacked fabrics with long mutual contacts, indicating that the present form is modified by diagenetic overgrowth. This may lead to weakly developed crystal faces, although the individuality of each micrite particle is respected. This relatively simple microfabric is similar to the overgrowth fabric of Folk (1965) and the "structure punctique" of Loreau (1972).
- Amoeboid crystals of microspar (Fig. 7) measuring 5 – 30 μm in diameter grow within the micro-voids and do not digest individual micrite grains. This may progressively envelope these elements, resulting in aggregates, the resulting porphyroblastic fabric being similar to that described by Folk (1965).

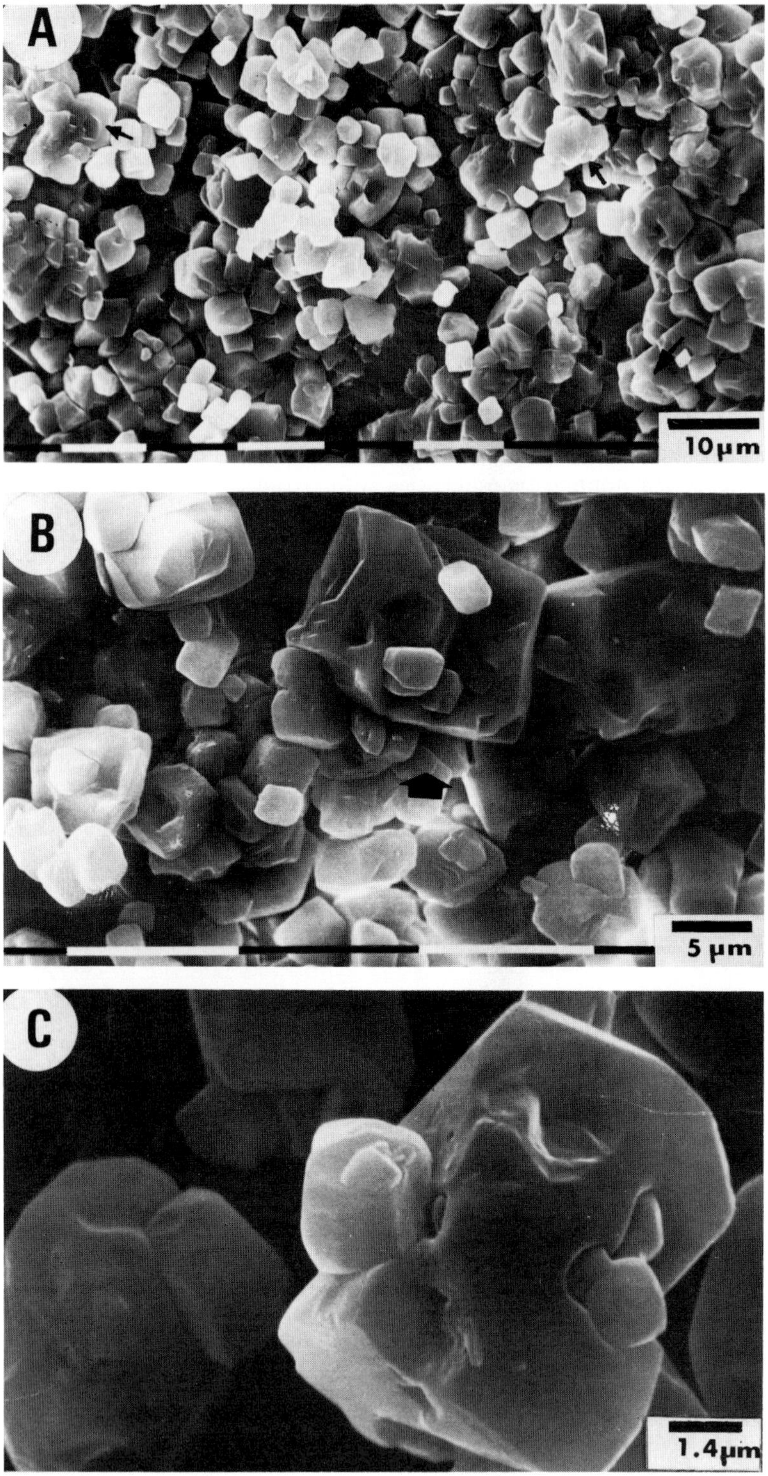

Fig. 6

These aggregates float within the micrite and, at least during the early stages of diagenesis, do not form a dense mosaic, the initial microspar not forming by dissolution-precipitation or by grain growth.

- Subhedral composite microspar crystals 5 – 30 μm in size are also common (Fig. 6). The faces of many crystals have rounded protuberances, suggesting that the crystal has envelopd (but not necessarily dissolved) micrite grains. This microspar thus has inclusions and probably represents a somewhat more advanced growth of amoeboid microspar.
- Euhedral microspar may form when crystals grow into voids (Fig. 7c). It probably forms also as the later stages of composite microspar development.

SEM study suggests that microspar in these Senonian reefs is formed mainly as cement develops between grains of micrite which are progressively enveloped. Thus, these grains probably remain as distinct inclusions with the crystals of microspar. The process of dissolution-precipitation often invoked under the term neomorphism by Folk (1974), although possibly existing, certainly is not the predominant process in microspar development.

Vertical and lateral distribution of microspar within the reef: Because the nature and intensity of microsparitization may help in understanding the origins and the relative age of this particular diagenesis, two profiles were sampled within a particular reef body (Fig. 2).

Profile A passes through the thickest part of the rudist mound and presumably includes the bathymetric culmination. In all samples, including those from the base, the micrite grains are overpacked and grain growth is clear. However, there is little microspar in the lower parts of the lens. Within the upper parts of the profile, richer in rudists and their debris, micrite grains are similarly interpenetrating. However, this fabric is associated with numerous microspar aggregates and euhedral microspar crystals.

Profile B, which traverses the flank of the same high, although including overpacked micrite throughout, has little microspar. It would seem therefore that microspar is best developed near the culmination of the rudist mound where its presence results in a hard, fine-grained, highly crystalline limestone. Its localization may reflect either a slightly coarser initial sedimentary fabric which favored microspar cementation, or, more likely, an intense calcite diagenesis possibly relating to the emergence of the reef.

3.2.2 The Micrites and the Microsparites Within the Conglomeratic Elements and Their Matrix

Comparative study of the diagenetic fabrics within the micritic gravels and their surrounding mud matrix enables the identification of the early diagenetic phase. Although the differences are not striking, the micrite composing most gravels

Fig. 6. A and **B** slightly overpacked textures composed of micrite grains and composite microspar (*arrow*) from the wacke-packstones of the rudist build-ups. **C** A composite, subhedral microspar crystal with micritic inclusions

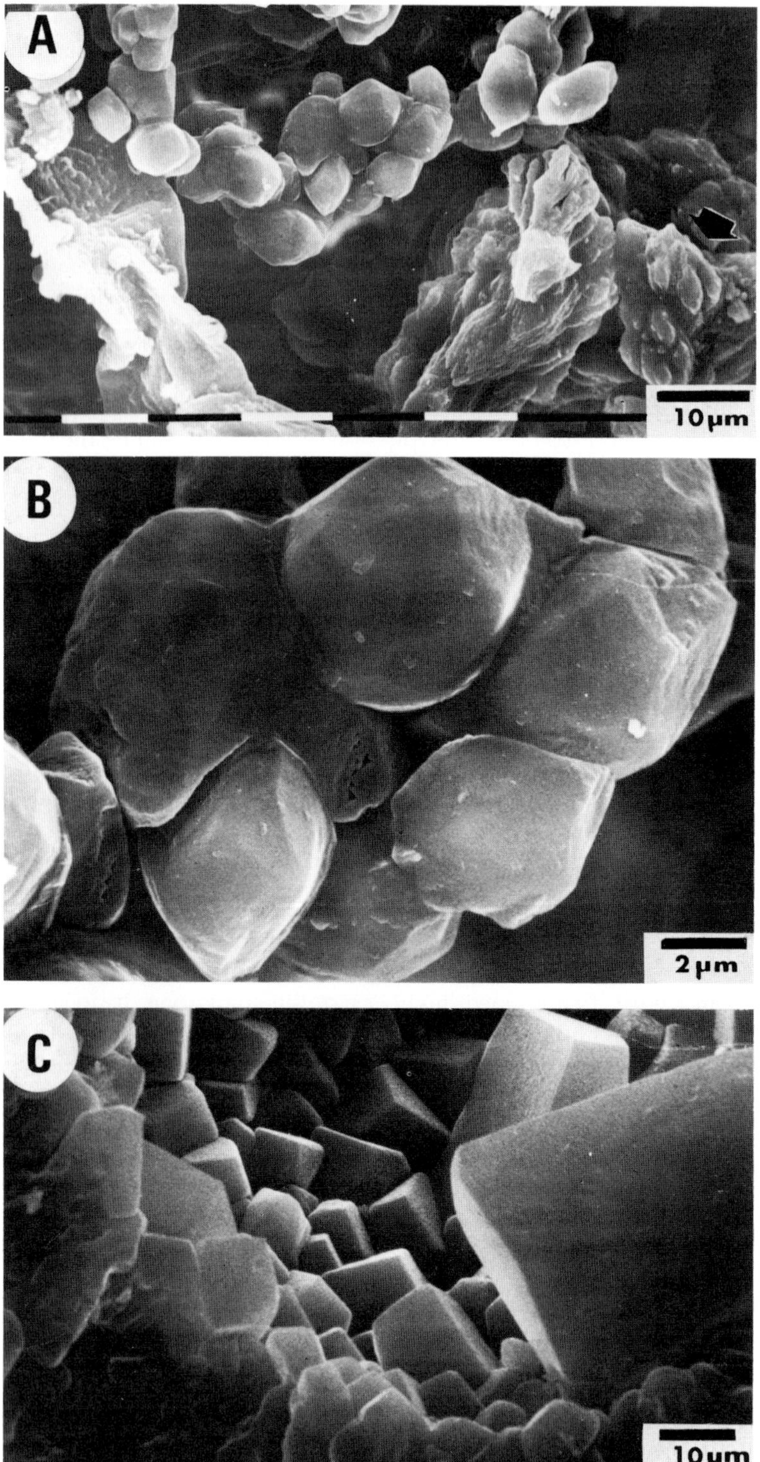

Fig. 7

includes both overpacked micrite grains *and* composite microspar crystals. The surrounding sediments, generally rich in coccoliths and pelagic foraminifera, while including overpacked micrite fabrics, have less microspar.

3.2.3 Strontium Concentration in Micrites

Sr has been measured both in the reef and in the overlying open marine carbonates at Jebel el Kebar; it was presumed that micrites affected by meteoric waters would have lower Sr concentrations. Results are not conclusive; while differences do exist, they are relatively minor. Sr in micrites from the reef body average 175 ppm, those in the overlying flank beds not affected by early diagenesis having 305 ppm. As the initial mineralogical composition of these two micrites is not known, these minor differences have only limited value.

3.2.4 Conclusion Concerning Micrite Diagenesis

This comparative study indicates that the diagenetic product responsible for the early lithification of the micrite within the rudist reefs, and therefore the overall consolidation of much of the reef body, is microspar exhibiting various types of fabric. The diverse microspar fabrics, while being concentrated near the top of the reef body, do not appear to have any clearly defined vertical or lateral distribution. It is possible, nevertheless, that aggregate microspar represents the early stages of microspar evolution, and that the progressive loss of porosity within the micrite reflects the progressive evolution towards precipitation of subhedral and euhedral microsparite.

3.3 Diagenesis of the Rudist Reef at Jebel Serraguia: Comparison with the Rudist Reef Complex at Jebel el Kebar

The diagenetic phenomena visible within the lenticular reef body, in spite of its relatively grainy textures, are not spectacular. They include ubiquitous dissolution of the aragonitic rudist components and the development of sparitic cements within the molds. Bioclastic sediments, both within the reefs and in the individual gravels surrounding the reef body, are locally cemented by a thin, isopachous fringe of blocky or somewhat bladed calcite, at least some of which may be marine in origin.

The diagenetic fabric within the micritic fraction, both in the reef body and within individual micritic gravels surrounding the reef, as at Jebel el Kebar, include all types of aggregate microspar crystals (Fig. 8). Microspar fabrics are

←

Fig. 7A–C. Microspar fabrics from the reef body at Jebel Serraguia. **A** Composite microspar (*arrow*) and aggregates associated with probable microbioclastic debris. **B** Detail of microspar aggregate; note that the microspar tends to surround the micrite grains. **C** Euhedral microspar gowing as a cement within a microvoid

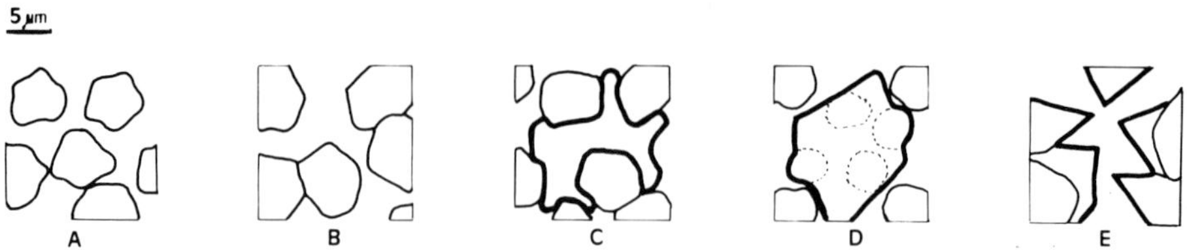

Fig. 8 A – E. Sketch of principal micrite and microsparite fabrics. **A** Equant grains of micrite with no overgrowth (see Fig. 6A). **B** Micrite showing long grain contacts and thus overgrowth (see Fig. 6A). **C** Amoeboid microspar forming an aggregate grain (see Fig. 7A, B). **D** Subhedral, composite microspar crystal with probable inclusions (see Fig. 6B, C). **E** Euhedral microspar crystals growing in microvoid (see Fig. 7C)

rare within the open marine micrites enveloping the conglomerates, suggesting that the microspar, preferentially developed in the conglomerates, is an early diagenetic product. As at Jebel el Kebar, it is the principal cement responsible for lithification of the reef body.

4 Discussion and Conclusions

The two rudist build-ups at Jebel el Kebar and at Jebel Serraguia, in spite of their somewhat different sedimentary properties, have both been affected by an important phase of diagenesis whose main expression is the reef-derived conglomerates flanking each reef. This early consolidation is also confirmed by the subvertical limits of certain parts of the reef.

This important erosion of both reef bodies suggests possible common hydrodynamic effects due to a relative lowering of sea level. Whatever the fundamental cause, both lithification and erosion are major factors determining the geometry of these build-ups. The principal cause of early lithification appears to be microspar, which is abundant both within the reef bodies and in their erosional products, the conglomerates.

This study, although inconclusive in certain respects, nevertheless shows that microsparitization may well record phases of reef emergence. Furthermore, the diagenesis of micrites influences the petrophysical properties of the reef body and thus its reservoir potential. Thus, in spite of the classical difficulties relating to the study of fine-grained sediment, their post-sedimentary evolution is an important part of reef diagenesis.

Acknowledgments. We would like to thank A. Saadallaoui, A. Raguideau and H. Hemissi for technical help, Dr. J. P. Loreau for fruitful discussion, and Dr. J. H. Schroeder for reviewing of the manuscript. This study was, in part, supported by the Ministère Tunisien de l'Enseignement Supérieur et de la Recherche Scientifique DRST program T4/82/FST and by the DRST-CNRS Convention.

References

Bernet-Rollande MC, Philip J (1981) Sur la découverte d'un récif à Caprinidés (rudistes) dans le Sénonien supérieur du Djebel Serraguia (Tunisie Centre Occidentale). C R Hebd Séance Acad Sci Paris 292 II:417–422

Bricker OP (ed) (1971) Carbonate cements. Johns Hopkins Univ, Stud Geol 19:376

Folk RL (1965) Some aspects of recrystallization in ancient limestones. In: Pray LC and Murray RC (eds) Dolomitization and Limestone Diagenesis, a symposium. Soc Econ Paleont Mineral Spec Publ 13:14–48

Folk RL (1974) The natural history of crystalline calcium carbonate: effect of magnesium content and salinity. J Sediment Petrol 44:40–53

Khessibi M (1978) Etudes géologiques du secteur de Maknassy-Mezzouna et du Jebel el Kebar (Tunisie Centrale). Thesis, Univ Lyon

Loreau JP (1972) Pétrographie de calcaires fins au microscope électronique à balayage: introduction à une classification des micrites. C R Hebd Séance Acad Sci

M'Rabet A, Sassi S, Tlatli M, Ben Ayed N (1985) Sedimentation and diagenesis of the Senonian rudist reef complex of Jebel Serraguia, Central Tunisia. In: Rosell J, Remacha F, Zamorano M (eds) Abstracts 6th European Regional Meeting Intern Ass Sedimentol Lerida, pp 313–316

Negra MH (1984) Paléoenvironnements et diagenèse des faciès récifaux à rudistes au Jebel el Kebar, Tunisie Centrale. Thesis, University of Paris Sud-Orsay

Schneidermann N, Harris PM (1985) Carbonate cements. Soc Econ Paleontol Miner Spec Publ 36:379

Diagenesis of Upper Jurassic Sponge-Algal Reefs in SW Germany

R. KOCH[1] and M. SCHORR[1]

1 Introduction

Sponge-algal reefs are characteristic of the Upper Jurassic of S.Germany where they occur as massive carbonates. Abundant data have been published concerning these "reefs" and their relationship to the adjacent, well-bedded "normal" facies (Gwinner 1976, Aldinger 1968, Wendt 1980). Most publications are concerned with stratigraphic and paleogeographic problems or with general petrographic analyses.

The massive facies with sponge-algal reefs is generally thicker than the equivalent bedded facies. This is mainly due to early cementation and to a higher carbonate production within the reefs, but also to later, differential compaction. The reef bodies are often dolomitized and form the so-called Frankendolomit within the Franconian Alb.

A fundamental interrelationship exists between dolomite and sponge-algal reefs. The so-called "saccaroid limestones" (dedolomite) are also related directly to dolomitization of the reef facies. The theories regarding dolomitization range from "descending dolomitization from the reefs" to the general remark that the dolomitization was "late diagenetic". Lang (1964) and Fritz (1965) describe the distribution of dolomite in the Swabian Alb in detail. Bausch (1963) defined its distribution in the Franconian Alb.

The marls and well-bedded "normal facies" are predominantly calcitic; only minor amounts of dolomite were noted. Dolomite in these sediments is generally abundant only at sites situated close to dolomitized sponge reefs, indicating that dolomitization progressed from these reefs into the adjacent facies (Lang 1964).

The two most important outcrops showing all significant diagenetic features are located near Tieringen and Schmiechen. Lang (1964) reports that dolomite and dedolomite occur as more or less coherent complexes throughout the Swabian Alb, although most abundant in the Middle Kimmeridgian. Fritz (1965) mentions that both dolomite and dedolomite occur in wells in the W Molasse Trough, as described by Schneider (1962). Dolomite was found in the Middle Kimmeridgian and Lowermost Tithonian in the Ehingen-1 and Gaisbeuren-1 wells. Dedolomite is present in the Gaisbeuren-1 and the Lauben-1 wells, whereas calcitic sponge-algal-reef complexes were found in the Scherstetten-1 and the Heimertingen-1 wells.

[1] Institut für Sedimentforschung der Universität Heidelberg, Im Neuenheimer Feld 236, P.O.B. 103020, 6900 Heidelberg, FRGermany

Reef Diagenesis
Edited by J. H. Schroeder and B. H. Purser

In view of hydrocarbon findings in the Entlebuch-1, Haimhausen-1 and Kinsau-1 wells over the past few years, limestones and dolomites of the Upper Jurassic sponge-algal reefs are currently of considerable interest (Betz and Wendt 1983).

As only limited data on the diagenesis of the sponge-algal reefs is available, the purpose of our paper is to demonstrate the diagenetic history of the reefs by describing the different types of cement and their geochemical properties. We have analyzed samples from reefs ranging from Upper Oxfordian to Lower Tithonian. Time and processes of dolomitization and dedolomitization have also been evaluated. These results permit a new interpretation concerning the water depths in which sponge-algal reefs developed, this interpretation contradicting earlier findings regarding reef environments and their time of possible emergence.

2 Area of Study; Paleogeography; Stratigraphic Framework

The Upper Jurassic (Oxfordian to Tithonian) in S. Germany is distributed over an area of more than 10000 km^2 (Fig. 1) where it is called the White Jurassic because of the light color of the rocks. Carbonate rocks of the Upper Jurassic are more resistant to erosion than the clays and marls of the underlying Middle Jurassic, and therefore form a high plateau with steep escarpments to the N and NW – the Swabian and Franconian Alb. This plateau dips to the SE and disappears under the sediments of the Molasse where it has been penetrated by wells in the "Molasse Trough". In both the Swabian and Franconian Alb, the Upper

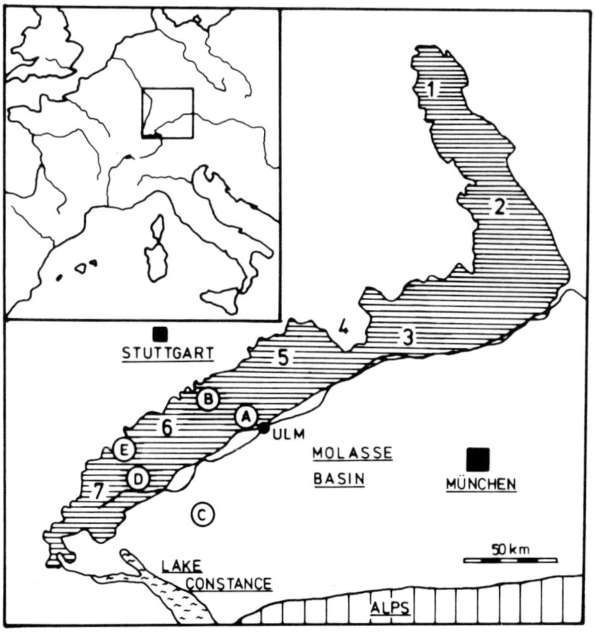

Fig. 1. Map of the Upper Jurassic in S Germany and location of outcrops analyzed (*A–E*). *1* Northern Franconian Alb; *2* Central Franconian Alb; *3* Southern Franconian Alb; *4* Ries; *5* Eastern Swabian Alb; *6* Central Swabian Alb; *7* Southwestern Swabian Alb. (After Ziegler 1977). *A* Valley of the river Blau near Ulm (3 outcrops); *B* Hochwanger Steige; *C* TB-3 Saulgau well; *D* Valley of the river Danube near Sigmaringen (2 outcrops); *E* Tieringen near Balingen (2 outcrops)

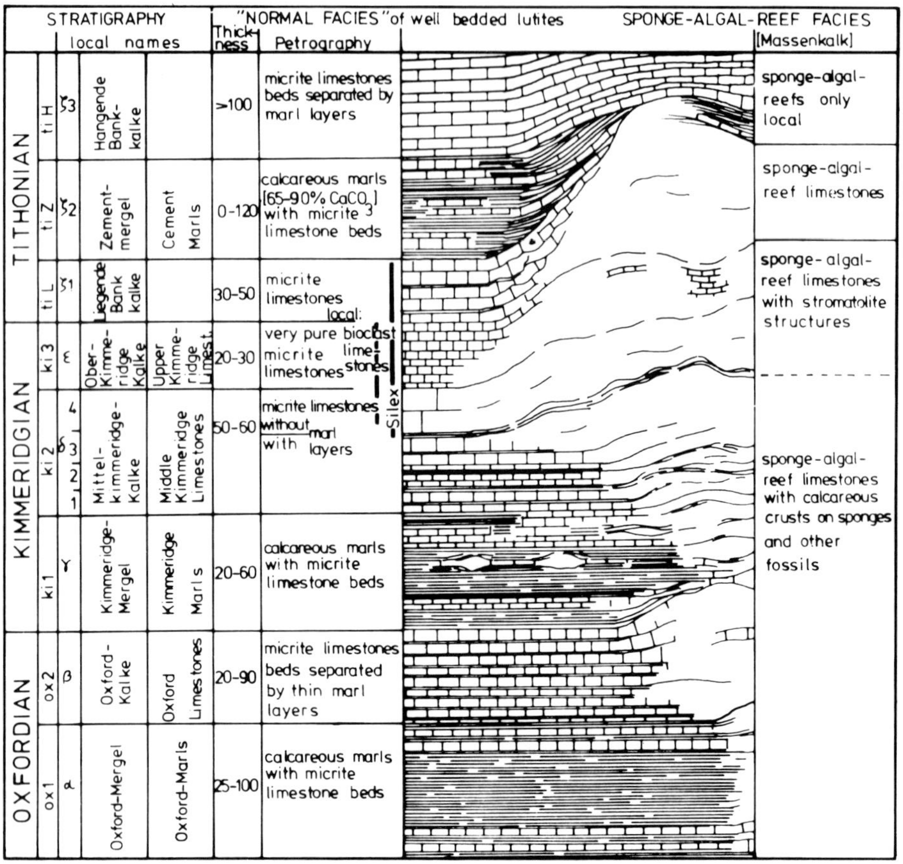

Fig. 2. Stratigraphy and facies relations of the Upper Jurassic of the Swabian Alb. (After Gwinner 1976 and Wendt 1980)

Jurassic shows a very similar development (Flügel and Steiger 1981). In general, three main lithologic types can be distinguished: (1) claystones and marls, (2) limestones of the well-bedded "normal facies", and (3) massive limestones and dolomites comprising the sponge-algal reefs (Massenkalk, Fig. 2).

The sponge-algal reefs started to develop in the Early Oxfordian in the SW of the Swabian Alb. They are most abundant in the Middle and Upper Kimmeridgian and grew on a wide shelf which extended over much of Central Europe. This shelf area was connected to the Tethyan ocean in the south following flooding of the so-called Vindelician Ridge during the Upper Jurassic (Wendt 1980).

The environment of deposition, i.e., the sites of formation of the sponge-algal reefs, the water depth and the submarine relief are problems which can be traced through the literature over the last 40 years.

In general, it has been agreed that the Upper Jurassic in W Europe was a period of a regressive sedimentation and that it shows a consequent shallowing from the Lower Oxfordian to the Upper Tithonian (Meyer 1975). Different water depths have been proposed as possible environments for sponge-algal reef development. Water depths exceeding 150 m and a relief above the sea floor of about 100 m are frequently postulated (Aldinger 1968, Gwinner 1976). On the other hand, Paulsen (1964) mentions a remarkable shallowing in the Lower Tithonian. Aldinger (1945) postulated that the reefs emerged above sea level (subaerial exposure) with possible periodic freshwater influx. This interpretation is supported by isotopic data presented by Fritz (1965). A similar interpretation was also proposed by Bausch (1963), who discussed cyclic subsidence.

3 Reef Morphology, Organic Communities and Vertical Zonation

In general, the sponge-algal reefs show lateral and vertical extension of some meters to some tens of meters. During their maximum development (Middle Kimmeridgian) they attained a thickness of up to 250 m covering several km^2 (Wendt 1980). Intercalated well-bedded facies reveals that these large sponge-algal reef complexes were composed of small individual sponge-algal reefs (Fig. 2).

Siliceous sponges (lithistids and hexactinellids) are characteristic frame-builders which formed a loose, open framework bound by horizontal and dome-shaped algal layers, thus reducing the open pore space to about 50% or less (Wendt 1980). Sponges constitute about 10% of the fauna (Wagenplast 1972), or more (Wendt 1980). They are plate- to saucer-shaped and may reach diameters of 1 m. In the sponge-algal reefs, the siliceous skeletons are usually replaced completely by calcite and a "pseudopelletoid-fabric", which preserves the gross outer shape of the sponges. Other components of the reef-dwelling fauna are terebratulid and rhynchonellid brachiopods and pelecypods (Wendt 1980).

In the adjacent well-bedded facies, ammonites and belemnites are abundant. Siliceous sponges found therein were generally transported from the nearby reefs (Wagenplast 1972).

The detailed study of sponge-algal reefs (Middle Kimmeridgian to Lower Tithonian) (50 to 60 m SW – NE elongation and 16 m high) mainly in a quarry near Ulm (Schorr and Koch 1985), demonstrated a clear vertical zonation (Fig. 3) within the reefs. This zonation is based mainly on the distribution of (1) ooids together with small intraclasts and pellets, (2) calcitic bryozonas, (3) primary aragonitic bryozoans, (4) microproblematicum-1, (5) sponge spicules, (6) finely dispersed clay patches and (7) geochemical data. Siliceous nodules often occur near the base of the reefs.

The vertical zonation was also found in reefs in other outcrops (Fig. 1) and in the TB-3 Saulgau well. In some reefs a possible facies zonation is already indicated by the distribution of dolomite (Fig. 4). Reefs characterized by facies zonation are also described in the Franconian Alb (Flügel and Steiger 1981).

The reefs examined are very similar to the ellipsoidal build-ups with a predominance of mud, described in Wilson's (1975) classification. It might therefore be more correct to call the reefs sponge-algal mud mounds. Their zonation is

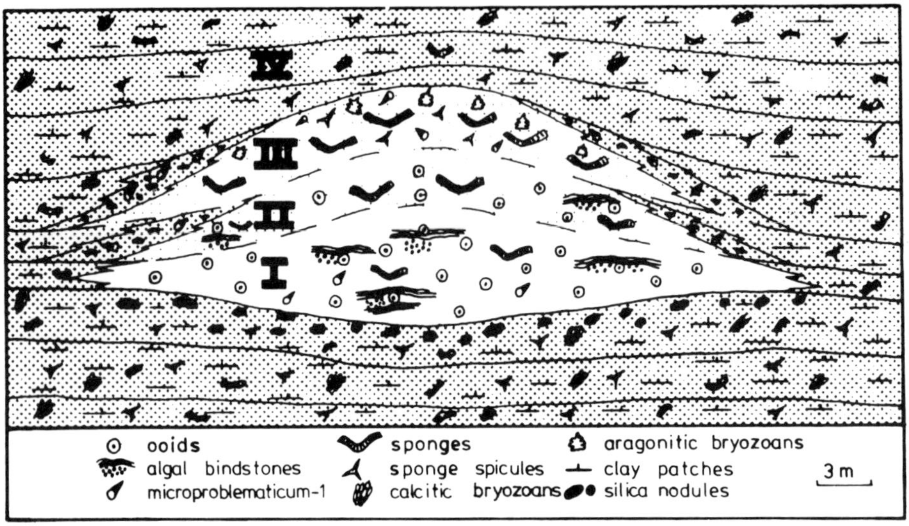

Fig. 3. Schematic zonation of a sponge-algal reef. Characteristic zones are *I* core facies; *II* transition facies; *III* outer facies; *IV* well-bedded facies. Only allochems important for the zonation are depicted. The diameter of the sponge-algal reef is about 50 to 60 m and thickness is about 16 m. (Cf. Schorr and Koch 1985)

comparable to the Silurian reefs described by Narbonne and Dixon (1984), and to many other samples given by Wilson (1975).

4 Diagenesis of Sponge-Algal Reef Limestones

4.1 Cementation

The "core facies" (Fig. 3) includes many types of primary cement whose localization is favored by a high intergranular porosity. At present all cements are calcitic. However, relics of precursor cements occur in the neomorphic calcites, suggesting original aragonite or Mg-calcite. They indicate the presense of an early marine diagenesis. Other clear cements lack relic fabrics, and these are interpreted to be of primary freshwateer origin. Cathodeluminescence study shows that all clear calcite cements have a well-defined, zonar luminescence whereas the early marine cements and all allochems do not luminesce. In the following descriptions, the numbers in *quotation marks* refer to the "diagenetic flowsheet", Fig. 10.

4.1.1 Marine Phreatic Cements

These are isopachous crusts whose two principal variations reveal differences in crystal size and in the fabric of relics.

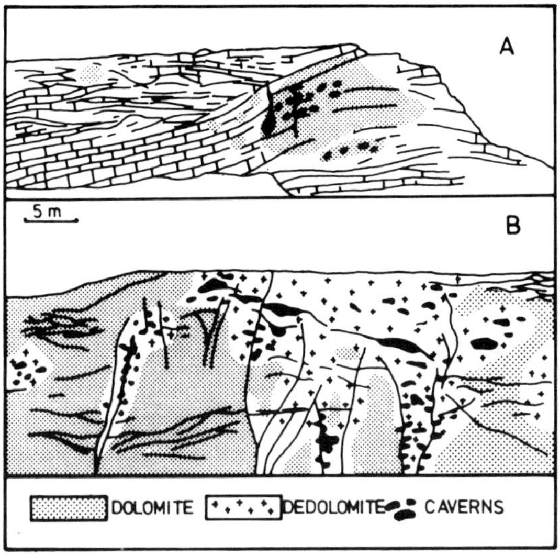

Fig. 4. A Relationship between well-bedded "normal facies" and sponge-algal reef dolomite at Tieringen quarry (Oxfordian). **B** Dolomite and dedolomite in the quarry of Schmiechen (Middle Kimmeridgian). Dedolomite is related to fractures and also to the present surface. (After Lang 1964)

a) Rim cements with aragonite needle precursors. The rims are 60 – 120 μm thick. Relics outline earlier needles oriented subperpendicular to the substrate and form an irregular outer boundary (Fig. 5 B, Fig. 10 *4*). Such needles characterize certain aragonite cements forming today in marine environments (Alexandersson 1972, Schroeder 1972). These rims now consist of microspar whose many individual crystals (< 10 μm) of variable optic orientation, form a finely crystalline, limpid mosaic.
b) Rim cements with bladed Mg-calcite precursors. These are 20 – 40 μm thick. Relic fabrics indicate earlier bladed crystals oriented perpendicular to the substrate gaving a regular outer limit (Fig. 5 A). Two to five blades now constitute one neomorphic crystal. This relic fabric if often characteristic of modern Mg-calcite cements.

4.1.2 Marine Vadose Cements

These occur on the roofs of voids and vugs (Fig. 5 C) and only randomly on the lower surfaces of particles. They exhibit fabrics similar to the rim cements with Mg-calcite precursors. They are thus interpreted to be neomorphically altered Mg-calcite cements which were probably formed within the intertidal zone (Purser 1978). The voids include pellet silt fillings similar to those described from Jurassic inter- and supratidal environments by Aissaoui and Purser (1983). Neomorphism of these calcite rims is thought to result from the overprinting of early marine cement by freshwater diagenetic processes analogous to those described by Folk (1973).

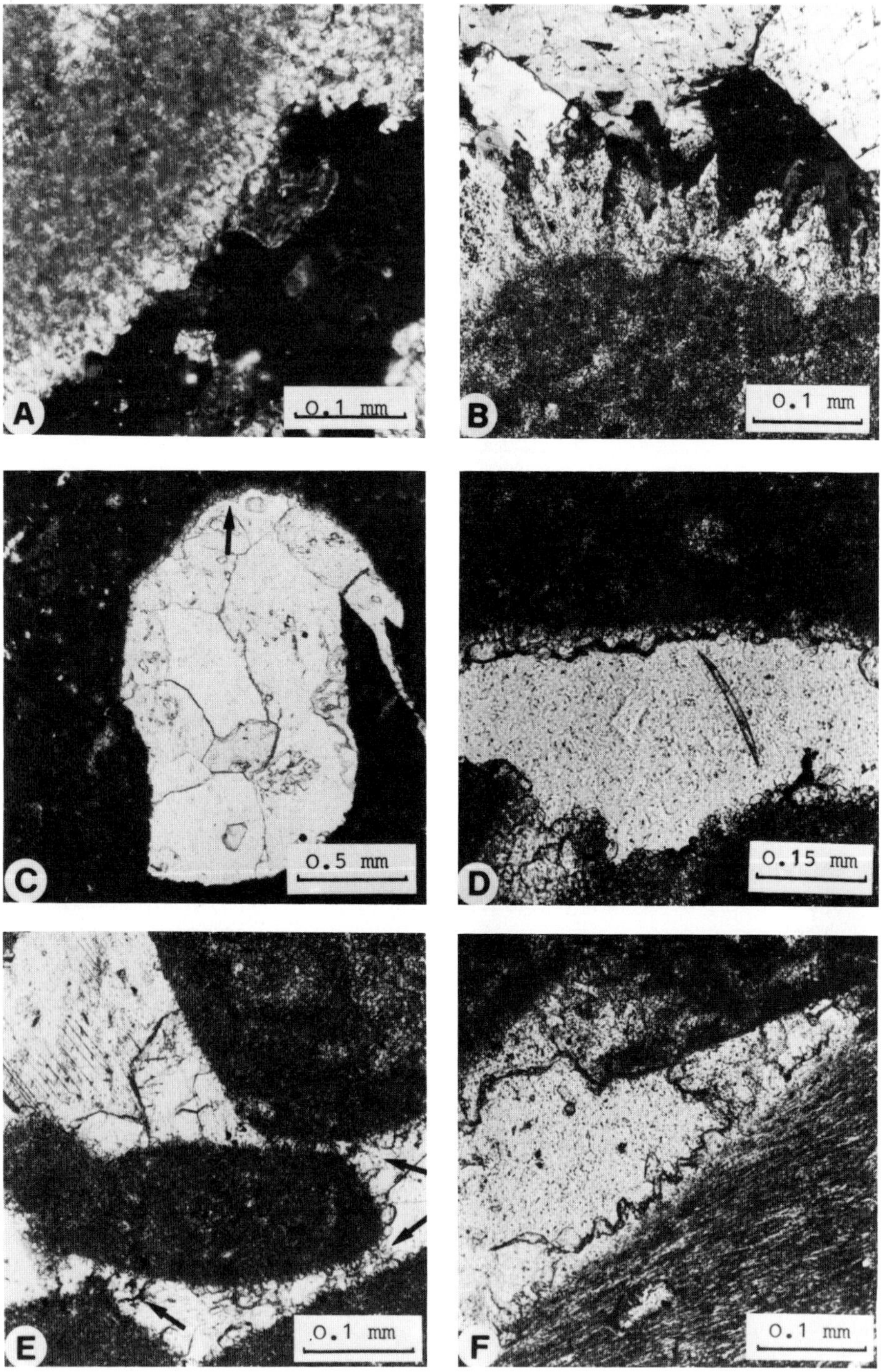

Fig. 5

4.1.3 Freshwater Vadose Cements

These are clear, granular crystals lacking relic fabrics. They are primary calcite and occur as gravitational and meniscus cements:

a) Gravitational cements (Müller 1971) (Fig. 5D) occur mainly in larger voids in the core facies where this is composed of pelsparites bound by algal layers (Wirsing and Koch 1986). They are composed of limpid crystals of 5 – 30 μm size.
b) Meniscus cements (Dunham 1971) are commonly present in packstones and grainstones in the reef core facies (Figs. 5E, 10 *2, 3*). They are composed of limpid, 5 – 50 μm crystals closely resembling the "pore-rounding" meniscus cement described by Harrison and Steinen (1978) from Holocene caliche crusts in Barbados. They are indicative of vadose conditions.

4.1.4 Freshwater Phreatic Cements

Relatively abundant, they include "dog-tooth", granular and drusy calcite cements, as well as poikilitopic and syntaxial overgrowth cements. On the basis of crystal size, limpid fabric and habit, these cements are interpreted to be of meteoric origin. They are similar to cements described by Schroeder (1979).

a) Dog-tooth cements (Fig. 5F) are developed locally. They comprise limpid, 20 – 60 μm crystals which are oriented subperpendicular to the substrate which has not influenced the texture of these crystals.
b) Granular cements are most common (Figs. 5E, 6A and 10 *3, 6, 7*). They generally fill residual pore space following marine or meteoric vadose cementation. Granular cements are composed of clear, 5 – 60 μm crystals.
c) Drusy mosaic comprises crystals which show an increasing size towards the centers of pores. Crystal components range from 10 to 50 μm in size.
d) Poikilitopic cements which enclose several grains (generally pellets) occur locally. They consists of large (up to 1 mm), clear crystals with twin-lamelli exhibiting the same luminescence features as freshwater cements. They are therefore interpreted to be of meteoric origin.
e) Syntaxial overgrowths (Evamy and Shearman 1965) occur commonly on echinoid fragments, especially where a solution cavity was formed around bioclasts embedded in micrite, and on fragments present in packstones within the core facies. No syntaxial overgrowth is are formed on echinoid fragments with micritized rims. Echinoid detritus with adjacent solution cavities are comparable to the "solution coronas" (Walkden and Berry 1984) interpreted to indicate subaerial emergence and meteoric diagenesis.

Fig. 5A – F. Early diagenetic fabrics. **A** Rim cement with relics suggesting a Mg calcite precursor. **B** Rim cement with relics suggesting an aragonite needle precursor. **C** Marine vadose cement (*arrow*) localized on the roof of a void. **D** Gravitational cement composed of limpid, granular crystals of meteoric origin. **E** Meniscus cement (*arrows*) developed at grain contacts. These consist of limpid, granular crystals. **F** "Dog-tooth" calcite growing on mollusc fragment and on pellets

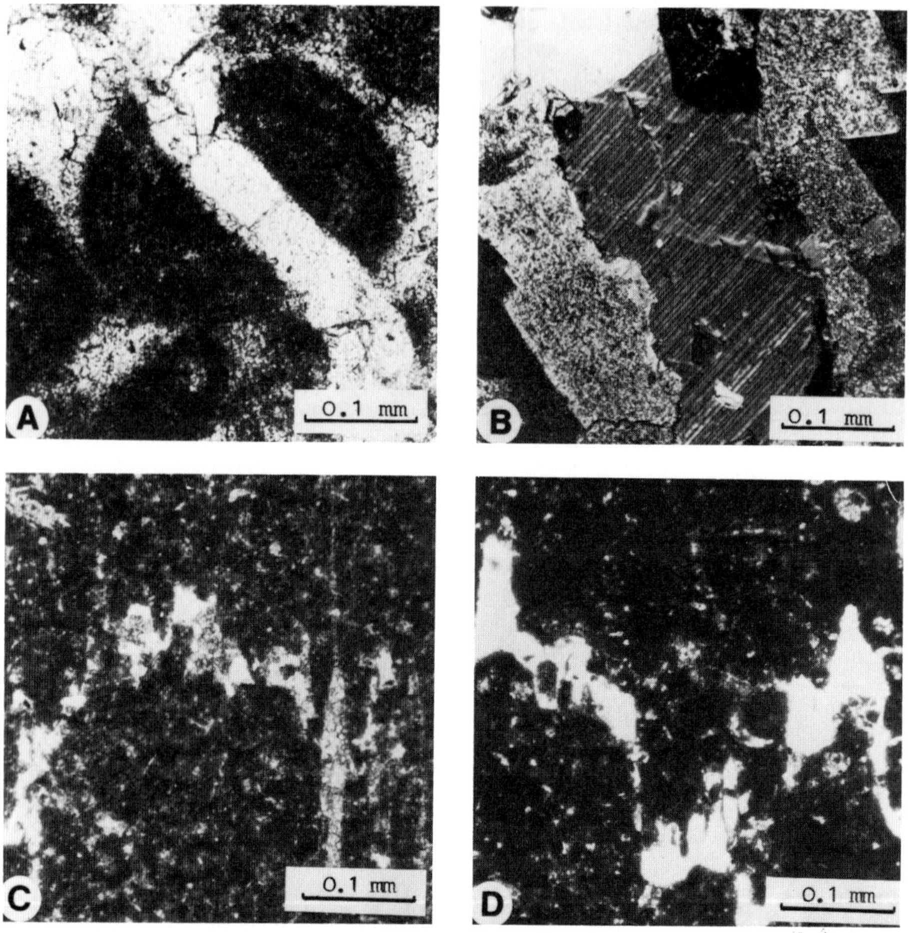

Fig. 6A – D. Fracturation and formation of microkarst. **A** Packstone from reef core. Following cementation by isopachous rim cement (Mg calcite precursor), the early lithified sediment was broken. The fractures and other residual pores were later cemented by limpid, granular to blocky calcite. **B** Dedolomitized rock is fractured. These fractures are filled by a blocky, limpid calcite formed during the second freshwater phase (see Fig. 10 *20*). **C** Formation of microkarst along stylolite. The lack of clay concentrations in stylolites indicates early formation. Freshwater influx led first to solution (microkarstification) in the vadose zone and, subsequently, to precipitation of equigranular calcite within the meteoric phreatic zone. **D** Renewed freshwater influx resulted in further dissolution along stylolites and their granular cements, resulting in the development of rounded stylolites

4.2 Solution Features and Microkarst

Diagenetic phenomena include stylolites which are often associated with dissolution and precipitation. Stylolites occur with or without clay seams.

In many examples it may be shown that solutions have invaded the stylolites forming 20 – 50 μm-sized microchannels (Figs. 6C and 10 *9, 10*). This must have occurred at a very early stage as these stylolites had not accumulated clay seams.

It is assumed that this stage was attained immediately after initial lithification, during shallow burial. In this respect, it is recalled that Bathurst (1971) has described stylolite formation in limestones under an overburden less than ca. 90 m. Ricken (1985) reports a minimum overburden of 40 m for Upper Oxfordian limestones from the Swabian Alb also exhibiting pressure-solution fabrics.

Rocks with very early stylolites lacking clay seams could have readily undergone later subaerial weathering. Subsequently, in the phreatic zone, the microkarst (channel porosity of Choquette and Pray 1970) was filled by equigranular, 5 – 10 μm-sized crystals (Fig. 6C). A second period of subaerial exposure sometimes results in an additional phase of leaching which has led to microkarst features with rounded stylolites (Figs. 6D, 10 *11*). An enlargement of the solution cavities to 20 – 100 μm is clearly visible.

Outcrops indicate that these features can also be caused by recent weathering. However, similar features are found in wells where they are interpreted to be of Upper Jurassic age.

4.3 Dolomite

Our studies suggest that dolomitization has generally begun within the most porous and permeable parts of the reefs, notably in the "core facies". Initial dolomite growth of rhombohedra 50 – 100 μm in size, occurs between cement crystals (Figs. 7A, 10 *13*). In spite of early cementation, these microvoids nevertheless remained conduits for dolomitizing pore waters during the early stages of diagenesis. Dolomitization also followed stylolites (Figs. 7B, 10 *14*) enlarged by freshwater influx showing that dolomitization followed permeability conduits into micritic sediments. These crystals have a size of 30 – 50 μm.

On a somewhat larger scale, dolomitization progressed laterally from the core facies, replacing adjacent micritic carbonates (Figs. 7C, 10 *17*). The transition zone between dolomite and limestone is only 5 to 50 cm in width, as also noted by Lang (1964) and by Fritz (1965). Within this dolomitization "front" the first dolomite crystals are idiomorphic rhombs floating in micrite but with increasing degree of dolomitization a hypidiomorphic to xenomorphic fabric is formed. These dolomites are generally calcium-rich ($Ca_{54}Mg_{46}$) and occur both at outcrop and in the subsurface.

Locally, larger fractures formed during early karstification were not closed completely by dolomite (Figs. 7D, 10 *12*) whose 250 – 500 μm crystals exhibit hypidiomorphic to xenomorphic textures. These dolomites are generally stoichiometric.

Cores in completely dolomitized bioherm complexes first penetrate an upper dolomite limit (front) and travers the completely dolomitized reef core some 10 m thick, as in the well TB-3-Saulgau described by Bach (1985), Huber (1985) and Meder (1985). In this well there is a progressive decrease in Sr concentration (bulk samples) from the dolomite limit (100 – 150 ppm) to the completely dolomitized core (<30 ppm). Parallel to this, the dolomites become less calcic and form larger, stoichiometric crystals. We interpret this to be the result of recrystallization of the precursor calcian dolomite, as described by Land (1983).

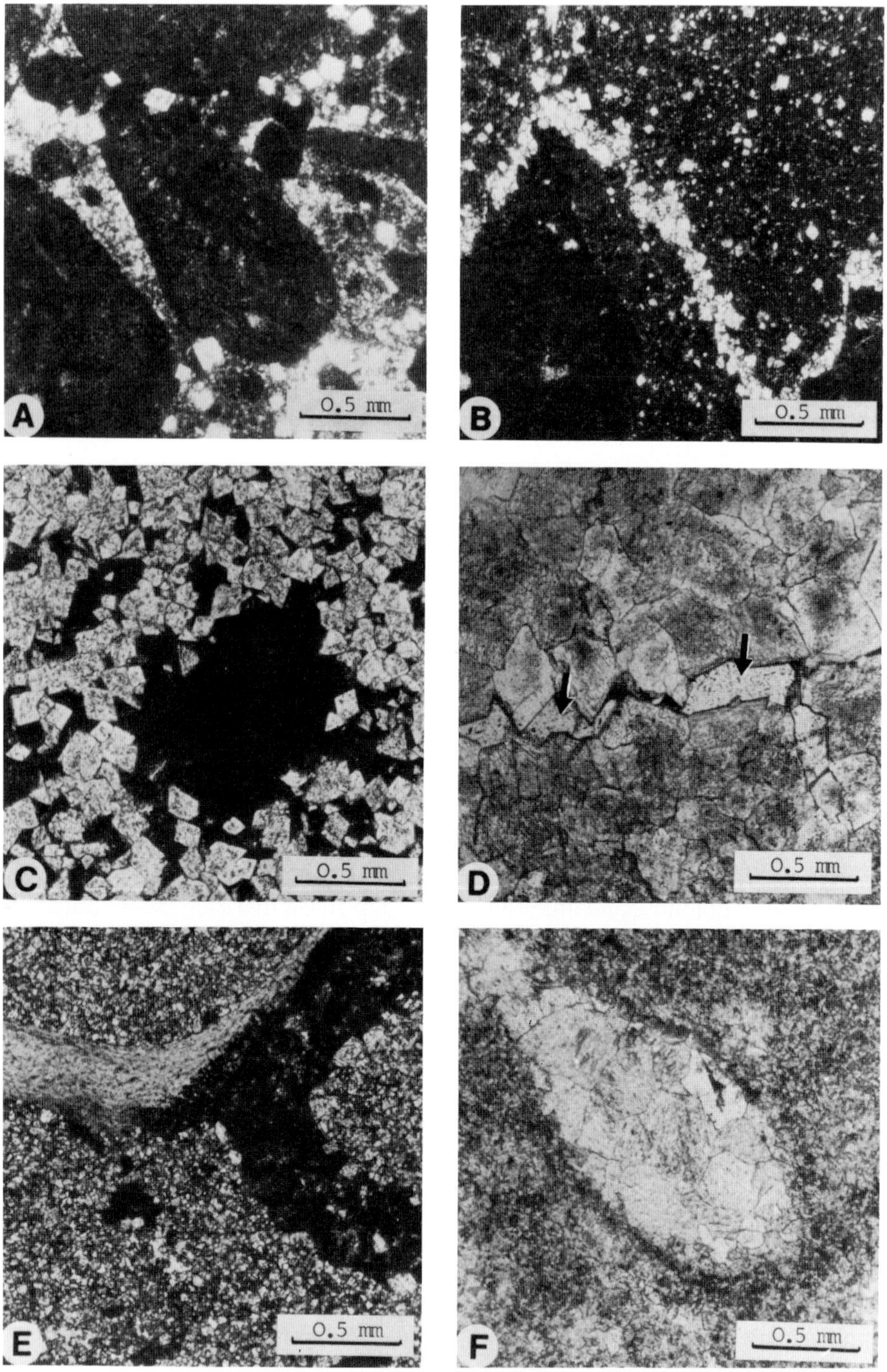

Fig. 7

Where dolomite replaced matrix, cements or allochems, calcitic grains are often preserved (Figs. 7E, 10 *15*). This is also true for grains composed initially of Mg calcite (e.g., echinoids) which were transformed to low Mg calcite during initial freshwater influx. Subsequently, these allochems were dissolved completely, probably during later freshwater flushing, the large molds still being preserved (Fig. 10 *21*). These secondary voids may also be cemented by a different type of dolomite. Whereas the earlier dolomite has many inclusions, this mold-filling dolomite consists of large, limpid crystals (up to 500 μm) which often show corrosion fabrics (Figs. 7F, 10 *29*).

The later, void-filling dolomite constitutes somewhat less than 15% of a completely dolomitized bioherm. Thus, the main dolomitization phase, as well as dedolomitization, occurred much earlier than the late, diagenetic (burial) stylolitization (Fig. 8F) associated with thick clay seams. At this time, dolomitization had probably ceased.

Petrographic and geochemical data (as outlined later) suggest that the sponge-algal reefs (especially the core facies) suffered a strong freshwater influx which probably mixed with seawater. The authors conclude that the main dolomitization of the reefs was due to the mixing of marine and freshwaters in a manner comparable to the "dorag" model (Badiozamani 1973). Recrystallization to stoichiometric dolomite probably occurred during burial.

4.4 Dedolomite

Three types of dedolomite were observed at outcrops, in the TB-3 Saulgau and in other drilling wells. The first type of dedolomite is comprised of 100–120 μm-sized rhombohedra which have been completely replaced by fine-crystalline calcite (Figs. 8A, 10 *19*). This is similar to the dedolomite described by Evamy (1967) and by Purser (1985), formed under near-surface conditions (De Groot 1967). Fritz has termed it centrifugal dedolomitization.

A second type of dedolomite corrodes external surfaces of dolomite crystals, extending along cleavage planes (Figs. 8B, 10 *26*). It is commonly developed in completely dolomitized rocks, notably in zoned rhombs. This type of dedolomitization developed laterally from fossil molds within the dolomite which are now filled with calcite (Fig. 10 *25*). The calcite replaces the dolomite in optical continuity and thus is the result of in situ dissolution and reprecipitation. The dolomite rhombohedra serve as a nucleation site for the newly precipitated calcite. The final product may include "calcite rhombohedra" showing relics of initial zonation (Figs. 8C, 10 *27*). Meder (1985) has described these features from the

Fig. 7A–F. Dolomitization phenomena. **A** Initial rhombohedra grow in intercrystalline pores of cement between grains (Fig. 10 *13*). **B** Dolomitization also advanced along stylolites enlarged by dissolution (Fig. 10 *14*). **C** Dolomite advancing from reef core replaced micritic carbonate by forming scattered rhombs (Fig. 10 *17*). **D** Locally, larger fractures and channels formed by microkarstification were not cemented completely by dolomite (*arrows*) (Fig. 10 *12*). **E** Where dolomite had completely replaced the matrix and cements, the calcitic allochems are often preserved (Fig. 10 *15*). **F** Calcitic allochems were often dissolved during a later stage of diagenesis and their molds filled by a subsequent phase of dolomite (Fig. 10 *29*)

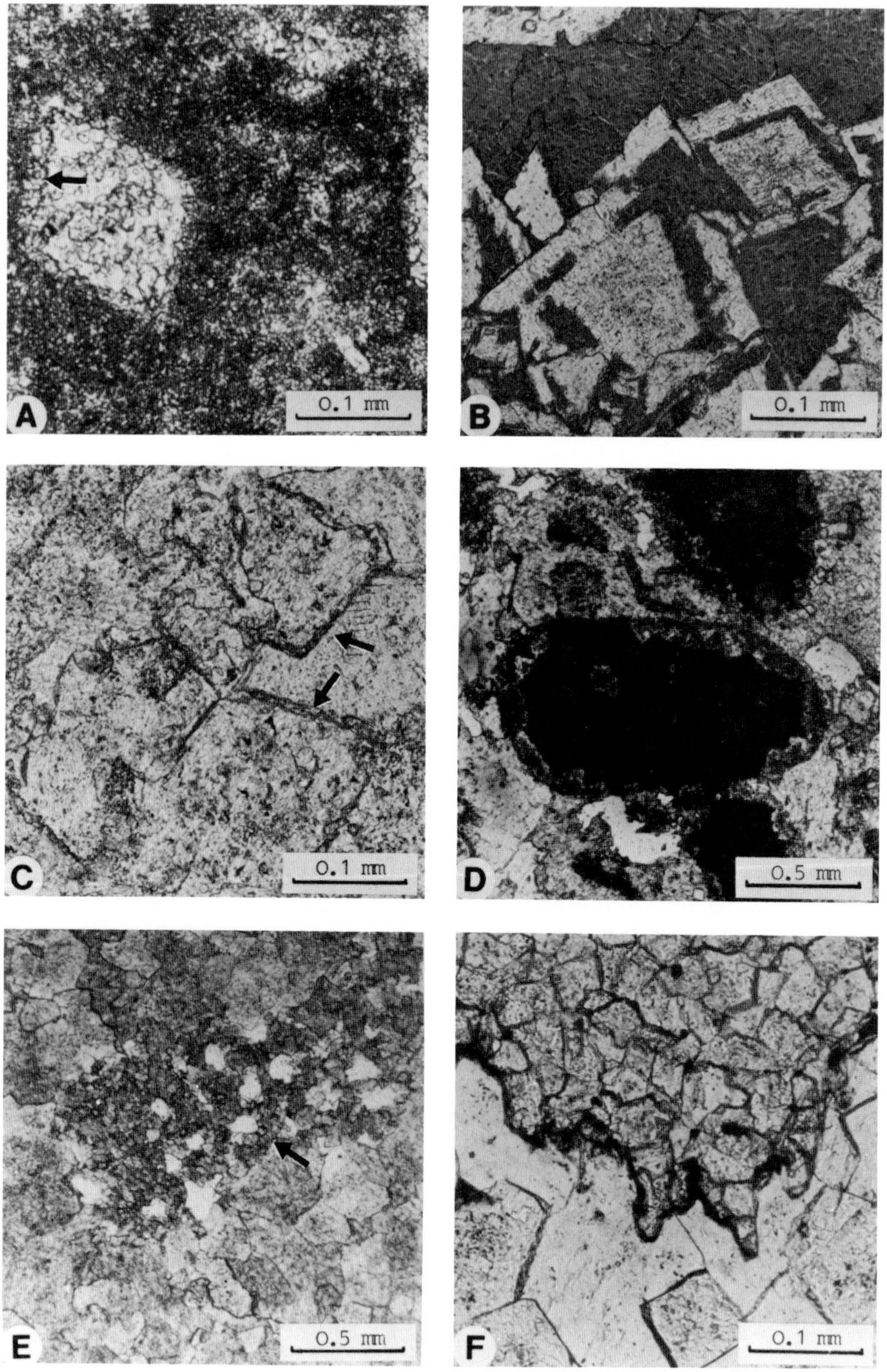

Fig. 8

TB-3 Saulgau core. Samples from the outcrop near Schmiechen, described by Fritz (1965) were found by the authors to exhibit also these features.

The third type of dedolomite (Figs. 8D, 10 *22*) preserves the gross primary structure of partly dolomitized allochems. Dolomitization, which begins within the intercrystalline spaces in the cement (as already noted), has just begun to replace the allochems. At this stage dolomitization ceased, possibly when the pore waters became excessively fresh. This led subsequently to dedolomitization of the grain periphery. Many solution cavities and corrosion of crystals are visible in these rocks. They indicate that dolomitization terminated at an early stage and was replaced by solutions favoring dedolomitization (Purser 1985). Lang (1964) and Fritz (1965) have concluded that dedolomitization was post-Jurassic, having occurred during Cretaceous to Recent freshwater influx. Dedolomite in the Middle Kimmeridgian and Lower Tithonian in wells located in the Western Molasse Trough (as already noted) tends to contradict this interpretation; both dolomitization and dedolomitization were probably completed mainly during Kimmeridgian and Tithonian times. The fact that these specific stratigraphic levels crop out extensively in the Swabian Alb has led to the possibly erroneous conclusion that dedolomitization must be post-Jurassic.

4.5 Internal Sediments

One of the interesting features, both at outcrop and in the subsurface, is the filling of secondary voids by internal sediment. These fillings include pelletoidal silt, vermicular kaolinite, and dolomite silt (Fig. 8E). These internal sediments, notably the dolomite silt, suggest an additional phase of freshwater influx. The dolomite silt is similar to the crystal silt of supratidal origin described by Aissaoui and Purser (1983). As noted by Lang (1964) in the Swabian Alb, the fabric of the dolomite is first destroyed, forming dolomite sand. We conclude that the same process took place during the Upper Jurassic, forming a dolomite detritus which filtered into the voids.

4.6 Geochemical Data

Geochemical data is based on the analysis of 250 samples of limestone and dolomite collected from sponge-algal reef facies and 50 samples from adjacent bedded facies (Schorr and Koch 1985, Bach 1985). The distribution of Sr and Mg

Fig. 8A – F. Dedolomitization phenomena. **A** Completely dedolomitized rhombs now composed of minute calcite crystals (Fig. 10 *19*). **B** In the upper parts of photo a fossil mold is filled completely by large poikilitopic crystals of calcite. Around this calcitization a dedolomitization affects zoned dolomite rhombs (Fig. 10 *26*). **C** Zoned dolomite rhombs completely replaced by calcite exhibiting ghost structure (*arrows*) (Fig. 10 *27*). **D** Partial dolomitization and dedolomitization of allochems whose subsequent dissolution has formed solution cavities (Fig. 10 *22*). **E** Large fossil molds in dolomite filled with dolomite crystal silt (Fig. 10 *24*). **F** Stylolitization with clay concentration which formed much later than the main dolomitization event. Differences in dolomite crystal size reflect initial textural variations (wackestone/grainstone)

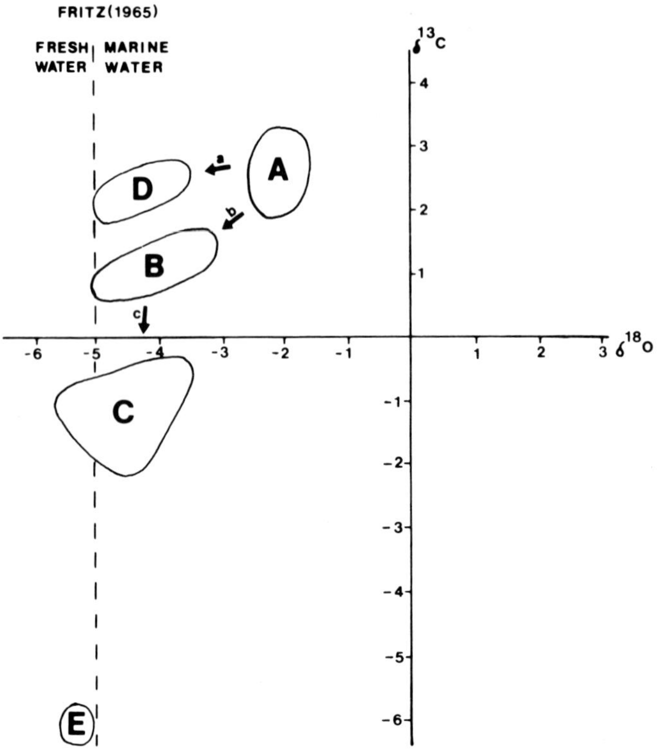

Fig. 9. Isotopic composition of selected samples. *A* Sponge-algal reef limestone from subsurface; $\delta^{13}C$ signal suggests formation in equilibrium with the atmospheric CO_2. *B* Reef limestone (subsurface) with strong freshwater effects confirmed by fossil molds and microkarst. *C* Reef limestone from outcrop. The more negative $\delta^{13}C$ indicates increasing freshwater influence. *D* Dolomitized reef from subsurface. Increased meteoric influence combined with marine mixing results in trend (*a*). *E* Dedolomite from outcrop. The negative ^{13}C reflects dedolomitization of former "mixed water" dolomites (see data in Fritz 1965)

was considered to be of special interest. In addition, 21 bulk samples were analysed for isotopic composition.

Limestones from the "reef core", which generally exhibit vadose diagenetic phenomena, have very low Sr (<40 ppm) and Mg (<1000 ppm) concentrations. Limestones from the "transition zone" and "outer zone" (Fig. 3) which contain more matrix, show slightly higher concentrations (Sr up to 80 ppm and Mg up to 1800 ppm). The surrounding bedded facies have about 120 ppm Sr and 2500 ppm Mg. The relatively low trace element concentrations, especially in the reef core, are interpreted as reflecting freshwater influence. Dolomitized sponge-algal reefs have even lower Sr concentrations (<30 ppm) suggesting an even stronger freshwater influx.

Isotopes. Fritz (1965) has interpreted oxygen isotope data ($\delta^{18}O$ of -5) as indicating a temperature of 25 °C. The trends *a* and *b* shown on Fig. 9 may be interpreted as indicating enhanced freshwater influence. The $\delta^{13}C$ signatures which average +3 are indicative of carbonate formed in equilibrium with the atmosphere. The more negative $\delta^{13}C$ values are caused probably by increased freshwater influence.

4.7 Sequences and Multiple Pathways of Diagenesis

All diagenetic features described can be expressed in a logical sequence presented in the form of a diagenesis flow-sheet, Fig. 10. This evolutionary series is similar to the "petrogenetogram" presented by Schroeder (1984). However, a difference concerns the three initial diagenetic stages (numbers *1, 4,* and *8* in Fig. 10) which are necessary to explain all phenomena observed within these Jurassic rocks. These three stages represent different bathymetric positions of the sponge-algal reefs (Fig. 11).

5 Discussion

Results suggest that the diagenetic history of the sponge-algal reefs is characterized by a succession comprising:

- Early lithification in the marine and/or meteoric environment;
- Dolomitization;
- Dedolomitization.

The multiple paths of diagenesis expressed in Fig. 10 show that diagenetic evolution of these Upper Jurassic carbonates is considerably more complex than previously thought.

A simple paleogeographic model (Fig. 11) expresses the variable depths at which these reefs developed, the variable mineral paragenesis depending on the setting of individual bodies; relative sea-level falls caused changes which differed according to the initial position of individual reefs.

- Sponge-algal reefs in greatest water depths remained in a submarine position and were affected only by submarine diagenesis (Fig. 11 B 3);
- Reefs living at intermediate water depths were located subsequently in the intertidal zone where they were susceptible to mixed water and meteoric diagenesis (Fig. 11 B 2);
- Reefs growing in very shallow waters were shifted into a supratidal environment where they were modified by meteoric vadose and phreatic effects (Fig. 11 B 1).

These three main diagenetic variants may overprint each other as the reefs evolved at different rates, implying that reefs which had undergone submarine cementation could have been affected later by freshwater or mixing water diagenesis as they grew into this depth range. The growth rate of reefs affected by freshwater cementation probably decreased as the binding organisms died. Different stages were attained by different sponge-algal reefs in the course of time. Reefs

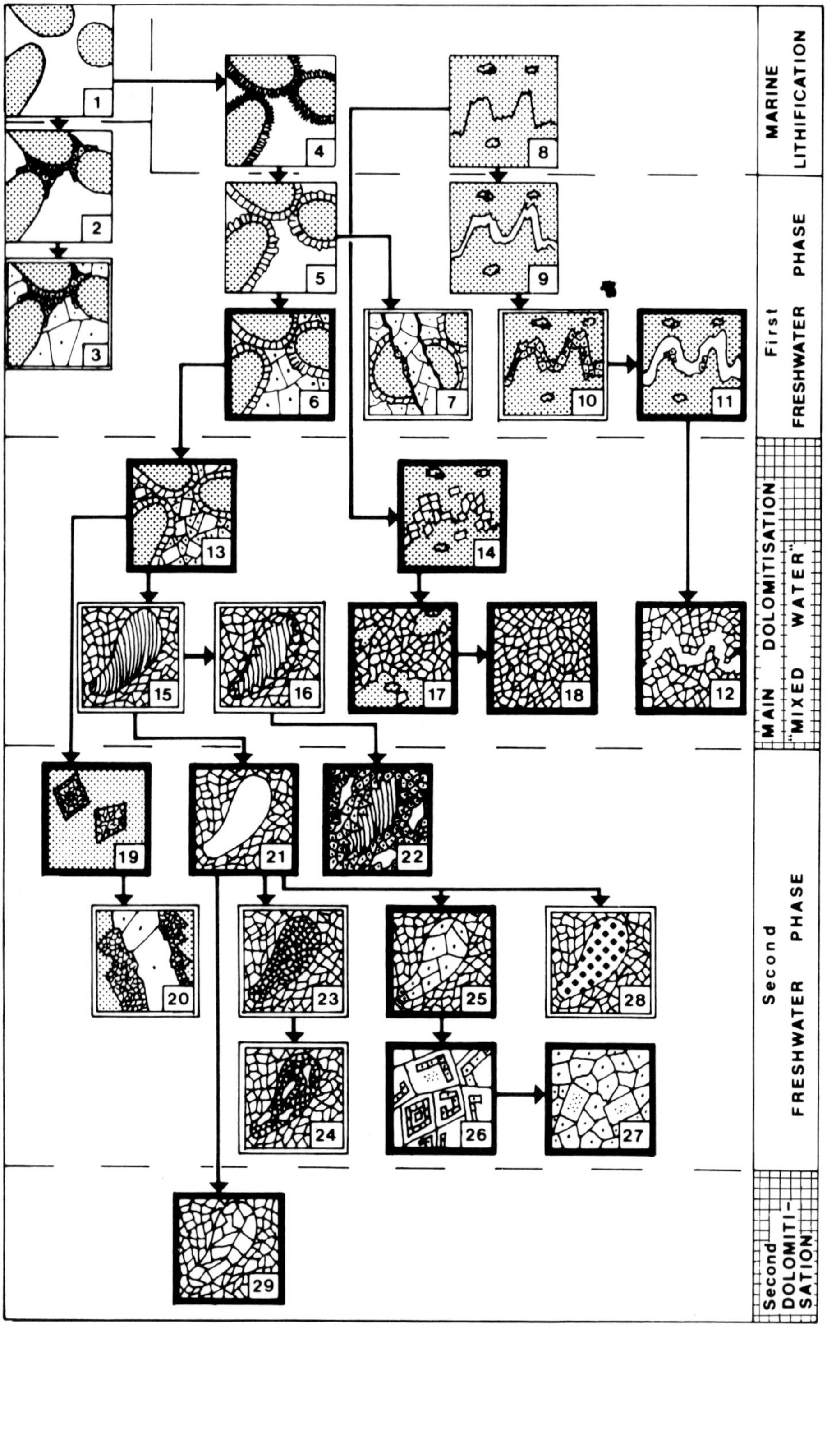
MARINE LITHIFICATION
First FRESHWATER PHASE
MAIN DOLOMITISATION "MIXED WATER"
Second FRESHWATER PHASE
Second DOLOMITI-SATION
1
2
3
4
5
6
7
8
9
10
11
12
13
14
15
16
17
18
19
20
21
22
23
24
25
26
27
28
29

Fig. 10

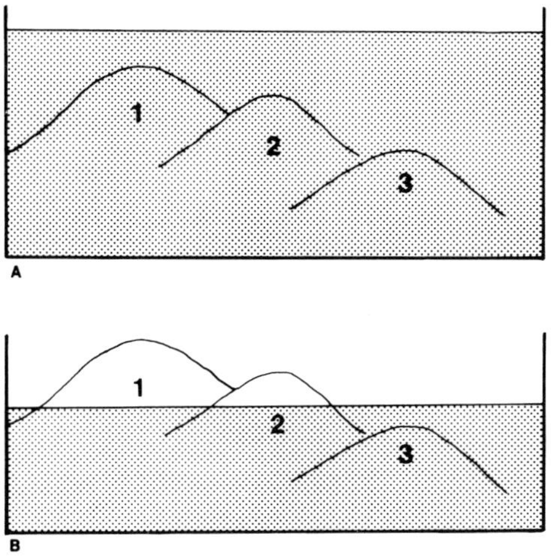

Fig. 11 A, B. Model showing relationships between reef bathymetry and subsequent diagenetic evolution. **A** High sea-level stage: Reefs grow at different depths and all are affected by submarine diagenesis. **B** Low sea-level stage: Reefs evolve through different diagenetic environments. Thus, in "reef *3*", submarine diagenesis affects the entire reef. In "reef *2*", the upper part is shifted into the mixing zone where dolomitization may occur. The top of "reef 3" is now in the meteoric zone where vadose and phreatic diagenesis may affect it. Below this zone mixed and normal marine waters result in characteristic fabrics relating to this environment. (After Wirsing, in preparation)

measuring several to tens of meters in thickness together formed a large composite body some 100 – 120 m in thickness, illustrated in Figs. 2 and 11. Within these masses both the nature and the diagenetic intensity (Fig. 10) are variable.

6 Conclusions

It has been shown that sponge-algal reefs were affected by early freshwater diagenesis following submarine cementation. "Mixing water dolomitization" occurred in many places during and after these early meteoric phases. A second, main freshwater phase let to important Upper Jurassic dedolomitization and other features related to surface and near-surface conditions.

Fig. 10. Diagenesis Flow Sheet. *Square with one line* = cannot be observed but must have been present as precursor; *squares with double lines* = can be observed; *squares with strong black line* = can be observed abundant. *1* Allochems without cement; *2* formation of granular meniscus cement (calcite); *3* complete cementation by granular to blocky freshwater cement; *4* marine cementation by isopachous rim (aragonite, Mg-calcite); *5* freshwater influx overprinting the precursor cements; *6* complete cementation by granular to blocky freshwater calcite; *7* fracturing of *5* and freshwater cementation; *8* weak lithification and formation of early stylolites without clay seams; *9* freshwater microkarst along early stylolites; *10* freshwater cementation in microkarst; *11* repeated karstification and formation of "rounded stylolites"; *12* complete dolomitization not closing former "rounded stylolites"; *13* mixed water dolomitization begins in the intercrystalline pores of the cements; *14* dolomitization along stylolites; *15* nearly completely dolomitized rock with some calcitic allochems; *16* dolomitization invades calcitic allochems; *17* dolomitization along stylolites increases; *18* completely dolomitized rock; *19* dedolomitization of single rhombohedra; *20* rhombohedra and rock are fractured and fractures cemented by blocky freshwater calcite; *21* dissolution of calcite allochems in dolomite; *22* dedolomitization of *16*; *23* fossil molds of *21* filled by fine "dolomite crystal silt"; *24* dissolution of parts of the crystal silt; *25* fossil mold of *21* filled by poikilitopic calcite; *26* dedolomitization replaces rhombs; *27* completely dedolomitized rock with "zonar" calcite crystals; *28* fossil molds of *21* filled with authigenic quartz or vermicular kaolinite; *29* fossil molds of *21* filled by second generation of dolomite

These are the principal diagenetic processes which affected the Upper Jurassic sponge-algal reefs examined. Because most of the phenomena are related to surface conditions, they shed new light on the problem of water depth in which these sponge-algal reefs have developed. It is clear that they emerged repeatedly during the Upper Jurassic. The possibility of a general shallowing of Upper Jurassic seas must also be considered. In our opinion, it is more likely that this Upper Jurassic sea covered a shallow, gently dipping shelf affected by cyclic sea-level changes. The slope of the shelf towards SW resulted in frequent intertidal environments in the N and NE, i.e. the Franconian Alb and the now eroded area north of the Swabian Alb. Consequently, these areas were more susceptible to dolomitization.

On slightly deeper parts of the shelf (area of W-Molasse-Trough) dolomites were formed only where sponge-algal reefs grew into shallow water and emerged subsequently into the mixed water zone. Here, most reefs underwent submarine diagenesis. The second phase of dolomitization which commonly cemented moldic voids occurred during deeper burial conditions, possibly relating to dewatering of underlying marine clays and marls. Most diagenetic modifications affecting Upper Jurassic sponge-algal reefs occurred within sedimentary and near-surface environments. These modifications are well preserved both at outcrop and in the subsurface.

Acknowledgments. We wish to thank E. Flügel, G. Müller, H. W. Pfefferkorn, and V. Schweizer for reading the manuscript and for numerous improvements and discussions. We thank the editors of this book for their patient and helpful corrections. We are very indebted to A. Bach, S. Huber, K. Meder, and G. Wirsing, who offered unpublished data concerning their research. We also thank R. Botz for isotopic analysis and W. Schmitz, M. Gastner and the colleagues of the Institut für Sedimentforschung for many discussions, and U. Kästner for the photoplates.

References

Aissaoui DM, Purser BH (1983) Nature of origins of internal sediments in Jurassic limestones of Burgundy (France) and Fnoud (Algeria). Sedimentology 30:273 – 283

Aldinger H (1945) Zur Stratigraphie des weißen Jura Delta in Württemberg. Jahresber Mitt Oberrhein Gel Ver NF 31:111 – 152

Aldinger H (1968) Ecology of algal-sponge reefs in the Upper Jurassic of the Schwäbische Alb, Germany. In: Recent developments in carbonate sedimentology in central Europe. Springer, Berlin Heidelberg New York, pp 250 – 253

Alexandersson T (1972) Intergranular growth of marine aragonite and Mg-calcite: evidence of precipitation from supersaturated sea water. J Sediment Petrol 42:441 – 460

Bach A (1985) Semiquantitative Röntgendiffraktometer-Analyse und geochemische Untersuchungen der Gesteine des Weißjura der TB-3 Saulgau (Oberschwaben) und ihre Beziehungen zu elektrischen Bohrlochmessungen. DiplArb Univ Heidelberg

Badiozamani K (1973) The Dorag dolomitization model – application to the Middle Ordovician of Wisconsin. J Sediment Petrol 43, 4:965 – 984

Bathurst R (1971) Carbonate Sediments and their Diagenesis. Elsevier, Amsterdam, p 632

Bausch W (1963) Der Obere Malm an der unteren Altmühl. Nebst Untersuchungen über das Riffproblem. Ausz Geol Bl NO-Bayern 2, 2:104 – 111

Betz D, Wendt A (1983) Neuere Ergebnisse der Aufschluß- und Gewinnungstätigkeit auf Erdöl und Erdgas in Süddeutschland. Bull Ver Schweiz Petrol Geol Ing 49, 117:9 – 36

Choquette P, Pray L (1970) Geologic nomenclature and classification of porosity in sedimentary carbonates. Am Assoc Petrol Geol Bull 54, 2:207–250

De Groot (1967) Experimental dedolomitization. J Sediment Petrol 37:1216–1220

Dunham RJ (1971) Meniscus cement. Johns Hopkins Univ Stud Geol 19:297–300

Evamy BD (1967) Dedolomitization and the development of rhombohedral pores in limestones. J Sediment Petrol 37, 4:1204–1215

Evamy BD, Shearman DJ (1965) The development of overgrowths from echinoderm fragments. Sedimentology 5:211–233

Flügel E, Steiger T (1981) An Upper Jurassic sponge-algal buildup from the northern Frankenalb, W-Germany. Soc Econ Paleontol Miner Spec Publ 30:371–397

Folk RL (1973) The natural history of crystalline calcium carbonate: effect of magnesium content and salinity. J Sediment Petrol 44:30–53

Fritz P (1965) Zur Genese von Dolomit und zuckerkörnigem Kalk im weißen Jura der Schwäbischen Alb (Württemberg). Arb Geol Paläontol Inst TH Stuttgart NF 50:100

Gwinner MP (1976) Origin of the Upper Jurassic of the Swabian Alb. Contrib Sediment 5:75

Harrison RS, Steinen RP (1978) Subaerial crusts, caliche profiles and breccia horizons: Comparison of some Holocene and Mississipian exposure surfaces, Barbados and Kentucky. Geol Soc Am Bull 89:385–396

Huber S (1985) Zur Mikrofazies und Diagenese des Oxford 1 – Kimmeridge 2,3 der TB-3 Saulgau (Oberschwaben). DiplArb Univ Heidelberg

Land LS (1983) Dolomitization. Am Assoc Petrol Geol, Education Course Note Ser 24:1–20

Lang HB (1964) Dolomit und zuckerkörniger Kalk im weißen Jura der mittleren Schwäbischen Alb. N Jahrb Geol Paläontol Abh 120:253–299

Meder K (1985) Zur Mikrofazies und Diagenese des Kimmeridge 2,4 – Tithon H der TB-3 Saulgau (Oberschwaben). Dipl Arb Univ Heidelberg

Meyer RKF (1975) Mikrofazielle Untersuchungen in Schwamm-Biohermen und -Biostromen des Malm Epsilon (Ober-Kimmeridge) und des obersten Malm Delta der Frankenalb. Geol Bl NO-Bayern 25/4:149–177

Müller G (1971) "Gravitational" cement: an indicator for the vadose zone of the subaerial diagenetic environments. Johns Hopkins Univ Stud Geol 19:301–302

Narbonne GM, Dixon OA (1984) Upper Silurian lithistid sponge reefs on Somerset island, arctic Canada. Sedimentology 31:25–50

Paulsen S (1964) Aufbau und Petrographie des Riffkomplexes von Arnegg im höheren weißen Jura der Schwäbischen Alb. Arb Geol Paläontol Inst TH Stuttgart NF 42:1–98

Purser BH (1978) Early diagenesis and the preservation of porosity in Jurassic limestones. J Petrol Geol 1:83–94

Purser BH (1985) Dedolomite porosity and reservoir properties of Middle Jurassic carbonates in the Paris Basin, France. In: Roehl PD, Choquette PD (eds) Carbonate Petroleum Reservoirs. Springer, Berlin Heidelberg New York, pp 343–355

Ricken W (1985) Epicontinental marl-limestone alternations: Event deposition and diagenetic bedding (Upper Jurassic, Southwest Germany). In: Bayer U, Seilacher A (eds) Sedimentary and evolutionary cycles. Lect Notes Earth Sci 1:127–162

Sandberg P (1985) Aragonite cements and their occurrences in ancient limestones. Soc Econ Paleontol Miner Spec Publ 36:33–57

Schneider J (1962) Der Jura in Erdölbohrungen des westlichen Molassetroges. Hermann Aldinger Festschrift, pp 163–172

Schorr M, Koch R (1985) Facies zonation of an Upper Jurassic algal-sponge bioherm (Swabian Alb, Germany). Facies 13:227–270

Schroeder JH (1972) Fabrics and sequences of submarine carbonate cements in Holocene Bermuda Cup reef. Geol Rundsch 61:708–730

Schroeder JH (1979) Carbonate diagenesis in Quaternary beachrock of Uyombo, Kenya: sequences of processes and coexistence of heterogenic products. Geol Rundsch 68:894–919

Schroeder JH (1984) The petrogenetogram of corals: spatial variations in diagenetic sequences. Palaeontogr Am 54:261–271

Wagenplast P (1972) Ökologische Untersuchung der Fauna aus Bank- und Schwammfazies des weißen Jura der Schwäbischen Alb. Arb Geol Paläontol Univ Stuttgart NF 67:1–99

Walkden GM, Berry JR (1984) Syntaxial overgrowth in muddy crinoidal limestones: cathodoluminescence sheds a new light on an old problem. Sedimentology 31:251–267

Wendt J (1980) Sponge reefs of the German Upper Jurassic. In: Hartmann WD, Wendt J, Wiedenmayer F (eds) Living and fossil sponges. Notes for a short course. Sedimenta 8:122 – 130
Wilson JL (1975) Carbonate facies in geologic history. Springer, Berlin Heidelberg New York, p 471
Wirsing G, Koch R (1986) Algen-Schwamm-Bioherme des Flachwasserbereiches (Schwäbische Alb, Weißjura Delta 3). Fazies 14:285 – 308
Ziegler B (1977) The "white" (Upper) Jurassic in S Germany. Stutt Beitr Naturk 26:1 – 79

Diagenesis of Upper Triassic Wetterstein Reefs of the Bavarian Alps

R. HENRICH[1] and H. ZANKL[2]

1 Introduction

Earlier studies of the Wetterstein series were concerned mainly with sedimentation (Sander 1936, Sarnthein 1965, 1966), or specific aspects of diagenesis (German 1968, 1969, Bechstädt 1975, 1979, Bechstädt and Dohler-Hirner 1983). The first modern facies concept, including ecological zonation, for a specific reef complex near Innsbruck, e.g., Hafelekar reef, was made by Brandner and Resch (1981).

More recent studies on the Wetterstein series in the middle part of the Bavarian Alps include sedimentation on platform margins (Henrich 1982), biocoenoses and ecology of different reef belts (Henrich 1983), facies and diagenesis of lagoonal carbonates (Henrich 1984).

The purpose of this study is to elucidate the specific biogenic, sedimentary and diagenetic fabrics of Early Upper Triassic Wetterstein reefs in the Bavarian Alps in order to provide an integrated picture of facies and diagenesis within these reefs.

1.1 Methods

This paper is based on facies mapping and observations of diagenetic features in the field as well as on careful petrographic analysis. Additionally, the following techniques were applied: SEM/EDAX analysis, determination of trace element composition of whole rock samples and selected diagenetic fabrics by atomic absorption spectroscopy, stable isotope compositions of cements and dolomite types analyzed with a Micromass VG 620 D mass spectrometer, and X-ray diffractometry of dolomites.

2 Location, Stratigraphy and Facies Distribution

2.1 Location

The area studied, the Hochstaufen-Rauschberg massif, is a mountain range in the eastern part of the Bavarian Alps situated near Salzburg (Fig. 1).

[1] Geologisch-Paläontologisches Institut, Universität Kiel, Olshausenstr. 40, 2300 Kiel 1, FRGermany
[2] Geologisch-Paläontologisches Institut, Universität Marburg, Lahnberge, 3550 Marburg, FRGermany

Reef Diagenesis
Edited by J. H. Schroeder and B. H. Purser

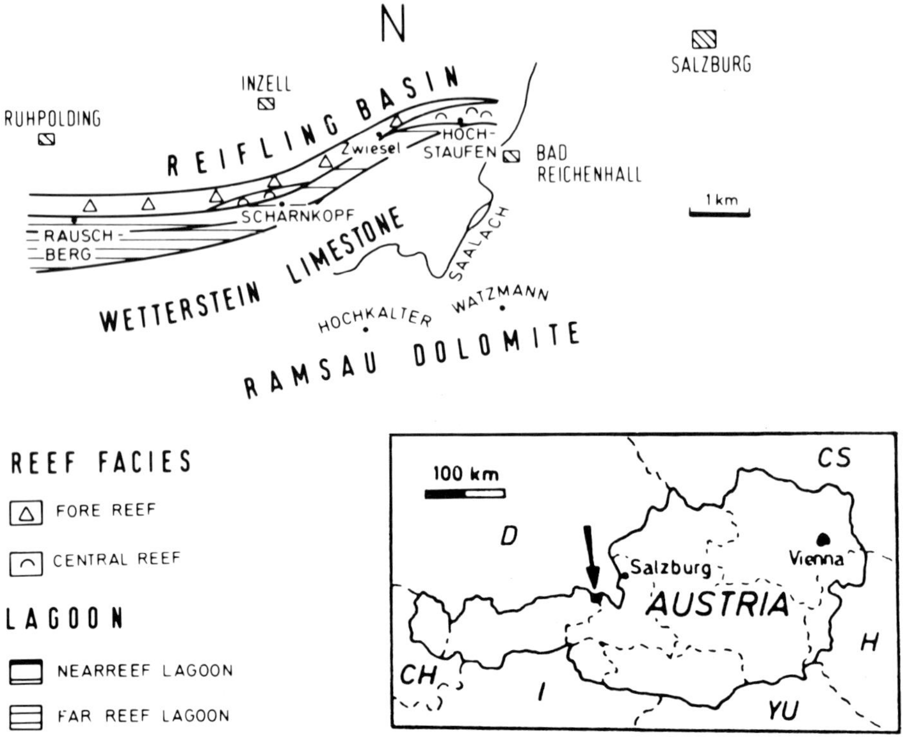

Fig. 1. Location map showing general facies distribution of Cordevolian carbonate series in the Tyrolian nappe along the middle part of the northern rim of the Calcareous Alps

2.2 Stratigraphy

Triassic sedimentation in the middle part of the Northern Limestone Alps took place in three regressive cycles (Table 1). Following transgression on the Permoskythian saline deposits (e.g., Haselgebirge), each cycle gradually built up a carbonate platform terminated abruptly.

The platforms in the Anisian comprise Reichenhall beds, the Wurstlkalke, and Steinalm Limestone; the Cordevolian comprises Wetterstein Limestone and Ramsaudolomite, while the Norian to Rhaetian include the Hauptdolomite and Dachstein Limestone. The platforms alternate with basins of varying depth, e.g., the Upper Anisian to Cordevolian Reifling Limestones and Partnach Marls, the Carnian Raibl beds, and the Norian to Rhaetian Hallstatt Limestones and Zlambach Marls.

Middle and Early Upper Triassic basinal sediments overlie the drowned platforms. Sedimentary framework in the Reifling and Partnach beds on top of the Steinalm Limestones records gradual subsidence of a lagoonal platform which was never subaerially exposed. In contrast, the transgressive, highly terrigenous Raibl beds overlying the Wetterstein Limestone were deposited after numerous periods of karst formation which interrupted sedimentation of the uppermost Wetterstein Limestone.

Table 1. Triassic development of the Hochstaufen-Rauschberg massif (Bavarian Alps)

MY	Formation		Thickness	Environment		Sea level
220	Norian	Hauptdolomit	1200 m	Dolomitic platform	Wide extended dolomitic platform with tidal flats and sabkhas	
	Jul-Tuvalian	Raibl beds	200 m to 300 m	Shallow-marine-basin	Shallow marine clastics, limestones and dolomites	
	Cordevolian	Wetterstein Limestone	350 m	Lagoon Wetterstein platform	Platform with tidal flats, lagoonal ponds and sabkhas	
			300 m 100 m	Reef complex	Central reef Fore reef and slope deposits	
		Reifling Limestone (upper group)	80 m		Fore slope	
		Partnach Marls	50 m	Basin	Basin subdivided in ridges and troughs	
229	Ladin	Reifling Limestone (lower group)	150 m			
233	Anisian	Steinalm Limestone	30 m	Shallow water platform	Platform margin Open platform	
		Succession of the "Wurstlkalke"	150 m		Oolite bars, restricted platform cycles	
		Reichenhall beds			Platform evaporitic facies	
239	Permoskythian "Haselgebirge"			Saline deposits		low high

2.3 Facies Distribution

2.3.1 General Facies Distribution Along the Wetterstein Platform

Regressive sedimentation is a typical feature of Wetterstein platforms. In the Cordevolian, the adjacent basins of argillaceous Partnach beds and Reifling Limestone were followed by carbonate platforms, and Wetterstein Limestone build-ups appear to have been initiated on topographic highs, platform configuration being partly predeterminated by the localization of earlier Steinalm platforms.

The Cordevolian to Upper Lower Carnian carbonate series of the Hochstaufen-Rauschberg massif reflect the last stage of progradation of a wide southern shallow water platform extending into the attached northern basin.

The regional paleogeography comprised a southern widespread dolomitic platform (Ramsau Dolomite) of at least several kilometers to tens of kilometers extension which graded northward into more lagoonal carbonates, e.g., Wetterstein Limestone (upper part). This Wetterstein lagoon is not exposed completely

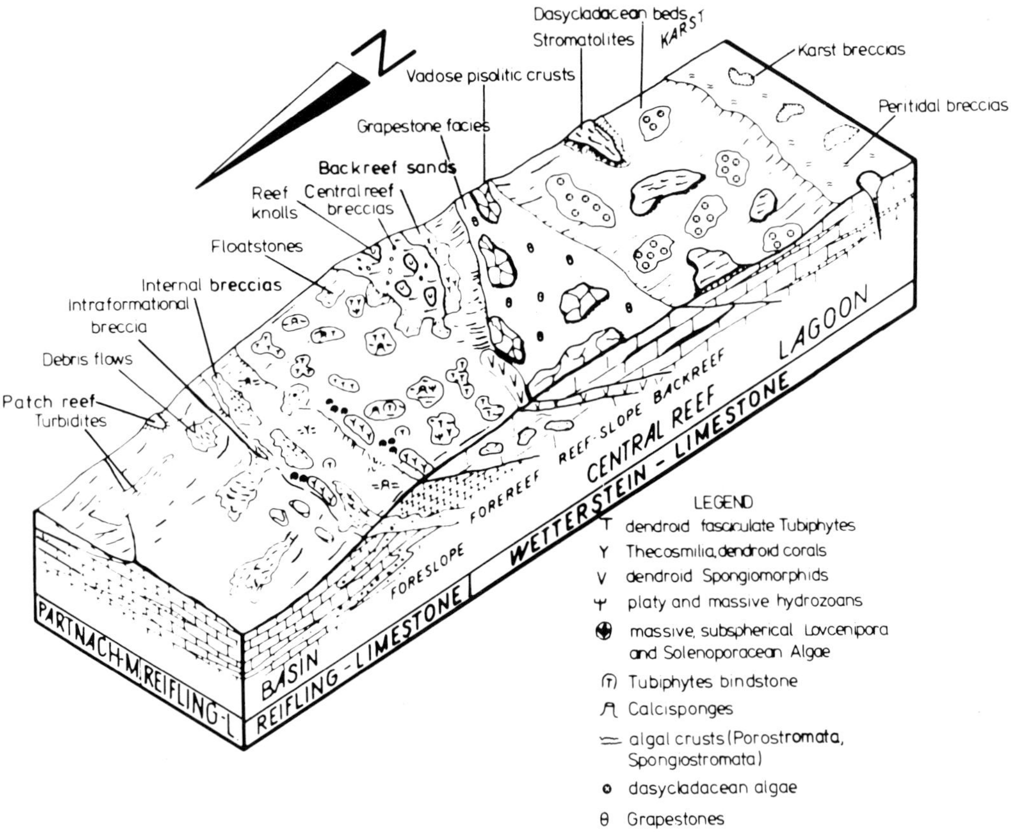

Fig. 2. Schematic block diagram of the different environments represented in the series of the Reifling Limestone, Partnach Marls, and Wetterstein Limestone

because of partial cover by the nappe pile; its extension is estimated to be of the order of a few kilometers. The detailed facies distribution of this platform and its northern marginal regions are illustrated in Fig. 2.

2.3.2 Reef Framework, Paleomorphology, Distribution and Dimension of Reef Bodies

Reef Framework. Like many fossil reefs, the Wetterstein reefs are not directly comparable to recent coral reefs; extensive wave-resistant frameworks are missing. The growth of the structures was controlled by stabilization and encrustation of debris, and most of the reef-building organisms had sediment trapping and sediment-binding functions. The reef builders were predominantly small. Encrusting organisms represented by a great number and variety of algae, e.g., abundant *Spongiostromata* crusts and small cushions, usually only a few centimeters in extension, comprised different porostromate algae, tufts of codiacean algae up to 20 cm in height, various genera and growth forms of sphinctozoa and inozoa, and various small problematica (e.g., *Ladinella porata, Baccanella floriformis, Thaumatoporella parvovesiculifera*) all participate in the construction of these rather small reef frameworks.

Most important primary framebuilders were *Tubiphytes obscurus.* They formed thickets several m^3, which included a great variety of calcisponges. Corals, especially the typical *Thecosmilia* heads of 1 to 2 m in height, and different forms of hydrozoans, e.g., *Spongiomorpha ramosa*, dominated certain sections of the reefs and formed specific biocoenoses (Fig. 2). Organisms of bulbous and cushion-like shape, e.g., solenoporacean algae and chaetetids, formed small patches up to some tens of centimeters in dimension within fine-grained debris and mud.

Because of the nature of the organisms involved, most of these frameworks did not reach significant elevations and were not able to persist in the surf zone. Primary cavities within the reef frameworks generally do not exceed several centimeters in size, but nevertheless constitute a significant proportion of the mass, especially where growth of the structure was very rapid, or extensive accretion of coarse reef debris occured, e.g., in the surf zone and on the reef flat. Reef cavities and depressions were inhabited by sessile foraminifers, catenulate hydrozoans, algae and various problematica.

Paleomorphology, Distribution and Dimension of Reef Bodies. The first reef structures, generally only a few meters in size, appeared within the basinal facies of the Reifling Limestone, where they developed on local debris sheets. The basal debris accumulation was stabilized initially by pioneering colonizers, mainly algae. *Tubiphytes* and encrusting sponges, and these were followed by 0.5 m high heads of fasiculate hydrozoan (e.g., *Holocoelia toulai*), forming a more rigid framework. Frequently, these reefs soon died in most cases as a result of rapid burial by debris.

In topographically higher positions on the fore-reef plate some larger structures developed. These generally included coral belts with closely spaced *Thecos-*

milia heads along the reef front, and cushions of solenoporacean algae and cheatetids.

The generally scarcely populated fore-reef was interspersed with occasional sponge-algal biostromes a few meter in thickness and single coral colonies. This relatively low diversity of organisms may have been due to persisting unfavorable bathymetric conditions and relatively high instability of the substrate.

Different central reef belts developed higher up on the slope in eastern and western positions. An eastern reef (Hochstaufenmassif, cf. Fig. 1) prograded slowly into a slightly deeper basin and subsequently grew to considerable thickness (350 m). It was clearly differentiated into three parts all of which commonly exhibit low diversity reef building organisms and a high proportion of reworked and brecciated material (Fig. 2).

The slope was characterized by calm water biocoenoses with thin sponge-algal biostromes developed on fine-grained substrate, frequently interspersed with unstable reef slope deposits (cf. Fig. 2). The high intertidal reef flat consisted of sediments continuously subjected to wave scour and thus numerous breccias with occasional autochthonous reef knolls a few meter in thickness prevailed. Early cementation and synsedimentary brecciation were the main processes. In the back reef a narrow belt of sand aprons formed on which locally coral patch reefs developed.

A western reef prograded relatively fast into shallow basinal environments. Higher sections of the reef rocks were periodically exposed subaerially and suffered vadose diagenesis. In the lower part, typical reef front sediments with an extensive belt of corals and cushions of small chaetetids (e.g., *Lovcenipora* sp.) and solenoporacean algae developed on fine-grained substrates. These were succeeded by a low angle slope showing continuous biotic zonation (Fig. 2). Sponge/hydrozoan/algal biostromes a few meters in thickness grew in sheltered environments. Corals, hydrozoans, and *Tubiphytes* constructed frameworks in more agitated water. In calm back reef positions *Tubiphytes* algal and *Spongiomorphid* beds developed. This reef, up to 150 m in thickness, is considered to have been built rapidly in sheltered positions along a continuous low angle slope (ramp type reef) and thus mainly reveals continuity of biocoenoses and lithofacies (for detail Henrich 1983).

2.3.3 Occurrence of Breccias in the Wetterstein Reef Complex

The numerous breccias in the Wetterstein rocks include a wide spectrum of reworked sediments along the platform margin, specific breccias formed on the reef flat of the eastern reef, and a variety of peritidal and karstic breccias intercalated in the lagoonal series.

As the Wetterstein platform expanded northwards into the basin, a monocline developed along its front. This flexure advanced with the platform front (Henrich 1982, 1983), and the resulting zone of weakness coincided with a great variety of breccias. The basic cause of the formation of the monocline and its progradation is related to greatly different rates of subsidence between the rapidly subsiding platform and the almost stable basin.

2.3.4 Vadose Pisolitic Crusts and Karst Surfaces (Fig. 4)

Areas of subaerial exposure on the Wetterstein platform include complex vadose pisolitic crusts in near-reef lagoonal sections and numerous karst surfaces developed in the inner, cyclic lagoonal deposits.

Near reef lagoonal carbonates reveal shallowing upward cycles of irregular thickness consisting of a thick sequence of marine limestone and a thin series of vadose pisolitic crusts. The well bedded vadose horizons of 0.5 to 2 m in thickness were intensively dolomitized and show many features comparable to the pisolite facies in the Permian Capitan reef complex (cf. Esteban and Pray 1983) including tepees up to meter sizes, layers, several tens of centimeters thick with inversely graded pisoids, and a set of cracks and fissures of up to several meters in extension filled by vadose cements. Laterally these vadose pisolitic crusts interfinger with limestone beds in grapestone and loferitic facies.

3 Diagenesis

3.1 Early Diagenetic Cement Fabrics

3.1.1 Cement Crusts Induced by Algal Growth

The frequent alternation and close association of algal films and mats with cement crusts illustrates the importance of algae in the construction of Wetterstein reef framworks. This importance is best illustrated by central reef breccias and slope floatstones which consist of multiple brecciated rock fragments whose clasts may be covered by several generations of organic crusts and radiaxial fibrous cements. Internal fillings include coarser sediments in fractures and spaces between these coarse clastic fabrics (Fig. 3 A, B). These breccias indicate intensive reworking in high energy, reef-flat environments and on unstable slopes together with very rapid cementation and lithification. Change in chemical microenvironment caused by excessive algal growth, e.g., crusts, tufts and cushions of *Tubiphytes, Pycnoporidium, Macrotubus, Girvanella* etc., favored cementation and calcification of organic films, and these helped to stabilize and harden the sediment.

In the Spongiostromata crusts micritic rims caused by these organic films frequently alternate with microsparitic cement layers (Fig. 3 B). The generally small crystal size of the microsparite layers suggests rapid crystallization, which seems to have been essentially triggered by a shift in the physicochemical microenvironment caused by photosynthetic removal of CO_2.

The association of algae and cement suggests that Spongiostromata is a true "organo-diagenetic" structure. However, many other partly calcified problematic algae, for example, *Baccanella floriformis* (Wurm 1982), complex *Tubiphytes* crusts and *Thaumatoporella parvovesiculifera*, might be better interpreted as "organo-diagenetic" structures rather than true biogenic growth forms (cf. discussion of *Epiphyton* and *Renalcis* by Pratt 1984).

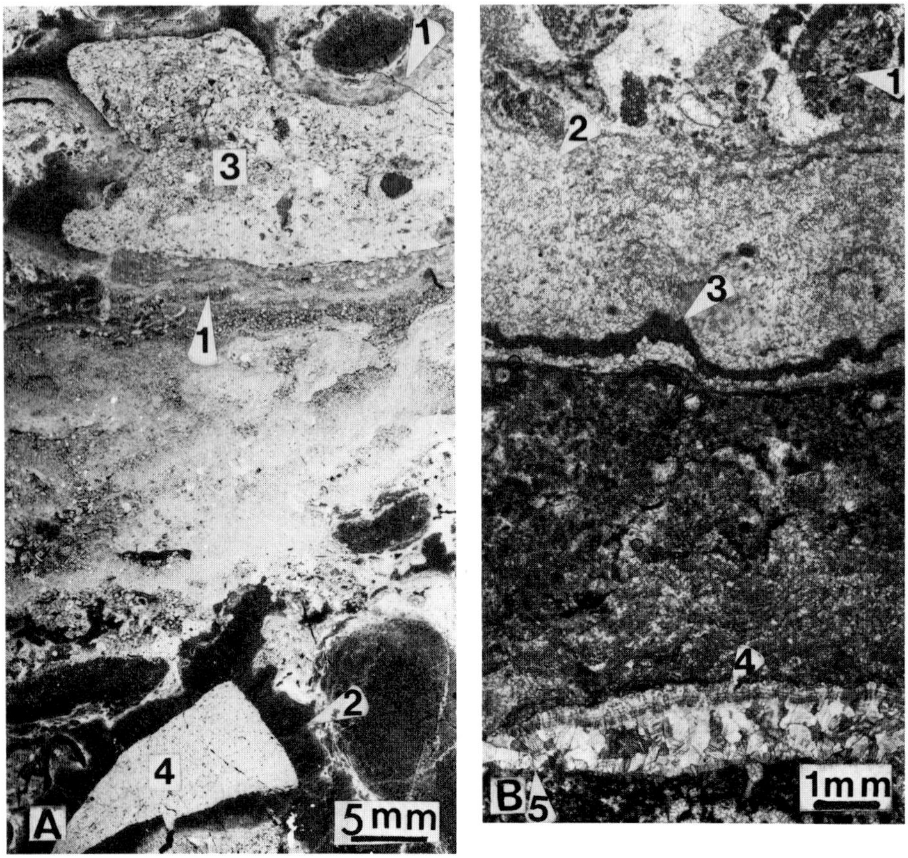

Fig. 3A, B. Cement fabrics in fore-reef floatstones and early crusts with cementation induced by algae. **A** Typical fore-reef floatstone, exhibiting synsedimentary slumping and differential compaction. Along shear planes (*1*) and within the newly formed cavities (*2*) radiaxial fibrous cement crusts precipitated. These cements are frequently intersected by diffuse micritic laminae (*1*) cut by younger sediment fills consisting of a mixture of fine-grained reef debris (*3*) and pelagic organisms predominantely, radiolarians and shell "filaments" (*4*). Fore-reef facies, sample No. 422 (negative print). **B** Detail of a complex algal cement crust within central reef breccia showing (*1*) coarse clasts including neomorphically altered pore filling cements; (*2*) Spongiostromata crusts with diffuse relict laminae of algae in a neomorphically altered cement; (*3*) dense micritic algal crusts intercalated within a layer of sparry cement; (*4*) younger dissolution fissure filled with radiaxial fibrous calcite, in part replaced by late saddle dolomite and blocky calcite (*5*) along a stylolite seam. Central reef facies, sample No. 220

3.1.2 Pseudofibrous Cement Crusts, Festooned Cellular Crustal Cement, Microstalactitic and Meniscus Cements

Characteristic cements in the Wetterstein lagoonal carbonates resemble much of the spectrum described by Assereto and Kendall (1977):

1. Pseudofibrous crustal cements consist of sparry calcite forming pseudomorphs after elongate fibrous crystals. The original crystal shapes are retained by micritic envelopes along the original margin of the crystals, which are hexagonal in cross-section. Different types of pseudofibrous crustal cements can be distinguished according to variations in morphology:

Mamillary crusts (Fig. 4D) consist of planar or slightly arched laminae (0.1 – 1 mm in thickness) formed of ghosts of radially oriented fibers that encrust the walls of fractures (Assereto and Kendall 1977, p. 169). These crusts usually attain thicknesses of several millimeters to centimeters and may extend laterally for several tens of centimeters. This cement, probably a calcitic pseudomorph of an initial aragonitic beach rock cement, is very common in the vadose horizons.

Ray crusts (Fig. 4F) are planar crustal cements arranged parallel to the walls of the fissures. The dimension of these crusts is comparable to mamillary crusts, but in addition they exhibit several intercalations of irregular, radial aggregates of elongate crystals several mm to cm in length. Crusts of this type often alternate with internal sediments and therefore must have been formed very early.

2. Festooned cellular crustal cement (Figs. 4D, F) consists of alternations of crinkled micritic laminae (20 – 500 μm in thickness) with admixtures of fine sediment particles and microsparitic laminae with relic radial fabrics. This type of cement closely follows all irregularities of the walls of the fissures and also alternates with internal sediments. In many cases, microstalactitic cements are included in the crusts, especially near the upper limits of the cracks. The cements were formed at a very early stage and may be compared with Recent beach rock cements of the supratidal environments of the Persian Gulf (Purser and Loreau 1973).

3. Cave popcorn, microstalactites, microstalagmites and meniscus cements (Fig. 4A, E) commonly occur on the floors and roofs of the cavities, in the spaces between pisoids and in festooned crustal cellular cement. They form irregularly laminated, lobate or arborescent gravitational cement layers which include agglutinated micrite. Due to their precipitation from percolating waters of variable chemistry in the upper part of the supratidal zone, they are both calcitic and dolomitic.

3.1.3 Radiaxial Fibrous Calcite Cement

According to their textural fabrics, two subtypes of radiaxial calcite may be distinguished (cf. Bathurst 1975). Type A are large, elongate crystals consisting of a set of very distinct subcrystals with slightly different extinctions (Fig. 5B). Type B shows vague subcrystals resulting in undulatory extinction of the whole crystal (Fig. 5A). In the material studied, type B appears to be richer in inclusions and is thought to represent a more altered replacement structure (cf. Kendall and Tucker 1973).

Radiaxial fibrous cements encrust both primary voids and secondary solution cavities, and thus are early precipitations somewhat modified by freshwater diagenesis. In many cases, relics of precursor cements, e.g., microstalactitic cements

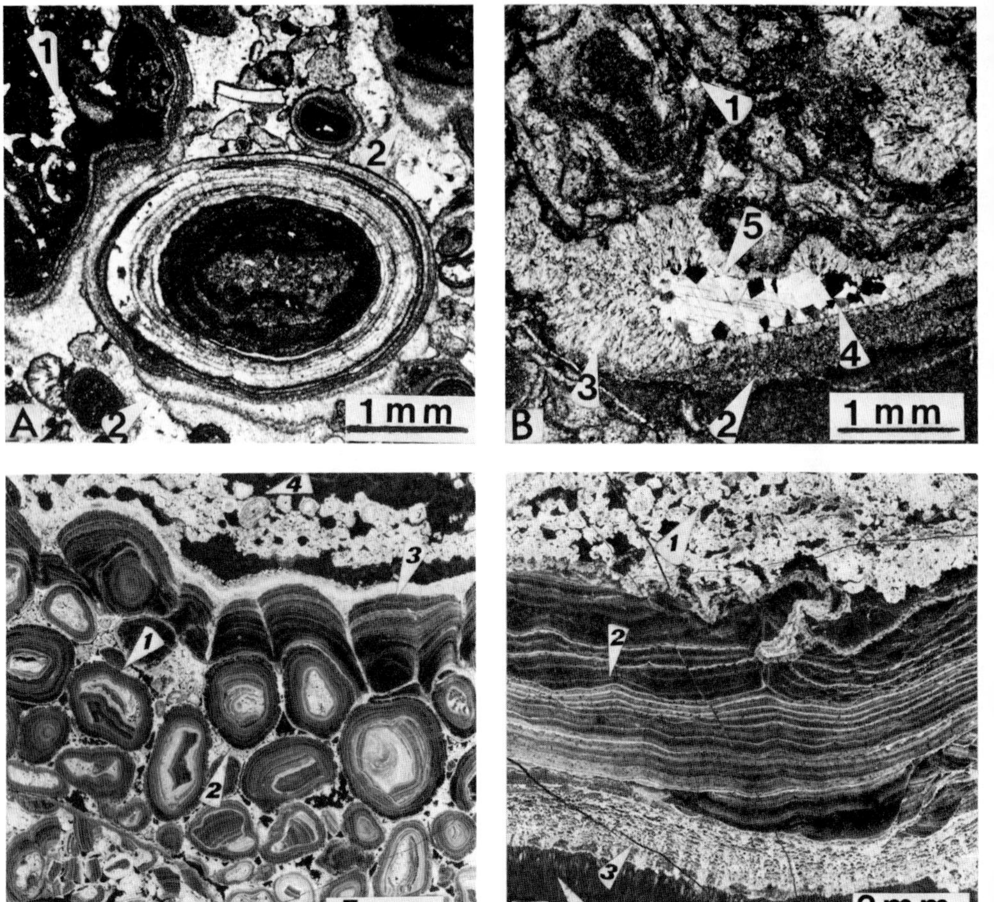
A
1
2
2
1 mm
B
1
5
4
3
2
1 mm
C
4
3
1
2
5mm
D
1
2
3
4
3mm
E
1
4
5
2
3
3mm
F
1
2
4
3
5
3mm

Fig. 4

and mamillary crusts, can be recognized (Fig. 6). In lagoonal sediments, especially in vadose pisolite horizons, radiaxial fibrous cement crusts of several centimeters thickness and several meters extension are cut by a set of younger cracks and fissures filled with cement and internal sediments. In reef rocks, radiaxial cements are either overlain by internal sediments or alternate with biogenic crusts (Figs. 3, 6). They also occur along secondary solution fissures in higher reef sections (Fig. 6).

In a recent reappraisal, Kendall (1985) rejects the replacement theory for the formation of radiaxial fibrous calcite and provides evidence of a primary precipitation in a marine phreatic environment, e.g., split-crystal growth from highly supersaturated solutions.

We conclude that the radiaxial fibrous calcite cement in the Wetterstein reef rocks, at least in part, is a replacement fabric of an initially different cement. Occurrences of radiaxial fibrous cement in the material from higher sections of the reef rocks supports this interpretation. Here the radiaxial fibrous fabric is developed both as linings of primary cavities and on the walls of secondary solution fissures, which show relics of precursory vadose cement.

It is also observed within primary cavities. Thus, formation of this radiaxial fibrous cement mosaic must have taken place during the progradation of lagoonal sediments, which bury reef rocks. Secondary dissolution fissures in the topmost sequences of the reef rocks indicate subaerial exposure with vadose and/or phreatic dissolution phases, prior to lagoonal sedimentation. The first vadose pisolite horizon developed in near-reef lagoonal sediments occurs 1 to 2 m above the reef top. It is possible that influx of lagoonal brines drained deeply into the reef and fore-reef rocks during initial burial stages, causing precipitation of fibrous precursor cements and their replacement by radiaxial fibrous cement crusts. Periodic sea-level fluctuations during platform progradation may have been an extremely effective mechanism for the discharge of lagoonal brines into the reef rocks.

These radiaxial fibrous crystals reveal a rather broad scatter of trace element and isotopic composition (cf. Table 2) supporting evidence of a precipitation

Fig. 4 A–F. Pseudofibrous crustal cements, festooned cellular cement and microstalactitic cement. **A** Pisoid with multiple concretionary growth (*1*) and microstalactitic cements (*2*). Lagoonal facies, sample No. 111. **B** Detail from vadose pisolitic crusts: (*1*) distorted mamillary crusts; (*2*) dolomitized basal sediment; (*3*) younger crack filled by radiaxial fibrous calcite; (*4*) dolosparite; (*5*) blocky calcite. Lagoonal facies, sample No. 111. **C** Inversely graded pisolite composed of pisoid fragments encrusted by younger pisoids (*1*). Some pisoids show radial cracks (*2*). The pisolite is covered by a polygonal crust (*3*). (*4*) Unaltered loferite on top. Lagoonal facies, sample No. 350 (negative print). **D** Large fissure oriented parallel to loferite bedding (*1*) filled by various vadose cements: (*2*) mamillary crusts; (*3*) festooned cellular cement; (*4*) radiaxial fibrous cement. Lagoonal facies, sample No. 268 (negative print). **E** Loferite with reworked algal chips (*1*), fine-arenitic organic infilling (*2*) and numerous sheet cracks with typical secondary solution features, notably bizarre solution rims (*3*) along the walls and floors of the cavities. Microstalactitic (*4*), radiaxial fibrous and final blocky calcite fillings (*5*) are common. Lagoonal facies, sample No. 479/1 (negative print). **F** Pisolitic crust with different vadose cements: (*1*) mamillary crusts and ray crusts; (*2*) diagenetically altered festooned cellular cement; (*3*) radiaxial fibrous calcite cement; (*4*) blocky calcite; (*5*) finely pelleted internal sediment. Lagoonal facies, sample No. 237 (negative print)

Fig. 5 A, B. Fabric variability in radiaxial fibrous cements. **A** Radiaxial fibrous calcite type A showing large elongate crystals consisting of a set of very distinct subcrystals each having slightly different extinction. Lagoonal facies – vadose pisolite horizon, sample No. 236. **B** Radiaxial fibrous calcite type B showing indistinct subcrystals resulting in undulatory extinction of the whole crystal. Lagoon – vadose pisolite horizon, sample No. 269a

from normal marine and lagoonal brines of variable compositions and latter modification by meteoric pore fluids.

3.2 Dolomite Types

Four major types of dolomite are distinguished in the Wetterstein series on the basis of the following criteria:

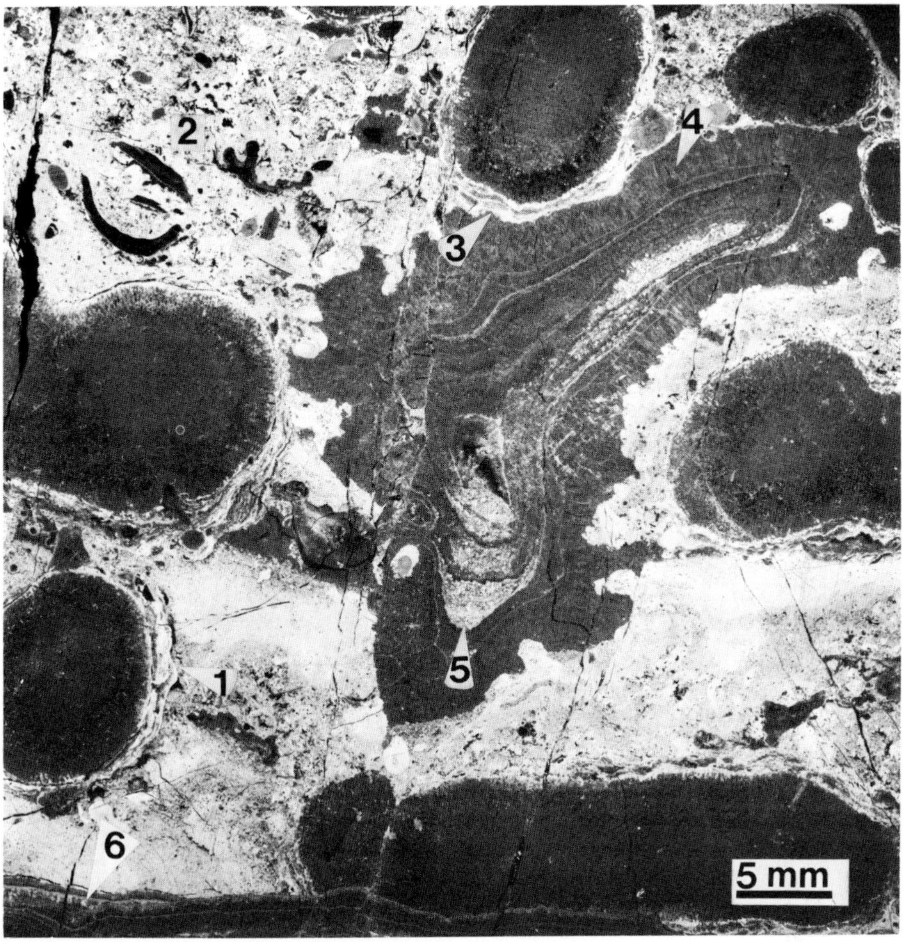

Fig. 6. Radiaxial fibrous calcite lining primary reef cavities and secondary solution fissures. Typical bafflestone fabric in the *Spongiomorphid* beds: Branches are encrusted by laminated algae (*1*). The sediment contains echinoderm and *Tubiphytes* fragments, gastropod and brachiopod shells (*2*). Primary cavities were lined with algal crusts (*3*) before the formation of the radiaxial fibrous cement (*4*) and filling by younger, distinctly layered internal sediments (*5*). A dissolution fissure parallel to bedding is lined with vadose mamillary crusts before being filled with radiaxial fibrous cements (*6*). Uppermost beds of the reef facies, sample No. 416 (negative print)

- sizes and habit of crystals
- shape and geometry of the dolomitized rock unit
- limitation to strata-bound sedimentary or diagenetic structures
- coexistence with other mineralizations
- geochemical and isotopic composition.

Table 2. Trace element and stable isotope composition of cement and dolomite types

Sample No.	Cement/ dolomite type	Facies	Mg %	Sr ppm	Na ppm	K ppm	Fe ppm	Mn ppm	$\delta^{18}O$ PDB	$\delta^{13}C$ PDB
329a	RFC	Central reef, cavity lining	1.68	149	183	52	73	25		
459a	RFC	Fore reef, floatstone, cavity lining	0.35	200	477	122	87	61	−3.79	+3.99
343a	RFC	Vadose pisolitic crust	0.44	195	495	233	485	26		
346a	RFC	Vadose pisolitic crust	0.47	142	201	75	246	37	−6.39	+3.33
489a	RFC	Filling of a karst fracture	1.07	167	413	322	905	30	−5.54	+3.64
213a	CBC	Final pore filling in reef cavity	0.97	133	249	103	136	40	−7.30	+2.69
213b	CBC	Final pore filling in reef cavity	0.94	123	195	74	210	42		
215	CBC	Final pore filling in reef cavity	0.27	185	326	217	353	142	−7.16	+2.96
489b	CBC	Final filling of karst fractures	0.18	107	225	93	62	27	−6.54	+3.77
346c	CBC	Final filling of vadose crust	0.37	115	282	67	118	38	−6.96	+3.47
378C	DM + matrix	Green emersion breccia	11.09	185	590	1020	7220	1074	−2.09	+3.89
378A	DM	Green emersion breccia	6.63	382	492	773	4520	1225	−3.70	−0.78
489C	DM + SD	Filling of karst fissure	6.97	245	721	1875	4930	53	−6.65	+2.44
378B		Calcite black pebble	1.15	2782	423	282	845	375	−4.11	+3.16
111	RD	Vadose pisolitic crust	10.08	75	398	568	660	37	−5.76	+3.87
241	RD	Vadose pisolitic crust	9.26	54	204	75	167	32	−6.29	+3.35
221	RD	Central reef	8.07	80	427	127	245	42	−5.70	+3.91
336	RD	Fore reef	9.16	66	167	66	82	43	−5.62	+4.06
206	SD	Fore reef	4.14	123	176	77	259	64	−7.92	−1.59
208	SD	Solution cavity	7.43	58	331	150	622	51	−7.67	+0.49
329b	SD	Final pore filling Central reef	7.05	103	189	85	125	55		
459b	SD	Central fore reef	10.0	82	682	410	250	77	−8.99	+2.76
343b	SD	Central vadose crust	4.97	152	269	363	988	43		

3.2.1 Syngenetic Dolomicrite

This dolomicrite consists of small idiomorphic (1 – 5 μm) rhombohedra (Fig. 7A) forming 1 – 2-m-thick inter- to supratidal dololaminite and stromatolite beds within cyclic inner lagoonal carbonates. Most of these dolomites display a slight excess of Ca^{2+} (Mg/Ca ratios 48 – 49/51 – 52). Various sedimentary and early diagenetic features indicate very rapid lithification and early dolomitization, e.g., numerous polygonal desiccation cracks, abundant shrinkage pores within algal mats, and characteristic peritidal breccias composed of multiply reworked, sharp-edged, early dolomitized stromatolite and dololaminite chips. Analogous early diagenetic dolomites are known from Recent sabkha environments. Various contraversial theories concerning the mechanism of this early dolomitization are discussed by Zenger and Dunham (1980).

Early dolomites also occur within vadose pisolitic crusts in near-reef cyclic lagoonal carbonates. Here minute rhombohedra of 1 – 5 μm size occur along microstalactitic cement seams and along 50 – 100 μm thick lamellae within the pisoids, where they alternate with 100 – 150-μm-thick calcitic laminations, some of which show a well-developed brick-like texture of 10 – 20-μm-thick and 40-μm-long calcite plates.

Assereto and Folk (1976) describe similar Triassic pisoids from Lombardy which exhibit alternating layers of radial rays and calcite plates with a brick-like texture, and thin dolomicrite levels. Their interpretation of these fabrics is based on a comparison with Recent equivalents from the supratidal environments of the Persian Gulf (Purser and Loreau 1973). They imply recrystallization and inversion of bundles of aragonite needles under hypersaline conditions, and a dolomitization of an organic-rich Mg calcite micrite under schizohaline conditions.

SEM/EDAX analysis of the Wetterstein pisoids records a wide range of preservation and several phases of dolomitization (cf. Henrich 1984). Some pisoids consist of thick micritic cortical layers alternating with thin micritic laminae. Here dolomite is confined to very distinct laminae (minute rhombs of 1 – 2 μm in diameter). EDAX concentric distribution patterns of sulfur which may suggest an influence of thin organic films in the formation of these concentric micritic textures. Other pisoids are composed of thick dolomicrite layers and thin calcimicrite levels affected by patchy dolomitization. In addition, entirely dolomitized pisoids, including coarser dolosparitic cores, were observed. All these dolomites coexist with rare diagenetic alkaline feldspars, reflecting the predominence of arid, hypersaline conditions.

Dolomitization started early (e.g., the distinct dolomicrite layers described by Assereto and Folk 1976) and persisted during early burial. A later secondary micritization affected the whole fabric, obscuring the initial textures (brick-like texture etc.). The exact mechanism involved are not completely understood, but are probably related to meteoric conditions.

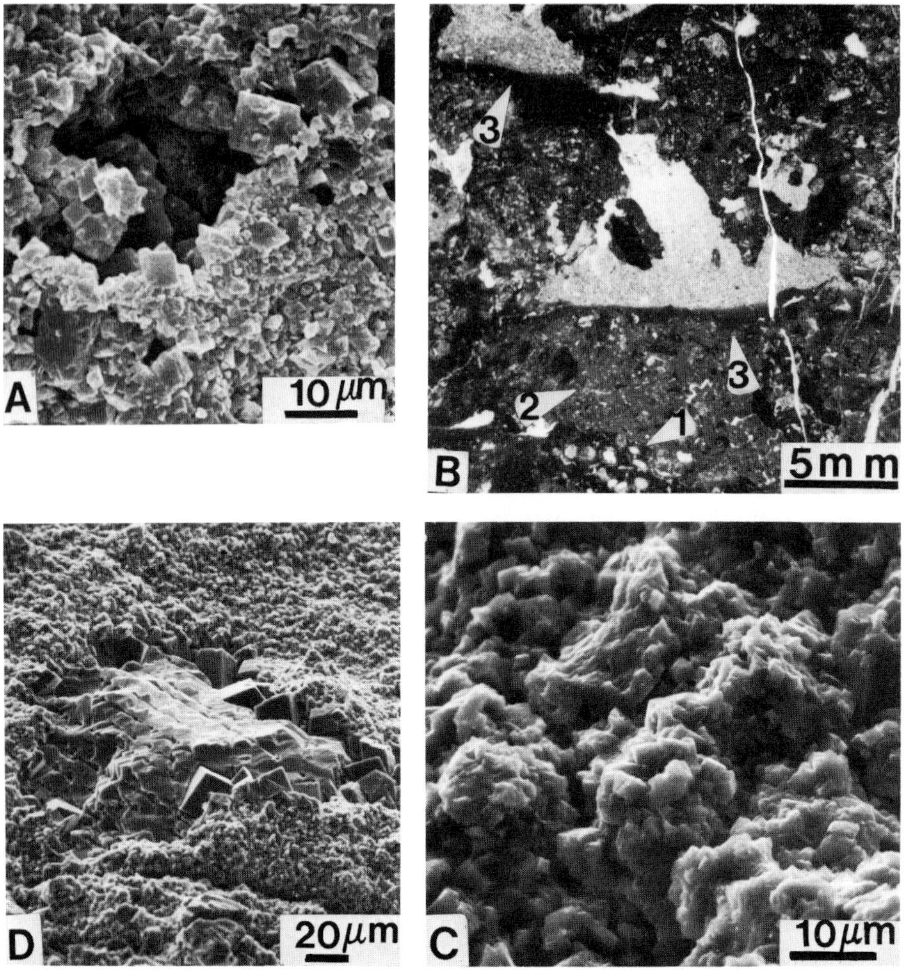

Fig. 7 A – D. Syngenetic dolomicrite and near-surface dolomicrosparite within karstic sediments. **A** Syngenetic dolomicrite in a reworked stromatolite with high secondary porosity and dolosparite in open pores. Lagoonal facies, sample No. 472. **B** Detail of karst filled with green marly limestone and black pebbles showing fragments of partly dissolved dasycladacean-foraminiferal limestone (*1*). The matrix (*2*) is affected by multiphased dissolution tubes and cavities filled with dense residual micrite and dolomicrosparite (*3*). Lagoonal facies, thin section No. 378/3. **C** Typical texture of near-surface dolomicrosparite exhibiting grape-like aggregates of dolomite. The somewhat indistinct contours of the aggregates are caused by thin amorphous films of silica (Si, Al, K, Fe). Lagoonal facies, sample No. 378. **D** Near-surface dolomicrosparite precipitated along a dissolution tube in the matrix of a karst breccia. This tube is lined with large idiomorphic dolomite and central clear blocky calcite. Lagoonal facies, sample No. 469

3.2.2 Near-Surface Dolomicrosparite Associated with Karst

A characteristic dolomicrosparite was precipitated within irregular dissolution tubes, voids and relictic pores of green emersion breccias. These breccias are thin, lense-shaped, marine-reworked paleosoil deposits hundreds of meters in extension, which are frequently intercalated in the uppermost lagoonal rocks of the Wetterstein series.

The precipitation of the dolomicrosparite is preceded by major phases of karstic dissolution intensively affecting the underlying dasycladacean limestone beds. Partly dissolved clasts of these host rocks were subsequently incorporated into the emersion breccia. Some of these clasts are impregnated by bituminous films and/or iron oxides, which envelope the surface of bioclasts. Residual voids were filled by a sparry mosaic of clear calcite, and subsequently reworked and incorporated with black pebbles into the green emersion breccia.

The matrix of these green, reworked paleosoils consists of dense micritic marls or marly limestones, which are rich in echinoid fragments and ostracods. These muds were often subjected to multiple phases of dissolution forming a network of irregular voids, tubes and channels of several millimeters in extension, all of which are marked by dense micritic rims and enrichment of insoluble residue. Some of these tubes may possibly be roots. Some of the voids collapsed and most of them were filled with mud, resulting in the patchy appearance of the matrix.

The youngest dissolution channels and voids (preferentially of stromatactis type) include geopetal fillings of very fine-grained residue with dolomicrosparite above (Fig. 7B). Often the boundary between geopetal sediment and overlying dolomicrosparite is gradual, and this may be interpreted either as the result of input of micrite during precipitation of the dolomite or as partial dolomitization of the geopetal filling.

SEM investigations show that the dolomite crystallized as grape-like aggregates of slightly hypidiomorphic rhombohedra (5 – 10 μm in diameter, Fig. 7C). Most of the dolomite aggregates are covered by thin films of amorphous silica and/or pyrite, as revealed by EDAX (Fig. 7C). The remaining pores were filled by large, zoned dolomite crystals and/or clear blocky calcite during the final stages of diagenesis (Fig. 7D).

The dolomicrosparites reveal a clear excess of Ca^{2+} (43 – 45 mol % $MgCO_3$; XRD) suggesting formation from nonevaporite pore fluids (cf. Choquette and Steinen 1980, Lumsden and Chimahusky 1980). The trace element composition, with extraordinary high contents of Fe, Mn, and K, supports earlier evidence by EDAX analysis. Stable isotope composition (cf. Table 2) records a moderately low $\delta^{18}O$ signal (-2.09 to -6.65) and a fluctuating $\delta^{13}C$ (-0.78 to $+3.89$), which may indicate fluctuating marine and meteoric conditions.

The dolomicrosparite fabrics and geochemical data suggest formation under reducing conditions within the mixing zone of karst systems, which are transitional to a marine phreatic environment.

3.2.3 Replacement Dolomite

Hypidiomorphic to idiomorphic dolomites were precipitated in pore spaces and lattices of completely lithified rocks and replaced sedimentary (e.g., matrix textures, bioclast fabrics, etc.) and early diagenetic structures (e.g., early diagenetic cement crusts). These dolomites are interpreted to have formed dur-

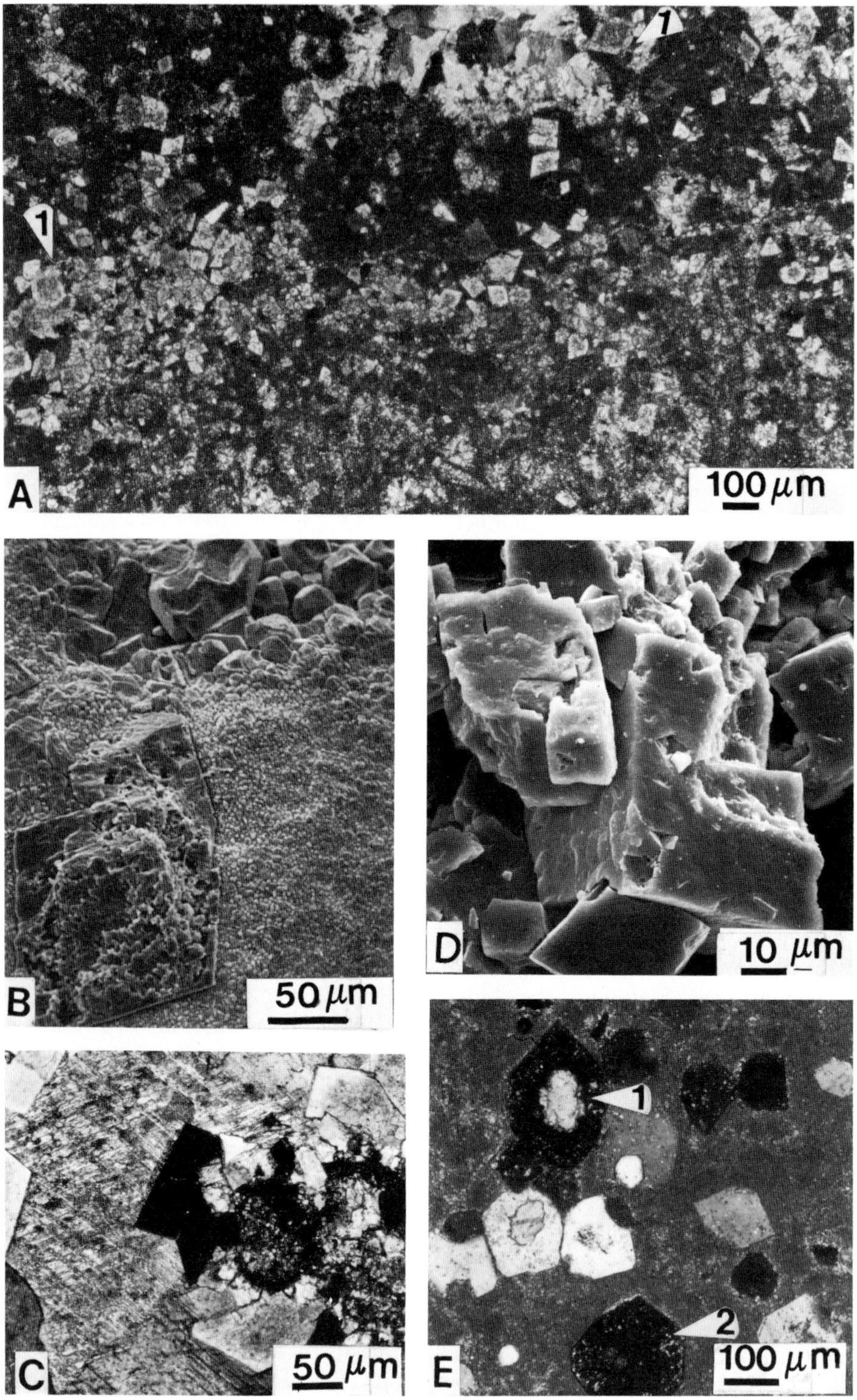

Fig. 8

ing early burial, since they, in turn, were affected by later intensive pressure solution.

Replacement dolomite affected reef rocks and near-reef lagoonal carbonates, the latter especially in the vicinity of the vadose pisolite horizons. These hypidiomorphic and idiomorphic dolomite rhombohedra, commonly 50 – 150 µm (maximum 250 µm) in diameter, exhibit clear outer rims and a dusty brownish microcrystalline inner core (Fig. 8 A). SEM indicates that these textures are a result of dedolomitization (Fig. 8 B). Undersaturated solutions affected the less stable inner parts of dolomite crystals via intracrystalline shear planes, and were most effective in the centers of the dolomite rhombs and these strongly corroded rhombohedra have not been refilled by calcite.

Most of these dedolomitized rocks are light brownish to yellowish in color, reflecting admixtures of iron oxides and hydroxides. Mg/Ca ratios of an average 48/52 were measured by X-ray diffractometry for these dolomites. Atomic absorption spectroscopy of whole rock samples revealed a significant increase in iron content correlated to increasing dolomitization and subsequent dedolomitization phenomena. As has been suggested by several authors (Katz 1971, Frank 1981), ferroan dolomite may have be dedolomitized under near-surface, oxidizing conditions, the iron oxides and hydroxides being by-products.

Stable isotope composition (Table 2) indicates a light $\delta^{18}O$ (–5.62 to –6.46) and a relatively heavy $\delta^{13}C$ (+3.35 to +4.06). These values show approximately the same scatter as those for radiaxial fibrous calcites precipitated within vadose pisolitic crusts. The high $\delta^{13}C$ may represent the relatively unchanged normal background of hypersaline environments, while the $\delta^{18}O$ may reflect significant dilution.

These observations suggest near-surface dolomitization in the mixing zone of meteoric and hypersaline brines (Badiozamani 1973, Land 1973, Folk and Land 1975, Sears and Lucia 1980, Dunham and Olson 1980) rather than by reflux of hypersaline brines (Adams and Rhodes 1960, Deffeyes et al. 1965). Frequent sea-level oscillations are thought to be the most effective mechanism in having pumped the large volumes of pore water needed for dolomitization.

This dolomitization persisted during early burial stages during deposition of inner lagoonal sediments, including the various periods of karstification. Freshwater flushing during the younger karst periods seem to have formed dedolomitization phenomena in the earlier dolomites.

However, relatively late dolomites generally have been little affected by dedolomitization. Typically, they consist of hypidiomorphic intergrowths of

Fig. 8 A – E. Replacement dolomite and dedolomitization. **A** Biogenic structures of central reef rocks, strongly modified by replacement dolomitization. A Clear outer rim and a dusty microcrystalline brownish core (*1*) of the rhombohedra reflect dedolomitization. Grumuleuse structure is visible in lower part. Central reef facies, sample No. 219. **B** Dolomite rhombs embedded in a dense matrix. The cores of dolomite crystals exhibit dissolution textures due to dedolomitization. Residual voids are filled with blocky calcite. Lagoonal facies, sample No. 275. **C** Zoned idiomorphic dolomite and central blocky calcite as late pore-filling cements in a dissolution cavity within a karst breccia. Lagoonal facies, sample No. 469. **D** Typical texture of replacement dolosparite with hypidiomorphic intergrowths and high intercrystalline porosity. Dolomitic matrix of a fore-reef breccia. Sample No. 105b. **E** Zoned authigenic quartz crystals with corroded carbonate core (*1*) and carbonate inclusions in outer zones (*2*), replacing saddle dolomite in reef rocks. Central reef facies, sample No. 122c

10 – 150-μm-sized rhombohedra with high intercrystalline porosity (Fig. 8C). The intensity of this later dolomitization varies regionally and probably depends on the duration of karstification which varied in different parts of the platform. The spectrum of diagenetic alterations include:

1. Lenses of several m^3 to tens of m^3 of epigenetic dolosparite bodies, often characterized by somewhat lighter colors relative to the surrounding rocks. The replacement of sedimentary and early diagenetic structures gives a sucrosic dolomite mosaic.
2. Selective dolomitization along horizons of greater porosity and permeability in the matrix of fore-reef internal breccias (Henrich 1982, 1983) and in the matrix of bioturbated sediments as well as in beds and lenses of coarse debris deposited in fore-reef areas.

3.2.4 Saddle Dolomite and Associated Authigenic Quartz Crystals

Saddle dolomite occurs as final fillings of intergranular pores and in residual spaces in reef cavities, as well as along younger tectonic fractures. In many cases,

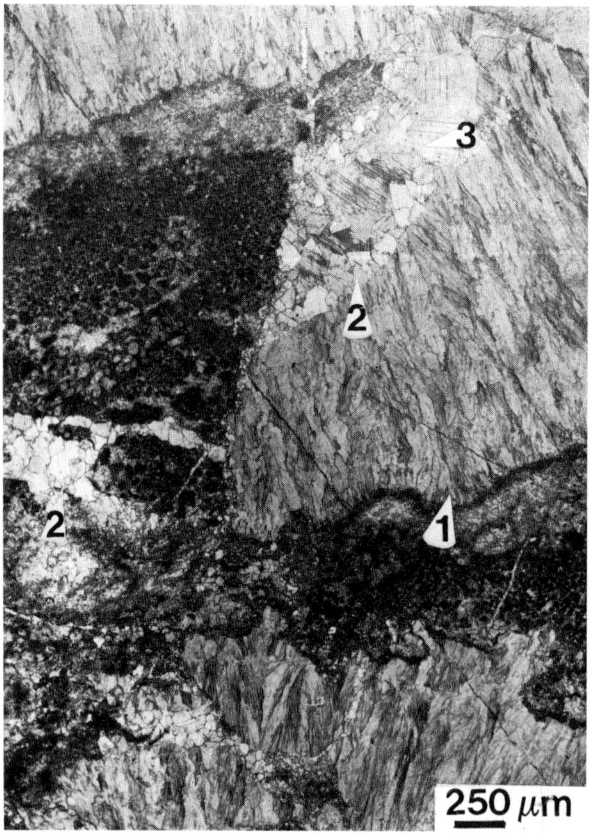

Fig. 9. Radiaxial fibrous cement replaced by deep burial saddle dolomite. Cavity filling in fore-reef rocks showing strong corrosion of radiaxial fibrous calcite crusts (*1*) and are replaced by coarse saddle dolomite (*2*) and late clear blocky calcite (*3*). Fore-reef facies, sample No. 453

radiaxial fibrous cements are intensively corroded and reworked before replacement by a coarse xenotopic (definition according to Gregg and Sibley 1984) saddle dolomite (Fig. 9A). These dolomite crystals are recognized in thin section by their sweeping extinction and curved edges leading to pointed terminations. Saddle dolomite occurs together with or replaces late clear iron rich blocky calcites (Fig. 9A).

Late dolomite cements are associated with epigenetic lead-zinc ore mineralizations along Upper Cretaceous/Early Tertiary faults and fractures, and are therefore regarded as late diagenetic mineralizations formed in deep-seated environments at moderately elevated temperatures. This assumption is supported by their chemical composition (iron content, high zinc concentrations) as measured with atomic absorption spectroscopy (Table 2) and by their light isotopic $\delta^{18}O$ (-7.67 to -7.92) and $\delta^{13}C$ (0.94 to -1.59). Gregg and Sibley (1984) suggest a minimum temperature of at least 50°C for the development of the xenomorphic dolomite texture. Radke and Matthis (1980) assume a temperature range of 60°C to 150°C for saddle dolomite.

Large, zoned authigenic quartz crystals 100 – 200 μm in diameter are closely associated with saddle dolomite and late diagenetic lead-zinc ore deposits. These quartz crystals contain various inclusions, e.g., irregular carbonate cores, and minute carbonate inclusions grouped within the peripheral zones of the crystal. They also include earlier dolomite rhombohedra (Fig. 8E).

3.3 Late Calcite Sparite Cements

A sequence of late diagenetic cements fills the remaining pore spaces and reef cavities. Some of the larger cavities, especially those of secondary origin, are not always filled with cement, but retain pores several millimeters to centimeters in size.

In lagoonal sediments a single blocky calcite crystal or a mosaic of clear equant calcite crystals often constitutes the final filling of intergranular voids. Stable isotope composition of these calcites reveal a low $\delta^{18}O$ (-6.54, -6.96) and a rather "unchanged" $\delta^{13}C$ ($+3.47$, $+3.77$), indicating meteoric formation of these calcites.

In the reef rocks, precipitation of a xenomorphic calcite microsparite with a coarsening of crystals towards the centres of the voids is generally succeeded by a final filling of clear blocky calcite with twin lamellae. Stable isotope composition of these late calcites indicate very light $\delta^{18}O$ (-7.16, -7.30) and $\delta^{13}C$ ($+2.69$, $+2.96$) suggesting precipitation from formation waters in deep burial environments (cf. Dickson and Coleman 1980).

3.4 Pressure Solution and Fracturing

These late diagenetic processes are summarized as follows:

1. Onset of pressure-solution resulting in microstylolites.

2. Intensive pressure solution subsequent to late calcite sparite cement precipitation (Sect. 3.3) leading to the formation of long stylolite seams.
3. Late tectonic fracturing. Along fractures and joints, precipitation of very clear blocky calcites without twins and clear slightly brownish dolomite, locally with ore mineralization, especially iron oxide seams (see Sect. 3.2.4).
4. Faulting and fracturing during Late Cretaceous and Early Tertiary nappe tectonics, when the greatest depth of burial of at least 3 km had been reached; along these fractures late saddle dolomite associated with lead-zinc ore deposits.

3.5 Late Karst

Late diagenetic features are related to deep mature karst systems, which were effective even during later stages of burial and compaction. For example, during sedimentation of Raibl, clastics and younger carbonates, water seems to have been injected downward along karstic channels. Karstification was most intensive along the unstable platform margin where synsedimentary tectonics and intensive reworking provided elevated permeabilities facilitating the drainage of karst fluids. As a result, nearly all prominent late diagenetic mineralizations were restricted to the close proximity of karst fabrics. The pore fluids migrating through these karstic channels evolved during burial of the reef rocks. Their evolution is reflected in the diagenetic products described above.

4 Porosity Development

Primary porosity in the Wetterstein reefs was generally low mainly because of the specific reef framework, constructed by small, often fragile and encrusting organisms with effective sediment baffling and binding functions. Furthermore, many primary voids were rapidly occluded during early cementation probably induced by photosynthetic activity.

On the other hand, sedimentary reworking on the unstable reef slope were ubiquitous and this resulted in very coarse fabrics composed mainly of clasts with partial infilling by younger sediments. These coarse sediments attained a high porosity and permeability. Although large volumes of pore space were filled rapidly with early cements, there remained considerable residual porosity.

During initial burial, meteoric waters increased the porosity by formation of dissolution cavities and fissures which, in turn, were partly filled with later cement crusts. Frequent sea-level oscillations stimulated dissolution/precipitation during a considerable period of time (cf. Brandner 1984). As a result, early diagenesis in the Wetterstein reef rocks finally resulted in parts of the sequence being completely cemented, while other parts remained highly porous and permeable, especially along sections subjected to intense synsedimentary reworking and meteoric dissolution.

During subsequent burial diagenesis the main water influx occurred along these permeable zones. Intensive karstification resulted in an overall increase in porosity and permeability; it also favored dolomitization, resulting in the formation of large dolosparite bodies, thus increasing porosity and permeability, especially in basal Wetterstein sections. The most porous rocks are developed near the unstable platform margin, where, in spite of intensive pressure solution and precipitation of late diagenetic cements, considerable porosity is preserved.

Acknowledgments. Dr. B. H. Purser, Dr. J. H. Schroeder, and Dr. J. Thiede read various drafts of the manuscript and made numerous valuable suggestions. The isotopic data was provided by ^{14}C Laboratory University of Kiel (Dr. H. Erlenkeuser). We thank J. Carstens (preparation of samples), M. Rössler (operation of the mass spectrometer), M. Burk and H. Münter (atomic adsorption spectroscopy), B. Rippe and E. Wettengl (draftings), and R. Dukat and B. Hecht (photographic assistance). We offer special thanks to Claudia Henrich for assistance in final drafting and criticism of the manuscript. Financial support was provided by the Deutsche Forschungsgemeinschaft.

References

Adams JE, Rhodes ML (1960) Dolomitization by seepage refluxion. Am Assoc Petrol Geol Bull 44:1912–1920

Assereto RL, Folk ML (1976) Brick-like texture and radial rays in Triassic pisolites of Lombardi, Italy: A clue to distinguish ancient aragonite pisolites. Sediment Geol 16:205–222

Assereto RL, Kendall CG (1977) Nature, origin and classification of peritidal tepee structures and related breccias. Sedimentology 24:153–210

Badiozamani K (1973) The Dorag dolomitization model – application to the Middle Ordovician of Wisconsin. J Sediment Petrol 43:965–984

Bathurst RGC (1975) Carbonate sediments and their diagenesis. Elsevier, Amsterdam, 658

Bechstädt T (1975) Lead-zinc ores dependent on cyclic sedimentation (Wetterstein-Limestone of Bleiberg-Kreuth, Carinthia, Austria). Miner Dep 10:234–248

Bechstädt T (1979) The lead-zinc deposit of Bleiberg-Kreuth (Carinthia, Austria): palinspastic situation, paleogeography and ore mineralization. Verh Geol Bundesanst 3:221–235

Bechstädt T, Dohler-Hirner B (1983) Lead-zinc deposits of Bleiberg-Kreuth. Am Assoc Petrol Geol Mem 33:55–63

Bosellini A (1984) Progradation geometries of carbonate platforms: examples from the Triassic of the Dolomites, northern Italy. Sedimentology 31:1–24

Brandner R (1984) Meeresspiegelschwankungen und Tektonik in der Trias der NW-Tethys. Jahrb Geol Bundesanst 126/4:435–475

Brandner R, Resch W (1981) Reef development in the Middle Triassic (Ladinian and Cordevolian) of the Northern Limestone Alps near Innsbruck, Austria. Soc Econ Paleontol Miner Spec Publ 30:203–231

Choquette PW, Steinen RP (1980) Mississippian non-supratidal dolomite, Ste. Geneviève Limestone, Illinois Basin: evidence for mixed-water dolomitization. Soc Econ Paleontol Miner Spec Publ 28:163–196

Dickson JAD, Coleman ML (1980) Changes in carbon and oxygen isotope composition during limestone diagenesis. Sedimentology 27:107–118

Deffeyes KS, Lucia FJ, Weyl PK (1965) Dolomitization of Recent and Plio-Pleistocene sediments by marine evaporite waters on Bonaire, Netherlands Antilles. Soc Econ Paleontol Miner Spec Publ 13:71–87

Dunham JB, Olson ER (1980) Shallow subsurface dolomitization of subtidally deposited carbonate sediments in the Hanson Creek Formation (Ordovician-Silurian) of central Nevada. Soc Econ Paleontol Miner Spec Publ 13:139–161

Esteban M, Pray LC (1983) Pisoids and pisolite facies (Permian), Guadalupe Mountains, New Mexico and West Texas. In: Peryt T (ed) Coated grains. Springer, Berlin Heidelberg New York, pp 503–537

Evamy BD (1967) Dedolomitization and the development of rhombohedral pores in limestones. J Sediment Petrol 37:1204 – 1215

Folk RL, Land LS (1975) Mg/Ca ratio and salinity: Two controls over crystallization of dolomite. Am Assoc Petrol Geol Bull 59:60 – 68

Frank JR (1981) Dedolomitization in the Taum Sauk Limestone (Upper Cambrian) southeast Missouri. J Sediment Petrol 51:7 – 18

German K (1968) Diagenetic pattern in the Wettersteinkalk (Ladinian, Middle Trias), Northern Limestone Alps, Bavaria and Tyrol. J Sediment Petrol 38:490 – 500

German K (1969) Reworked dolomite crusts in the Wettersteinkalk (Ladinian, Alpine Triassic) as indicators of early supratidal dolomitization and lithification. Sedimentology 12:257 – 277

Gregg JM, Sibley DF (1984) Epigenetic dolomitization and the origin of xenotopic texture. J Sediment Petrol 54:908 – 931

Henrich R (1982) Middle Triassic Carbonate Margin development: Hochstaufen-Zwieselmassif, Northern Calcareous Alps, Germany. Facies 6:85 – 106

Henrich R (1983) Der Wettersteinkalk am Nordwestrand des tirolischen Bogens in den Nördlichen Kalkalpen: der jüngste Vorstoß einer Flachwasserplattform am Beginn der Obertrias. Geol Palaeontol 17:137 – 177

Henrich R (1984) Facies, dolomitization and karstification of lagoonal carbonates: Triassic of the Northern Alps. Facies 11:109 – 156

Katz A (1971) Zoned dolomite crystals. J Geol 79:38 – 51

Kendall AG (1985) Radiaxial fibrous calcite: a reappraisal. Soc Econ Paleontol Miner Spec Publ 36:59 – 77

Kendall AG, Tucker ME (1973) Radiaxial fibrous calcite: as a replacement after acicular carbonate. Sedimentology 20:365 – 389

Land LS (1973) Contemporaneous dolomitization of Middle Pleistocene reefs by meteoric water, North Jamaica. Bull Marine Sci 23:64 – 92

Land LS (1973) Holocene meteoric dolomitization of Pleistocene limestones, North Jamaica. Sedimentology 20:411 – 424

Longmann MW (1980) Carbonate diagenetic textures from nearsurface diagenetic environments. Am Assoc Petrol Geol Bull 64:461 – 487

Lumsden DN, Chimahusky JS (1980) Relationship between dolomite nonstoichiometry and carbonate facies parameters. Soc Econ Paleontol Miner Spec Publ 28:123 – 137

Pratt B (1984) Epiphyton and renalcis-diagenetic microfossils from calcification of coccoid bluegreen algae. J Sediment Petrol 54:948 – 971

Purser BH, Loreau JP (1973) Aragonitic, supratidal encrustations on the Trucial Coast, Persian Gulf. In: Purser BH (ed) The Persian Gulf. Springer, Berlin Heidelberg New York

Radke BM, Matthis RL (1980) On the formation and occurrence of saddle dolomite. J Sediment Petrol 50:1149 – 1168

Sander B (1936) Beiträge zur Kenntnis der Anlagerungsgefüge (Rhythmische Kalke und Dolomite aus der Trias). Min Petrogr Mitt 48:27 – 139

Sarnthein M (1965) Sedimentologische Profilreihen aus den mitteltriadischen Kalkgesteinen der Kalkalpen nördlich und südlich von Innsbruck. Verh Geol Bundesanst 1965/1-2:119 – 162

Sarnthein M (1966) Sedimentologische Profilreihen aus den mitteltriadischen Karbonatgesteinen der Kalkalpen nördlich und südlich von Innsbruck. 1. Fortsetzung. Ber Nat Med Ver Innsbruck 54:33 – 39

Sears SO, Lucia FJ (1980) Dolomitization of northern Michigan Niagara Reefs by brine reflexion and freshwater/seawater mixing. Soc Econ Paleontol Miner Spec Publ 28:215 – 235

Wurm D (1982) Mikrofazies, Paläontologie und Palökologie der Dachsteinriffkalke (Nor) des Gosaukammes, Österreich. Facies 6:203 – 296

Zenger DH, Dunham JB (1980) Concepts and models of dolomitization. An introduction. Soc Econ Paleontol Miner Spec Publ 28:1 – 9

III Paleozoic Reefs

The Upper Permian Reef Complex (EZ1) of North East England: Diagenesis in a Marine to Evaporitic Setting

M. E. TUCKER and N. T. J. HOLLINGWORTH[1]

1 Introduction

The Zechstein of North East England is a complex sequence of carbonate evaporite cycles of Late Permian age. It includes the Permian Magnesian Limestone of North East England which was deposited around the margin of the Zechstein Sea which covered much of northwest Europe some 230 – 240 million years ago. The Magnesian Limestone includes a 100-m-thick reef complex (part of the Ford Formation) which trended north-south in County Durham and Tyne and Wear (Fig. 1). The growth of the reef (Tunstall Reef Member) resulted in the development of an extensive lagoon to the west, towards the shoreline. There has been much discussion on how the reef was formed and on the roles played by organisms, especially bryozoans and algae, but pervasive dolomitization has obliterated much of the evidence and preservation of the fauna is generally poor. Dolomitization has also destroyed many of the early diagenetic textures of these rocks. However, recent discoveries of undolomitized reef rock have provided beautifully preserved fossils and allowed the recognition of submarine cements. Evidence for the latter is now being found in dolomitized parts of the reef too. This paper documents the diagenesis of the Zechstein reef of NE England in the light of data revealed in the reefal limestones, and against a background of major changes in sea level, exposure and evaporite precipitation.

2 Stratigraphy

Stratigraphic correlation and nomenclature for the English Zechstein has been revised by Smith et al. (1974), Harwood et al. (1982) and Taylor (1984) (see Fig. 2), and the new names have been used in this paper where appropriate. The English Zechstein can be divided into two depositional provinces separated by the Cleveland High, the Durham province to the north and the Yorkshire province to the south. In the Durham province, three cycles can be distinguished, each ending with an evaporitic phase when gypsum-anhydrite and/or halite were precipitated. Although there are five cycles in total, the last two are best known from the subsurface of Yorkshire and the North Sea. The main control of Zechstein deposition and formation of the carbonate-evaporite cycles is considered to be eustatic changes in sea level (Smith 1980, Clark and Tallbacka 1980).

[1] Department of Geological Sciences, University of Durham, Durham DH1 3LE, England

Reef Diagenesis
Edited by J. H. Schroeder and B. H. Purser

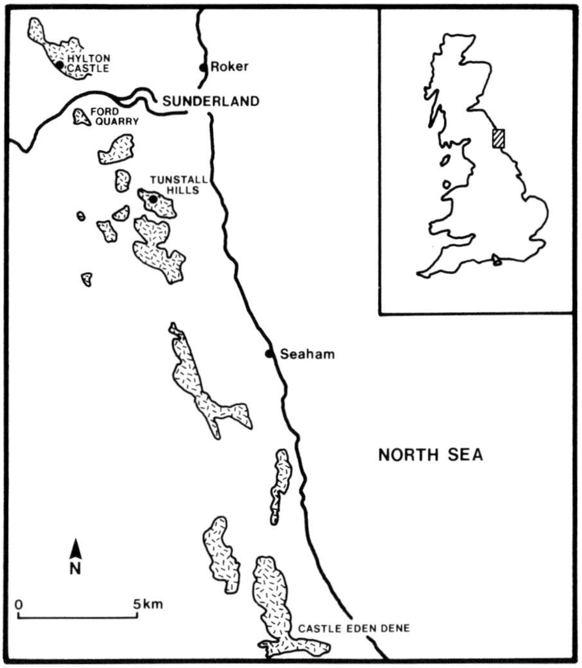

Fig. 1. Location and outcrop of the reef facies (Tunstall Member) of the Ford Formation in County Durham and Tyne and Wear

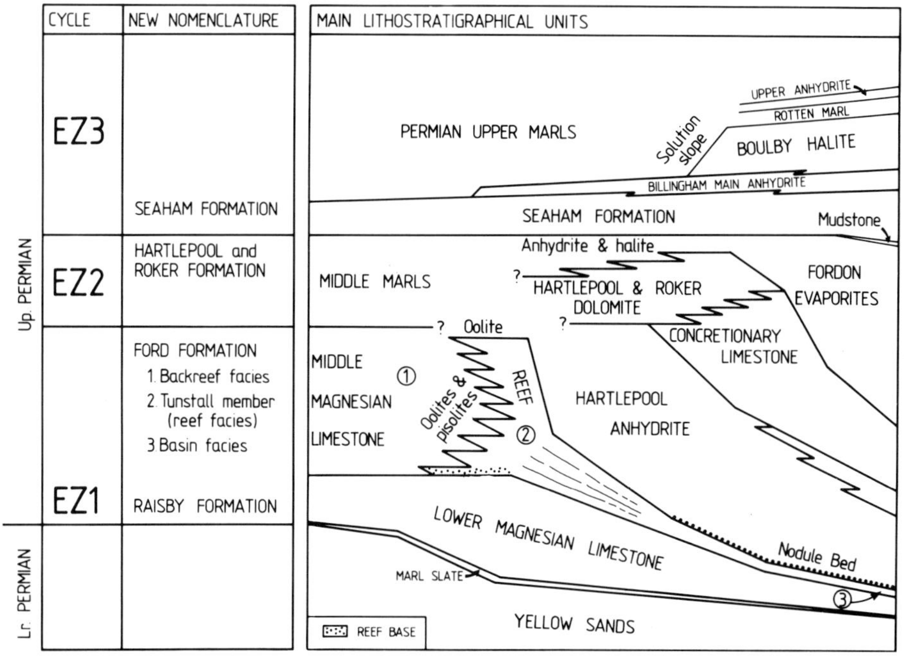

Fig. 2. Stratigraphic cross-section and nomenclature of the Zechstein sequence in the Durham Province

Initial flooding of the Zechstein basin led to the deposition of the Marl Slate (Kupferschiefer) which passes upwards into marine carbonates of the Raisby Formation (Lower Magnesian Limestone). In the Durham Province, the succeeding Ford Formation comprises lagoonal and basin deposits separated by a shelf edge reef (the Tunstall Member, Fig. 2) which reaches a maximum thickness of 100 m. The back reef facies mostly consists of dolomicrites, with local cross-bedded oolitic and pisolitic dolostones. Widespread regression and evaporative drawdown led to the termination of reef growth and precipitation of the Hartlepool Anhydrite in the Zechstein Basin (Smith 1980) and against the steep outer face of the reef. A eustatic sea-level rise and transgression over the shelf deposited the Hartlepool and Roker Formation, a sequence of shallow water lime mudstones and oolites. The end of this second cycle is marked by deposition of the Fordon Evaporites, which filled most of the Zechstein basin. The succeeding Seaham Formation is the last major carbonate unit of the European Zechstein, and was buried by the Billingham Anhydrite.

3 Evolution and Paleoecology of the Zechstein Reef

The reef of the Tunstall Member contains a diverse and abundant fauna described by King (1850), Kirkby (1858), and Trechmann (1913, 1925, 1945). A decrease in abundance of stenohaline species and faunal size has been used as evidence for increasing salinity during reef growth (Trechmann 1925). However, recent work by one of the authors (NTJH) suggests that salinity was a major factor in causing faunal extinction only during the latest stages of reef development. The distinct faunal succession was probably controlled by environmental factors associated with reef evolution, leading to the creation of numerous ecological niches within the reef framework (cf. Walker and Alberstadt 1975).

The reef is founded upon a lenticular shelly rudstone (the "basal coquina" of Smith 1958, 1981) which contains a varied assemblage of bryozoans, bivalves, brachiopods and gastropods. The reef base community is dominated by epifaunal forms; an infaunal and quasi-infaunal shelly fauna accounts for only a small percentage of the total. *Dielasma* is the most abundant brachiopod and it commonly occurs in nests. Bryozoans are also abundant, but many are fragmented. *Acanthocladia* formed delicate, bushlike colonies. Gastropods were mostly herbivores feeding on algae. The coquina bank accumulated in an agitated, shallow, normal marine environment, where water depths were mostly less than 2 to 3 m, with occasional local emergence. The precipitation of marine cements between bioclasts and the binding and baffling actions of bryozoan frameworks led to the establishment and growth of the reef itself.

Distinct communities did not evolve until the reef was fully developed. Early reef core lithofacies display a wide degree of community overlap but at higher levels, the reef shows distinct reef core, fore reef and reef flat communities. The reef core community is dominated by bryozoans, bivalves and brachiopods. Faunal variations across the reef are shown in Fig. 3.

The reef migrated rapidly basinwards over its own talus. Probable periodic subaerial exposure of reef flat areas led to increased environmental stress and a

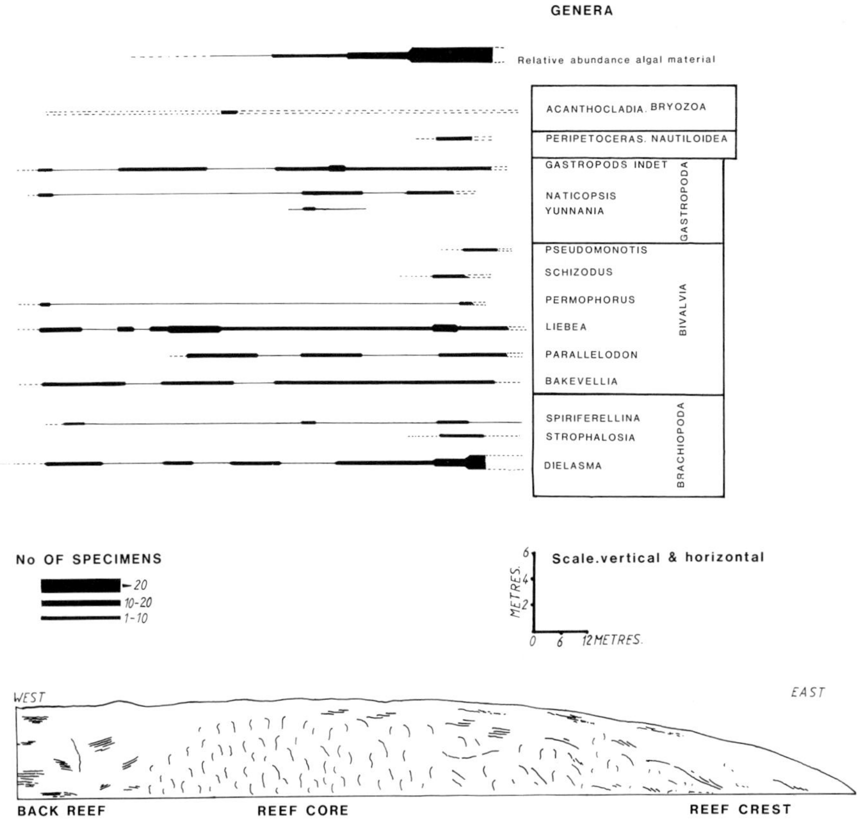

Fig. 3. Cross-section of upper part of reef as seen at Ford Quarry, Sunderland, showing faunal distribution and abundance

community with a restricted range of species. The highest faunal diversity and rate of organic productivity were confined to a narrow zone along the reef crest and upper fore reef slope, where environmental conditions were optimum (Fig. 3). The latest stages of reef growth are represented by the development of an algal-dominated reef flat associated with an increase in the abundance of algal-grazing gastropods and a relative decrease in abundance of strictly stenohaline genera. Termination of reef growth has been ascribed to an increase in salinity (Trechmann 1925), connected with evaporative drawdown of the Zechstein sea, which terminated cycle 1 carbonate sedimentation.

4 Diagenesis of the Tunstall Reef

The Zechstein carbonates and the Tunstall Member in particular have been subjected to extremes of diagenesis. From the undolomitized reefal limestones, the early diagenesis is expressed as extensive marine cementation. Aragonite and high Mg calcite were precipitated on the seafloor and calcite fan crystals during very shallow burial. The next major event was pervasive dolomitization which took place before any compaction. Sulphate was precipitated at a later stage, and then after compaction, during deeper burial, calcite was precipitated as a cement and as a replacement of early cements, sulphate and some dolomite. An important late diagenetic event was dedolomitization during Tertiary uplift.

4.1 Early Diagenesis: Marine Cements

Undolomitized Tunstall reefal facies occur in the "basal coquina" and in isolated patches in the lower reef core of the Tunstall Hills (Fig. 1). These limestones contain an abundance of marine cements and three main types are recognized: (a) aragonite crusts and botryoids, since replaced by calcite, (b) isopachous layers and fans of acicular calcite, interpreted as originally high Mg calcite, and (c) calcite fan crystals.

4.1.1 Calcitized Aragonite Cements

These cements are composed of coarse equant to columnar to irregular calcite crystals which commonly cross-cut original fabrics of the primary aragonite precipitates. In many cases, the calcite is cloudy with inclusions and these have a common optic orientation (Fig. 4). Under the SEM these are seen as elongate

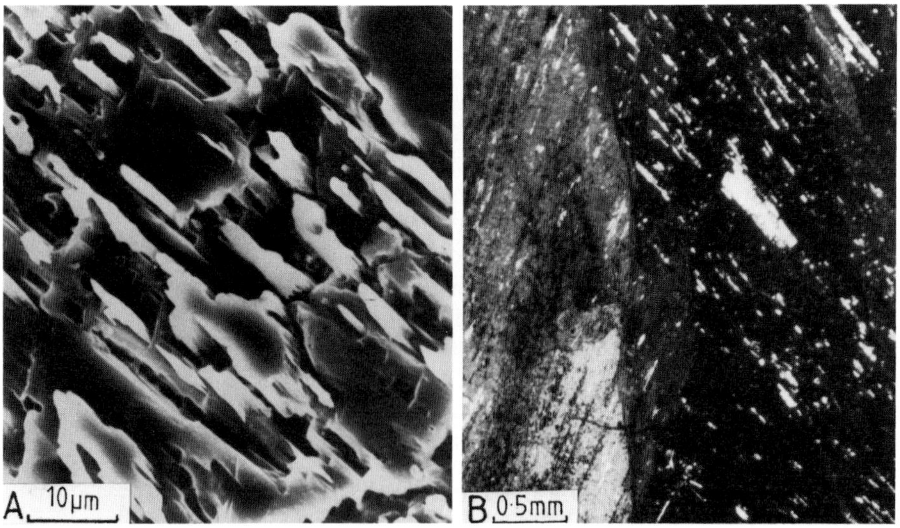

Fig. 4. A Aragonite relics in columnar calcite seen with the SEM. **B** Junction of two former aragonite fans (seen in *top right hand corner* of Fig. 5 a), replaced by columnar calcite crystals containing many aragonite relics (also seen in SEM) with a common orientation. The boundary between the two columnar calcites does not follow the original junction of the fans. Crossed polars

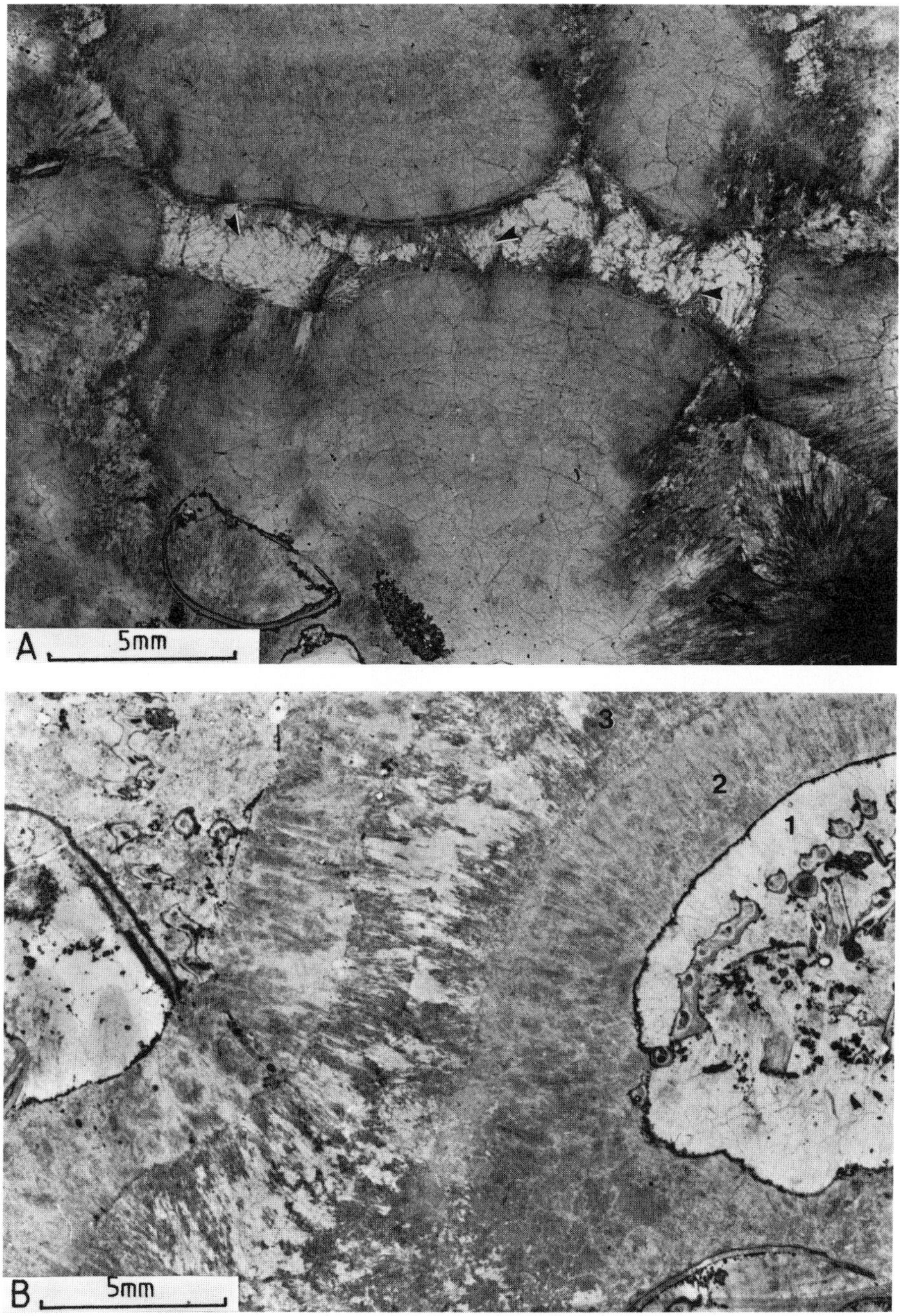

Fig. 5. A Thick cement crust, with aragonite relics (see with SEM), around bryozoans and brachiopods. Notice the local development of fibrous fabrics within the crusts. Calcite fan crystals (*arrows*) occur upon the surface of the former aragonite, in the central void. **B** Three layers (*1, 2, 3*) of calcite after aragonite around bioclasts (mainly bryozoans). Notice the sharp but locally indented boundary between the first and second generations and the variety of calcite replacement fabrics

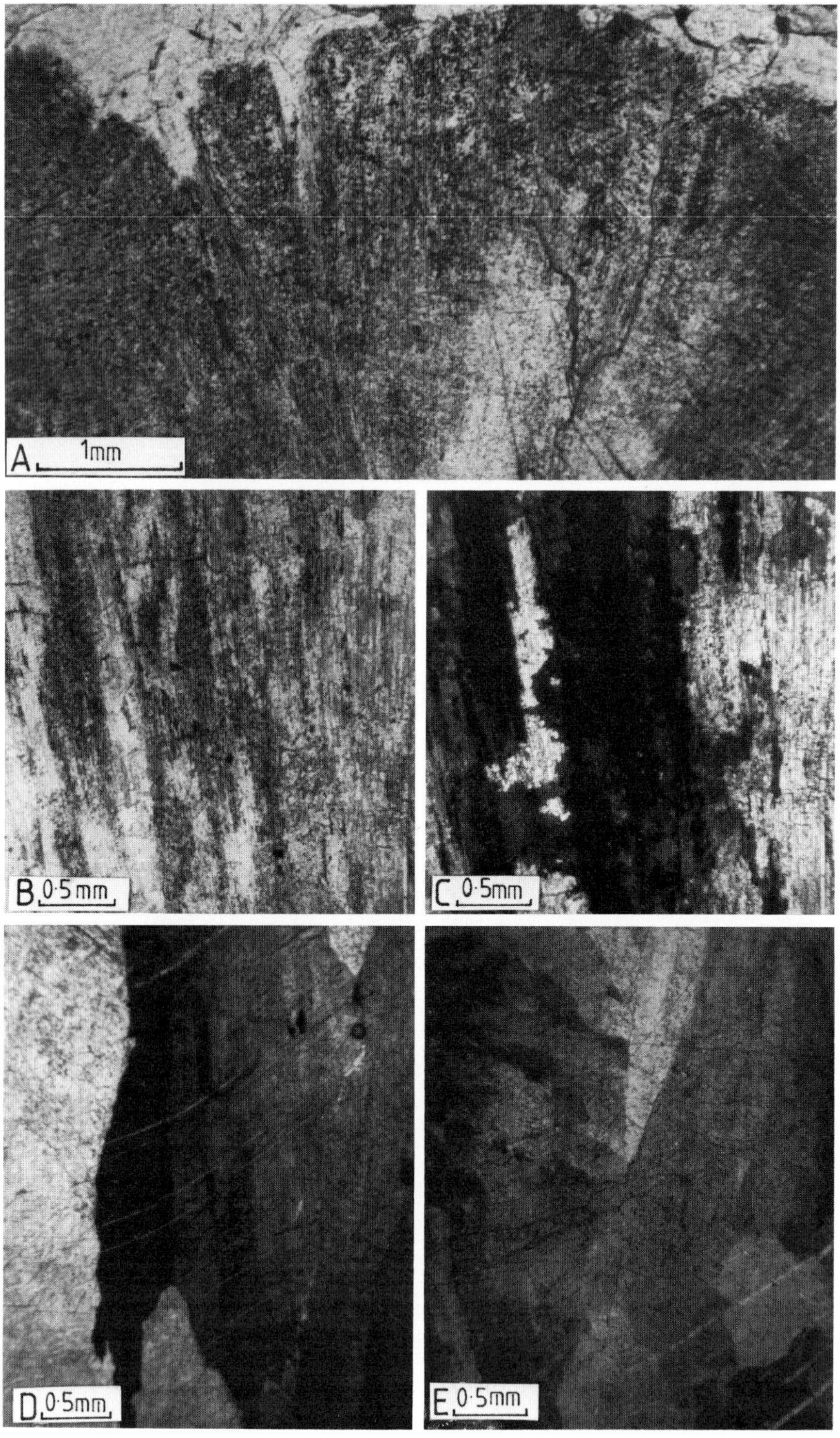

Fig. 6

mineral inclusions with a preferred orientation (Fig. 4). Where mineral relics are absent, there are often many holes. Use of EDAX reveals that the relics have a strong Ca peak, and a very small Sr peak. With XRD, a very small aragonite reflection is obtained with some samples. The relics are interpreted as aragonite and support the interpretation of these calcite mosaics as replacements of original aragonite cements.

The original aragonitic cements occur in two main growth morphologies, crusts and botryoids. Most common are thick (up to 30 mm) crusts around bioclasts (Fig. 5A, B). Some crusts are isopachous, and the original aragonite crystallites, as revealed by inclusion patterns had a parallel orientation. Crystallites may constitute bundles with a square-ended "termination" (Fig. 6A). Some thick crusts have mammillated outer surfaces (Fig. 5A), many with a delicate dentate margin, and crystallites fan away from the substrate. A growth banding is common. Some of these cement crusts have grown out from a substrate to envelop other skeletal grains. In some limestones two or three layers of these aragonite crusts were precipitated with irregular, indented surfaces between (Fig. 5B).

The former aragonite crusts are now composed of a distinctive, honey-coloured calcite which shows a great range of fabrics. In the inner part of some crusts, the calcite has unit or slightly undulose extinction, and twin planes are not prominent. The inclusion trains defining the former aragonite crystallites are commonly absent in the inner part of the crust (Fig. 6A). Farther away from the substrate, the crystals are generally elongate and columnar, normal to the original substrate. The extinction ranges from fascicular-optic (divergent optic axes) to unit to radiaxial (convergent optic axes) (Fig. 6D, E). Patchy extinction is common in some crystals, through the presence of subcrystals (Fig. 6E), and many twin planes are concave away from the substrate (Fig. 6D). Some columnar crystals are oblique to the linear inclusion patterns. Elsewhere, the calcite after aragonite is a typical pseudospar of irregular to equant crystals, with some crystal boundaries following crystallite junctions (Fig. 6B, C). Extinction is random, but linear inclusion patterns are well-defined. This pseudospar is similar to that described by Mazzullo and Cys (1979) as a replacement of aragonite in the Permian carbonates of Texas. Within thick crusts, the whole range of calcite fabrics can be seen. For example, in Fig. 5A, areas of crust of columnar crystals pass into pseudospar with a prominent original crystallite pattern. The latter is particularly well seen where crusts from different substrates interfere with each other or where there is a strongly convex substrate.

Also within the formerly aragonite cement, there are patches of clear calcite without inclusion patterns. These are places where wholesale dissolution of the aragonite occurred, rather than calcitization.

Fig. 6. A Square-ended bundles of inclusion-rich former aragonite crystallites as part of two large fans. Dark inclusions are absent from the inner part of one fan. **B** Inclusion-defined crystallites of former aragonite. **C** As **B** under crossed polars showing neomorphic calcite partly following crystallite boundaries. **D** Columnar calcite after aragonite (inclusion-defined crystallites just visible) with curved twin planes, irregular crystal boundaries and radiaxial (convergent) extinction. **E** Columnar calcite crystal after aragonite (inclusion-defined crystallites just visible) with curved twin planes and complex subcrystals

Aragonite was also precipitated in the form of botryoids. These reach 20 mm in diameter and occur as discrete mamelons (Fig. 7A, B). In some places they originally consisted of coarse columnar crystals, radially disposed, with a little sediment between (Fig. 7A). Square-ended and pointed terminations are present. Others consisted of acicular crystallites closely packed together, with a dentate outer margin (Fig. 7B). Botryoids occur most commonly within articulated brachiopods or upon bivalve shells. The calcite replacing the botryoids varies from equant crystals with no relic inclusion patterns to irregular-columnar

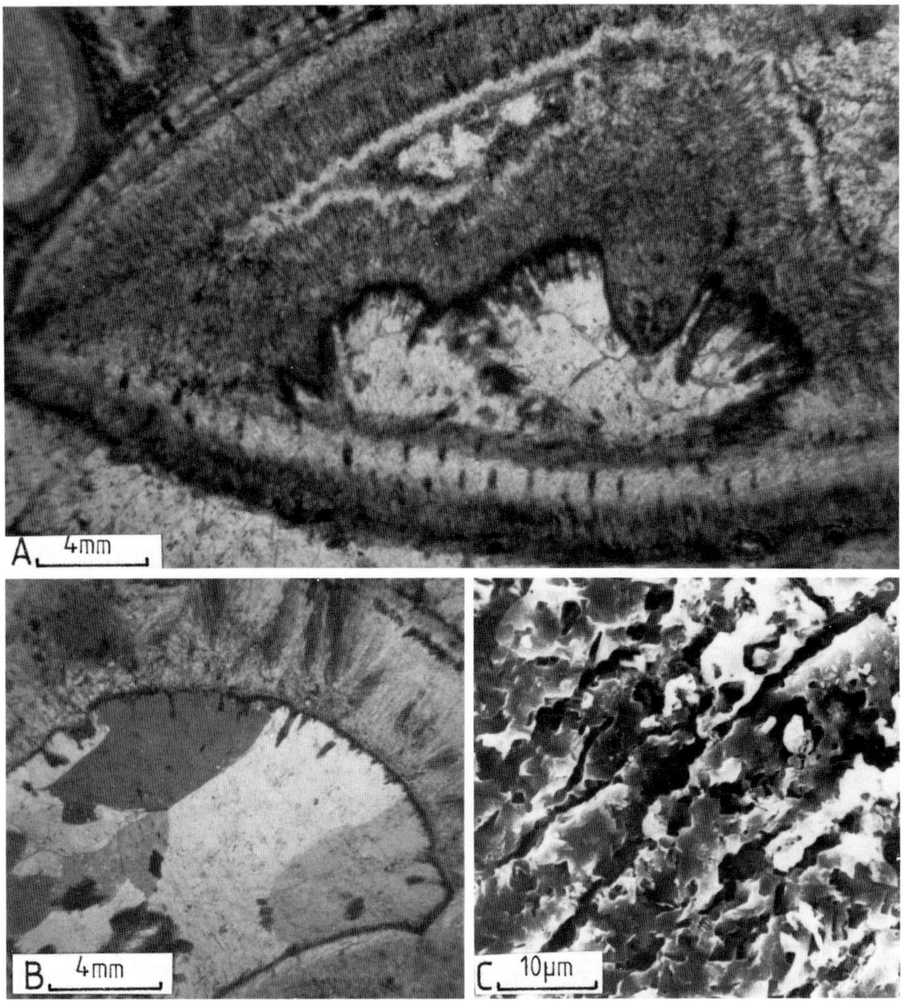

Fig. 7. A Small botryoids of former aragonite in punctate brachiopod shell, overlain by acicular calcite, which also coats the roof of the cavity. Some dissolution of the right-hand botryoid occurred before the acicular calcite was precipitated. **B** Botryoid of former aragonite with dentate outer margin, replaced by coarse calcite and occurring within a brachiopod shell. The botryoid is overlain by acicular calcite with small fans. Crossed polars. **C** Microdolomite inclusions in acicular calcite. SEM

crystals with preservation of the original acicular crystallites and abundant aragonite relics.

The calcite after aragonite mostly has a dull cathodoluminescene although there are bright, irregular patches. Areas within the former aragonite crusts occupied by clear, colourless calcite spar have a bright pink to orange luminescence with growth zones, indicating a void fill origin.

Most of the calcite after aragonite has replaced the precursor without an intervening void stage. The typical product of this calcitization is equant to irregular pseudospar or neomorphic spar and this is well known from calcitized aragonite shells and ooids (e.g. Sandberg 1975, Sandberg and Hudson 1983, Tucker 1985). However, some of the neomorphic spar described here is unusual in showing columnar crystals with complex extinction patterns. In some cases there was a control exerted by the original aragonite crystallites on the replacement crystals: the long edges of the columnar crystals may follow crystallite boundaries, but elsewhere there seems to be no relationship between the two, and the complex replacement crystals cut across the inclusion patterns.

Botryoids of aragonite are well known from several modern reefs (e.g. Belize, James and Ginsburg 1979) and they occur in Pleistocene reefs of New Caledonia and the Red Sea (Aissaoui 1985). Crusts of acicular aragonite are a common cement in modern reefs, although they are not as thick as those described here. Ancient aragonite cements are recorded from the Triassic reefs and associated carbonates of the Alps (Scherer 1977, Assereto and Folk 1980), a Permian reef in Yugoslavia (Flügel et al. 1984) and from the Permian Capitan reef of Texas (Babcock 1977, Mazzullo and Cys 1979).

4.1.2 Isopachous Layers and Fans of Acicular Calcite

Isopachous layers of closely-packed acicular calcite crystals occur around many bioclasts, and also upon formerly aragonitic botryoids (Fig. 7A, B). They are generally less than 2 mm thick, although there may be several generations. The cement layers are usually brown in colour, due to the weathering of iron within the calcite lattice. Small fans of this cement type are also common. These fans are much smaller than those originally of aragonite described above. Fine skeletal debris is commonly tightly circumcrusted by this acicular calcite cement, and polygonal boundaries occur where the crusts around grains meet. Associated with these acicular calcite cements are peloids of similar needle-like crystals. They are most clearly seen beneath umbrella cavities and within articulated brachiopod shells.

Under the SEM, the acicular nature of the cement is revealed (Fig. 7C), and small microdolomite inclusions can be seen and identified with EDAX. This cement type is non-luminescent.

These acicular cements are interpreted as originally high Mg calcite precipitates. The fabrics are regarded as primary, although the original cement has lost its Mg to the formation of the microdolomite inclusions (cf. Lohmann and Meyers 1977), and there has been an uptake of Fe. High Mg calcite is a common cement in modern reefs (e.g. Schroeder 1972, James and Ginsburg 1979) and a number of ancient fibrous calcite cements have been interpreted as originally high Mg calcite (e.g. Lohmann and Meyers 1977, Marshall 1981). The peloids of acicular calcite may also be a precipitate. Similar bodies are well known from modern reefs (e.g. MacIntyre 1985).

4.1.3 Calcite Fan Crystals

This cement type occurs upon the thick, formerly aragonitic crusts and consists of calcite crystals up to 2 mm long with distinctive fanning subcrystals (Fig. 8). The extinction is radial-fibrous, with each subcrystal having unit extinction. In some fan crystals, there are sharp boundaries between the subcrystals. In others, dark inclusions occur along subcrystal boundaries. Although mostly growing into cavities, and nucleating upon the former aragonite, there are places where the outer part of a fan crystal has replaced the aragonite.

These calcite crystals are interpreted as primary cements except where they have replaced aragonite. The fabrics are quite unlike those of calcite after aragonite where calcite mosaics cross-cut an original fabric. The fan crystals could be the result of split growth, as envisaged by Kendall (1985) for radiaxial fibrous calcite.

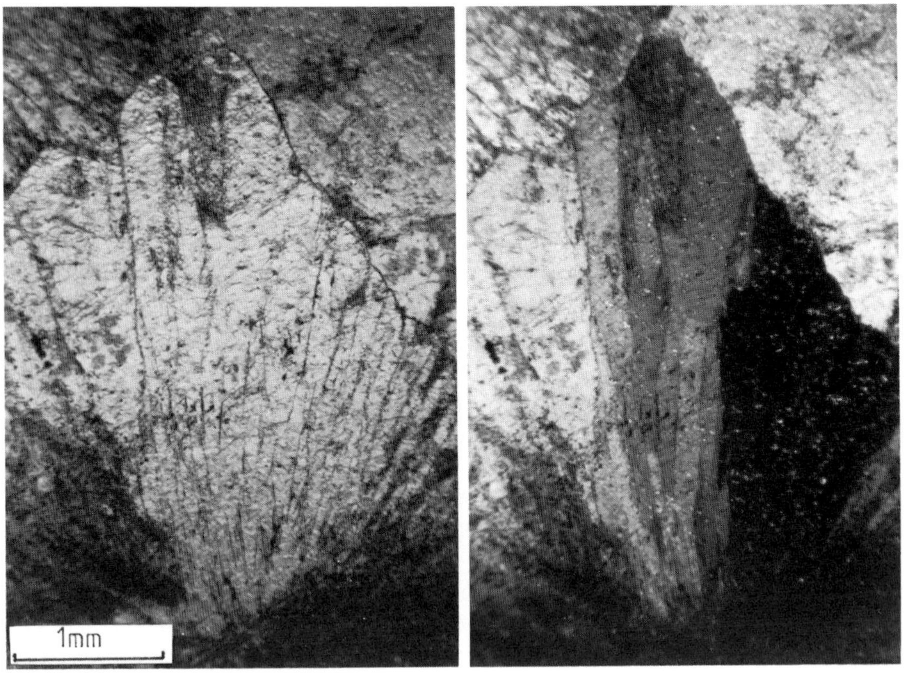

Fig. 8. Calcite fan crystal developed upon surface of former aragonite crust (seen in Fig. 5A) under ppl and xp

4.1.4 Cement Sequences, Evidence for Emergence and Discussion

In the reef limestones, the originally aragonite and high Mg calcite cements may occur in any order. However, isopachous acicular calcite is commonly a first layer on shells, and discrete aragonite botryoids may occur within and upon this cement. The thick aragonite crusts are often a second generation, filling much void space, and several layers may occur (Fig. 5B). In the reef-base skeletal limestones quite commonly there is a sharp and pitted junction between the first marine cement layer and a second generation, usually of the thick aragonite crusts (Figs. 5B, 9A). The boundary between the two may cut across the first layer, so that the latter is of uneven thickness (Fig. 9A). This feature is interpreted as the result of a phase of dissolution during a short period of emergence.

Evidence for marine vadose conditions in the basal reef coquina is also provided by local asymmetries in the acicular calcite cement. Where this cement occurs between shells, the cement fringe may thin away, suggesting the effect of meniscus films of water (Fig. 9B). Such cement morphologies are well known from beachrocks in the higher intertidal zone (Taylor and Illing 1969).

The calcite fan crystals are the last phase of early cementation and it is likely that they were precipitated during very shallow burial, when pore waters were still marine.

The evidence from the local pockets of undolomitized reef shows that seafloor cementation was ubiquitous. The skeletal sand and gravel which forms the base of the reef was cemented in the very shallow subtidal, and locally the

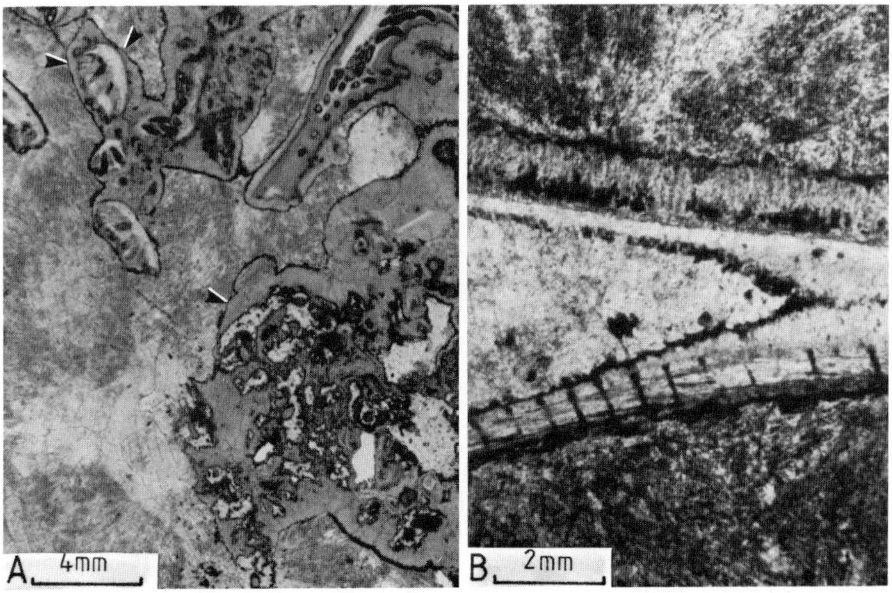

Fig. 9. A Acicular calcite coating grains and lighter calcite-after-aragonite filling remaining voids. There is an irregular and cross-cutting boundary (*arrows*) between these two cement generations. **B** Asymmetric acicular calcite cement layer between two brachiopod shells. Calcite after aragonite fills voids

shell bank grew above sea level so that marine vadose conditions were established. Small islands of reef rubble, as occur at present along the Florida shelf-edge, would have existed, with cements producing beachrock. Dissolution surfaces within the cements could indicate a meteoric influence, but the absence of wholesale dissolution suggests only the temporary establishment of meteoric lenses.

With the development of the bryozoan-algal reef above the basal coquina, marine cements continued to be precipitated in abundance. The best evidence for this is seen in scattered occurrences of undolomitized reef, but dolomitized marine cements, especially the botryoids, have been found in the reef core and reef flat facies (e.g. at Castle Eden Dene, south Durham, Fig. 10). Thus, in spite of the mostly fabric-destructive dolomitization of the Tunstall Member, there is sufficient evidence to demonstrate the widespread precipitation of marine cements. They appear to have been precipitated in all reef facies developed along the shelf margin, but they have yet to be found in the back-reef facies. The cements may have been essential to the development of the reef, contributing towards the stability of the basal coquina for colonization by bryozoans and epifaunal brachiopods and bivalves. Marine cementation then helped to maintain the reef itself as a structure able to withstand wave attack and grow up to 90 m above the basin floor.

4.2 Later Diagenesis

Later diagenetic events have considerably affected the EZI reef of N.E. England, with pervasive dolomitization being the most important. Two phases of dolomitization can be distinguished and related to other events. Also of note is the precipitation of gypsum-anhydrite in the reef carbonates. The calcitization of the early cements, where they were not dolomitized, and the precipitation of calcite spar in voids were both burial processes. Dedolomitization is a common phenomenon, affecting the dolomites during burial and on later uplift. Later diagenetic processes in the Durham reef can be divided into three categories: *pre-compaction events:* namely pervasive dolomitization and gypsum-anhydrite precipitation; *post-compaction events:* calcitization and calcite spar precipitation, dedolomitization, dolomite void and fracture fills and stylolitization; and *on uplift:* the principal phase of dedolomitization.

4.2.1 Pre-Compaction Diagenesis

Pervasive dolomitization is seen throughout the Tunstall reef carbonates, apart from the mostly reef-base limestones described earlier, but even in these, dolomite rhombohedra are common. Reef dolostones consist mostly of quite well preserved fossils in a xenotopic mosaic of decimicron-sized anhedral-subhedral dolomite crystals. Only locally does the reef consist of massive dolomite with no skeletal relics. Bryozoans, one of the most common faunal elements in the reef, originally had finely crystalline high Mg calcite skeletons and dolomite replacement has mostly retained the fine crystal size, so that much of the textural detail

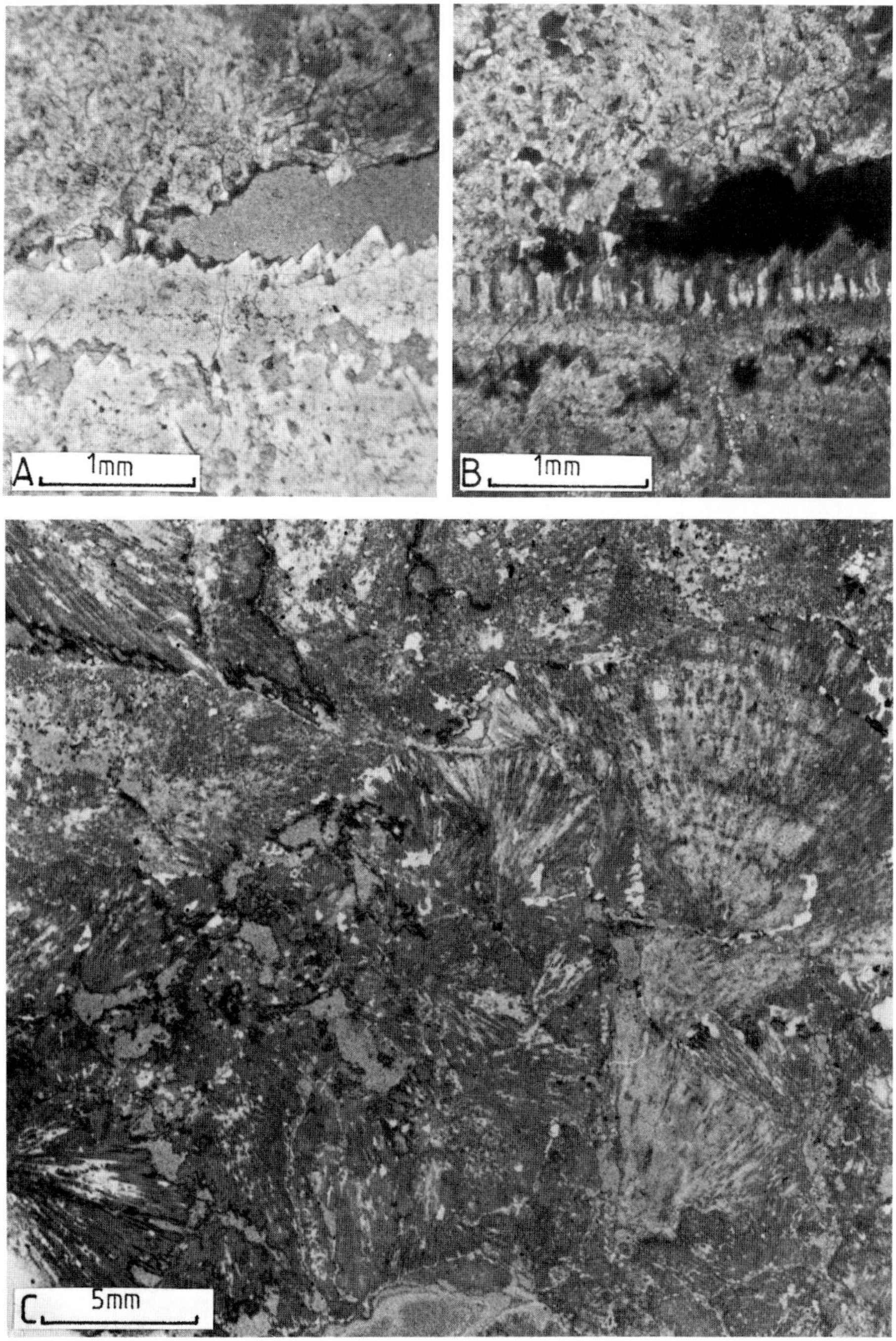

Fig. 10. A Dolomitized crinoid with vugs and dolomitized matrix. **B** As **A** under crossed polars showing columnar dolomite after marine cement coating crinoid and undulatory dolomite in matrix. **C** Dolomitized fans of former aragonite, with much vuggy porosity

is preserved at the microscopic level. The same type of dolomite occurs within brachiopods and foraminifera, both of which had an original microcrystalline fabric (although of low Mg calcite). Crinoid fragments, originally large single high Mg calcite crystals, are replaced by a single dolomite crystal, but often replacement has been incomplete, and there are voids lined by rhomb faces of the large crystal (Fig. 10A). Between bioclasts there occur anhedral to subhedral dolomite crystals which are mostly cloudy. Commonly crystals are unit-extinguishing but undulose-extinguishing dolomite is also present. With use of the diffuser and the eye of faith, areas of undulatory dolomite appear to be replacements of fans and crusts which were probably aragonite originally. Crinoid columnals are commonly surrounded by a layer of columnar undulatory dolomite crystals (Fig. 10B) which are a replacement of a marine cement (?high Mg calcite). Where obvious aragonite fans have been dolomitized (Fig. 10C), the replacement crystals are dark (cloudy) and $5-20$ µm in size. The radiating arrangement of the original crystallites is still seen and square-ended terminations on bundles of crystallites are well developed. Within the fans there are often patches of coarse dolospar where there is no preservation of the original cement texture. In the higher parts of the reef, algal encrustations are common (Smith 1981), and these are mostly replaced by dolomicrite. Throughout the reef dolostone there are patches of euhedral dolomite rhombs with much intercrystalline pore space. In general the dolostone is very porous ($10-30$%) with many small vugs on the millimetre scale. Dolomite crystals around the vugs have rhomb faces directed into these pores.

In the reefal limestones dolomite occurs as scattered replacement rhombohedra within the marine cements but it is more common within the cavities. Dolomite crystals line brachiopod shells and occur upon the cement crusts. In many cases the dolomite rhombs have been replaced by calcite indicating a prespar origin (Fig. 11A). The calcite spar is post-compaction, so that a pre-compaction, post-marine cement timing is suggested for the dolomitization, i.e. relatively early during shallow burial diagenesis.

Two mechanisms may account for the dolomitization: seepage reflux and meteoric-marine mixing. Seepage reflux was invoked by Smith (1981) and Clark (1980a). The occurrence of undolomitized limestones towards the base of the reef complex would be consistent with downward-moving fluids. During deposition of the Tunstall Member, oolites and skeletal sands and muds accumulated in the back reef lagoon. There is some evidence, in the form of gypsum pseudomorphs, for evaporites being deposited in this lagoon or in shoreline sabkhas, but much of the landward part of this backreef sequence has been removed by erosion. Mg-rich brines could have formed in this backreef area and descended into the carbonates during the sea-level fall which terminated reef growth (Fig. 12). In the next Zechstein cycle (EZ2), the succeeding Middle Marls were deposited on the inner shelf area above the marginal wedge of EZ1 carbonates, while the Hartlepool and Roker Dolomite (EZ2) was deposited at the shelf margin (Fig. 2). The marls contain lagoonal and sabkha evaporites which could again have been a source of dolomitizing solutions in a seepage-reflux mechanism. However, an alternative system involving mixing of meteoric water with marine porewaters in the carbonate wedge, as sea level fell after reef growth, cannot be entirely excluded.

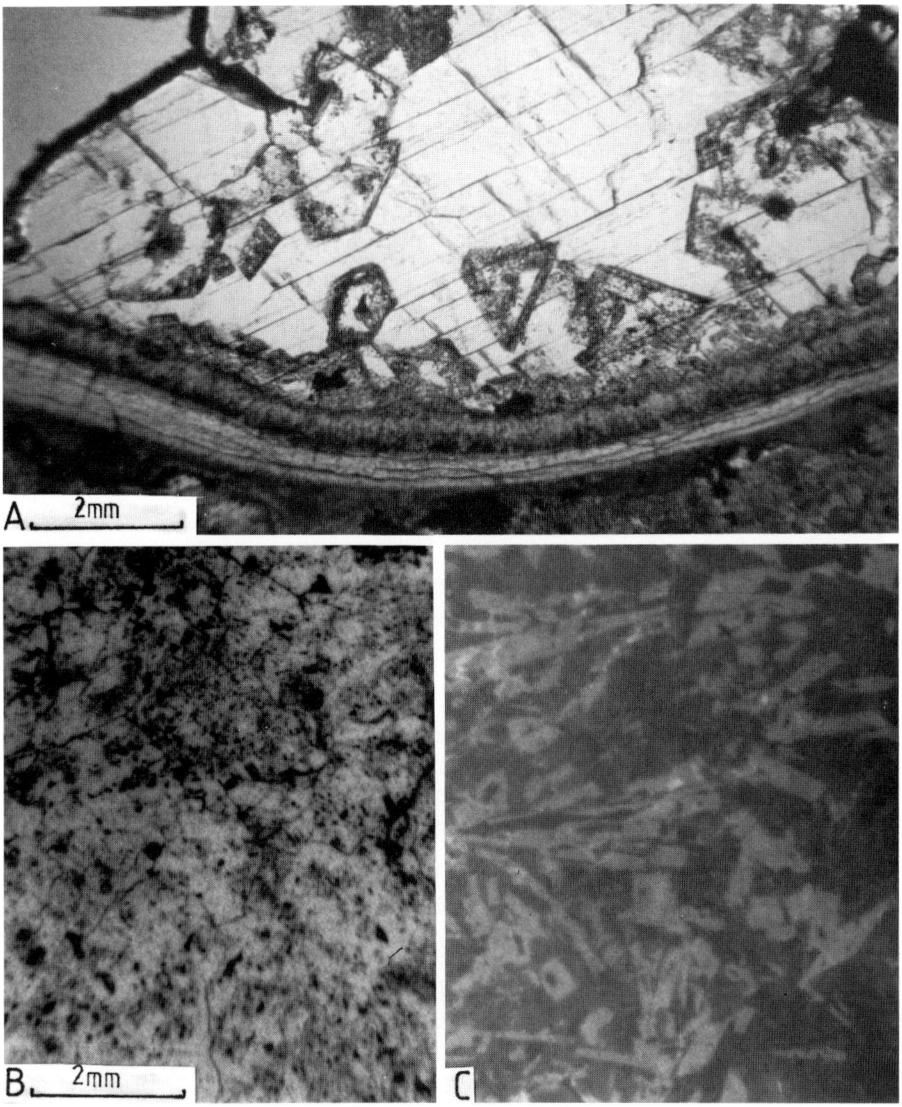

Fig. 11. A Calcitized dolomite rhombs within late coarse calcite spar partly filling brachiopod shell lined by isopachous acicular calcite fringe. **B** Pseudospar calcite after aragonite, ppl. **C** Same field of view as **B** under cathodoluminescence, showing relic gypsum crystal shapes

Gypsum-anhydrite occurs within the reef dolostones and there is evidence for the former presence of gypsum within the reefal limestones. In the dolostones from surface outcrop, occasional gypsum partly fills the vugs. It is likely that this is a relic of much more extensive sulphate cementation. Anhydrite is common in subsurface Zechstein carbonates generally (Clark 1980a), where it occurs as a cement and as a replacement of dolomite, and Smith (1981) noted its occurrence in the Tunstall Member reef dolostone.

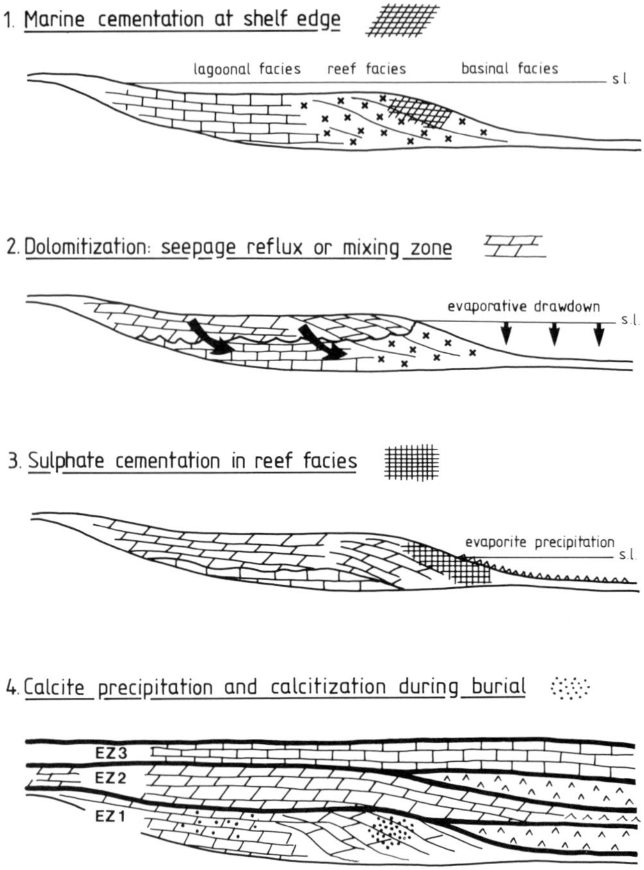

Fig. 12. Summary of diagenesis of Zechstein reef of North East England

In the reefal limestones clear lozenge and rectangular shapes are seen within the cloudy calcitized aragonite and within the calcite spar. These shapes are very clearly seen under cathodoluminescence, where they have a bright pink colour relative to the surrounding dull hue (Fig. 11 B, C). The shapes of these luminescent areas strongly suggest that they are pseudomorphs after gypsum. Under SEM, small inclusions with the calcite crystals are often composed of Ca and S (EDAX analysis), again suggesting the occurrence of gypsum.

The precipitation of gypsum-anhydrite in the reefal carbonates, after the dolomitization, probably took place during deposition of the Hartlepool Anhydrite (Smith 1981). This evaporite was deposited adjacent to and below the reef in a subaqueous and sabkha setting (Fig. 12). Pore fluids in the reef at this time would probably also have been saline.

4.2.2 Compaction

The effects of overburden pressure are seen in the form of broken bioclasts within the reefal carbonates. In the limestones, broken shells show that the calcite spar was precipitated after this compactive fracturing. In the dolostone, broken bryozoans demonstrate that the major phase of dolomitization took place before grain fracture and that there was also a post-compaction phase of dolomite precipitation.

4.2.3 Post-Compaction Diagenesis

The main process affecting the reefal carbonates after compaction was the precipitation of calcite.

Calcitization of aragonite, described earlier, led to neomorphic spar with a range of crystal fabrics. The timing of this replacement is shown by cement crusts which were broken before calcitization (cf. Tucker 1985).

Calcitization of dolomite (dedolomite) is common in the reefal limestones. Dolomite rhombs have been replaced by calcite, mostly as part of large spar crystals, and cleavage and twin planes are seen traversing the dolomite shapes (Fig. 11 A).

Dedolomites are common in the reefal dolostones, and these take the form of nodules and aggregates of radiating fibrous brown calcite crystals (known locally as cannon balls, although the best ones are developed in a higher EZ2 carbonate, the Cannon Ball Limestone). Less regular masses of coarse calcite crystals occur throughout the dolostone. In spite of the mineralogical changes, fossils within the dedolomites are often well preserved, although in calcite again, but areas between bioclasts consist of coarse, clear calcite crystals with scattered dolomite relics. Smith (1981) noted that the dolomite is most common in reef dolostones closest to the residue of the Hartlepool Anhydrite.

Zechstein dedolomites have been discussed at length by Clark (1980a), who distinguished between those formed during burial and those related to subsequent uplift. Smith (1981) considered that the dedolomitization was brought about by meteoric groundwaters during gypsification and dissolution of the Hartlepool Anhydrite. This probably took place during major Tertiary uplift of the Zechstein sequence. Notwithstanding this, some burial dedolomitization is indicated by the replaced dolomite rhombs in the reefal limestones.

Calcite spar was precipitated in remaining voids during the late burial of the reefal limestones. It shows all the characteristic features of a cavity fill spar cement and is mostly brightly luminescent with growth zones. Some poikilotopic calcite also occurs.

The origin of the solutions for the burial precipitation of calcite spar and the calcitization is unclear. Dedolomitization apparently takes place when pore-waters have a high Ca/Mg ratio and there is a low $p\,CO_2$ (de Groot 1969, Clark 1980a). It is generally believed that calcite spar is precipitated from low Mg^{2+}

waters (Folk and Land 1975). These could be meteoric waters or connate (marine) waters which have been depleted in Mg^{2+}. It seems unlikely that freshwater could penetrate deeply into the thick Zechstein carbonate-evaporite package during burial diagenesis to displace marine and hypersaline porewaters. The extensive precipitation of dolomite would have removed much Mg^{2+} from groundwaters, perhaps leaving sufficient Ca^{2+} for calcite precipitation. A further possibility is the dehydration of gypsum. Much of the Zechstein sulphate, such as the Hartlepool Anhydrite, was deposited as gypsum (e.g. Clark 1980b) and this would have dehydrated to anhydrite on burial with the release of water, presumably containing much Ca^{2+} and SO_4^{2-}. Sulphate is known to be an inhibitor of dolomite precipitation (Baker and Kastner 1981), so that the expulsion of this fluid from the evaporite beds of the Zechstein, perhaps mixing with Mg^{2+}-depleted connate water, could have led to calcite precipitation, dedolomitization and calcitization of aragonite.

Within the reefal dolostones, a second stage of *dolomite precipitation* is seen where dolostones have been fractured. The fractures are filled with clear dolomite rhombs, and vugs along the line of breakage are filled with dolomite spar cement. This clear dolomite cement is only of minor occurrence and could simply be of local origin, with ions derived from pressure solution. *Stylolites* are common in the reefal limestones and dolomites.

5 Discussion and Conclusion

The early, marine diagenesis of the Upper Permian reef of N.E. England has much in common with modern reefs, where extensive precipitation of aragonite and high Mg calcite is producing solid reef rock. The cements appear to be preferentially precipitated in the shelf-edge reef facies, and in the reef base coquina, deposited in turbulent shallow waters. The identification of metastable carbonate minerals in the Permian is consistent with the fluctuations in the mineralogy of marine precipitates deduced by Sandberg (1983) for the Phanerozoic. The Upper Carboniferous to Triassic is considered to be a time of aragonite and high Mg calcite cements (and ooids), like the Recent, whereas the Ordovician to Lower Carboniferous and Jurassic-Cretaceous seas favoured calcite.

There are some significant features of the Permian marine cements described here. The calcite fan crystals are unlike modern marine cements. However, they are interpreted as primary precipitates and probably formed during very shallow burial, while pore waters were still marine. The columnar calcite with radiaxial fabrics that has replaced aragonite is of interest in view of the discussion concerning radiaxial fibrous calcite, a common and quite specific type of cement in Paleozoic limestones. Formerly regarded as a replacement of an acicular precursor (Kendall and Tucker 1973), it is now being re-interpreted as a primary cement whose peculiar fabrics are due to split growth (Kendall 1985). However, the occurrence in the Permian Tunstall Member of aragonite crusts replaced by calcite ranging from equant through to columnar neomorphic spar suggests that radiaxial fabrics may occur in replacement calcites. However, the gross features of these Permian calcites are quite different from "classic" radiaxial fibrous calcite.

The later diagenesis of the Tunstall reef follows a pattern consistent with pore water evolution in an evaporitic basin (Fig. 12). Evaporative drawdown of water in the Zechstein basin, as a result of a eustatic sea-level fall in the world ocean (Clark and Tallbacka 1980, Smith 1980), terminated reef growth. This permitted the dolomitization of the marginal carbonate bank by groundwater either in a seepage-reflux or mixing-zone model. Undolomitized limestone occurs mainly in the lower part of the reef complex. With the precipitation of sulphates in the basin adjacent to the reef, gypsum and/or anhydrite were also deposited in cavities in the dolomitized reef. With continued sedimentation, some fracturing occurred due to increased overburden, followed by calcitization of dolomite, gypsum and aragonite and calcite spar precipitation. On uplift in the Tertiary, dissolution of the evaporites led to dedolomitization of the carbonates, particularly those along the former shelf margin.

Acknowledgments. Sincere thanks are due to Ron Lambert and George Randall for thin sections, Gerry Dresser for photography, Andy Reid for artwork, Carole Blair for word processing, and Denys Smith for helpful criticism of the script.

We are also grateful to Mr. and Mrs. Appleton of Rock Cottage for allowing access to outcrops. Neville Hollingworth acknowledges this work under the tenure of a grant from the Natural Environment Research Council.

References

Aissaoui DM (1985) Botryoidal aragonite and its diagenesis. Sedimentology 32:345 – 361

Assereto R, Folk RL (1980) Diagenetic fabrics of aragonite, calcite, and dolomite in an ancient peritidal-spelean environment: Triassic Calcare Rosso, Lombardia, Italy. J Sediment Petrol 50:371 – 394

Babcock JA (1977) Calcareous algae, organic boundstones, and the genesis of the Upper Capitan Limestone (Permian, Guadalupian), Guadalupe Mountains, West Texas and New Mexico. Soc Econ Paleontol Mineral Permian Basin Sect Field Conf Guidebook, pp 3 – 44

Baker PA, Kastner M (1981) Constraints on the formation of sedimentary dolomite. Science 213:214 – 216

Clark DN (1980a) The sedimentology of the Zechstein carbonate formation of eastern Drenthe, The Netherlands. Contr Sediment 9:131 – 165

Clark DN (1980b) The diagenesis of Zechstein carbonate sediments. Contr Sediment 9:167 – 203

Clark DN, Tallbacka L (1980) The Zechstein deposits of southern Denmark. Contr Sediment 9:205 – 231

Flugel E, Kochansky D, Ramovs L (1984) A Middle Permian calci-sponge/algal/cement reef: Straza near Bled, Slovenia. Facies 10:179 – 256

Folk RL, Land LS (1975) Mg/Ca ratio and salinity: two controls over crystallization of dolomite. Bull Am Assoc Petrol Geol 59(1):60 – 68

Groot De K (1969) Experimental dedolomitisation. J Sediment Petrol 37:1216 – 1220

Harwood GM, Smith DB, Pattison J, Pettigrew T (1982) Field excursion guide EZ82. Symp English Zechstein, Leeds Univ 1982

James NP, Ginsburg RN (1979) The seaward margin of Belize Barrier and Atoll Reefs. Int Assoc Sediment Spec Publ 3:1 – 191

Kendall AC (1985) Radiaxial fibrous calcite: a reconsideration. In: Schneidermann N, Harris PM (eds) Carbonate cements. Soc Econ Paleontol Miner Spec Publ 36:59 – 77

Kendall AC, Tucker ME (1973) Radiaxial fibrous calcite: a replacement after acicular carbonate. Sedimentology 20:365 – 389

King W (1850) A monograph of the Permian fossils in England. Palaeontrogr Soc Monogr

Kirkby JW (1858) On some Permian fossils from Durham. Trans Tyneside Nat Fld Club 3:286–297
Lohmann KC, Meyers WJ (1977) Microdolomite inclusions in cloudy prismatic calcites: A proposed criterion for former high magnesium calcites. J Sediment Petrol 47:1078–1088
MacIntrye IG (1985) Submarine cements – the peloidal question. In: Schneidermann N, Harris PM (eds) Carbonate cements. Soc Econ Paleontol Miner Spec Publ 36:109–116
Marshall JD (1981) Zoned calcites in Jurassic ammonite chambers: trace elements, isotopes and neomorphic origin. Sedimentology 28:867–887
Mazzulo SJ, Cys JM (1979) Marine aragonite sea-floor growths and cements in Permian phylloid algal mounds, Sacramento Mountains, New Mexico. J Sedimentol Petrol 49:917–936
Sandberg PA (1975) New interpretations of Great Salt Lake ooids and of ancient non-skeletal carbonate mineralogy. Sedimentology 22:673–702
Sandberg PA (1983) An oscillating trend in Phanerozoic non-skeletal carbonate mineralogy. Nature 305:19–22
Sandberg PA, Hudson JD (1983) Aragonite relic preservation in Jurassic calcite replaced bivalves. Sedimentology 30:879–892
Scherer M (1977) Preservation, alteration and multiple cementation of aragonitic skeletons from the Cassian beds (U. Triassic, southern Alps): petrographic and geochemical evidence. N Jahrb Geol Palaeontol Abh 154:213–262
Schroeder JH (1972) Fabrics and sequences of submarine carbonate cements in Holocene Bermuda cup reefs. Geol Rundsch 61:708–730
Smith DB (1958) Observations on the Magnesian Limestone reefs of northeastern Durham. Gr Br Geol Surv Bull 15:17–84
Smith DB (1970) Submarine slumping and sliding in the Lower Magnesian Limestone of Northumberland and Durham. Proc Yorks Geol Soc 38:1–36
Smith DB (1980) The evolution of the English Zechstein basin. Contr Sediment 9:7–34
Smith DB (1981) The Magnesian Limestone (Upper Permian) Reef Complex of Northeastern England. Soc Econ Paleontol Miner Spec Publ 30:187–202
Smith DB, Brunstrom RGW, Manning PI, Simpson S, Shotton FW (1974) Correlation of the Permian rocks of the British Isles. Spec Rep Geol Soc Lond 5
Taylor JCM, Illing LV (1969) Holocene intertidal calcium carbonate cementation. Qatar, Persian Gulf. Sedimentology 12:69–107
Taylor JCM (1984) Late Permian-Zechstein. In: Glennie KW (ed) Introduction to the petroleum geology of the North Sea. Blackwell Scientific, London, pp 61–83
Trechmann CT (1913) On a mass of anhydrite in the Magnesian Limestone at Hartlepool, and on the Permian of south-eastern Durham. Q J Geol Soc Lond 69:184–218
Trechmann CT (1925) The Permian formation in Durham. Proc Geol Assoc Lond 36:135–145
Trechmann CT (1945) On some new Permian fossils from the magnesian limestone near Sunderland. Q J Geol Soc Lond 100:333–354
Tucker ME (1985) Calcitized aragonitic ooids and cements from the Late Precambrian of southern Norway. Sediment Geol 43:67–84
Walker KR, Alberstadt LP (1975) Ecological succession as an aspect of structure in fossil communities. Palaeobiology 1:238–257

Diagenesis of Aragonitic Sponges from Permian Patch Reefs of Southern Tunisia

M. Scherer[1]

1 Introduction and Geological Setting

Djebel Tebaga in southern Tunisia is the only known outcrop of marine Upper Permian on the African continent (Fig. 1). Throughout the Djebel Tebaga a variety of patch reefs, metres to tens of metres in size, are embedded in crinoidal and fusulinid limestones and marls (Fig. 2). Beds directly underlying the reefs are commonly reduced in thickness resulting from early compaction. Besides coralline algae, the important frame-builders are calcareous sponges represented especially by *Inocoa, Sclerospongia, Disjectoporida*, and less common *Sphinctozoa.* Close to extinction at the Permo-Triassic time boundary, rugose corals are rare and do not contribute to the reef framework.

During the latest Permian-earliest Triassic regression, these patch reefs were subjected to an arid to semi-arid coastal climate, as indicated by onlapping Triassic beach and dune sands showing occasional dinosaur footprints. The Triassic to Jurassic sequence consists mainly of red sandstones, shales and marls, representing coastal desert environments with occasional shallow marine incursions (Busson 1967, Assereto and Benelli 1971). The Upper Triassic Azizia Limestone described from Libya (Assereto and Benelli 1971) was not encountered in the area and is laterally replaced by sandstone (Busson 1967). The Permian to possibly Middle Jurassic sediments are locally unconformably overlain by coastal/marine Lower Cretaceous clastics and carbonates which, following a Late Cretaceous to Tertiary regression and uplift, were eroded down to the Permian in the centre of Djebel Tebaga.

A previous study has shown that the Permian sponge skeletons are still partly preserved as aragonite (Wendt 1977). The calcareous sponges of Djebel Tebaga therefore offer a unique opportunity to analyze the early diagenetic, subsequent burial and weathering products and the overall paragenetic sequence.

2 Methods

In addition to petrographical studies supported by point counting, staining techniques, UV-light and cathodoluminescence microscopy, trace elements and stable oxygen and carbon isotopes were analyzed. The particular type of trace element has been selected, firstly, to deduce the environment in which biominer-

[1] Shell U.K. EXPRO, Shell-Mex House, Strand, London, WC2, England

Reef Diagenesis
Edited by J. H. Schroeder and B. H. Purser

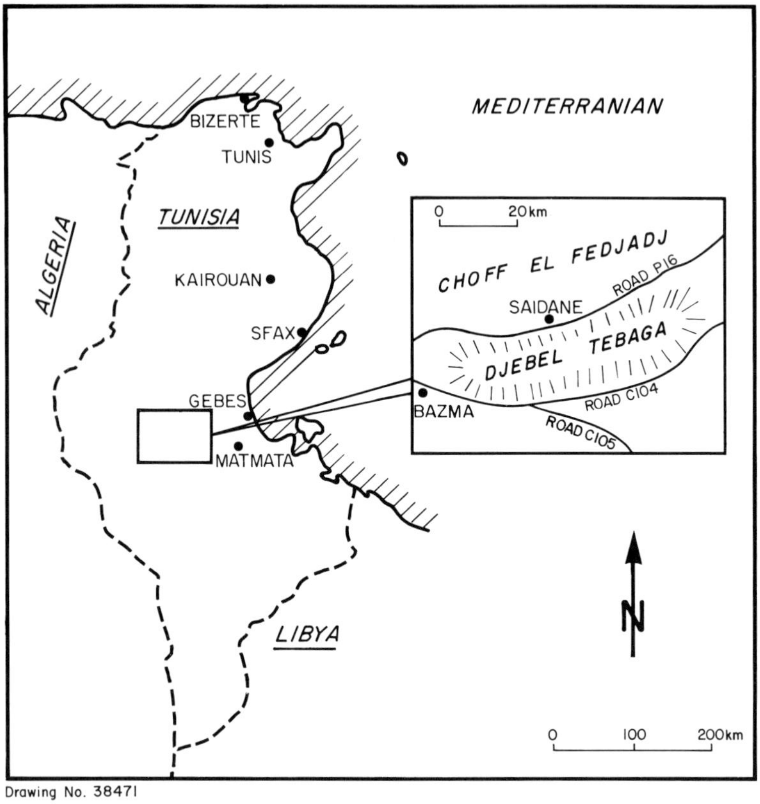

Fig. 1. Location of the area studied

alization took place and, secondly, to interpret the conditions during diagenesis: sodium, strontium, and magnesium were used following suggestions by Land and Hoops (1973), Usdowski (1973), Siegel (1960), and Kinsman (1969). Ferrous iron was also selected because it is a good indicator of reducing conditions and, by negative evidence, of oxidizing conditions, if iron is available in the system.

Spot analyses were carried out on thin sections by energy dispersive X-ray analysis (EDAX). Since the contents are often too low to be effectively measured by EDAX, and elements such as sodium and, to a certain extent, also magnesium, have a very low radiation energy, the separating window between sample and detector was removed. Atomic absorption analyses were performed on total samples using a Perkin Elmer Model 460 because of higher resolution to precise the information obtained from point measurements.

Carbonate phases in total samples were determined by X-ray diffraction using Cu K alpha radiation. Quantitative compositions were calculated from comparison of the integrated main peak areas of the samples with those of known mixtures of aragonite and calcite of Holocene and Pleistocene corals and with Triassic dolomitic reef material from the Alps. From repetitive runs a standard deviation of 5% was obtained. For stable isotope analysis crushed samples were

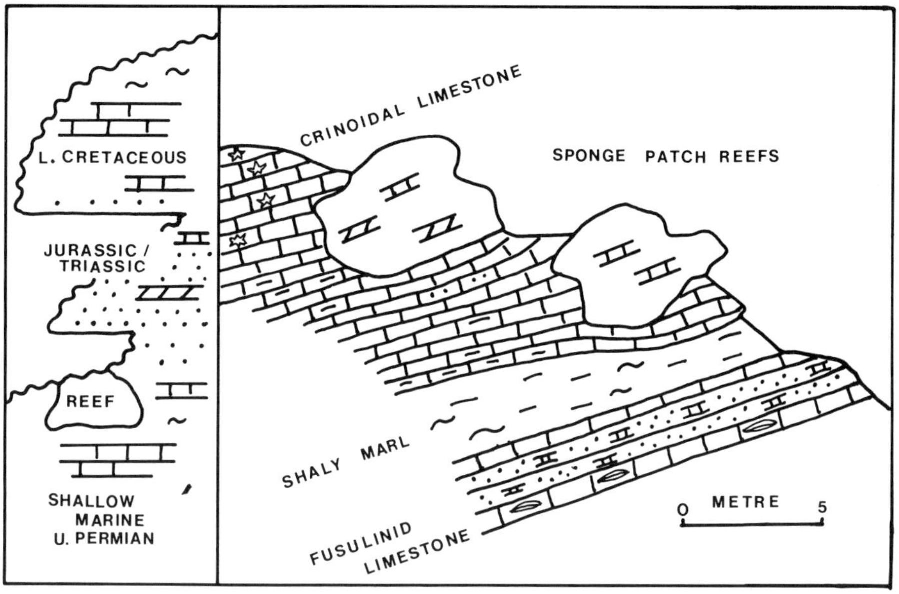

Fig. 2. Schematic stratigraphic column and typical cross-section of patch reefs with surrounding sediments

treated in a solution of sodium hypochlorite to remove organic matter and washed and dried samples of known mineral composition were treated with 100% phosphoric acid at 25 °C (McCrea 1950) to obtain carbon dioxide. Sample T26/1 was separated into aragonite and calcite fractions by heavy liquids and both phases were then treated as above. Analysis of stable isotopes was carried out on the CO_2 gas with a 60°, 15 cm single focussing mass-spectrometer.

For the statistical treatment of normalized data purpose-written programmes were applied for curve fitting based on the Gauss-Jordan elimination method, least-squares routines, and the Simplex algorithm. The functions obtained are shown in Table 2. To estimate most likely values of trace elements and stable isotopes in the various carbonate phases a Monte Carlo error propagation was used (Anderson 1976). For further informtion on the statistical methods, the reader is referred to Daniel and Wood (1980) and Caceci and Cacheris (1984).

Samples have been deposited with the University Tübingen, Institut für Geologie und Paläontologie, D-7400 Tübingen, Sigwartstrasse 10, Federal Republic of Germany.

3 Results

3.1 Chemical Evidence

Analytical results performed on total samples are given in Table 1. The most likely values of stable isotopes and trace elements in the various carbonate phases are summarized in Table 3.

Table 1. Analytical data of Permian calcareous sponges from Tunisia, Djebel Tebaga

No.	Name	$\delta^{13}C$ ppt PDB	$\delta^{18}O$ ppt PDB	Aragonite %	Dolomite %	Calcite %	Mg ppm	Sr ppm	Na ppm	Fe ppm	Residue %	Mn ppm	Ca^{c} %
1	T2/1						6232	1883	399	4841			38.7
2	T2/2	+0.50	−5.77	0	2	98	26134	265	210	8766		1363	36.3
3	T2/3			40	0	60	3762	3327	1073	5655			38.6
4	T2/5			44	0	56	2211	4746	833	6103		500	38.6
5	T3/1	+1.65	−4.94	0	0	100	5234	1597	174	3640		155	39.0
6	T4/1			0	0	100	2681	439	162	2830			39.4
7	T5/3			30	0	70	5632	2887	345	8857			38.2
8	T7/1	+5.28	−5.39	19	0	81	7593	2576	366	6252	2.55		35.8
9	T7/4	+2.89	−5.81	0	1	99	3100	1051	405	4052		270	39.1
10	T7/5			0	5	95	8615	1697	338	4586			38.5
11	T8/1						17026	1362	315	5121			37.6
12	T8/2						10930	2788	242	6970			37.9
13	T13/1			5	0	95	2801	1600	294	5610			38.9
14	T14	+3.33	−7.75	0	0	100							
15	T19/1	+5.82	−2.30	1	0	99	2229	497	371	7671	2.25		36.7
16	T19/2	+5.70	−3.65	9	0	91	1788	1546		5770	1.47		37.6
17	T19/3			0	0	100							
18	T19/4	+4.98	−2.19	1	0	99	3920	543		3102	1.81		37.4
19	T20/1	+5.56	−5.22	0	2	98	6726	554		7053	1.70		36.9
20	T20/2	+5.73	−5.28	1	0	99	2503	819		6146	1.52		37.5
21	T20/3			1	0	99	4518	817		6951	1.57		37.2
22	T20/4	+5.24	−4.46	9	0	91	6942	650		8123	1.45		36.9
23	T20/5	−0.09	−6.29	0	0	100	2058	427		7287	1.97		37.1
24	T20/6	+5.45	−4.63	20	0	80	5289	969	274	6465	1.65		37.1
25	T20/7			0	0	100	6635	627		5839	1.73		37.0
26	T20/8			2	0	98	6391	903	233	6178			38.6
27	T20/11			0	0	99	7214	988	220	10710			39.1
28	T20/12	+3.77	−5.46	0	0	100	7473	875	285	8792		1503	38.1
29	T21/1	+5.21	−5.47	13	0	87	4459	1365		4297	1.62		37.4
30	T21/2	+5.61	−4.21	15	0	85	5510	1003		6747	1.18		37.6
31	T21/3	+5.37	−3.99	0	0	100	6629	863		7014	0.14		38.4
32	T25/2			31	0	69	4079	3352	327	9997			38.2

33	T25/3	+3.75	−5.73	0	0	100							
34	T26/1	+4.99	−5.11	33	0	67	3245	4498	684	6746	1.69		36.8
35	T26/1[a]	+1.36	−5.97	100	0	0							
36	T26/2[b]	+3.27	−5.44	0	2	98	8399	951		3577	4.06		34.7
37	T26/2[b]	+4.28	−4.68	0	100	0			$\delta^{18}O$ corrected by −0.8				
38	T26/3			0	2	98	8415	1163	426	3912		213	38.6
39	T26/4	+2.89	−5.18	40	0	60	6442	3267	839	12200		1540	37.6
40	T28/1			0	0	100	4286	575		7960	4.16		34.6
41	T28/2			0	0	100	2973	901		8576	1.83		36.9
42	T28/3			0	0	100	4141	962		8177	4.74		33.9
43	T28/4	+4.75	−5.19	20	0	80	6585	2106		7171	0.80		37.6
44	T30/2			0	0	100	4441	513	522	13062		1045	38.1
45	T30/3			0	2	98	7997	2325	238	15284			37.4
46	T31/1			15	0	85	6565	2157		9002	2.91		35.3
47	T31/2	+3.71	−6.15	10	0	90	4764	2460		7920	1.37		37.1
48	T31/3	+4.29	−5.85	10	0	90	4370	1345	220	7053	2.03		36.7
49	T31/4	+3.38	−5.76	0	0	100	3623	935	405	7579	1.45	607	37.2
50	T31/5	+3.44	−4.71	16	0	84	5777	3612		8483	1.95		36.3
51	T31/6	+3.69	−5.64	29	0	71	5497	3366	238	7705	3.32		35.0
52	T31/7			11	0	89	3163	1874	387	4582	0.97		38.0
53	T31/8	+3.09	−6.39	20	0	80	4206	2722	244	10604	2.73		35.5
54	T31/10			10	0	90	4419	1957	433	11091			38.2
55	T31/22	+4.06	−5.45	15	0	85	5698	2262	279	15364		1094	37.5
56	T31/27			31	0	69	4715	2987	236	2055		472	38.9
57	T31/28	+2.26	−5.26	2	0	98	4685	1160	320	6110		879	38.7
58	T31/29			5	0	95	5017	1489	393	8460			38.5
59	T31/31	+4.70	−5.59	0	0	100							
60	T31/32	+2.34	−5.55	0	0	100							
61	T31/60	+3.67	−5.63	0	0	100							

[a] Aragonite separated by heavy liquids.
[b] No. 36 measured after 1 h CO_2 development. No. 37 measured after 48 h CO_2 development.
[c] Calcium content calculated.

Table 2. Functions fitted to data

Element/mineral	Equation	Std. deviation	Correlation	Cases	Remarks
m Mg/m Ca vs. $\delta^{13}C$	$Y = 1.209 + 1.836X$	1.25	0.85	5	In 100% calcite samples
m Sr/m Ca vs. $\delta^{18}O$	$Y = 7.756 - 41.017/X$	0.84	0.89	6	In 100% calcite samples
m Mg/m Ca vs. $\delta^{18}O$	$Y = -4.915 - 1.206/X$	0.29	0.80	5	In 100% calcite samples
m Sr/m Na vs. aragonite	No correlation			14	
Calcite vs. Na	$Y = -129.59 + 42913.94/X$	171.7	0.53	25	
Calcite vs. Sr	$Y = 9410.07 - 87.61X$	772.0	−0.82	35	
Aragonite vs. Sr	$Y = 791.55 + 81.24X$	571.0	0.88	31	
Calcite vs. Mg	No correlation			31	
Aragonite, calcite vs. Na	$Z = -123.64 + 0.57X + 4.17Y$	178.2	0.57	25	
Aragonite vs. Na	$Y = 120.34 + 11.98X$	211.6	0.55	14	Only samples above 10% aragonite
m Sr/m Na vs. $\delta^{13}C$	No correlation			14	
m Sr/m Na vs. $\delta^{18}O$	$Y = -4.72 - 0.371X$	0.40	−0.66	8	
Fe vs. Mn	$Y = -4836.39 + 641.905 \ln X$	323.60	0.77	12	
m Sr/m Na vs. m Mg/m Ca	$Y = 1.19 \cdot 1.27X$	0.69	0.61	15	Only samples above 10% aragonite
m Mg/m Ca vs. m Sr/m Na	$Y = 0.61 \cdot 1.65X$	1.02	0.67	15	
m Sr/m Na vs. m Mg/m Ca	$Y = 0.82 + 1.73X$	0.66	0.59	7	
Aragonite vs. Fe	No correlation				
m Sr/m Ca vs. m Mg/m Ca	No correlation				
$\delta^{13}C$ vs. $\delta^{18}O$	$Y = -7.32 + 0.48X$	0.59	0.60	16	

m = molar.

Table 3. Results after multivariate analysis and error propagation

Mineral phase	$\delta^{13}C$ ppt	$\delta^{18}O$ ppt	Mg ppm	Sr ppm	Na ppm	Fe ppm	Mn ppm	Mg/Ca molar 10^2	Sr/Ca molar 10^4
Aragonite skeleton	$+1.38 \pm 0.30$	-5.97 ± 0.20	260 (noise)	8998 ± 580	1933 ± 80	trace	trace	0.11	108.31
Aragonite needles				9850 ± 220					
Ferroan dolomite	$+3.27 \pm 0.30$	-5.44 ± 0.20							
Fe-calcite	$+5.72 \pm 0.40$	-5.23 ± 0.35	8700 ± 850	935 ± 60	358 ± 80	4900 ± 580	800 ± 95	3.77	11.26
Calcite	-0.09 ± 0.40	-6.29 ± 0.40	2058 ± 250	427 ± 30	180 ± 100	trace	trace	0.92	5.27

Note: Point measurements were included in the statistical analysis.

3.2 Petrographic Characteristics

Two processes generally reduce pore space:

- deposition of sediment particles in skeletal pores, and
- cementation, which is volumetrically the more important (2 – 5% versus 45 – 50%).

During the initial stage of diagenesis both processes seem to have been influenced by mean pore diameters. The trapping of sediment particles commonly occurred in voids larger than 0.5 mm, whereas early cement occurred preferentially in pores smaller than 0.2 mm. The majority of cements consists of aragonite, dolomite and calcite. However, based on sequential arrangement and chemical composition, these can be divided further into a sequence of six carbonates and three non-carbonates. The first cement commonly consists of still preserved aragonite needles (Fig. 3A, B, D) 20 to 150 µm in length with a mean length of 50 µm. The needles are similar in geometry and strontium content (about 1% wt; Table 3) to those described by Schroeder (1972) from the Holocene and by Scherer (1975, 1977) from the Pleistocene and the Upper Triassic.

The aragonite needles are succeeded by and are often incorporated into, non-ferrous dolomite rhombohedra of 20 to 50 µm in size (Fig. 3B). Like the aragonite cement, they are commonly present in voids smaller than 0.2 mm. Clusters of framboidal pyrite, often geopetally arranged, follow the dolomite (Fig. 3C).

A subsequent zoned ferroan dolomite rarely survived weathering but can still be recognized as dedolomite with limonite stained zones. Such zones are commonly collapsed, indicating that the dedolomitization went through a dissolution phase (Fig. 4A). The formation of ferroan dolomite was possibly contemporaneous with the dolomitization of internal sediments (Fig. 6E).

Volumetrically most important was the precipitation of blocky Fe-calcite cement, which entirely fills residual pore space. Fifty to 300 micrometers in size, crystals form a cement mosaic enclosing all earlier void fills (Fig. 3). The cement mosaic is fairly uniform, and sparry rim cement, or an increase in crystal size towards the centre of a pore, were not observed.

Following formation of Fe-calcite, a major change in the environment led to both oxidation of ferrous iron in the carbonate phase and dissolution of precursor cements. However, permeabilities must have been sufficiently low because parts of the previous diagenetic phases were preserved. Fe-calcite cement was leached, as indicated by rounded crystal faces and dissolution vugs (Fig. 3E). The resulting enlarged pore space was cemented by non-ferroan, low-magnesian calcite which almost completely replaced the ferroan dolomite (Fig. 4A). Locally, in strongly altered samples, aragonite needle cement was replaced by elongate, bladed crystals of non-ferroan calcite (Fig. 4B). Such crystals have also been described from Lower Proterozoic dolomites by Grotzinger and Reed (1983), who postulate that the observed fibrous dolomite cement represents a replaced precursor aragonite-needle cement. The calcite pseudomorphs described above tend to support this hypothesis.

Although non-ferroan calcite does not show zoning when stained, cathodoluminescence occasionally reveals an additional zone (Fig. 3F).

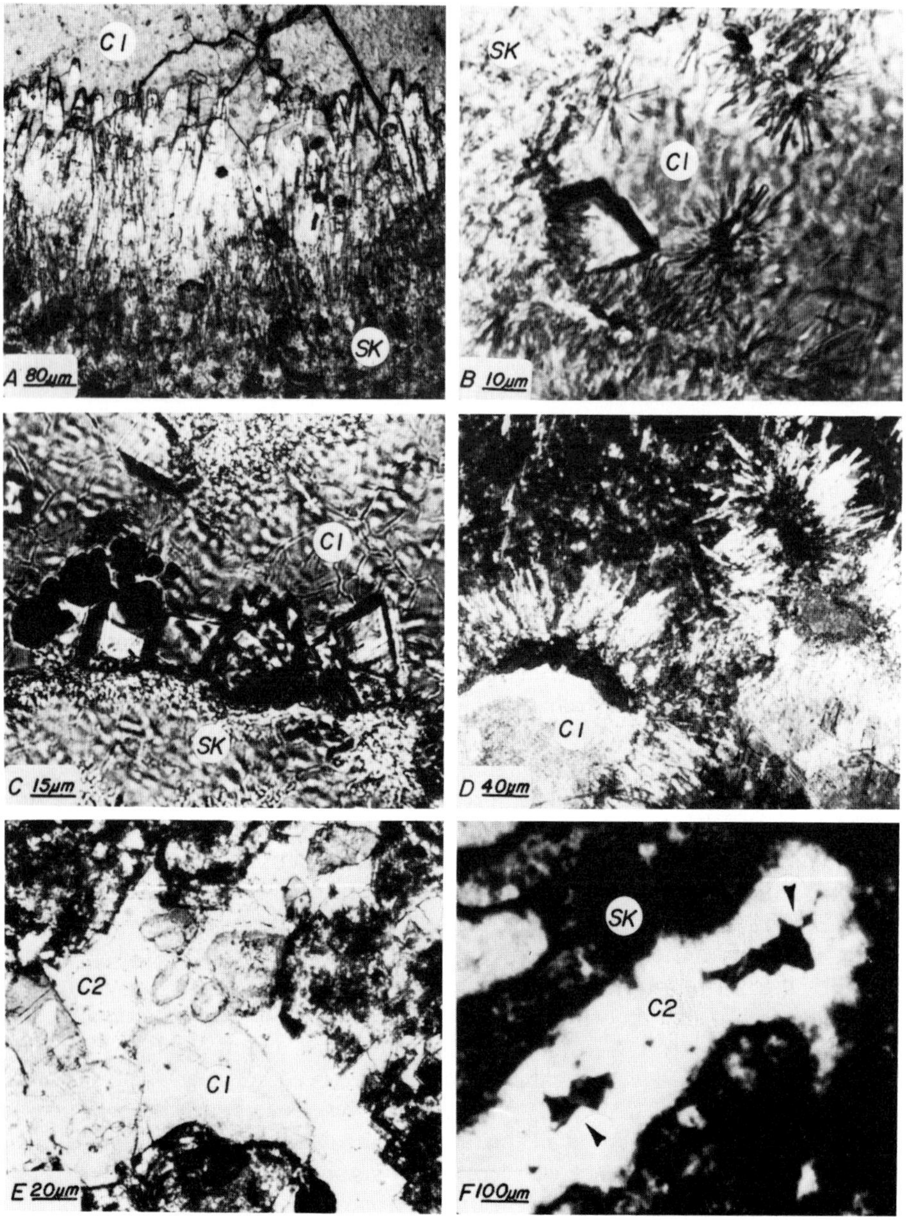

Fig. 3 A – F. Cement sequences I. **A** Aragonite needle cement growing on skeletal wall (*SK*) followed by Fe-calcite (*C1*); sample T7/4. **B** Non-ferroan dolomite rhombohedran incorporating some aragonite needles in sample T2/1 (Skeleton = *SK*). The rest of the void is completely filled by further aragonite needles and Fe-calcite. **C** Cluster of framboidal pyrite on dolomite rhombs (T2/1). **D** Aragonite needle cement completely preserved, while the aragonitic skeleton is replaced by Fe-calcite (T26/4). **E** Dissolution of Fe-calcite (*C1*). Secondary porosity occupied by non-ferroan calcite (*C2*). Skeleton dedolomitized and also replaced by non-ferroam calcite (T2/2). *Dark patches* are organic residue and limonite. **F** Cathodoluminescence photograph showing a darker calcite cement zone (*arrows*) succeeding non-ferroan calcite (*C2*) ion sample T32/28

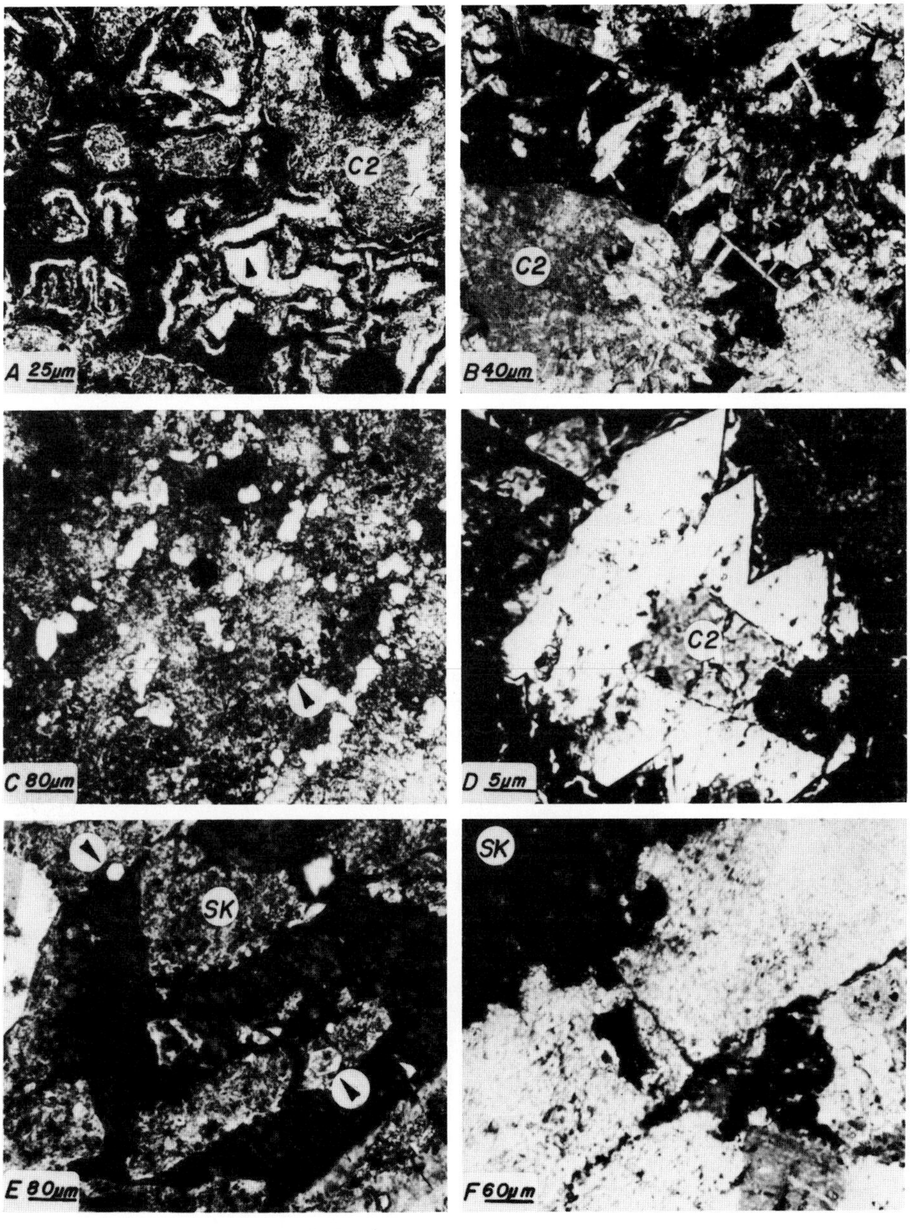

Fig. 4 A – F. Cement sequences II. **A** Dedolomitized ferroan dolomite. The presently limonitic zoning (*arrows*) is collapsed. Dedolomite and remainder of void-fill consist of non-ferroan calcite (*C2*). Sample T2/2. **B** Part of sample T7/8 completely replaced by non-ferroan calcite. Note calcite pseudomorph after aragonite needle cement. **C** Growth of iodiomorphic quartz crystals in sample T2/1. Framboidal pyrite (*arrow*) transformed into limonite. **D** Enlarged view of group of quartz crystals incorporating non-ferroan calcite (*C2*) in sample T2/2. **E** Large barite crystals (*dark*) incorporating broken pieces of skeleton (*SK*) and quartz crystals (*arrows*) in sample T2/5. **F** Barite (*white*) replacing the skeleton (*SK*) which was previously replaced by Fe-calcite (Sample T2/5)

In strongly weathered samples, idiomorphic quartz occurs which sometimes incorporates non-ferroan calcite thereby clearly indicating its sequential position (Fig. 4C, D). The final diagenetic process was the rare precipitation of barite which occurs as large crystals 100 to 500 micrometers in size. They often enclose carbonate fragments and quarz crystals (Figs. 4E, F).

3.3 Alteration of Skeletons

The first alteration observed in skeletons is bio-erosion due to boring, probably by filamentous algae or fungi which formed dark tubules about 2–5 μm in length composed of siderite and organic matter (Fig. 5). Although bio-erosion can be an important process during lithification (Schroeder and Zankl 1974), dissolution and precipitation seem to play a more dominant part in the diagenetic evolution of sponge skeletons at the Djebel Tebaga. Because the internal fabric of the aragonite skeleton is a determining factor in these processes, the microstructures are briefly described.

The most common structure is of trabecular type, consisting of subparallel columns of aragonite needles radiating from a "centre of calcification". Such microstructure is still common today in scleractinian corals. The "centres of calcification" are filled by organic matter during life and form small hollow tubules after decay of the organic fill. Other less common microstructures are described by Scherer (1977). This tubular system seems to be the preferred site for dissolution following removal of the organic matter. The first mineral transformation pro-

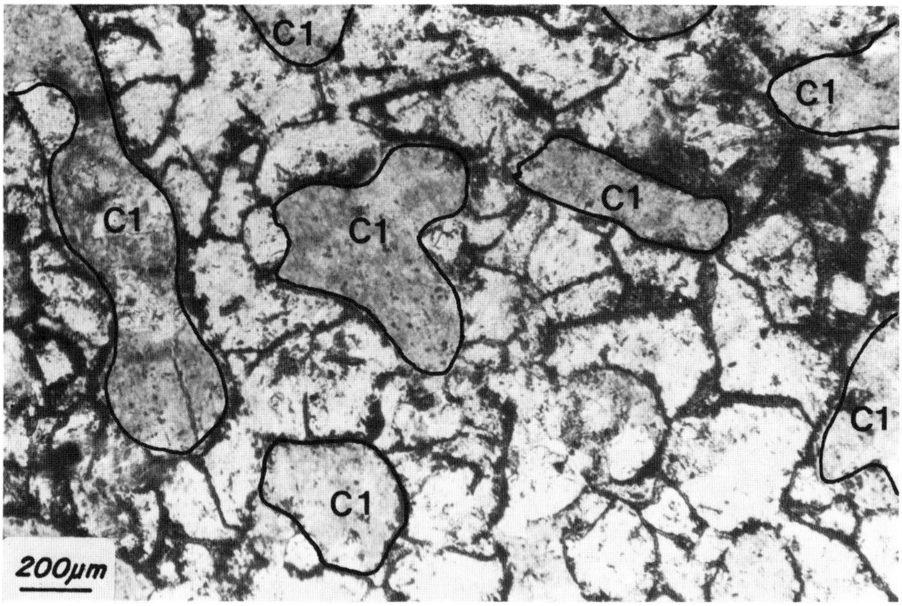

Fig. 5. Sponge skeleton (*white*) strongly bored by filamentous algae or fungi. Voids are cemented by Fe-calcite (*C1*; voids outlined by pen)

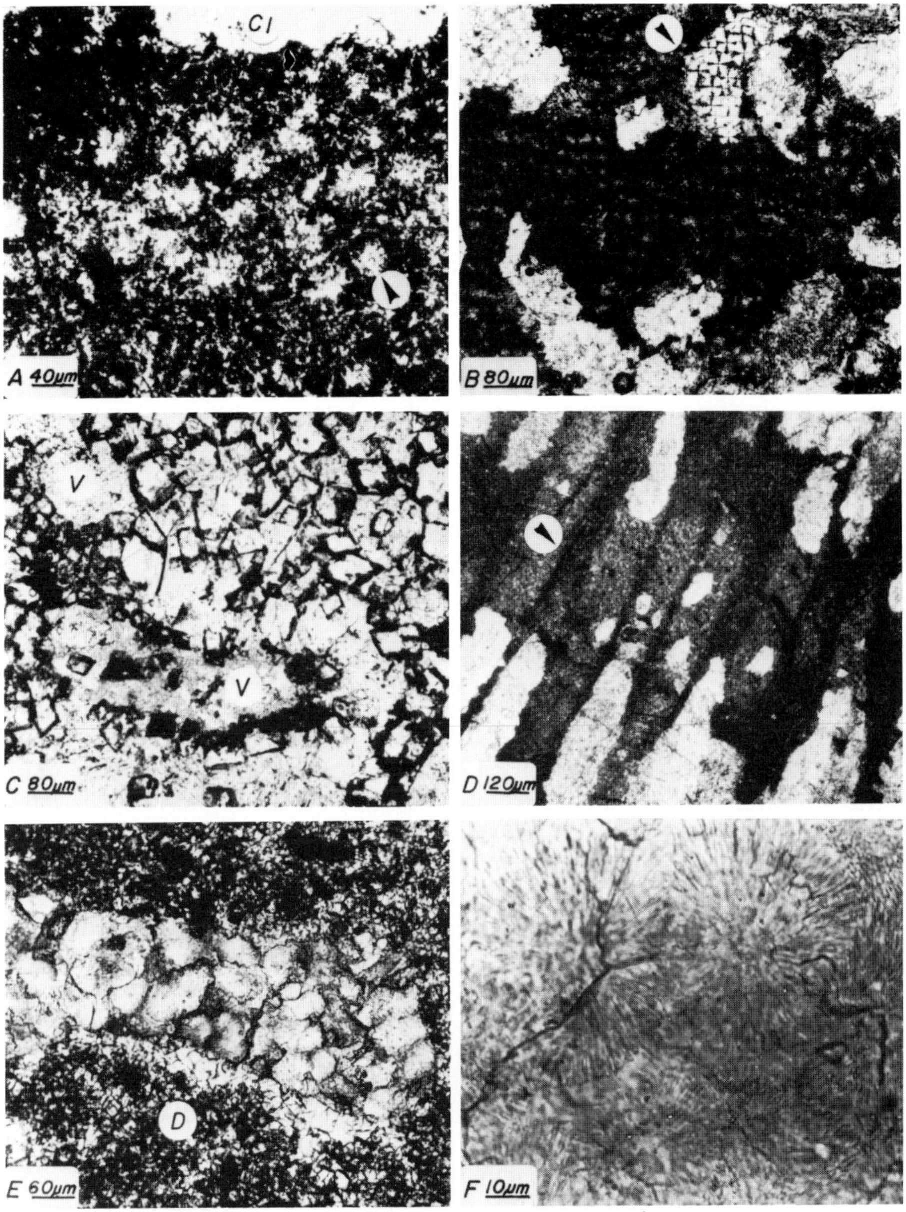

Fig. 6 A – F. Alteration of sponge skeletons. **A** Sample T2/3 stained with Feigl's solution (aragonite: *black*) shows replacement of skeletal aragonite by Fe-calcite (*white*) starting from the centre of calcification (*arrow*). Void filled by Fe-calcite (*Cl*). **B** Pseudo-extinction crosses in trabecular cross-sections of sample T25/9. *Arrow* indicates an area of preserved aragonitic skeleton. Areas replaced by Fe-calcite show blurred but still recognizable pseudo-extinction crosses (crossed polarizers). **C** De-dolomitized skeleton in sample T20/4. Dolomite rhombohedra remains as ghosts. Skeleton and void fill (*V*) replaced by non-ferroan calcite. **D** Sample T26/6 consisting only of non-ferroan calcite. The skeleton (*arrow*) remains as a ghost but seems to have influenced the orientation of calcite crystals. Note presence of quartz crystals near centre. **E** Sample T26/3. Trabecular cross-sections and radiating fibres are still visible after two phases of diagenetic replacement (Fe-calcite to non-ferroan calcite). The dolomitized internal sediment (*D*) is replaced by non-ferroan calcite. **F** Details of T26/3 showing the replacement of the trabecular microstructure after the second calcite replacement

ceeding from such centres is dolomitization of skeletal aragonite, the skeletons being locally dolomitized with complete loss of internal structure (Fig. 6C). The subsequent Fe-calcite has often replaced skeletons completely, by extending from these centres (Fig. 6A). Using crossed polarizers, pseudo-extinction crosses caused by the radial arrangement of aragonite crystals in the trabecular column remain visible, though blurred, in skeletons replaced completely by Fe-calcite. Extinction crosses in Fe-calcite are commonly rotated a few degrees from the normal north-south, east-west orientation compared to coexisting aragonite. This may indicate a slight reorientation of crystal axes.

During the formation of non-ferroan calcite, skeletons are completely dedolomitized and only the ghosts of dolomite rhombohedra remain embedded in large crystals of non-ferroan calcite (Fig. 6C). Skeletons survive either as ghosts within large crystals obviously influencing their orientation (Fig. 6D), or were delicately replaced, exhibiting trabecular column outlines and microstructures (Figs. 6E, F). However, such replacements of microstructures no longer exhibit pseudo-extinction crosses.

3.4 Porosity Development

Initial porosities range from 48 to 55%, and include intraskeletal pores measuring tens of micrometers to several millimetres. Such a range will clearly permit circulation of seawater through bigger pores whereas smaller ones may favour restricted microenvironments. Thus, selective water movement may explain internal sediments mainly in pores larger than 0.5 mm, and the abundance of aragonite needle cement and early dolomite rhombohedra better developed in pores smaller 0.2 mm.

Although internal sediments and first cements such as aragonite needles, dolomite and framboidal pyrite clearly reduce porosity, their volume, approximately 10%, has only a relatively small overall effect. However Fe-calcite cement, ca. 35%, considerably reduces porosities to an estimated value of 5%, while permeabilities were probably reduced to only some millidarcies and thus probably effectively hampered further replacement of aragonite.

The formation of non-ferroan calcite was preceded by a dissolution phase (Fig. 3E) which took place in an oxidizing environment as suggested by the change of ferrous to ferric iron now present in the form of limonite (Fig. 4A). Secondary porosity increased to about 15 to 20% before it was again reduced by non-ferroan calcite to the previous low level.

3.5 Generation of Hydrocarbons

Samples with highest aragonite content were studied using UV-light which revealed that generation and initial expulsion of a light yellow hydrocarbon had taken place from structureless organic matter (brown fluorescence) corresponding to Tissot and Welte's (1978) Type II.

4 Discussions

4.1 Estimation of Burial Depth and Temperature from Degree of Organic Metamorphism

Initial expulsion of hydrocarbons will have taken place following a low degree of organic metamorphism (Saxby 1982, Fig. 1). Using the age/temperature relationship of Lopatin (Waples 1980), such a degree is reached in the samples analyzed at a maximum depth of burial of 1500 to 2000 m assuming an average geothermal gradient of 3°C/100 m (45° to 60°C).

4.2 Temperature and Salinity

According to Allan and Matthews (1982), carbonates that have undergone mineralogical stabilization under meteoric water conditions tend to have a relatively low range of $\delta^{18}O$, whereas $\delta^{13}C$ shows a rather wide range partly inherited from the replaced carbonate. This also seems to be the case for this study (Fig. 7). However, compared with data from Allan and Matthews (1982, Fig. 1E) the Permian samples show strongly positive $\delta^{13}C$ values of up to +5.82 and fairly negative $\delta^{18}O$ data, with the bulk at around −5.5. Similarly high $\delta^{13}C$ values were found by Brand (1982) in Carboniferous limestones of Oklahoma and New Mexico, such values being considered not to be the result of isotopic equilibrium with ambient seawater. The cause of the $\delta^{13}C$ enrichment is not known.

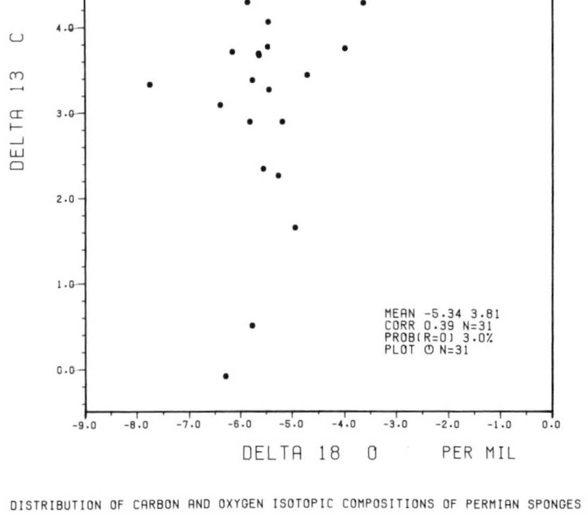

Fig. 7. Cross-plot of $\delta^{18}O$ versus $\delta^{13}C$ derived from total sample. Note the contrast between high variability of $\delta^{13}C$ and relative homogeneity of $\delta^{18}O$, which is interpreted by Allan and Matthews (1982) as being the result of meteoric water diagenesis

Negative $\delta^{18}O$ values, however, can also be the result of fractionation at higher temperatures. A more detailed estimation of salinity and temperature during the formation of the various carbonate phases can be achieved by applying a number of methods. These use the oxygen isotope composition and the trace element content, particularly of strontium and sodium, in the carbonate phase (Epstein et al. 1953, Land and Hoops 1973, Veizer and Demovic 1974, etc.). However, two assumptions have to be made, namely that the pore solutions were derived from seawater, and that Permian seawater had a composition similar to present-day shallow tropical seawater.

The following equations were then used to estimate temperature and salinity of pore fluids from stable oxygen isotope values and trace element composition of the various carbonate phases:

1. The oxygen isotope/temperature relationship from Shackleton and Kennett (1975).
2. An oxygen isotope/salinity function for seawater derived fluids calculated from data of Epstein and Lowenstam (1953), Epstein and Mayeda (1953), Dansgaard (1964), Lloyd (1965), Mook (1970), Fleischer et al. (1976) and Aharon et al. (1977): $\delta^{18}O$ water $= 7.18 + 0.24$ salinity $(g\,l^{-1}) - 0.81 \times 10^{-3}$ (salinity)^2 ($R = 0.94$).
3. The Doerner-Hoskins formula to estimate uptake of strontium into the carbonate lattice (Kinsman and Holland 1969) utilizing Kinsman's (1969) molar Sr/Ca ratio of 0.84×10^{-2} for present-day tropical seawater.
4. A temperature dependance of the strontium distribution coefficient for calcite ($k^{Sr}_{calcite}$) for a temperature range of 25 °C to 100 °C (calculated from data of Kinsman 1969, 1971; Wedepohl 1969; Usdowski 1973; Jacobson and Usdowski 1976): $k^{Sr}_{calcite} = 7.50 \times 10^{-2} \times T°C/(T°C - 11.529)$.

Of the five carbonate phases studied (Table 3), only two yielded reasonable results. The $\delta^{18}O$ value in skeletal aragonite of -1.38 is too negative to be precipitated in equilibrium with ambient seawater. Besides possible fractionation by the organism, a re-equilibration with later (non-marine) pore solutions is suggested. This is supported also by point analysis (EDAX) of Sr in skeletal aragonite. A decrease of Sr in the aragonite phase was found with increasing diagenetic alteration. However, in samples rich in aragonite, the Sr content of 9000 ppm in skeletal aragonite suggests growth in "normal seawater". The sodium content of 1706 ppm, stoichiometrically unbalanced by chloride (350 ppm) shows a deficiency of approximately 850 ppm compared to aragonite in equilibrium with seawater (White 1977). This deficiency, plus an Mg content below the limit of detection, may well be the result of incongruent dissolution as suggested by Schroeder (1969) and Otis and Müller (1985).

Insufficient material precluded measurement of isotope compositions of aragonite needle cement. However, the Sr content of approximately 1% slightly higher than that in the skeletal aragonite, is comparable with values found in Triassic, Pleistocene and Holocene equivalents (Scherer 1977, 1975, 1974).

The ferroan dolomite has a strongly negative $\delta^{18}O$ value of -5.44. Because of strong dedolomitization during weathering it is suggested that this also reflects re-equilibration.

The volumetrically important Fe-calcite cement, however, has a fairly constant composition from sample to sample, which is believed to reflect the original temperature and salinity fractionation. Applying the Doerner-Hoskins formula, the Sr content of 935 ppm in Fe-calcite is best reproduced with distribution coefficients at temperatures between 27° and 30°C for $(\mathrm{m\ Ca^{2+}})_{\mathrm{solid}}$ of 0.97. A calcite precipitated from seawater at 24 °C should contain 1100 ppm of Sr (Kinsman 1969). The salinity of the parent solution of Fe-calcite could therefore have been slightly lower, at about 85% of modern seawater salinity. Because present-day shallow seawater has a salinity of 35 to 40 $\mathrm{g\,l^{-1}}$, the salinity of the calcite parent solution could have been in the order of 30 to 34 $\mathrm{g\,l^{-1}}$. If the distribution coefficient increases with a fall in salinity, as suggested by Jacobson and Usdowski (1976), this calcite could have been formed at temperatures slightly higher than 30 °C.

That the salinity was slightly lower than that of seawater was confirmed from the sodium content. 358 ± 80 ppm of Na in Fe-calcite is only partly stoichiometrically balanced by chloride (235 ± 35 ppm/$\mathrm{Cl^-}$). An error propagation revealed the most likely value to be 206 ppm of sodium incorporated into the carbonate phase. Compared to calcite precipitated from seawater (230 ppm Na after Veizer et al. 1978) there is a 10% deficiency in Fe-calcite, thus suggesting a value of 90% of present seawater salinity; this corresponds to the results obtained from the Sr distribution. The magnesium content of 8700 ppm seems to be much lower than that expected for such a salinity. Using Lowenstam's (1961) Mg/Ca molar ratio for seawater of 5.43 and a distribution coefficient of 0.02 (Winland 1969) the parent solution of Fe-calcite probably had a salinity of only 12 to 14 $\mathrm{g\,l^{-1}}$. Applying these salinities of 12 to 14 $\mathrm{g\,l^{-1}}$ into the formula of the evaporation curve (2) oxygen isotope values of −4.42 to 0.05 are obtained for the parent solution.

From Eq. (1) temperatures of 21° to 47 °C are calculated for the formation of Fe-calcite.

It therefore seems, regardless of the wide range, that Fe-calcite was precipitated under slightly elevated temperatures under reducing conditions, as indicated by the presence of ferrous iron. Assuming an average geothermal gradient of 3 °C/100 m and a mean surface temperature of 15 °C, a burial depth of 200 to 1000 m is estimated for the major diagenetic process in the Tunisian sponge reefs. The cementation probably took place during the initial burial phase and seems to have stopped before reaching its maximum burial of 1500 to 2000 m, as indicated by the degree of metamorphism of the organic matter still enclosed in the skeletons (see Sect. 4.1).

Non-ferroan calcite contains considerably smaller amounts of trace elements and the isotope composition is more negative than in Fe-calcite. The absence of ferrous iron in the calcite lattice was most probably caused by oxidizing conditions. These factors suggest mixing of pore solutions with meteoric waters, probably close to the surface. From Sr and Na contents, again fairly consistent in the samples, values can be calculated as being about half the salinity of the Fe-calcite solution. Consequently, lower temperatures of 16° to 29 °C are obtained by applying Eq. (1), which would indicate the formation of this calcite that took place during the last phase of the Cretaceous to Tertiary uplift from depths of 400 m to surface.

4.3 Incongruent Dissolution

Schroeder (1969), and Oti and Müller (1985) have shown that biogenic carbonates are depleted of strontium and magnesium at different rates, and have proposed a process of incongruent dissolution. Walls et al. (1977) go further by stating that the incongruent dissolution of minerals with different solubilities cannot entirely explain the observed dissolution rates of Sr, Mg, and Ca in biogenic carbonates. Walls and co-workers suggested that an "ionic mobility" is the reason for early diagenetic equilibrium with the surrounding chemical environment.

Aragonitic skeletons of Permian sponges seem to be suitable material for testing the model, which had been developed in the laboratory and which has lacked geological confirmation. Furthermore, it also offers the possibility of testing the position of sodium in the sequence of preferential dissolution.

As shown in Table 3, Mg, Sr, and Na were found at quite different concentrations in the skeletons; while the magnesium content is close to the detection limit of 250 ppm, strontium in aragonitic skeletons of samples with high aragonite content attains 9000 ± 580 ppm, and sodium approximately 1700 ± 80 ppm. However, point analyses show that both strontium and sodium in the skeletal aragonite decrease below these values in more strongly altered samples, thereby indicating a relationship between depletion and diagenetic equilibrium.

The incorporation of trace elements by aragonitic sponges does not seem to be influenced by metabolic fractionation if one compares Triassic and Holocene calcareous sponges (Scherer 1977). Thus the difference between the calculated trace element content of aragonite in equilibrium with seawater and the concentration of the same elements in Permian sponges should be a measure of incongruent dissolution and/or ionic mobility.

Based on values of 8500 to 9000 ppm Sr (Kinsman 1969), 2600 ppm Na (White 1977) and 1000 – 1500 ppm Mg (Scherer 1977, Fig. 17) incorporated into aragonite in equilibrium with seawater, the loss in Permian sponge skeletons is approximately 5%, 35%, and 80% respectively. The order of preference is therefore Mg > Na > Sr which is similar to that detected by Schroeder (1969) and Walls et al. (1977) for Mg and Sr. Sodium, though less resistant than strontium, nevertheless retains a significant part of the original concentration. The use of sodium incorporated into the carbonate lattice as a relative paleosalinity indicator in ancient carbonate rocks (Veizer et al. 1977) therefore seems to be a valid method.

The small needles of skeletal aragonite have a large surface area relative to that of blocky Fe-calcite. The skeleton therefore is more susceptible to incongruent dissolution and the exchange of trace elements in lattice positions at, or close to, the crystal surface. In contrast, Fe-calcite crystals have a large volume to surface ratio and, for this reason, will retain their original composition longer after a change in diagenetic environment, and are thus more reliable for paleo-environment analysis.

4.4 The Petrogenetogram

Introduced by Schroeder (1984) the petrogenetogram is a schematic illustration of spatial variations and temporal sequences of genetic/diagenetic products.

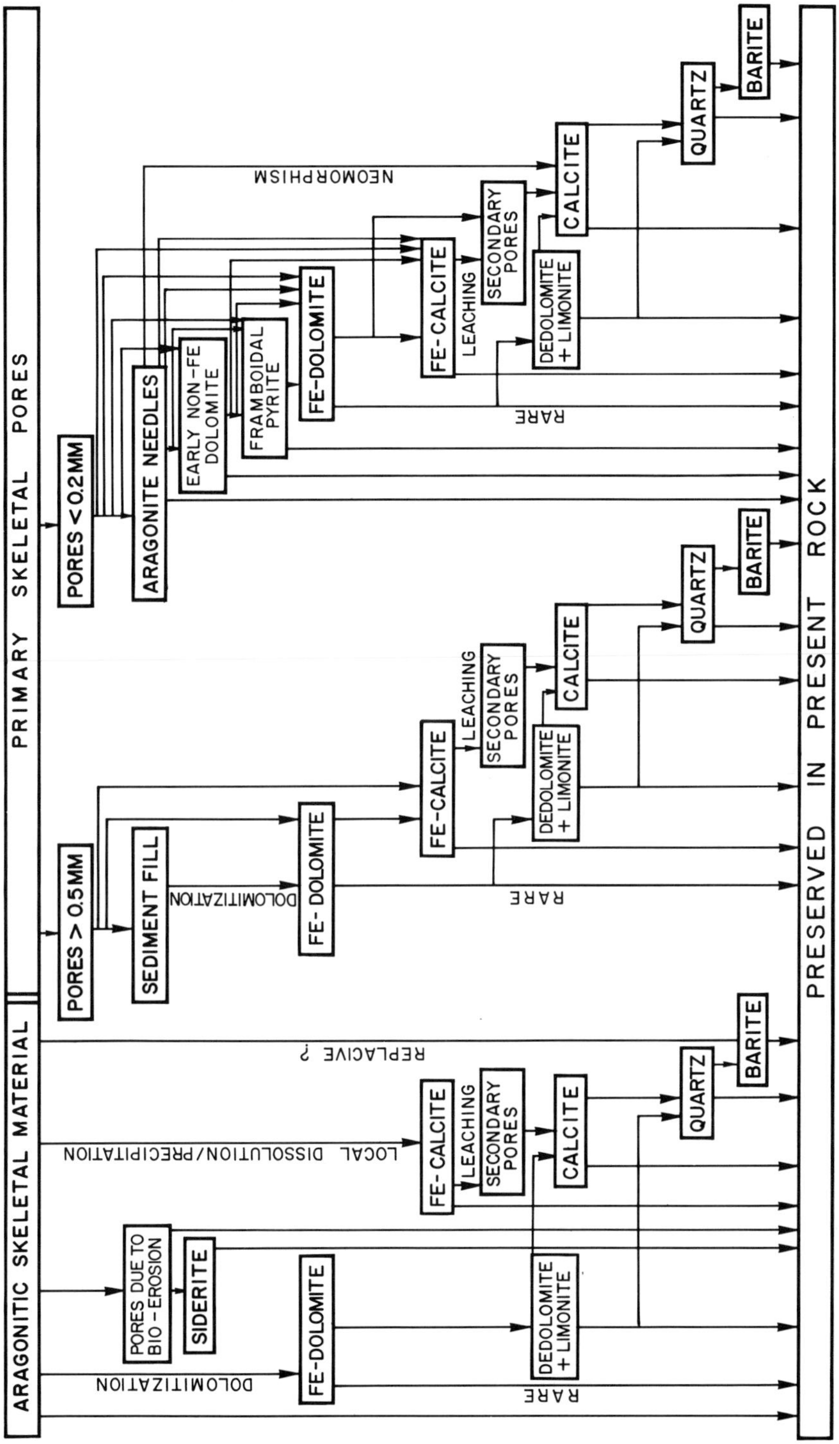

Fig. 8. The petrogenetogram of Permian calcareous sponges from the Djebel Tebaga, Southern Tunisia

In Fig. 8, an attempt is made to visualize various paths or parallel diagenetic sequences. The petrogenetogram demonstrates the intimate relationships between pore systems and skeletons on one side and diagenetic products on the other. The initial pore distribution influenced diagenetic products and sequential arrangements, especially during early diagenesis. Later, however, with more drastic changes in the pore-water chemistry, intensive cementation of the total pore system and hence reduction of overall porosity is more obvious. A number of dissolution phases can be recognized. They varied in intensity and even affected previous cement sequences during the Late Cretaceous to Tertiary period of uplift and weathering.

5 Conclusions

Principle conclusions of this study are:

a) The primary mineralogy of the reef-building calcareous sponges is aragonite of which relicts are still present (up to 44%).
b) Skeletal microstructures can be preserved in detail under certain conditions in spite of several stages of neomorphism. Thus, without supporting chemical analysis, their presence is not necessarily an indication for a calcite primary mineralogy.
c) The preservation of Permian aragonite in the Djebel Tebaga is probably the result of several factors:
   - Rapid reduction of porosity and permeability during Fe-calcite cementation which hampered access of fluid undersaturated in respect of aragonite,
   - Pore-fluid conditions close to aragonite stability caused by arid conditions and marine incursions during Triassic to Jurassic times,
   - Rapid uplift during Tertiary to Holocene above a meteoric groundwater table to the surface where wheathering is slowed down by poor permeabilities and an arid present-day climate.
d) The range of initial pore sizes in skeletons influenced diagenetic products and sequential arrangements.
e) Incongruent dissolution and variations in ionic mobility change the trace element content and isotopic composition of skeletal aragonite. Strontium, however, and, to a certain extent, sodium, are relatively stable in blocky carbonate cement and can be used as environmental indicators although sodium does not substitute for Ca^{2+} in the calcite lattice according to Okumura and Kitano (1986).
f) Close to extinction of rugose corals at the Permo-Triassic boundary and the onset of scleractinian corals in the Triassic, calcareous sponges did not succeed in building large reef structures.

Acknowledgments. I wish to thank J. Wendt, Tübingen, for inviting me to participate in a field trip to Southern Tunisia and for the determination of species. The author is also grateful to Mrs. D. Kost for analytical work. Stable isotope determinations were performed by W. G. Mook, Groningen, and H. Friedrichsen, Tübingen. The study was part of the Special Research Project Paleoecology (SFB 53) and was funded by the German Research Foundation (DFG). Thanks are also given to the Tunisian Goverment for permission to work and to export the samples. Draughting and typing were carried out at Shell UK, Exploration and Production, London. The author also thanks Shell Internationale Petroleum Maatschapij BV, The Hague, for permission to publish.

References

Aharon P, Kolodny Y, Sass E (1977) Recent hot brine dolomitization in the "Solar Lake", Gulf of Elat, isotopic, chemical, and mineralogical study. J Geol 85:27 – 48

Allan JR, Matthews RR (1982) Isotope signatures associated with early meteoric diagenesis. Sedimentology 29:797 – 817

Anderson GM (1976) Error propagation by the Monte Carlo method in geochemical calculations. Geochim Cosmochim Acta 41:511 – 525

Assereto R, Benelli F (1971) Sedimentology of the Pre-Cenomanian Formations of the Jebel Gharian, Libya. In: Gray C (ed) Symposium on the geology of Libya, 14 – 18 April 1969, Tripoli. Univ Libya, pp 37 – 83

Brand U (1982) The oxygen and carbon isotope composition of Carboniferous fossil components: seawater effects. Sedimentology 29:139 – 147

Busson G (1967) Mesozoic of Southern Tunisia. In: Martin L (ed) Guidebook to the geology and history of Tunisia. Petrol Expl Soc Libya, 9th Ann Field Conf 1967:131 – 151

Caceci MS, Cacheris WP (1984) Fitting curves to data – the Simplex algorithm is the answer. BYTE May 1984:340 – 362

Daniel C, Wood FS (1980) Fitting equations to data. 2nd ed. Wiley, New York

Dansgaard W (1964) Stable isotopes in precipitation. Tellus 16:436 – 468

Epstein S, Lowenstam HA (1953) Temperature-shell-growth relation of recent and interglacial Pleistocene shoal-water biota from Bermuda. J Geol 61:424 – 438

Epstein S, Mayeda TK (1953) Variations of the $^{18}O/^{16}O$ ratio in natural waters. Geochim Cosmochim Acta 4:213 – 224

Epstein S, Buchsbaum R, Lowenstam HA, Urey HC (1953) Carbonate water isotopic temperature scale. Geol Soc Am Bull 62:417 – 426

Fleischer E, Goldberg M, Gat JR, Magaritz M (1976) Isotopic composition of formation waters from deep drillings in southern Israel. Geochim Cosmochim Acta 41:511 – 525

Grotzinger JP, Reed JF (1983) Evidence for primary aragonite precipitation, Lower Proterozoic (1.9 Ga). Rocknest dolomite, Wopmay orogen, northwest Canada. Geology (Boulder) 11:710 – 713

Jacobson RL, Usdowski HE (1976) Partitioning of strontium between calcite, dolomite and liquids. Contr Miner Petrol 59:171 – 185

Kinsman DJJ (1969) Interpretation of Sr^{2+} concentrations in carbonate minerals and rocks. J Sediment Petrol 39:486 – 508

Kinsman DJJ (1971) Diagenetic history of limestones determined from Sr^{2+} distribution. In: Bricker OP (ed) Carbonate cements. Johns Hopkins Univ Stud Geol 19:259 – 263

Kinsman DJJ, Holland HD (1969) The coprecipitation of cations with $CaCO_3$ IV. The coprecipitation of Sr^{2+} with aragonite between 16° and 96 °C. Geochim Cosmochim Acta 33:1 – 17

Land LS, Hoops GK (1973) Sodium in carbonate sediments and rocks: a possible index to the salinity of diagenetic solutions. J Sediment Petrol 43:614 – 617

Lloyd RM (1965) Oxygen isotope enrichment of seawater by evaporation. Geochim Cosmochim Acta 30:801 – 814

Lowenstam HA (1961) Mineralogy, $^{18}O/^{16}O$ ratios, and strontium and magnesium contents of recent and fossil brachiopods and their bearing on the history of the oceans. J Geol 69:241 – 260

McCrea JM (1950) The isotopic chemistry of carbonates and a paleotemperature scale. J Chem Phys 18:849 – 857

Mook WG (1970) Stable carbon and oxygen isotopes of natural waters in the Netherlands. IAE A-SM-129/12:163 – 190

Okumura M, Kitano Y (1986) Coprecipitation of alkali metal/ions with calcium carbonate. Geochim Cosmochim Acta 50:49 – 58

Oti M, Müller G (1985) Textural and mineralogical changes in coralline algae during meteoric diagenesis: an experimental approach. N Jahrb Miner Abh 151:163 – 195

Saxby JD (1982) A reassessment of the range of kerogen maturities in which hydrocarbons are generated. J Petrol Geol 5:117 – 128

Scherer M (1974) Submarine recrystallization of a coral skeleton in a Holocene Bahamian reef. Geology (Boulder) 2:499 – 500

Scherer M (1975) Cementation and replacement of Pleistocene corals from the Bahamas and Florida. N Jahrb Geol Paläontol Abh 149:259 – 285

Scherer M (1977) Preservation, alteration and multiple cementation of aragonitic skeletons from the Cassian Beds (U. Triassic, Southern Alps): Petrographic and geochemical evidence. N Jahrb Geol Paläontol Abh 154:213 – 262

Schroeder JH (1969) Experimental dissolution of calcium, magnesium, and strontium from recent biogenic carbonates: a model of diagenesis. J Sediment Petrol 39:1057 – 1073

Schroeder JH (1972) Fabrics and sequences of submarine carbonate cements in Holocene Bermuda cup reefs. Geol Rundsch 61:708 – 730

Schroeder JH (1984) The petrogenetogram of corals: spatial variations in diagenetic sequences. Palaeontogr Am 54:261 – 271

Schroeder JH, Zankl H (1974) Dynamic reef formation: a sedimentological concept based on studies of recent Bermuda and Bahama reefs. Proc 2nd Int Coral Reef Symp 2:413 – 428

Shackleton NJ, Kennett JP (1975) Paleotemperature history of the Cenozoic and the initiation of Antarctic glaciation: oxygen and carbon isotopic analysis in DSDP sites 277, 279 and 281. In: Kennett JP, Houtz RE (eds) Initial reports of the Deep Sea Drilling Project. US Government Printing Office, Washington, p 743

Siegel FR (1960) The effect of strontium on the aragonite-calcite ratios of Pleistocene corals. J Sediment Petrol 30:297 – 304

Tissot BP, Welte DH (1978) Petroleum formation and occurrence. Springer, Berlin Heidelberg New York

Usdowski HE (1973) Das geochemische Verhalten des Strontiums bei der Genese und Diagenese von Ca-Karbonat und Ca-Sulfat-Mineralen. Contr Miner Petrol 38:177 – 195

Veizer J, Demovic R (1974) Strontium as a tool in facies analysis. J Sediment Petrol 44:93 – 115

Veizer J, Lemieux J, Jones B, Gibling MR, Savelle J (1977) Sodium: Paleosalinity indicator in ancient carbonate rocks. Geology (Boulder) 5:177 – 179

Veizer J, Lemieux J, Jones B, Gibling MR, Savelle J (1978) Paleosalinity and dolomitization of a Lower Paleozoic carbonate sequence, Somerset and Prince of Wales Island, Arctic Canada. Can J Earth Sci 15:1448 – 1461

Walls RA, Ragland PC, Crisp EL (1977) Experimental and natural early diagenetic mobility of Sr and Mg in biogenic carbonates. Geochim Cosmochim Acta 41:1731 – 1737

Waples DW (1980) Time and temperature in petroleum formation: application of Lopatin's method to petroleum exploration. Am Assoc Petrol Geol Bull 64:916 – 926

Wedepohl KH (1969) Primäre and diagenetische Strontiumgehalte von Karbonatgesteinen. Ber Dtsch Ges Geol Wiss B Miner Lagerstättenf 14:17 – 23

Wendt J (1977) Aragonite in Permian reefs. Nature 267:335 – 337

White AF (1977) Sodium and potassium coprecipitation in aragonite. Geochim Cosmochim Acta 41:613 – 625

Winland HD (1969) Stability of calcium carbonate polymorphs in warm, shallow seawater. J Sediment Petrol 39:1579 – 1587

Facies Relationships and Diagenesis in Waulsortian Mudmounds from the Lower Carboniferous of Ireland and N. England

J. MILLER[1]

1 Introduction

Waulsortian buildups are dominated by lime mud, and conspicuously lack frame-building organisms such as corals, calcareous algae and stromatoporoids. They were first described by Dupont (1863) from the Dinantian (Lower Carboniferous) of the Waulsort area in Belgium. Dupont had already described certain sparry masses ("Stromatactis") from Devonian mud mounds as stromatoporoids. In 1883, he recognized similar structures in the Waulsortian limestones, and considered these massive buildups to be reefs of stromatoporoids and fenestellid bryozoans. Nowadays, the sparry fabrics are considered to be diagenetic features. Lacking a skeletal framework, Waulsortian buildups are best regarded as carbonate banks rather than true reefs (Lees 1964).

Apart from the type area in the Dinant Synclinorium of Belgium (Lees et al. 1977), the main Waulsortian outcrops are in south and central Ireland (Lees 1964, Sevastopulo 1982) and central and northwest England (Lees and Miller 1985). In North America, examples of Waulsortian buildups include those of the Sacramento Mountains, New Mexico (Pray 1958) and the Big Snowy Mountains, Wyoming (Cotter 1966). Buildups may show evidence of considerable depositional relief above the contemporaneous sea floor, as at Furfooz, Belgium (Lees et al. 1985) and Muleshoe, New Mexico (Jackson and DeKeyser 1984). Such buildups are composite structures, composed of smaller, superimposed lenticular banks with off-set growth centres. Elsewhere, notably in Ireland, the Waulsortian developed as complexes of considerable thickness (over 800 m) and regional spread.

Recent detailed petrographic analyses of Waulsortian buildups have revealed the existence of four depth-related Phases: *A, B, C,* and *D*, representing a range of marine environments from Phase *A*, the deepest (probably more than 300 m), to Phase *D* in the photic zone, about 120 m or shallower (Lees et al. 1985, Lees and Miller 1985). The paleogeographical situation of Waulsortian facies appears to be on the distal portion of carbonate ramps (Lees 1982, Miller and Grayson 1982).

The origin of the lime mud and its ability to sustain steep buildup slopes in the absence of a skeletal framework has been much debated. Lees and Miller (1985) found that muds from Waulsortian Phases *B, C,* and *D* contained abundant moulds of filamentous organisms, possibly cyanobacteria, as well as recognizable blue-green algal genera such as *Girvanella*. However, no remains of

[1] Grant Institute of Geology, West Mains Road, Edinburgh EH9 3JW, Scotland

Reef Diagenesis
Edited by J. H. Schroeder and B. H. Purser

filamentous organisms were found in buildup muds of Phase *A* banks, demonstrating the independence of the fundamental mud-making mechanism from the preserved biota.

Most work on the origin of the Waulsortian facies has concentrated on the mudmound rocks themselves. However, many clues to the origins and diagenetic history of the mudmounds must lie in the relationship of the banks to their enclosing facies, particularly in the details of bank to off-bank transitions. General descriptions of lateral facies relationships were produced by Lees (1964) for the Irish Waulsortian, for buildups in Northern England by Miller and Grayson (1972) and by Lees et al. (1977) for the Belgian Waulsortian. Lees (1964) also described the cement sequences and geopetal sediment successions in Irish buildups. More detailed diagenetic studies of buildups from the Dublin Basin were made by King (1984), but hitherto no systematic comparison of diagenesis in bank facies and off-bank facies has been made.

The Belgian mounds are generally too tectonized for detailed study of lateral facies changes, but many Waulsortian off-bank to bank transitions from Ireland and Northern England have now been examined, results being summarized in this chapter. Despite variations in the nature of the off-bank sediments and the phase of the associated buildup facies, lateral and basal transitions are found to be remarkably consistent, as are many features of their early diagenesis.

2 Material, Methods, and Terminology

2.1 Material

Lateral facies transitions were sampled from Waulsortian banks at Hill 707, near Loughrea, Co. Galway, Ireland. This material is described in detail, and comparisons made where relevant with other localities, as the transition sequences are remarkably consistent. Apart from Hill 707, other Irish localities included Knockadrum, Marble Hill and Knockastia. In Northern England, transitions were studied at Coplow, Great Dunnow, Hall Hill, Knoll Wood, Newsholme, Sugar Loaf and The Ridge. Particulars of these localities are given in Appendix 1 of Lees and Miller (1985).

2.2 Methods

Thin sections were polished on both sides (Miller 1986) and mounted with epoxy resin for cathodoluminescence (CL) examination in a Nuclide ELM2B Luminoscope. The interpretation of CL in carbonates has been discussed by Fairchild (1983), Amieux (1982), Meyers (1974) and Sommer (1972).

Samples for SEM study were smoothed with 1000 grade Carborundum before being etched for 2 – 10 min with a saturated solution of disodium EDTA at room temperature. The chelating action of this solution was found by Miller and Clarkson (1980) to reveal delicate ultrastructure in calcite skeletal material without inducing artifacts due to excessive dissolution. All SEM material was sputter-coated with gold.

Thin sections and acetate peels were stained with Alizarin Red S and potassium ferricyanide using a modification of Dickson's (1969) technique to indicate the distribution of Fe in calcites.

2.3 Facies Terminology

Waulsortian buildups have an envelope of peri-Waulsortian facies. Lees (1964) and Miller and Grayson (1972) preferred the term *bank facies* for the Waulsortian buildup rocks themselves. Peri-Waulsortian facies include *flank*, and *inter-bank* facies, following the usage of Miller and Grayson (1972). Bank facies are often well-bedded, and Lees (1964) introduced the term *bank bed* for the discrete layered units within the banks. At bank margins, outwardly dipping bank beds pass into thin-bedded flank facies, and the transitional zone is the bank *toe* (Fig. 2), a term coined by Whitbread (1963).

3 Bank Margin Transitions, Hill 707, Loughrea, Co. Galway, Ireland

3.1 Bank Margin Geometry

There are four, partly superimposed Waulsortian banks exposed on Hill 707. Their growth centres and growth vectors were mapped by Lees (1964, pp. 495 – 497). The prograding margin of Bank 4 (Fig. 1 A) shows bundles or sets of bank beds steepening away from the growth centre or core, where bedding is less clearly defined. Judging from internal geopetal evidence, many of these marginal beds were deposited on slopes up to 40°. Transitions to thin-bedded crinoidal flank facies occur within distances of 1 – 3 m. Bundles of flank beds are separated by shale partings, and the individual limestone units are themselves bounded by thin shale films. Towards bank toes, shale films are lost and the flank beds progressively amalgamate (Fig. 1 B). The flank : bank bed thickness ratio varies between 1 : 6 and 1 : 20.

3.2 Lithological Changes

At Hill 707, inter-bank facies are thinly bedded, argillaceous spicular wackestones (Fig. 1 C), extensively replaced by chert and dolomite. Approaching the bank, there is a drastic decrease in clay content and an increase in quantity and grain size of bioclastic debris in the flank beds. Flank facies beds are thin (1 – 3 cm) wavy-bedded packstones and grainstones (Figs. 1 B, 2 a). Crinoidal debris is dominant, and there are abundant shreds of fenestellid bryozoans, fragmented "stick" bryozoans, ostracods, moravamminid sponges, crushed mol-

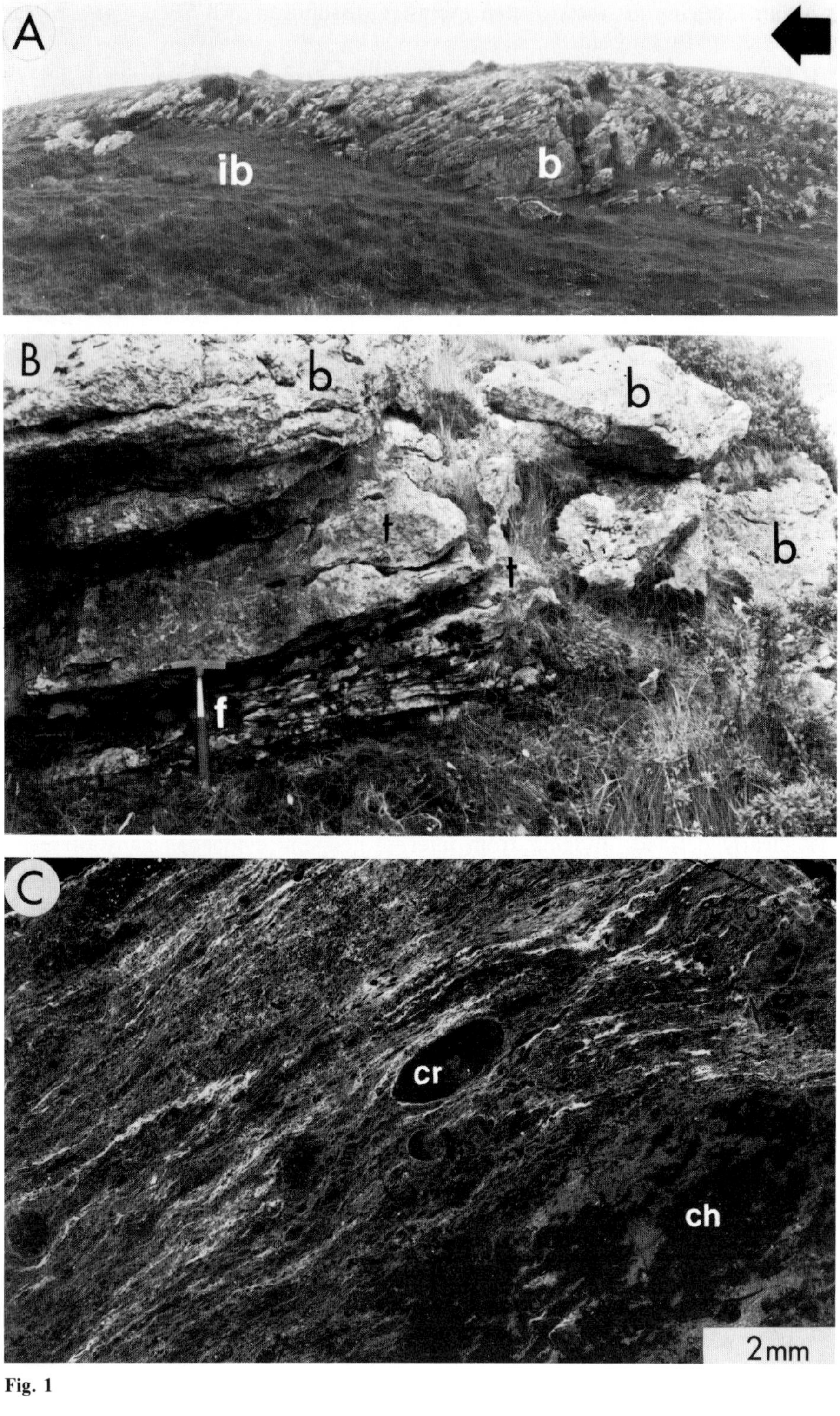

Fig. 1

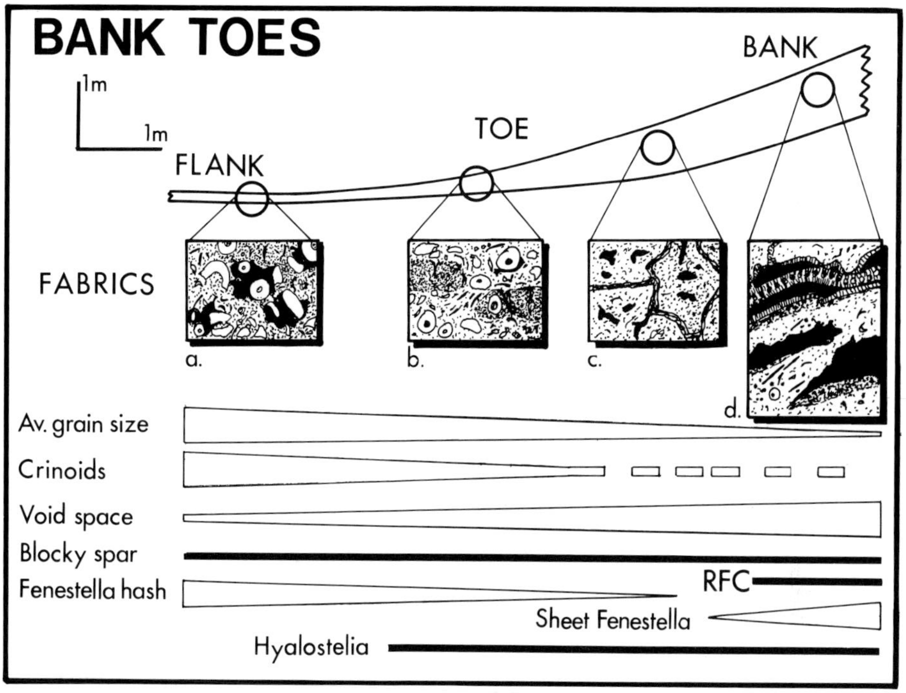

Fig. 2. Main petrographic characters of Waulsortian lateral facies changes. Insets *a – d* drawn from thin sections. Range *boxes* and *bars* represent trends only; not to scale. *RFC* radiaxial fibrous calcite cement, clear spars *black*

luscan and brachiopod shells, together with comminuted rugose coral debris (Fig. 3 B). Bioclasts are poorly sorted and somewhat abraded. Crushing of fossils and distortion of thin shale wisps indicates compaction, and there are abundant stylolites.

Passing from the flank towards the toe of each bank bed, there is a progressive increase of fine-grained comminuted bioclastic debris at the expense of coarser bioclasts (Fig. 2). Within the proximal flank beds, there are discrete, lenticular masses of optically very dense micrite, from a few mm across up to several cm (Figs. 2b, 3 A). These mud lenses are much less fossiliferous than the matrix and often contain moulds of hyalosteliid sponge spicules (Fig. 3 A). Petrographically, the lenses are closely similar to the primary micrites of the bank facies, which Lees (1964) termed the M1 mud generation (Fig. 9). There are no signs that they are detached clasts from the bank.

Fig. 1. A Southwest face of Hill 707, Co. Galway. Waulsortian bank (*b*) prograding in direction of *arrow*. Regional dip negligible, so dips are depositional. Scale figure on right (A. Lees) 2 m. **B** Detail of lateral facies passage from **A**, hammer 25 cm: *b* bank facies passing into *t*, toe and then *f*, flank facies. Note upward sequence from interbank facies (covered, below hammer) through flank, toe and bank facies as bank prograded from right to left. **C** Interbank facies argillaceous spicular wackestone, Hill 707: negative print of thin section; *cr* crinoid; *ch* replacement chert

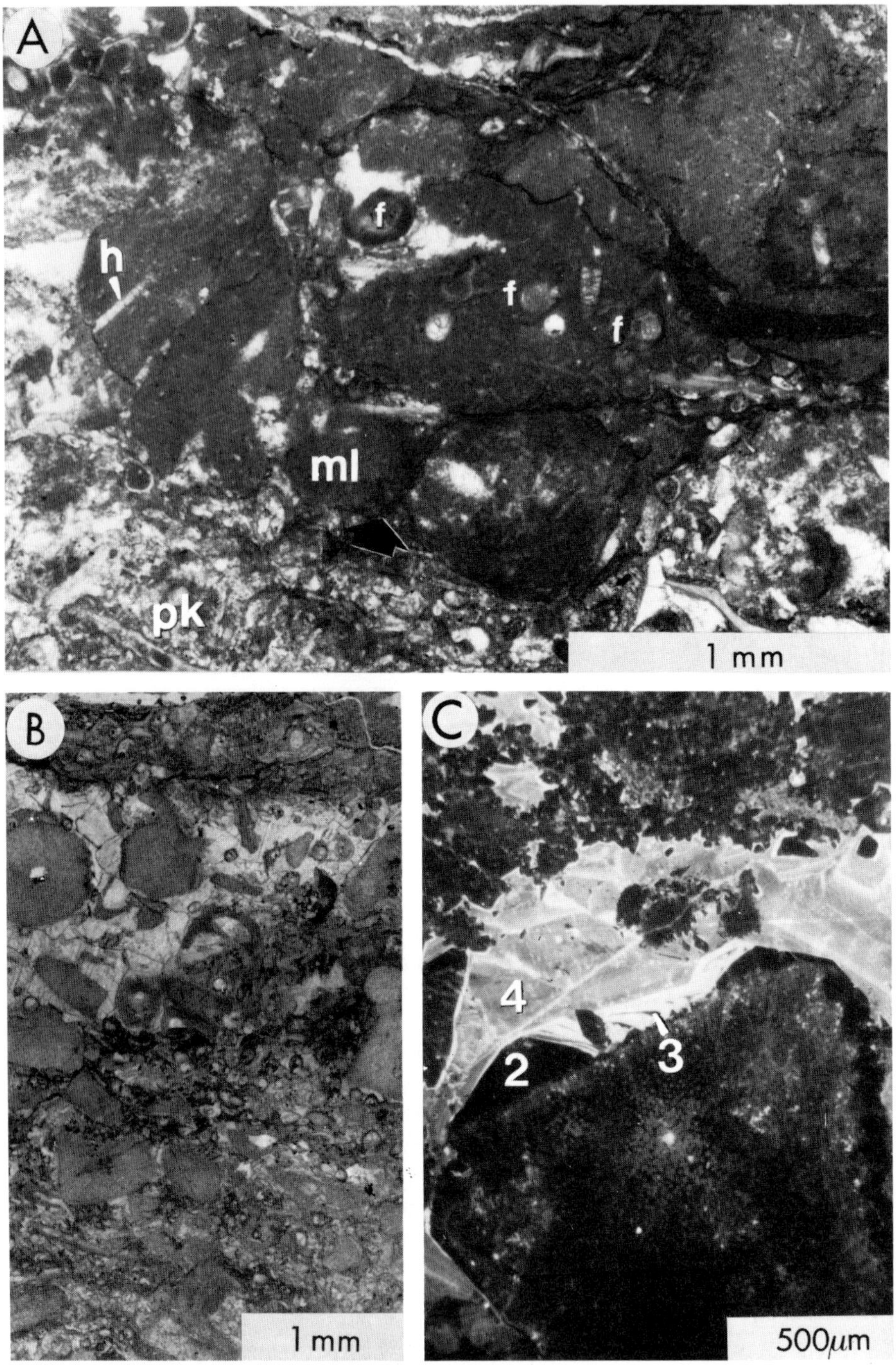

Fig. 3A–C. Flank facies photomicrographs, Hill 707. **A** Mud lens (*ml*) in crinoid-bryozoan packstone, with dense micrite and clear sparry patches, showing sharp margin (partly fossil-controlled), fenestellid bryozoans (*f*) and hyalosteliid sponge spicule moulds (*h*). **B** General view of flank facies in transmitted light, showing areas of extensive dissolution and cementation by clear calcite. **C** CL image of echinoid plate with syntaxial calcite cements, stages numbered. Note the Stage 4 cement in the stereome pores, and some dissolution of the RH margin of the plate

The lenses are mostly sharp-edged (sometimes bounded by stylolites), although soft sediment mixing with the matrix has taken place locally. There is a pronounced compactional drape of bioclastic matrix about them, indicating that the lenses were firm before compaction. Commonly, there are re-fittable crumbly-edged fractures or tears opening on to their outer surface (Figs. 4B, C). These are attributed to pre-compactional dewatering in a firm gel-like substrate, as (1) the fracture walls have torn away from bioclasts rather than across them, (2) the walls are re-fittable, (3) the tears open towards the exterior of the lenses, and (4) bioclasts from the matrix were pulled down into the fractures. The lenses also contain small, irregular, sparry patches (Fig. 3A).

Passing from the toe up the slope of the bank beds, the proportion of mud lenses increases rapidly at the expense of the bioclastic matrix. The lenses become larger and more irregular in shape as they lie in mutual contact. They begin to coalesce (Fig. 2c), producing a characteristic pseudo-brecciated appearance in the field. Thin zones of clay and fine bioclastic material were squeezed up between coalesced mud lenses (Fig. 2d). Again, the generally sharp margins of the mud areas and their crumbly-edged tear fractures indicate that they were firm before compaction and dewatering of the softer clay-rich matrix. Clear, spar-filled voids of irregular shape and variable size are common, but these contain no geopetal sediments.

In bank bed toes, bioclasts may be very much rarer than in the flank and bank beds. However, fenestellid material is much more complete, and hyalosteliid spicule moulds become abundant. Crinoids occur only sporadically.

Further up the bank slopes, the toe pseudo-breccias pass rapidly into more homogeneous wackestones and mudstones. Initially, there is much crumbly-edged fracturing (Fig. 4A), producing re-fittable pull-apart structures probably associated with dewatering of a former gel-consistency substrate. Such pull-aparts may have been influenced, at least partially, by the steep depositional slope. Fenestellid colonies commonly bridge the fractures but are not themselves broken, pointing to the early origin of the fractures in a substrate which was firm but not brittle. The fracture system was subsequently exploited and enlarged by dissolution (see below). Some internal re-organization of the fabric appears to have taken place after the initial fracturing phase: bioclastic calcisilt geopetal sediments (M2) rest on bridges in the primary fracture system. Some of these geopetal fills themselves have crumbly-edged tears (Fig. 4A) indicative of further dewatering. Remaining void space is filled with a succession of cloudy, fibrous and clear blocky calcite cements or with blocky cements alone.

The bank facies wackestones are locally rich in small bioclasts (Figs. 2, 5A). Hyalosteliid sponge spicule moulds are extremely common in the M1 sediment, and are associated with spar-filled moulds of curved filaments 30 – 40 μm in diameter, of unknown origin, but common in Waulsortian micrites from Phases *B – D* (Lees and Miller 1985). In addition to the fenestellid bryozoans, there are ramose bryozoans, trilobites, crinoids, gastropods and ostracods. Clastic material, particularly clay, is extremely rare in the bank beds (apart from shale films between bank beds), in contrast to the abundance of shale in the inter-bank and flank facies. Scattered, very small plagioclase crystals are present, which luminesce bright blue and thus are not authigenic (Kastner 1971). Much rarer

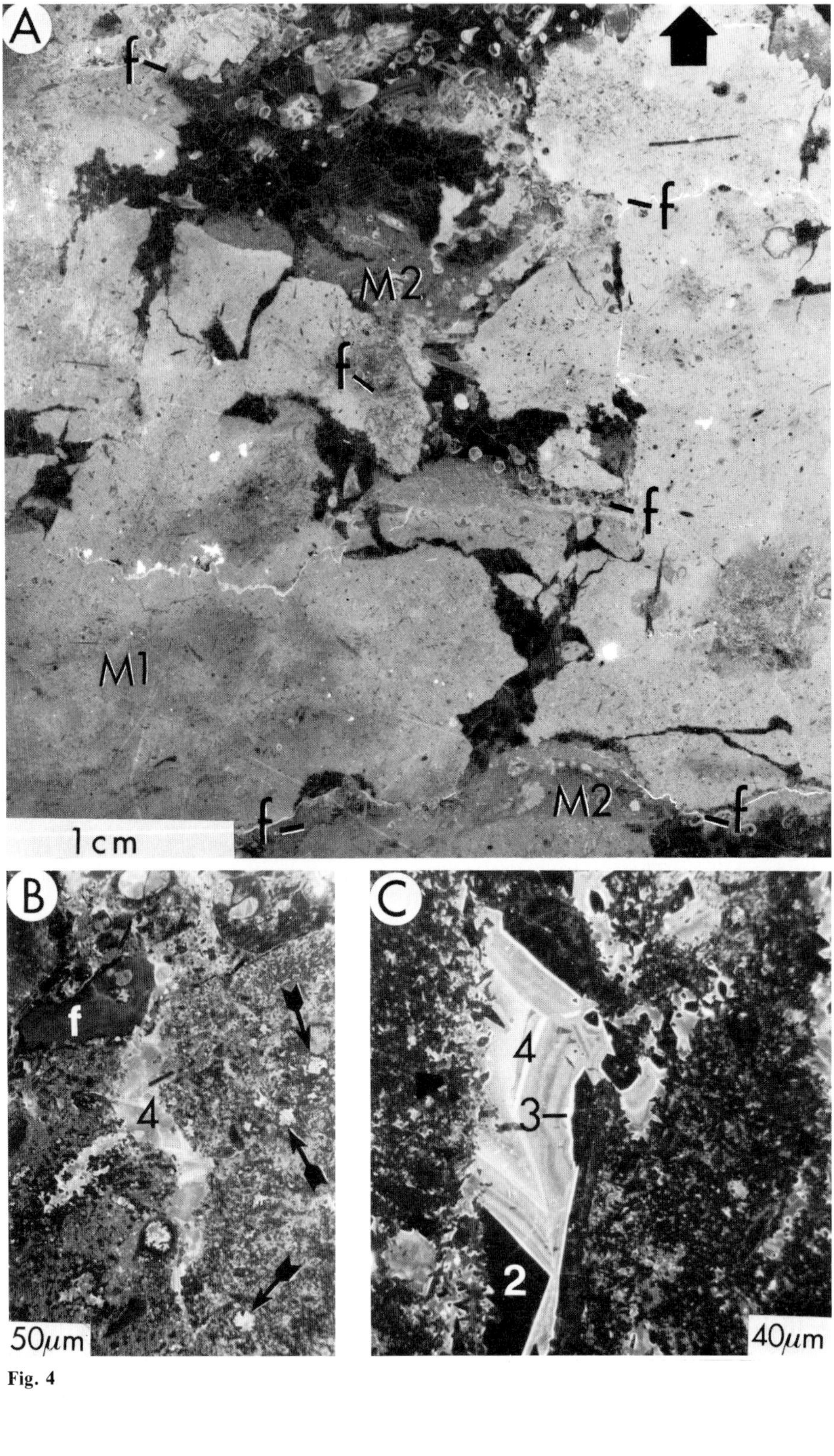

Fig. 4

detrital apatite grains present a bright yellow CL. Otherwise, bank limestones are very pure carbonate.

In the bank facies proper, beds at the prograding bank margin show fabrics that reflect the inclined depositional slope (Fig. 5 A) (Lees 1964, pp. 523 – 527). Spar-filled cavities are prominent features of the bank facies, and passing bankwards from the toes, they increase in size, abundance and complexity (spar may account for 50 – 90% by volume of bank facies). The cavities are often highly irregular, with complex digitate roofs and geopetal floors. Many cavities and any fenestellid sheets crossing them, were centripetally encrusted with radiaxial fibrous calcite (RFC) cements (Figs. 2d, 5 A), before final filling with blocky spars. Individual bank beds often show a regular succession of sparry fabrics and fauna from bottom to top of the bed. These rhythmically developed sequences will be described elsewhere.

3.3 Diagenetic Stratigraphy

3.3.1 Cement Sequence

Using CL and staining, a detailed sequence of Hill 707 cements was established. This sequence provided a framework for relative timing of other diagenetic processes, extending Meyers' (1974) concept of "cement stratigraphy". The cement sequence is divided into a series of stages, each of which represents a pronounced change in crystal habit, origin, mineralogy or CL characteristics. Cement stages may also be demarcated by consistent discontinuities, such as dissolution surfaces.

The most complete cement sequence, comprising six stages, occurs in spar-filled voids from the Hill 707 bank facies (Fig. 6).

Stage 1 consists of isopachous, radiaxial fibrous calcite with very dull and blotchy luminescence. At Hill 707, this cement has been recrystallized. It is slightly cloudy in transmitted light, but seen with SEM (Fig. 7 B), the crystals lack the fibrous appearance, abundant inclusions and curved twin planes of less altered radiaxial fibrous calcite found in nearby buildups (Fig. 7 A). Stage 1 does not normally occur in intraskeletal voids, such as ostracods, brachiopods or sponge spicule moulds, but only in larger cavities within the M1 sediment. This is not due to substrate specificity, as radiaxial fibrous calcite encrusts M1 micrites, geopetal sediments of different grain sizes, and various kinds of skeletal materi-

---

Fig. 4 A – C. Petrography of flank and toe facies, Hill 707. **A** Negative print of thin section from proximal toe facies, *arrow* shows way up. Primary mud (*M1*) with re-fittable crumbly-edged tear system, pulled away from delicate fenestellid bryozoan fronds (*f – f*). Note abundant hyalosteliid sponge spicule moulds (*dark circles* and *ellipses*). Dissolution exploited and extended fracture system. Geopetal sediment (*M2*) rests on bridges. Cements are all clear prismatic or blocky spars (no RFC). Pyrite white. **B** CL image of sinous tear opening onto surface of M1 lensoid in flank facies, mainly filled by Stage 4 cement. Note pervasive dissolution of M1, with voids lined by dark and bright Stages 2 and 3 cement, especially in hyalosteliid spicule moulds (*arrowed*). Top quadrant shows bryozoan packstone matrix of lens, including fenestellids (*f*). **C** CL image of crumbly-edged fracture in M1 lens of bank toe, showing dissolution of wall and penetration of cement stages 2 – 4 into mud matrix

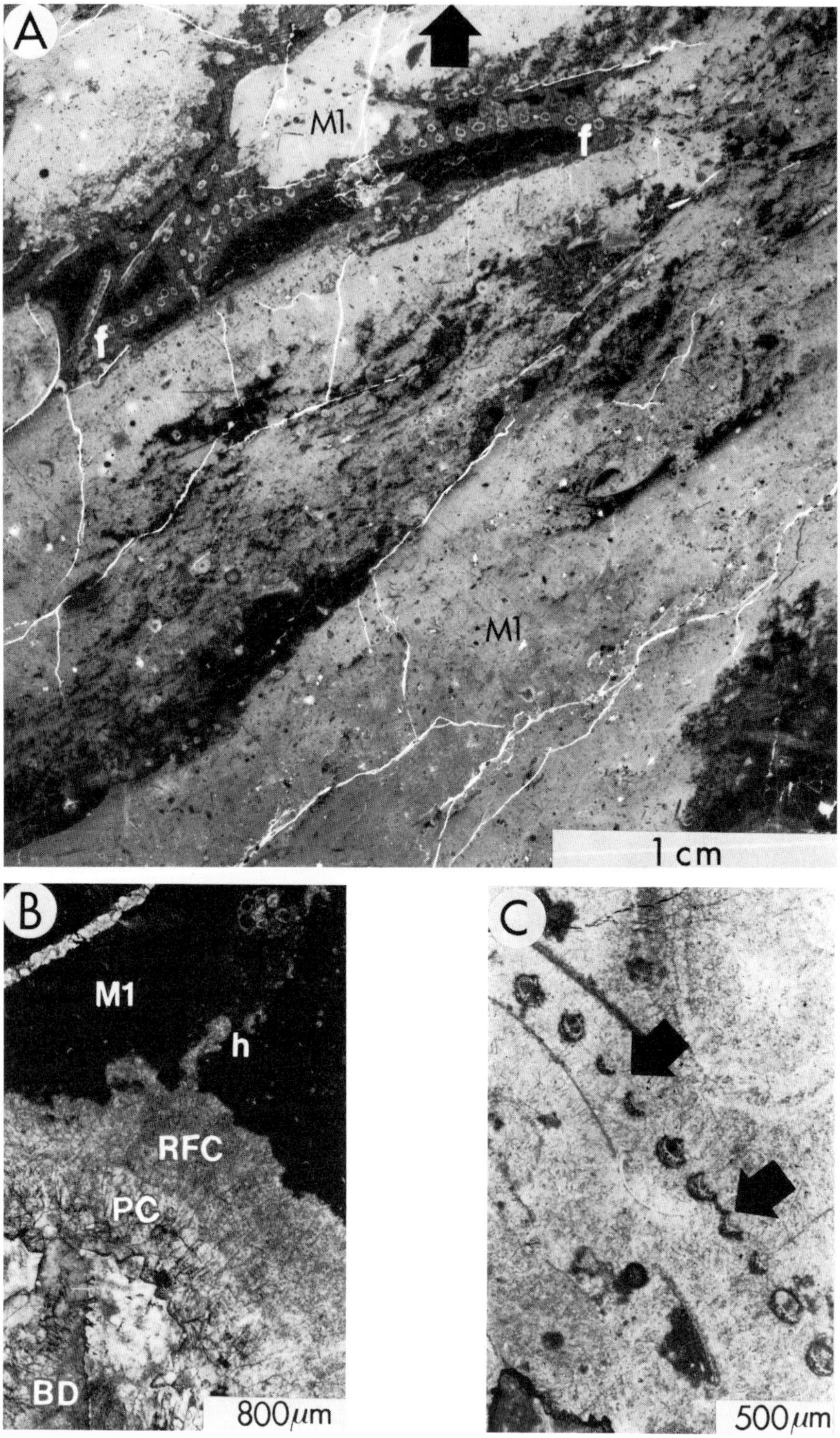

Fig. 5

CEMENT STAGE	CL CHARACTERS	TL CHARACTERS	INTERPRETATION
6	Brick-red, medium luminescent	Baroque ferroan dolomite (BD)	Vein-fed, warm reducing brines, deep burial
5	Non-luminescent	Quartz (QZ)	Vein-fed, deep burial
4	Medium orange-brown zoned	Clear, blocky, ferroan calcite	Vein-fed, reducing zone, burial stage
corrosion			
3	V. bright orange-yellow may be zoned	Clear calcite, syntaxial with 2	Marine phreatic precipitate, reducing zone
corrosion			
2	Non-luminescent	Clear, prismatic low-Mg calcite (PC)	Marine phreatic precipitate, oxidizing zone
1	Non-luminescent or dull, blotchy	Cloudy RFC (radiaxial fibrous calcite)	High-Mg marine phreatic, oxidizing zone

Fig. 6. Cement sequence, Hill 707, Co. Galway

al. Furthermore, Stage 1 radiaxial fibrous calcite cements are not found outside the bank facies: they have never been observed in any Waulsortian toe, flank or interbank lithology, and are thus facies-specific. Locally, geopetal sediments may lie directly on Stage 1 cements (e.g. Lees 1964, p. 526), showing that void-fill cementation and sedimentation processes were often coeval.

Cement Stage 2 consists of clear, inclusion-free, prismatic, non-luminescent non-ferroan calcite. The crystals are generally euhedral, with rhombohedral and scalenohedral terminations. In flank facies, Stage 2 cements are most strikingly developed as syntaxial overgrowths on echinoderm material (Figs. 3B, C). Stage 2 cements were locally corroded and etched (Fig. 7E) before deposition of Stage

Fig. 5A–C. Bank facies petrography, Hill 707. **A** Negative print of thin section, *arrow* indicating way up, showing primary depositional fabric. Primary mud (*M1*) contains voids supported by fenestellids (*f–f*), centripetally lined with RFC (*grey*) and filled with blocky spars (*black*). Many other voids are dissolution or collapse features and lack Stage 1 RFC. **B** Photomicrograph of void wall in M1 showing scalloped dissolution surface, opening into hyalosteliid spicule mould (*h*), filled with RFC, prismatic calcite (*pc*) and Stage 6 baroque dolomite (*bd*). **C** Photomicrograph showing dissolution of fenestellid frond (*arrowed*) and bivalve shell traversing void in bank facies; all encrusted with Stage 1 RFC

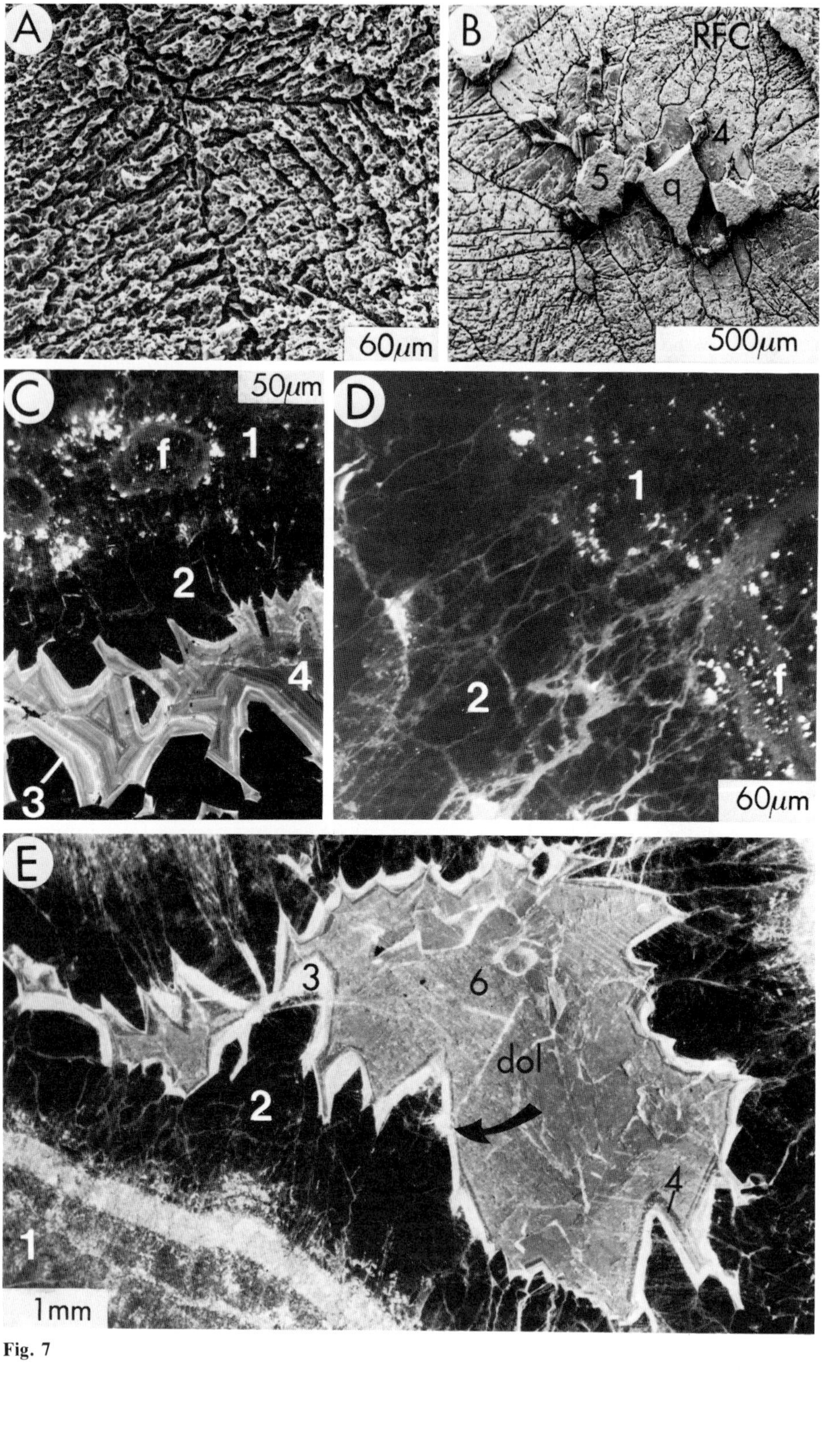

Fig. 7

3. Stage 3 is syntaxial with the Stage 2 prisms and is distinguished by its very bright yellow-orange CL. Framboidal pyrite is often associated with the Stage 2–3 boundary. Where Stage 3 is thickest, it comprises several bright and dull zones (Fig. 3C). However, Stage 3 is often very thin and locally absent (Fig. 7C), probably removed in the significant corrosion which preceded Stage 4 cementation.

Stage 4 is the first cement to be introduced by veins. It has a medium bright orange-brown to dull brown CL typical of Fe-quenching, and is conspicuously zoned (Fig. 7C). The veins which carried the Stage 4 fluids anastomose (Fig. 7D), invading intercrystal boundaries and cleavage planes of earlier cement stages, resulting in local brecciation or recrystallization, especially of the Stage 1 radiaxial fibrous calcite.

Stage 5 is a quartz cement (Fig. 7B). Stage 6 consists of baroque dolomite, with a medium bright brick-red CL (Figs. 5B, 7E). Both are vein-fed.

This carbonate cement sequence conforms to the models proposed by Frank et al. (1982) and Grover and Read (1983), in which cement stages reflect decreasing redox potential of pore fluids with time and burial. Inclusion-rich Stage 1 radiaxial fibrous calcite cements are interpreted as marine phreatic precipitates within the oxidizing zone (Kendall 1985). Many of the inclusions in these cements are microdolomites (Fig. 7A), suggested by Lohmann and Meyers (1977) to be indicative of former high-Mg marine calcites. Inclusion-free, low-Mg Stage 2 calcite crystals are non-luminescent, indicating precipitation within the oxidizing zone, where Mn^{3+} and Mn^{4+} ions could not be incorporated into the lattice. A change to reducing conditions in the stagnant pore waters is signified by the very bright luminescence of Stage 3 cement, where Mn^{2+} was incorporated as CL activator into the calcite lattice, but without quenching by Fe^{2+}. Blue-stained ferroan Stage 4 cements record more strongly reducing conditions (Grover and Read 1983). The fluids from which Stage 4 cements precipitated were injected via brittle fractures, probably late in the burial history. These cements incorporate Mn^{2+}, but the activator is quenched by Fe^{2+}. The fine zonation of Stage 4 cements suggests fluctuating Eh/pH conditions during precipitation. Evidently the injected fluids were also initially undersaturated with respect to carbonate, as there is marked corrosion of underlying Stage 3 or earlier cement (Fig. 7D).

Minor silicification occurs in the M1 sediments of the flank lenses, toes and bank beds, where silica also replaces bioclasts. Replacement chert is extensive in inter-bank facies (Fig. 1C). The timing of this silicification apparently corresponds with the Stage 5 quartz cement (Fig. 8). The source of the silica cannot

Fig. 7A–E. Photomicrographs of bank facies cements. **A** SEM view of etched RFC from Marble Hill, Co. Galway, showing fibrous nature, abundant inclusion pits and curved cleavage planes. **B** SEM view of etched void from Hill 707 showing recrystallized RFC followed by Stage 4 cement and final void-fill of quartz (*q*) Stage 5. **C** Cross-section of void in CL, showing fenestellid (*f*), blotchy Stage 1 RFC, prismatic Stage 2 and 3 and zoned Stage 4. **D** CL view of Stage 4 vein-fed cement forcing its way along inter-crystal boundaries of Stages 1 and 2, causing fracture and dissolution of these earlier cements. Bright spots are clusters of small dolomite rhombs. **E** CL image of void cements, showing blotchy Stage 1 RFC, much invaded by fine veins of Stage 3 and traversed by a vein of Stage 4. Note dissolution of Stage 2 by Stage 3 (*arrowed*), and penetration of Stage 3 between crystals of earlier cements. Zoned Stage 4 very thin, probably dissolved by Stage 6 baroque dolomite (*dol*) fluids

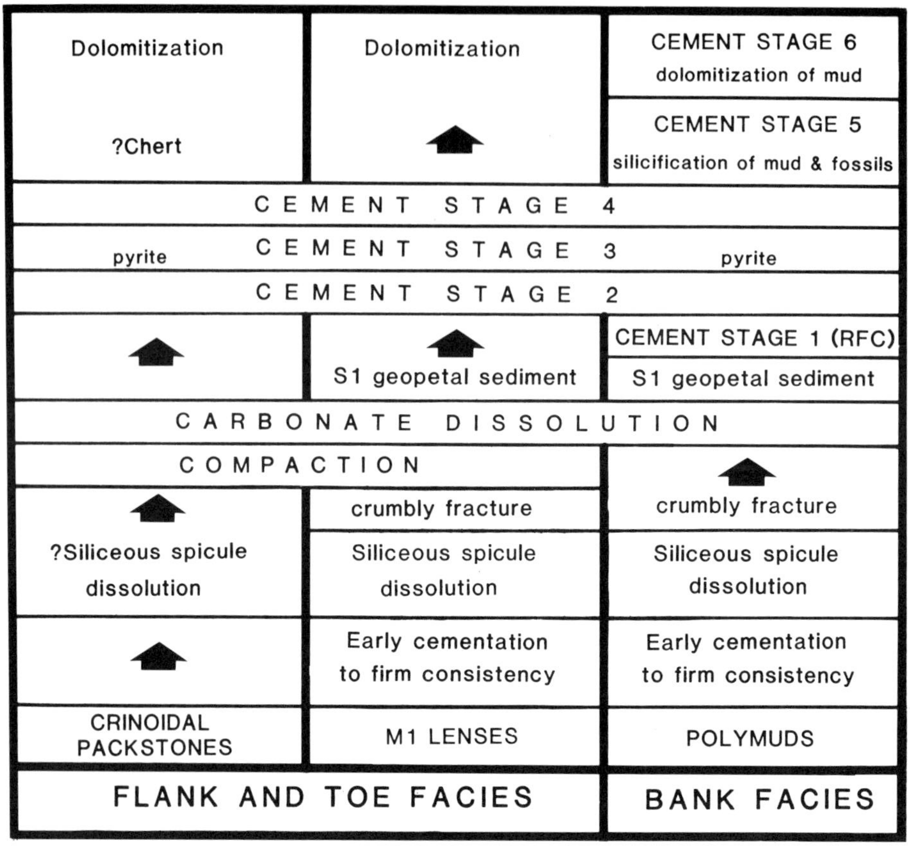

Fig. 8. Simplified diagenetic stratigraphy for flank, toe and bank facies, Hill 707, Co. Galway. Stylolitization and later phases of neomorphism of M1 muds omitted for clarity

have been sponge spicules, dissolution of which took place before Stage 1 cement.

Baroque dolomites (Stage 6 cement) have been interpreted by Choquette (1971), Radke and Mathis (1981) and Grover and Read (1983) as products of warm, reducing brines: together with the quartz cement and silicification, they are considered here to be products of burial diagenesis. Small, brightly luminescent dolomite rhombs are scattered in the micrite matrix of the bank facies (Fig. 9C) and they also probably developed from the same source brines as the baroque dolomite. The cherty, argillaceous inter-bank facies at Hill 707 are also extensively affected by patches of dull-medium luminescent dolomite.

Later stages of fracturing and injection of reducing fluids at Hill 707 are recorded by fine, medium-luminescent, blue-staining ferroan veinlets, which ramify through the muds and invade intercrystal boundaries and cleavage planes (Figs. 7D, 9B). These veins cross-cut all cement stages (except perhaps the quartz and baroque dolomite), but do not give rise to a distinct cement generation, as no unfilled voids remained at the time of their emplacement.

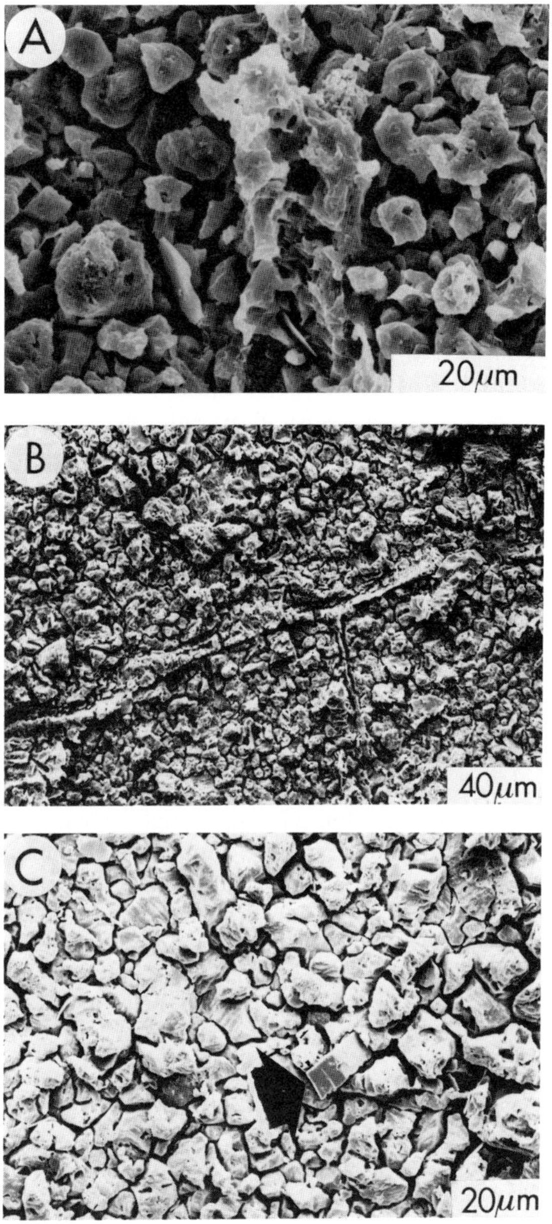

Fig. 9A–C. SEM photomicrographs of etched M1 sediments, Hill 707, Co. Galway. **A** M1 from mud lens, proximal flank facies; note inclusions in many grains and central vein of calcite cement. **B** M1 from toe, showing small inclusion-rich grains and larger, irregular cement patches with higher relief. Anastomosing vein of Stage 4 cement. **C** Crystal mosaic of M1 from bank facies, showing small inclusion-rich grains and irregular, cleaved cement blebs. Late replacement dolomite crystal *arrowed*

3.3.2 Micrite Diagenesis

There is little direct evidence for the original nature of any of the M1 muds, or on the earliest phases of their cementation. The muds are too fine-grained for high-magnification CL study with a cold cathode luminescence machine, and in any case they have been patchily altered by dissolution and neomorphism. SEM reveals that the M1 sediment consists of an interlocking mosaic of crystals from 2 – 20 μm across. Grain size and crystal shape of M1 from flank/toe lensoids and from bank facies are very similar (Fig. 9). The smaller, equant crystals are rich in blocky inclusions, probably microdolomites, suggesting that the M1 crystal aggregate is not solely a neomorphic product. Cowan (1980) found similar inclusions in M1 sediments from Waulsortian buildups in New Mexico, and King (1984) reported them from Waulsortian banks of the Dublin Basin. Possibly, these smaller inclusion-rich M1 crystals represent former high-Mg precursor grains. Larger, irregular crystals (Fig. 9C) have the character of blocky spar cements, which appear from CL to be Stages 2, 3, and 4. These cements filled small voids which developed in the M1 mud during a pervasive pre-Stage 1 dissolution phase (Fig. 4B, C), and are therefore indicative of a secondary porosity. Further cementation and neomorphism of M1 mud in the vicinity of burial stage veins is demonstrated by CL. The patchy luminescence of M1 can thus be partly related in colour and intensity to cement Stages 2 – 5, following a complex set of processes including dissolution as well as local neomorphism of micrite grains during their equilibration with ambient fluid compositions. Bioclasts show comparable variations in CL which can similarly be related to the cement stratigraphy. Dolomite rhombs grew in the muds at a late stage (Fig. 9C), and there are also patches of replacement silica, although true chert is absent from bank and toe beds. Significantly, the muds remained diagenetically "open" enough for these late stage fluids to affect them, as observed in New Mexico bioherms by Meyers et al. (1982).

The diagenetic overprinting of the primary Waulsortian muds makes it impossible to distinguish any synsedimentary marine cements which may have contributed to the initial firmness of M1 lenses and steep bank bed slopes. Evidence for synsedimentary cementation is thus circumstantial, relying on textural features as well as the evident stability of these bank margin slopes. Cowan (1982) and King (1984) also failed in their attempts to establish the nature and degree of early marine cementation in Waulsortian M1 sediments by direct petrographic means. However, Meyers et al. (1982) performed stable isotope analysis on New Mexico bioherm muds and detected a signature indicative of marine cementation.

3.3.3 Early Dissolution Phases

A most significant aspect of the diagenetic stratigraphy (Fig. 8) is the presence of two periods of dissolution (one of silica, one of carbonate) which pre-date the first recorded stage of cementation (Stage 1). Hyalosteliid sponges, common in M1, are prominently represented by calcite-filled moulds of their large supporting rod-bundle spicules, up to 15 cm long and over 1 mm in diameter. There are also patches of microscleres from the main body of the sponge. The rod-bundle spicules were formerly siliceous, bearing a central canal. This mouldic preservation occurs in all the Waulsortian so far examined. It contrasts with the situation observed in non-Waulsortian facies, where replacement of the silica spicules by calcite is the rule, the outline of the central canal remaining. In the toe bed micrite lensoids, the spicule moulds contain Stage 2 and 3 cements (Figs. 4B, C). Where subsequent carbonate dissolution caused moulds to open into large cavities in the bank beds, they may be filled with Stage 1 radiaxial fibrous calcite cement (Fig. 5B). Silica dissolution thus took place before Stage 1 cementation. The large spicule moulds also show that the M1 matrix was firm enough to support the voids at the time of silica dissolution. Thus the silica dissolution post-dates any presumed synsedimentary cementation of the M1 sediment, but pre-dates the Stage 1 radiaxial fibrous calcite (Fig. 7). Land (1976) noted a similarly early dissolution of opaline spicules in Recent coral reefs, but these spicules were very much smaller than those of *Hyalostelia.* Such early dissolution of silica may be attributable to a pH increase during initial diagenesis of organic matter (Birnbaum and Wireman 1984). However, Friedman et al. (1976) attributed early dissolution of quartz in some Recent carbonate sediments to pH fluctuations caused by algal metabolism.

The next dissolution phase involved carbonate, including aragonitic molluscan shells, the moulds of which are filled with Stage 2 and 3 cements. Significant volumes of secondary porosity were created in both bank bed micrites and flank facies. The dissolution followed the development of the crumbly edged dewatering fractures in the bank bed micrites and flank lensoids, and to some extent enlarged and exploited these existing fracture systems (Fig. 4). It also enlarged fenestellid shelter voids and commonly attacked *Fenestella* fronds where they crossed such voids (Fig. 5C), as well as truncating other bioclasts. The dissolution voids are characterized by the exploitation of existing fabrics (Figs. 4B, C) and their frequently scalloped and deeply embayed roofs (Fig. 5B). Often, these secondary cavities lack internal geopetal sediments. The long hyalosteliid rod-bundle spicule moulds often appear to have acted as conduits for fluids involved in the dissolution, and later allowed ingress of the fluids from which Stage 2 and 3 cements subsequently precipitated. The patchy nature of the intergranular and void space thus created suggests that less well-cemented areas of the micrites were selectively enlarged.

Although this early dissolution phase is an extremely common phenomenon in the Waulsortian, it has not been recorded. It always predates the earliest recognizable sparry cement stages (Fig. 4B, C), and demonstrates that the early diagenesis of Waulsortian facies was not merely a history of progressive cementation, as previously imagined. In order to effect the dissolution, the marine pore waters within the buildup and flanking sediments must have either become undersaturated for carbonate, which is unlikely, or were lowered in pH. Changes in pH could have been a further consequence of bacterial degradation of certain kinds of organic matter within the micrites (Birnbaum and Wireman 1984).

In the case of Hill 707, it is not possible to trace the effects of early silica or carbonate dissolution into inter- or sub-bank facies owing to the high degree of compaction and chertification of these sediments. Elsewhere, for example in the North of England, interbank facies do not show signs of early carbonate dissolution. Furthermore, interbank hyalosteliid spicules are preserved by calcite replacement rather than being mouldic, and both early silica and carbonate dissolution phases were seemingly restricted to the banks and their immediate vicinity. Land (1976) noted a similar restriction of opal dissolution to Recent coral reefs as compared with off-reef areas.

In the diagenetic stratigraphy (Fig. 8), the deposition and early consolidation of M1 muds is separated from Stage 1 radiaxial fibrous calcite cementation by two dissolution phases, internal fracturing, void formation and some internal sediment generation. Therefore, Stage 1 cement was not contemporaneous with the synsedimentary cement which is assumed to have provided initial firmness of buildup muds. Radiaxial spars thus had little significance in providing rigidity to the bank beds during their accretion phases, as suggested by Pray (1958), among others.

4 Lateral Transitions Elsewhere

Lateral transitions were examined in other Irish Waulsortian buildups (Knockadrum, Marble Hill and Knockastia), which, together with Hill 707, are assigned to Waulsortian Phases *A* and *B* (Lees and Miller 1985, Fig. 6). In the North of England, lateral transitions were studied at Coplow Quarry, The Ridge, Hall Hill, Sugar Loaf, Great Dunnow, and Newsholme: these buildups have the characteristic component assemblages of Waulsortian Phases *C* and *D*. They are considered to have been deposited in somewhat shallower water than the Irish examples cited; Waulsortian Phase *D* apparently originating in the photic zone (see Lees and Miller 1985).

Despite the different phases in these buildups, lateral facies transitions of prograding buildups are remarkably consistent. They closely follow the pattern seen at Hill 707, in their development of M1 lenses in proximal flank facies and coalescence of those lenses in the toe-bank passage. Differences lie mainly in the abundance and diversity of fossils in the associated flank and interbank facies, with diversity being greatest in the higher phases (i.e. *C* and *D*). There is also some variation in the distance over which the transitions from flank to bank take place, ranging from several metres to extremely abrupt transitions within 1 m. Nevertheless, there is confirmation for the conclusion of Lees and Miller (1985) that the fundamental mud-generating mechanism of Waulsortian banks was largely independent of the phase and thus the depth at which the mounds formed.

Early diagenetic trends were also closely comparable to those of Hill 707, including the early silica and carbonate dissolution phases. The main differences involve the complexity of post-RFC cement sequences, where some cement stages may have several substages (D. Gillies, personal communication), and in details of the burial history. Such late diagenetic histories record regional events and have no direct relationship with Waulsortian bank origins.

In Phase *D* banks, there is often a cement stage not seen at Hill 707. It predates the radiaxial fibrous calcite cement (which is always the first stage in other phases). Botryoidal, fibrous layers of cloudy calcite form successions of isopachous growths on bryozoans and cavity walls (Fig. 10B). Each cement layer is micritized and microbored, and sometimes encrusted with blue-green algal filaments, indicating that the void system opened on the buildup surface to admit the encrusting microbiota. Radiaxial fibrous calcite cement always follows the

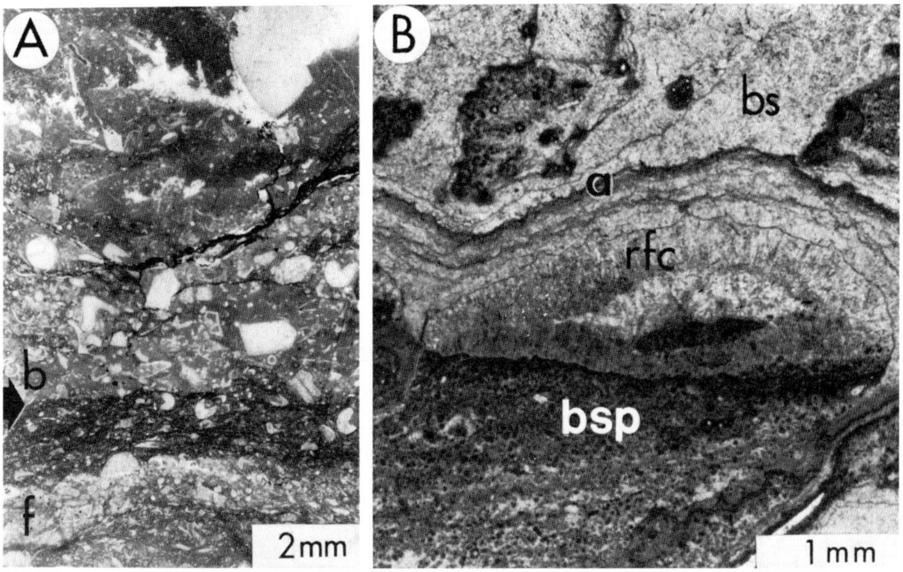

Fig. 10. A Photomicrograph of basal transition from sub-bank facies (*s*) to bank facies (*b*), Coplow Quarry, N. England. Note upward reduction of crinoid debris and development of void fabric near top. **B** Photomicrograph of Phase *D* void cement sequence, Knoll Wood, N. England. Botryoidal spars (*bs*) with micritized surfaces are crusted with algae (*a*) and followed by radiaxial fibrous calcite (*rfc*). At the base of the cavity, botryoidal spars (*bsp*) include abundant pellets

botryoidal cement; it is never micritized or algally coated. This simplies that radiaxial fibrous calcite cement was formed only in void systems which, although interconnected, had no large-scale openings on the buildup surface, contrary to the open framework model of cemented crusts postulated for buildup structure by Pratt (1982). There are no traces of the botryoidal cements or their correlatives in the associated Phase *D* interbank, flank or toe beds, where the earliest cements are clear low-Mg calcites equivalent to those stages which immediately post-date radiaxial fibrous calcite cement as in other buildups.

5 Basal Facies Transitions

Three transitions from sub-bank to bank facies have been studied petrographically (Knockadrum, Coplow Quarry and Newsholme Quarry). They are all remarkably abrupt. Sub-bank facies are often lithologically similar to interbank facies, and are usually argillaceous. Recapitulating the general trends shown in lateral passages (Fig. 2), shale decreases towards the buildup base, while individual beds become paler in colour and similar to flank facies, i.e. crinoid/bryozoan grainstones, packstones or wackestones. M1 muds appear very suddenly, usually with a noticeable colour change, as small blebs packed together with bioclasts (Fig. 10A). Within a few millimetres, shale is completely excluded, bioclast abundance is reduced and fully developed M1 muds appear, containing clear spar-filled dissolution voids (Fig. 10A). In a further few millimetres or centimetres, fenestellid sheets arrive and begin to control the geometry of voids (cf. Lees 1964, Fig. 12F), which are lined with centripetal radiaxial fibrous calcite crusts. The sub-bank to bank transition is completed within a metre, but the appearance of M1 is extraordinarily sudden, and confined to a single horizon. Basal transitions, like lateral transitions, show the same diagenetic trends and do not appear to vary with the Waulsortian Phase.

6 Depositional Environments and Buildup Origins

Lateral facies relationships give some clues as to buildup environments. Poorly sorted bioclastic packstones and grainstones with comminuted and fragmented fossils in the flank facies indicate mobile substrates with moderate current activity and frequent reworking, at least in the periphery of the buildups. Fine sediment, both clay and micrite, was mostly winnowed out from the coarser beds, remaining in protected sites in the lee of large skeletons and as sediment fills in skeletal voids. These periods of fairly high energy alternated with quieter episodes, represented by thin shale layers, upon which bryozoan colonies (fenestrates and ramose types) grew, together with hyalosteliid sponges (the latter absent from Phase *A*). These fossils were buried in situ, probably by suspension fall-out, and are preserved intact (apart from compression) in the shales. Pulsatory sedimentation is reflected not only in the shale-grainstone alternations, but also in the tendency towards "bundling" of bank beds at the advancing bank margin (Fig. 1A, B).

Towards bank margins, M1 lenses developed in the skeletal grainstones. They appear to represent localized centres of lime mud formation and are petrographically comparable to the bank bed M1 micrites. The optically dense, formerly high-Mg mud of the lenses is quite different from associated biomicrite patches within the flank grainstones. The bankwards increase in number of lenses and their coalescence in bank toes indicates that they must have been sites of active production of carbonate mud within the host sediment, possible by non-filamentous cyanobacteria. At bank toes, M1 muds became dominant, altering the nature of the substrate, partially excluding crinoids and some of the other typical flank biota but encouraging increased populations of hyalosteliid sponges. This localized high carbonate productivity presumably was responsible for originating the toe slopes and thus inducing bank margin advance.

Monty et al. (1982) reviewed petrographic and experimental evidence bearing on the origin of Frasnian (Devonian) carbonate mudmounds from Belgium, and concluded that the lime mud of these buildups was microbial in origin, with surficial organic films stabilizing the surface and producing gel textures. A similar origin has been postulated for Waulsortian lime-muds (Lees and Miller 1985), who ruled out other origins such as trapping and binding by fenestrates (Pray 1958, Philcox 1971) or by cryptalgal mats (Pratt 1982). The generation and amalgamation of micrite lenses in Waulsortian flank-bank transitions refutes the mere trapping and binding of detrital components as the main means of bank accretion, but instead emphasized in situ production of a distinctive M1 sediment restricted to the lenses and bank facies. It is noteworthy that the high shale content of interbank facies is abruptly curtailed close to the banks, and there is never any detrital clay within cavities in the bank facies, precluding a simple trapping mechanism.

Bathurst (1980, 1982) envisaged physicochemical cement precipitation as being important in mudmound growth, forming a succession of lithified crusts with intervening layers of soft sediment. The unconsolidated mud was supposed to have been winnowed, leaving cavities. This, of course, still begs the question of the origin of the initial mud. It is difficult to reconcile Bathurst's model with the lateral facies transitions described herein, and the absence in the diagenetic sequence of any evidence for the postulated lithified crusts. Textural indications (differential compaction of micrite lensoids, lack of micrite flow deformation on bank slopes, unfilled skeletal and interskeletal voids, paucity of bioturbation or mechanical reworking, predominance of crumbly-edged tears rather than brittle fractures) all point to a firm gel substrate consistency for M1 muds. Such consistencies are associated with fine-grained carbonates having very low clay contents (Zankl 1969) and need not imply initial cement contents of more than a few percent by volume. Brittle fracturing did not take place until late in diagenesis, and is associated with veins and inclined stylolites.

There is a complete lack of intraclasts from bank facies in adjacent flank or interbank facies, no true sedimentary breccias at the foot of steep bank margins, and remarkably little evidence of penecontemporaneous erosion of bank sediments. Slumping or sediment flow did not take place down the inclined bank slopes, so the accreting bank sediments were more cohesive and firm than the plastic consistency of highly compacted flank and interbank sediments.

The circumstantial, but strong, evidence for a degree of penecontemporaneous cementation of bank facies and M1 lenses in flank beds has to be reconciled with the lack therein of recognizable submarine cement stages equivalent to (or earlier than) the cloudy, formerly high-Mg calcites in the bank facies. Correlatives of these cement types are missing from interbank, flank and toe beds in all Waulsortian transitions so far examined. I conclude that environmental factors, such as temperature and pressure, were unfavourable for marine physicochemical cement precipitation at the depths where Waulsortian facies formed. Today, deep water situations may favour precipitation of clear low-Mg calcite rather than high-Mg calcite and aragonite, typical of shallower waters. Stow and Miller (1985) recorded some luminescent clear low-Mg marine overgrowth cements from Cretaceous calci-turbidites near the Walvis Ridge. However, in the Waulsortian case, cement stratigraphy shows that the clear cements in the off-bank sediments always post-date the earliest marine cements (cloudy botryoidal spar and radiaxial fibrous calcite) in the adjacent banks.

Van Laer and Monty (1984) reported finding well-preserved flabellate microbial filaments within radiaxial fibrous spars in unspecified Paleozoic mud mounds. Possibly, the restriction of the volumetrically abundant radiaxial fibrous calcite marine cements to void systems essentially unconnected to the lime bank surfaces reflects the environmental preference of a similar nonphotosynthetic microbial population. Biogenic carbonates are well-known: for example, bacterial construction of calcite crystals was reported by Chafetz and Folk (1984) from travertines. Further (indirect) evidence of interstitial organic activity in Waulsortian M1 muds may lie in the early silica and carbonate dissolution stages, which require drastic fluctuations of pH and/or alkalinity, most easily explained by direct biological acitivity or degradation of certain types of organic matter.

7 Summary and Conclusions

Interbank to bank facies transitions provide evidence concerning the origin of the primary lime muds in Waulsortian buildups, and comparison of early diagenetic histories from the adjacent facies gives important evidence bearing on the nature and timing of buildup lithification:

1. Close to bank margins, lenticular masses of micrite developed within flank facies grainstones and packstones. They represent in situ production of carbonate mud similar to the M1 primary muds of the banks, and were most likely the product of non-filamentous, non-photosynthetic cyanobacterial activity within the sediment. The lenses had an initial firm gel consistency indicative of a degree of early cementation prior to compaction of the matrix.
2. At bank toes, the M1 lenses coalesced, producing a pseudo-brecciated appearance, then merged into the fenestrate bryozoan-rich polymuds of typical bank facies. This indicates that the primary buildup muds and those of the flank lenses had a common origin.
3. Siliceous sponge spicules dissolved very early in diagenesis, and the silica dissolution was followed by carbonate dissolution, which opened up intergranular and void-space secondary porosity in bank, flank, and toe sediments.
4. Other than in Phase *D* buildups, the first directly observable stage of marine cementation is cloudy radiaxial fibrous calcite which is always absent from flank, interbank and toe facies. The earliest cement stages in these off-bank facies are clear calcites of Stages 2 and 3, probably representing progressive anaerobiosis of confined marine pore waters. All later cement stages were introduced by tectonic veins and brittle fractures, and are considered to be products of burial diagenesis. Bank bed micrites remained diagenetically open enough for these later pore fluids to produce patchy neomorphism, dissolution and cementation, obscuring the nature of the original lime muds and the earliest phases of their diagenesis.
5. In off-bank sediments, the absence of any cements which pre-date or correlate with Stage 1 radiaxial fibrous calcite (or botryoidal, micritized cements in Phase *D* buildups) is considered to reflect unfavourable physicochemical conditions for synsedimentary cement precipitation in the deep water context of the Waulsortian.
6. Early stages of cementation in M1 muds may have been mediated by microbial populations, either directly or as a by-product of favourable micro-environ-

ments produced during degradation of associated organic material. Cyanobacteria may also have been responsible for subsequent production of large volumes of radiaxial fibrous calcite cements in interconnected void systems within the banks.

7. Lateral and vertical facies changes appear to be independent of Waulsortian Phase.

Non-skeletal, mud-rich buildups, including those of Waulsortian type, are apparently restricted to the Paleozoic. Their absence in post-Paleozoic rocks may point to extinction of a specialized deep water, non-photosynthetic microbiota which was responsible for mud mound production. Whatever the reason for their demise, these buildups offer a fascinating deep water counterpoint to studies of evolution and diagenesis in shallow water reefs.

Acknowledgments. I am indebted to A. Lees for guidance in Ireland, for critically reading a draft of this paper and for continuing discussion during our joint study of the Waulsortian problem. Doug Gillies provided stained peels and discussion of cement sequences from N. England: he and Julian Andrews also suggested improvements on the MS. The University of Edinburgh generously defrayed travel expenses.

References

Amieux P (1982) La cathodoluminescence: méthode d'étude sédimentologique des carbonates. Bull Centres Rech Explor Elf-Aquitaine 6:437 – 483

Bathurst RGC (1980) Stromatactis – origin related to submarine-cemented crusts in Paleozoic mud mounds. Geology (Boulder) 8:131 – 134

Bathurst RGC (1982) Genesis of stromatactis cavities between submarine crusts in Paleozoic carbonate mud buildups. J Geol Soc Lond 139:165 – 181

Birnbaum SJ, Wireman JW (1984) Bacterial sulphate reduction and pH: implications for early diagenesis. Chem Geol 43:143 – 149

Chafetz HS, Folk RL (1984) Travertine: depositional morphology and the bacterially constructed constituents. J Sediment Petrol 54:289 – 316

Choquette P (1971) Late ferroan dolomite cement. Mississippian carbonates, Illinois Basin, USA. In: Bricker OP (ed) Carbonate cements. Johns Hopkins Stud Univ Geol 19:339 – 348

Cotter E (1966) Limestone diagenesis and dolomitization in Mississippian carbonate banks in Montana. J Sediment Petrol 36:764 – 774

Cowan P (1980) Diagenesis of lime mud, Mississippian-age bioherms, Sacramento Mountains, New Mexico. MS Thesis, State Univ New York at Stony Brook

Dickson JAD (1966) Carbonate identification and genesis as revealed by staining. J Sediment Petrol 36:491 – 505

Dupont E (1863) Sur le calcaire carbonifère de la Belgique et du Hainaut français. Bull Acad R Sc Belg 2e Sér 15:86

Dupont E (1883) Sur les orgines du Calcaire Carbonifère de la Belgique. Bull Acad R Sc Belg 3e Sér 5:211 – 229

Fairchild IJ (1983) Chemical controls of cathodoluminescence of natural dolomites and calcites: new data and review. Sedimentology 30:579 – 583

Frank JR, Carpenter AB, Oglesby TW (1982) Cathodoluminescence and composition of calcite cement in the Taum Sauk Limestone (Upper Cambrian), southeast Missouri. J Sediment Petrol 52:631 – 638

Friedman GM, Ali SA, Krinsley DH (1976) Dissolution of quartz accompanying carbonate precipitation and cementation in reefs: example from the Red Sea. J Sediment Petrol 46:970 – 973

Grover G Jr, Read JE (1983) Paleoaquifer and deep burial related cements defined by regional cathodoluminescent patterns, Middle Ordovician carbonates, Virginia. Am Assoc Petrol Geol Bull 67:1275 – 1303

Jackson WD, DeKeyser T (1984) Microfacies analysis of Muleshoe Mound (Early Mississippian), Sacramento Mountains, New Mexico: a point-source depositional model I and II. Bull West Texas Geol Soc 23(5,6):6 – 10, 6 – 9

Kastner M (1971) Authigenic feldspars in carbonate rocks. Am Miner 56:1403 – 1442

Kendall AC (1985) Radiaxial fibrous calcite. In: Schneidermann N, Harris PM (eds) Carbonate cements. Soc Econ Paleontol Miner Spec Publ 36:59 – 77

King DA (1984) A diagenetic history of the Waulsortian limestones, Eastern Midlands and Dublin Basin, Republic of Ireland. MS Thesis, State Univ New York at Stony Brook

Land LS (1976) Early dissolution of sponge spicules from reef sediments, North Jamaica. J Sediment Petrol 46:967 – 969

Lees A (1964) The structure and origin of the Waulsortian (Lower Carboniferous) "reefs" of west-central Eire. Philos Trans R Soc Lond B 247:483 – 531

Lees A (1982) The paleoenvironmental setting and distribution of the Waulsortian facies of Belgium and southern Britain. In: Bolton K, Lane HR, LeMone DV (eds) Symp Environmental Setting and Distribution of the Waulsortian Facies, El Paso Geol Soc and Univ Texas at El Paso, pp 1 – 16

Lees A, Miller J (1985) Facies variation in Waulsortian buildups II. Mid-Dinantian buildups from Europe and North America. Geol J 20:159 – 180

Lees A, Hallet V, Hibo D (1985) Facies variation in Waulsortian buildups I. A model from Belgium. Geol J 20:133 – 158

Lees A, Noel B, Bouw P (1977) The Waulsortian "reefs" of Belgium: a progress report. Mem Inst Géol Univ Louvain 29:289 – 315

Lohmann KC, Meyers WJ (1977) Microdolomite inclusions in cloudy prismatic calcites: A proposed criterion of former high-magnesium calcites. J Sediment Petrol 47:1078 – 1088

Meyers WJ (1974) Carbonate cement stratigraphy of the Lake Valley Formation (Mississippian), Sacramento Mts., New Mexico. J Sediment Petrol 44:837 – 861

Meyers WJ, Cowan P, Lohmann KC (1982) Diagenesis of Mississippian skeletal limestones and bioherm muds, New Mexico. In: Bolton K, Lane HR, LeMone DV (eds) Symp Environmental Setting and Distribution of the Waulsortian Facies. El Paso Geol Soc and Univ Texas at El Paso, pp 80 – 95

Miller J (1986) Microscopical techniques I. Slices, slides, stains and peels. In: Tucker M (ed) Sedimentary techniques. Blackwell, London (in press)

Miller J, Clarkson ENK (1980) The post-ecdysial development of the cuticle and the eye of the Devonian trilobite *Phacops rana milleri* Stewart 1927. Philos Trans R Soc Edinb B 288:461 – 480

Miller J, Grayson RF (1972) Origin and structure of the Lower Visean "reef" limestones near Clitheroe, Lancashire. Proc Yorks Geol Soc 38:607 – 638

Miller J, Grayson RF (1982) The regional context of Waulsortian facies in northern England. In: Bolton K, Lane HR, LeMone DV (eds) Symp Environmental Setting and Distribution of the Waulsortian Facies. El Paso Geol Soc and Univ Texas at El Paso, pp 17 – 33

Monty CLV, Bernet-Rollande MC, Maurin AF (1982) Re-interpretation of the Frasnian classical "reefs" of the southern Ardennes, Belgium. Ann Soc Geol Belg 105:339 – 341

Parkinson D (1957) Lower Carboniferous reefs of northern England. Am Assoc Petrol Geol Bull 41:511 – 537

Philcox ME (1971) A Waulsortian bryozoan reef ('cumulative biostrome') and its off-reef equivalents, Ballybeg, Ireland. Compt Rend 6e Congr Int Strat Geol Carbonif Sheffield 1967 4:1359 – 1372

Pratt BR (1982) Stromatolitic framework of carbonate mud-mounds. J Sediment Petrol 52: 1203 – 1227

Pray LC (1958) Fenestrate bryozoan core facies, Mississippian bioherms, southwestern United States. J Sediment Petrol 28:261 – 273

Radke BM, Mathis RL (1980) On the formation and occurrence of saddle dolomite. J Sediment Petrol 50:1149 – 1168

Sevastopulo GD (1982) The age and depositional setting of Waulsortian limestones in Ireland. In: Bolton K, Lane HR, LeMone DV (eds) Symp Environmental Setting and Distribution of the Waulsortian Facies. El Paso Geol Soc and Univ Texas at El Paso, pp 65 – 79

Sommer SE (1972) Cathodoluminescence of carbonates I, II. Chem Geol 9:257 – 273, 275 – 284
Stow DAV, Miller J (1984) Mineralogy, petrology and diagenesis of sediments at Site 530, southeast Angola Basin. Init Rep DSDP 75:857 – 873 (US Government Printing Office, Washington)
Van Laer P, Monty CLV (1984) The cementation of mud mound cavities by microbial spars. 5ème Congr Eur Sedimentol (abstr), pp 30 – 31
Whitbread DR (1963) Facies changes in the Carboniferous successions of South Ireland. Ph D Thesis, Univ Reading UK
Zankl H (1969) Structural and textural evidence of early lithification in limestones. Sedimentology 21:241 – 256

Early Lithification, Dolomitization, and Anhydritization of Upper Devonian Nisku Buildups, Subsurface of Alberta, Canada

H.-G. MACHEL[1]

1 Introduction

Carbonates and evaporites of the Upper Devonian Nisku Formation were deposited on an extensive cratonic shelf that extended more than 1300 km southeastward from central Alberta into western Wyoming and South Dakota. The study area is located in the subsurface of central Alberta between townships 46 to 52 and ranges 5W5 to 16W5 (Figs. 1 and 2). In this area, the average depth of the Nisku is about 2000 m in the northeast and about 3600 m in the southwest. The strata form a structural homocline which dips towards the southwest at an average angle of 0.5 degrees.

Carbonate buildups within the Nisku Formation have undergone complicated and remarkably heterogeneous diagenesis. Some buildups are predominantly limestone, others are completely dolomitized and anhydritized. Many buildups contain oil, others contain only gas, with or without H_2S, and the distribution of several other features (e.g., saddle dolomites, celestite, various types of anhydrite, replacive calcite) is also heterogeneous. This paper treats those diagenetic features that pre-date oil migration. Importantly, dolomitization and anhydritization are interpreted to have taken place at several hundred meters of depth, in spite of the facts that the strata were deposited in an evaporitic province, and that the Nisku Formation consists of several shallowing-upward cycles and was temporarily exposed.

2 Facies and Stratigraphy

Facies and stratigraphy of the Nisku are extensively documented in Machel (1983, 1984, 1985a), Anderson (1985), and Anderson and Machel (in preparation), and are only briefly summarized below.

In the study area, the Nisku Formation was deposited on the flat-lying Ireton Formation, which is approximately 200 m thick and consists mainly of illitic shales and marls. On the basis of seismic data, the Nisku is subdivided into inner shelf, outer shelf (bank), slope, and Winterburn Basin, and geographically into the Brazeau, Pembina and Bigoray areas (Fig. 2). Reefal (Zeta Lake) facies occurs as isolated, roughly circular buildups on the slope, and as an elongated buildup fringing the bank-edge (Fig. 3). The buildups on the slope are about

[1] Basin Research Institute, Louisiana State University, Baton Rouge, LA 70803-4101, USA

Reef Diagenesis
Edited by J. H. Schroeder and B. H. Purser

Fig. 1. The study area (*SA*) is located on top of the Ireton shale (*oblique ruling*). Ireton-equivalent (*IPA*) and Nisku primary anhydrite deposits (*NPA*) occur some 200 km E and SE of SA. Beaverhill Lake anhydrite (*BHL*) occurs below the Ireton close to SA

1.5 km across and about 100 m thick. The off-reef facies consists of the Lobstick Member, the Bigoray Member, and the couplet Cynthia Member/Wolf Lake Member (Fig. 4), each of which forms a shallowing-upward cycle. The Nisku Formation is transgressively overlain by the silty/marly Calmar Formation. This facies grades into reefal and then intertidal facies of the Blue Ridge Member some 30 to 40 m above the top of the Nisku.

Most Nisku buildups are fossiliferous mudmounds. Intertidal facies and features indicative of exposure were recognized at or near the top of only a few buildups. The off-reef and bank (Dismal Creek) facies are approximately contemporaneous with the buildups, and there was little relief between the various facies domains.

3 Diagenesis

The Nisku Formation was affected by more than 20 diagenetic processes, which are comprehensively discussed in Anderson (1985) and Machel (1985a). The

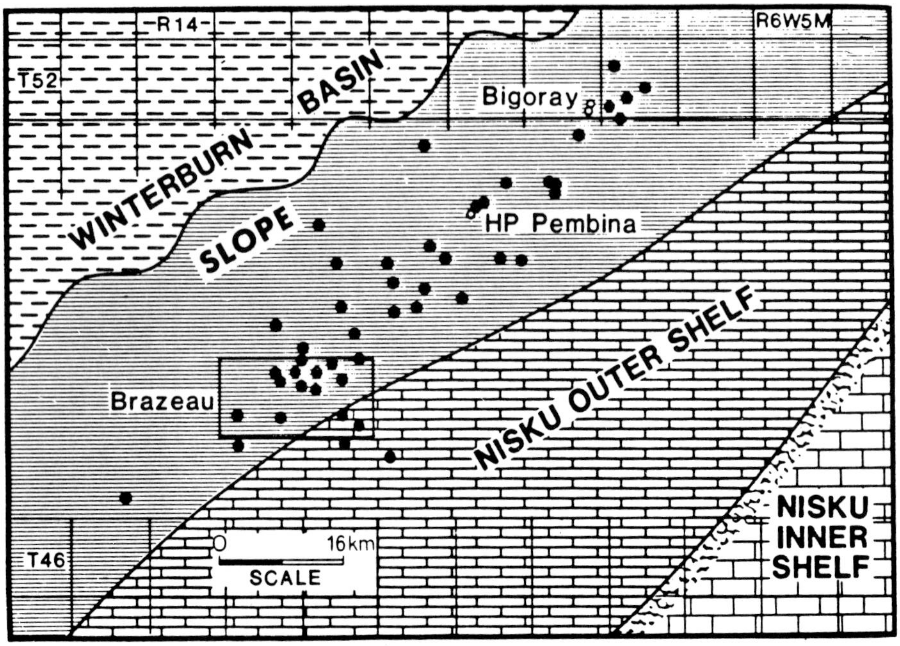

Fig. 2. Generalized subsurface map of the Nisku facies in the Brazeau, Pembina, and Bigoray areas (SA in Fig. 1)

present paper is focused on the limestone diagenesis, dolomitization, and anhydritization.

3.1 Calcite Cementation and Dissolution

There are no completely undolomitized buildups in the study area. Limestone buildups with only minor dolomitization occur almost exclusively in the up-dip Bigoray area (Fig. 2). Diagenetic features similar to those in the Bigoray area also occur in the *B-, D-,* and *I*-Pools of the down-dip Brazeau area (Fig. 3). Hence, it is likely that the limestone diagenesis was similar across the entire study area.

3.1.1 Petrography

The paragenetic sequence is illustrated in Fig. 5. Petrographic features are labeled with their respective phase number.

Phase 1. Cream to brown, prismatic calcite is abundant in some thin intervals. It occurs mainly in interparticle primary voids (Fig. 5 A), and is interstratified with marine internal sediment in some places. Composite crystal length is 1 to several cm. The crystals consist of innumerable fibrous subcrystals with sweeping, often

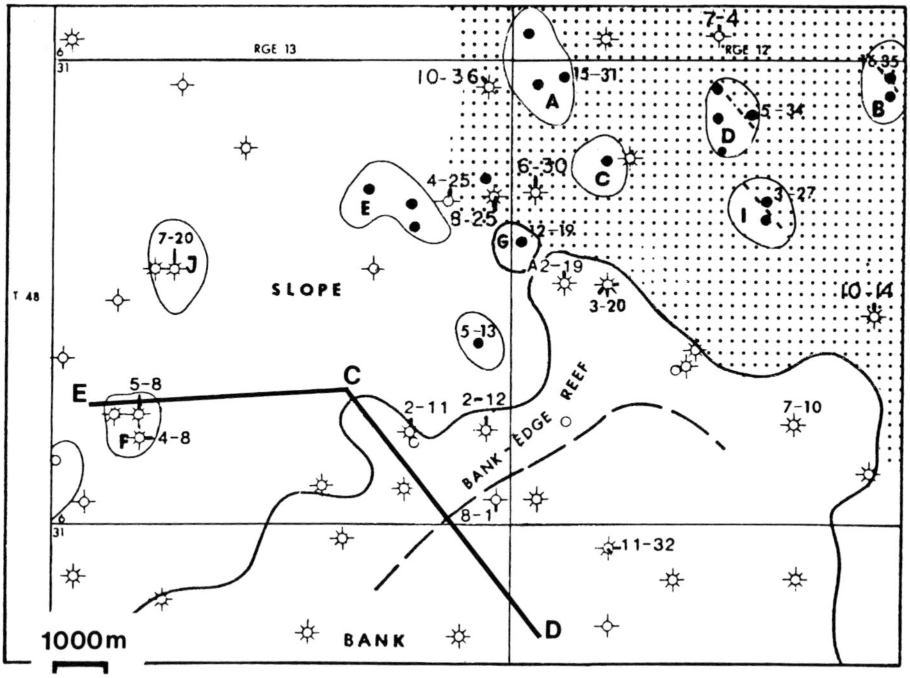

Fig. 3. Facies distribution in the Brazeau area (see Fig. 2 for location). *Filled dots* mark oil wells, *open circles* mark water or gas wells. *Line ECD* see Fig. 4. Cores in the *stippled area* consist predominantly of limestones

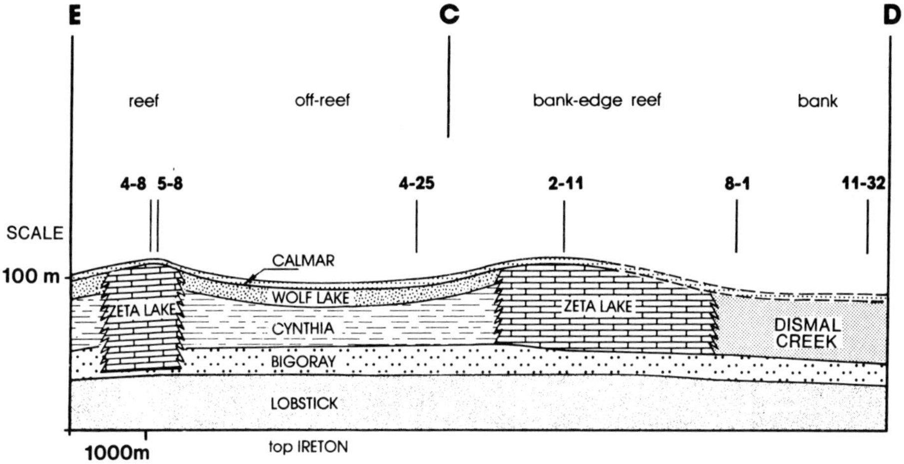

Fig. 4. Cross-section through all Nisku facies types in the Brazeau area. All wells are projected onto line ECD (see Fig. 3 for location). The top of the Ireton was taken as datum

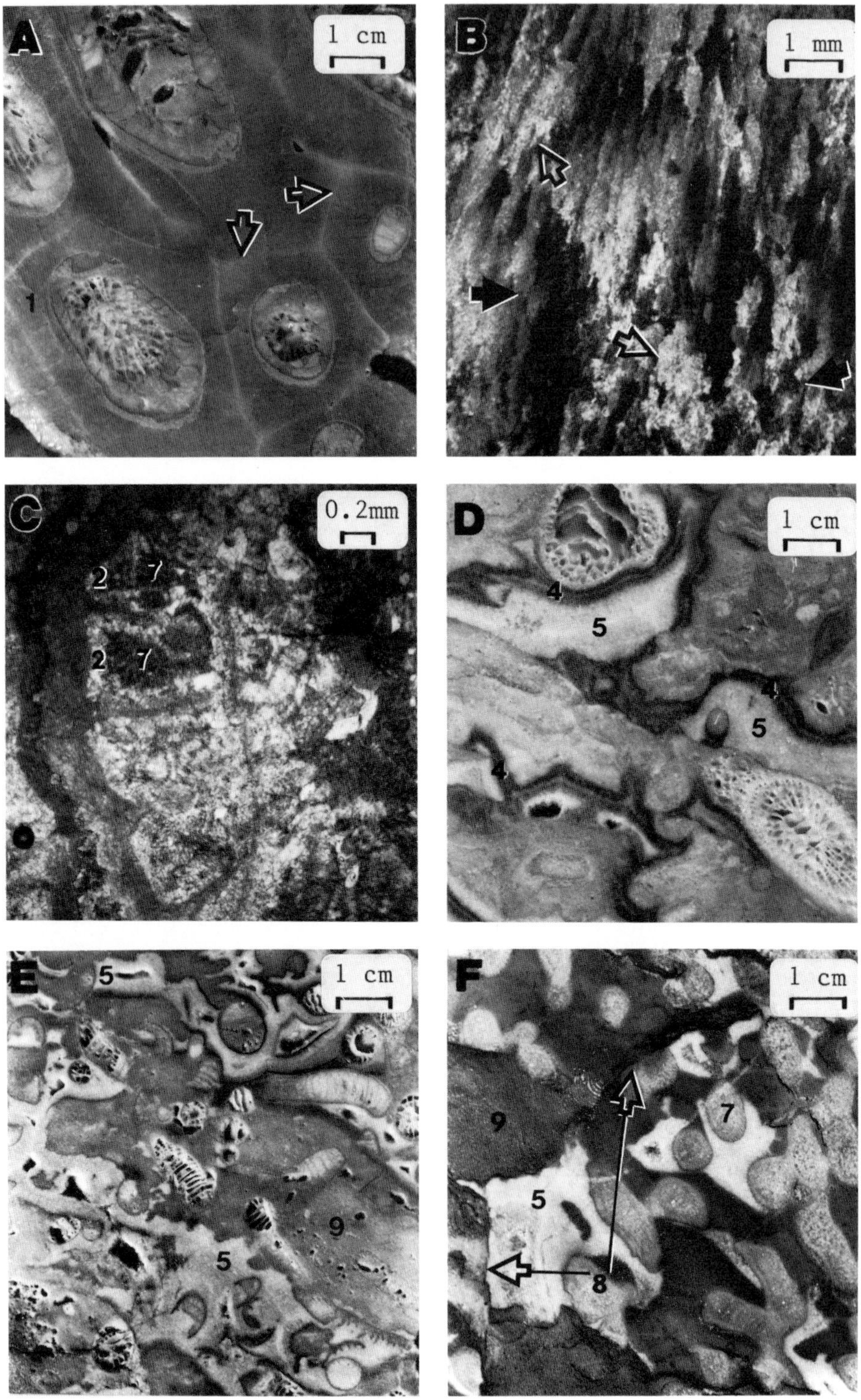

Fig. 5

radiaxial extinction, and they are partially recrystallized (Fig. 5B). Microdolomite inclusions occur.

Phase 2. Isopachous, scalenohedral calcite cement occurs throughout the buildups (and also in off-reef facies) and forms thin linings (200 – 500 microns) in primary inter- and intraparticle voids (Fig. 5C). This cement locally overgrew the radiaxial cements.

Phase 3. A rare dissolution phase enlarged some shelter and interparticle voids and removed some aragonitic fossils (gastropods, bivalves). The molds were filled in phases 4 and 5.

Phase 4. Banded fibrous calcite cement (Fig. 5D) occurs in the upper part of certain buildups (e.g., the upper 27 m of the Bigoray *C*-Pool) and mostly in interparticle and shelter voids. The crystals are pendant (microstalactitic), sometimes with multiple brown/beige-white banding. Average crystal length is 500 – 2000 microns.

Phase 5. White to cream, fibrous, isopachous calcite cement (Fig. 5D, E, F) is very abundant throughout the buildups. It post-dates the pendant calcite cement where the two occur together (Fig. 5D). The isopachous crystals partially or completely fill most primary inter- and intraparticle porosity and stromatactoid voids. Average crystal length is 1000 – 2500 microns. Some crystals are radiaxial.

Phase 6. Syndepositional/shallow burial matrix cementation, neomorphism, and physical compaction lithified the mud and sand matrix to micrite and microspar, and altered it to pseudospar. Textures indicative of syndepositional/shallow burial matrix lithification are (a) angular clasts of reefal material that occur in gravity flow deposits flanking some buildups; (b) in situ breccias of partially lithified mud, and (c) neptunian dykes with marine internal sediment within several buildups. Cements of phases 1 and 5 display neomorphic fabrics (such as irregularly shaped subcrystals with variable luminescence patterns, and anhedral neomorphic subcrystals: Fig. 5B).

Phase 7. Ferroan, limpid to translucent calcite spar (Fig. 5C, F) occurs mainly in the lowest parts of the buildups, and is not very abundant [according to

Fig. 5A – F. Calcite cements. Features are labeled with the number of their respective phase. **A** Polished slab of typical radiaxial fibrous calcite (phase 1), filling primary voids between corals. The crystals nucleated on the corals, then grew centrifugally into the interparticle voids and met at planar, polygonal compromise boundaries (*arrows*). **B** Thin section (crossed nicols) of sample shown in **A**. The radiaxial fibrous calcite is recrystallized. The radiaxial bundles are not well defined in parts of the sample (*filled arrows*), and some irregularly shaped crystals formed at the expense of several adjacent bundles (*open arrows*). **C** Thin section (plain polarized light) of fringes of scalenohedral calcite (phase 2) and ferroan calcite (phase 7, *stained dark*) in fasciculate coral (*circle at lower left* is bubble in epoxy). **D** Polished slab from upper part of the Bigoray *C*-Pool, with dark pendant cement (phase 4) and white isopachous cement (phase 5) in shelter voids. Note that the intracoral porosity is not affected by these cements. **E** Polished slab from the lower part of the Bigoray *C*-Pool, showing phase 5-cement in shelter, interparticle, and stromatactoid voids. Note lack of phase 4-cement in lower part of the buildup. Matrix is partially dolomitized (phase 9). **F** Polished slab of white isopachous calcite (phase 5) and translucent ferroan calcite in corals (phase 7) occluding all primary porosity. Stylolites (phase 8) truncate these cements. The matrix is partially dolomitized (phase 9)

Anderson (1985) this type of cement occurs also in the flanks of the Bigoray *I*-Pool and in off-reef intracoral voids]. The ferroan calcite occurs either as overgrowths on cements of previous phases or, most commonly, as equant rhombohedral spar in remnant intracoral voids with or without ferroan/nonferroan zonations (revealed by staining with potassium ferricyanide). Most voids affected by this type of cement were completely occluded (Fig. 5 F). Average crystal size is several hundreds of microns.

Phase 8. Ubiquitous chemical compaction (pressure solution) formed clay seams and stylolites (generation I) with average amplitudes ranging from a few mm to about 2 cm (Fig. 5 F). Residues consist mainly of illite, with some kaolinite and quartz. Stylolites of generation I have relatively sharp boundaries in dolomite-free intervals, but are somewhat more diffuse in the dolomitized parts. A second generation of stylolites (phase 11, Machel 1985a) offsets generation I.

3.1.2 Diagenetic Pore Waters

Calcites (brachiopods and rugose corals, micrite and microspar matrix, cements of phases 1, 4, 5, and 7) from the Bigoray *C*-Pool were analyzed for trace elements and isotopes. The samples were collected with a dental drill from material that was homogeneous under the light and cathodoluminescence microscopes. Submicroscopic domains of variable chemical composition are not relevant for this study. The analytical procedures and results are comprehensively discussed in Machel (1985a and 1986).

Petrographic evidence suggests that skeletal calcites, matrix, and cements of phases 1 and 5 are recrystallized and/or neomorphosed marine calcites. In *meteoric* waters, strontium, calcium and manganese contents of recrystallized and neomorphosed marine aragonite (A), marine high-Mg-calcite (HMC), and marine low-Mg-calcite (LMC) *should* plot in the respective "diagenetic stabilization fields" of Fig. 6 [after Brand and Veizer (1980), note: the Nisku data are not separated into petrographic groups because they form a very tight cluster]. The Nisku data (dots) plot in a zone where the A- and HMC-stabilization fields overlap, which has three important implications: (1) Since none of the Nisku data plot in the fields of marine A, HMC, and LMC (boxes), the present compositions of all Nisku calcites are diagenetic. (2) LMC cements were precipitated from solutions with Sr/Ca and Mn concentrations similar to those fluids in which HMC recrystallized to LMC. (3) The composition of the diagenetic solution *cannot* be inferred, because neomorphism and recrystallization take place also in marine phreatic systems, which may lead to trace element changes similar to those in meteoric systems (Machel 1985a and 1986).

The Sr-contents of the various petrographic groups scatter erratically if plotted against depth (Fig. 7). It is unlikely that these marked trace element variations within petrographically coeval calcites are due to different diagenetic pore waters. Rather, the observed variations are due to variable degrees of chemical reequilibration caused by different diffusion rates across crystal surface reaction zones with variable thicknesses and compositions (as discussed by

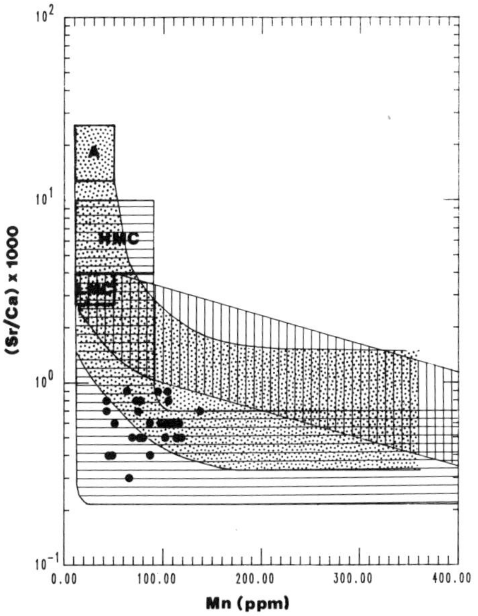

Fig. 6

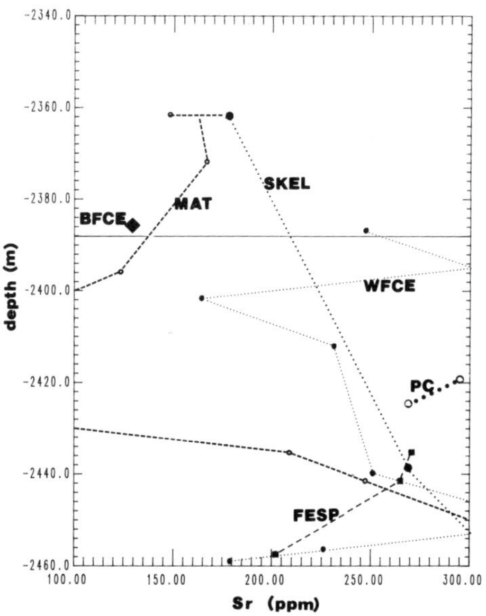

Fig. 7

Fig. 6. Diagenetic "stabilization fields" for aragonite (*A: stippled*), high-Mg-calcite (*HMC: horizontal ruling*), and low-Mg-calcite (*LMC: vertical ruling*) in meteoric water. (After Brand and Veizer 1980). Most Nisku calcites (*filled circles*) plot in a zone where the stabilization fields of A and HMC overlap

Fig. 7. Sr/depth plot of calcites from the Bigoray *C*-Pool. *SKEL* skeletal calcite; *MAT* matrix; *PC* prismatic calcite cement (phase 1); *BFCE* banded fibrous calcite cement (phase 4); *WFCE* white fibrous calcite cement (phase 5); *FESP* ferroan calcite spar (phase 7)

Pingitore 1982, Veizer 1983). Furthermore, the observed variations have no apparent relationship to depth. Plots of 26 other trace elements versus depth are similarly erratic. Apparently, no former vadose, meteoric phreatic, or marine phreatic zones can be recognized on the basis of trace elements.

Stable isotope compositions of the various calcites from the Bigoray *C*-Pool are plotted in Fig. 8. Three groups are visible on the basis of the oxygen isotope composition: matrix, ferroan spar, and all other types which cluster around a $\delta^{18}O$ of -5.8 permil PDB. The matrix samples are from the uppermost part of the pool, and their relatively heavy oxygen isotopes may be due to precipitation and/or neomorphism in evaporated groundwater, either as vadose micritic cement, or as meteorically neomorphosed marine mud. The $\delta^{18}O$ of -5.8 permil PDB of the pendant phase 4-cement and isopachous phase 5-cement probably characterizes the previously discussed phase of recrystallization and/or neomorphism (phase 6). Using the equation of Friedman and O'Neil (1977) and a temperature of 25° – 30°C, calcite with -5.8 permil PDB would precipitate from water with -4.0 to -3.0 permil PDB. This is precisely within the range of the best determination to date of the oxygen isotope composition of Upper Devonian *seawater* [-3.5 ± 2 permil SMOW; Popp et al. (1986)]. Hence neomorphism and recrystallization probably took place in a *marine phreatic* and/or

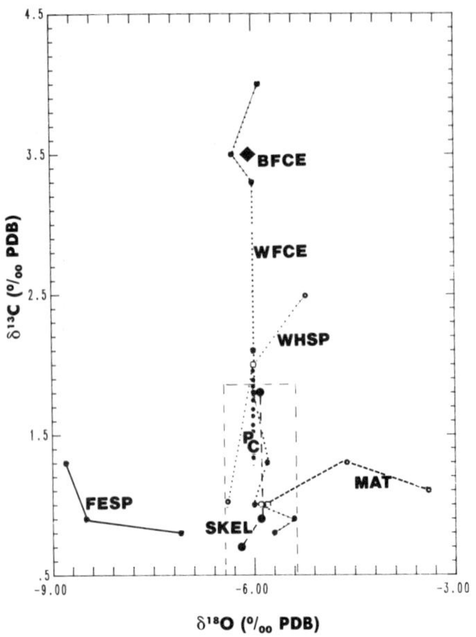

Fig. 8

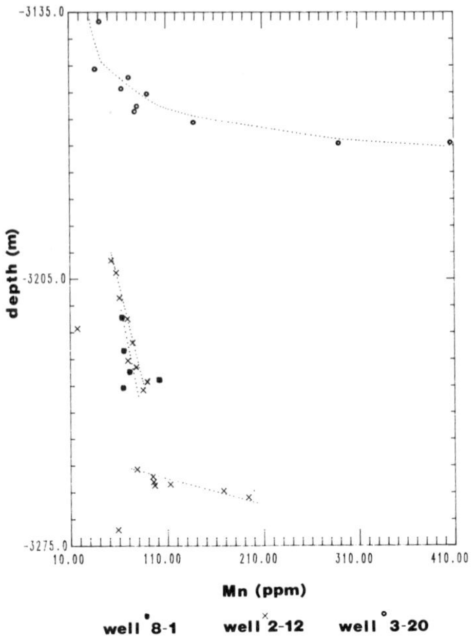

Fig. 9

Fig. 8. Plot of carbon/oxygen isotope ratios of calcites from the Bigoray *C*-Pool (abbreviations as in Fig. 7). MAT and FESP form distinct groups on the basis of oxygen isotope ratios, the other calcites display large variations in their carbon isotope ratios but a narrow range for the oxygen ratios

Fig. 9. Mn/depth plot of gray matrix dolomite (see Fig. 3 for well locations). Mn decreases upward in the bank (well 8 – 1) and in the bank-edge reef (wells 2 – 12 and 3 – 20), and the trends cross-cut the cycle boundaries. Pools on the slope have similar trends

shallow burial environment. The ferroan spar, which is primary LMC and not recrystallized, could be the meteoric phreatic time-equivalent to the ^{18}O-enriched matrix samples, or it could be of a deeper burial origin. Its mean $\delta^{18}O$ of -8.2 permil PDB could be due to precipitation from meteoric or brackish water, or at higher temperatures, or both.

The carbon isotope ratios of all samples in the lower part of the Bigoray *C*-Pool, and matrix throughout the core, are close to $+1.0$ permil PDB, indicating a system dominated by carbon derived from (Devonian) marine carbonate. Other carbon values are much higher (Fig. 8), which may be due to (a) photosynthetic uptake of ^{12}C by organic matter with a corresponding isotopic enrichment of the remaining carbon reservoir and/or (b) carbon derived from fermentation of organic matter. In either case, the observed carbon values could be inherited from the metastable calcium carbonate precursors, or they could have been acquired during recrystallization. Samples from the top of the Bigoray *C*-Pool are not depleted in ^{13}C as predicted by the model of Allen and Matthews (1977), probably because there was no soil at the exposure surface.

3.1.3 Discussion and Interpretation

On the basis of petrographic evidence, radiaxial cements of phase 1 formed syndepositionally during periods of slow lime mud production and/or sedimentation. Their trace element characteristics indicate precipitation and/or neomorphism in a marine phreatic environment. The thin scalenohedral rims of phase 2 also formed syndepositionally and/or soon after, but mainly in voids sheltered from currents and sedimentation, such as those within corals.

Following initial submarine cementation, the upper part of the Biogray *C*-Pool was exposed. The facies of this buildup displays a shallowing-upward sequence in the uppermost meters, with fenestral fabrics and some stromatolites. Dissolution in a vadose and/or mixed meteoric marine phreatic environment (phase 3) resulted in enlargement of some shelter and interparticle voids, and partial or complete dissolution of some aragonitic fossils. The pendant calcite cements (phase 4) may have precipitated from meteoric or marine water in the vadose zone.

The most extensive cementation occurred in a marine phreatic and/or shallow burial environment during phases 5 and 6. Most interparticle and stromatactoid voids were occluded by phase 5-cements, which were probably formed diachronously from the bottom to the top of the buildups, in the upper few (1 – 3) meters of the sediments at any given time during progressive sediment accumulation. Matrix lithification and physical compaction (phase 6) occurred over a fairly large depth range (perhaps a few centimeters to several tens of meters). This is indicated by (a) the heavy (vadose) oxygen isotope signal in the Bigoray *C*-Pool, (b) syndepositional angular mudstone clasts in gravity flow deposits flanking some buildups, (c) neptunian dykes, and (d) breccias of lithified mudclasts in a more coarse-grained matrix, which are attributed to differential cementation with subsequent physical compaction (Machel 1984, 1985a).

Ferroan spar (phase 7) may have formed in a meteoric phreatic environment contemporaneously with exposure of the top of the buildup. However, the geochemical characteristics are also compatible with a later, deeper burial origin. Chemical compaction (phase 8) resulted mainly in the formation of stylolites and diagenetic clay seams, which crosscut textures of all previous phases. Chemical compaction continued throughout dolomitization and anhydritization.

The above features are typical for most Nisku limestone buildups. However, the features indicative of subaerial exposure occur only in very few buildups, suggesting that the other buildups (1) did not experience a period of subaerial exposure long enough to produce recognizable fabrics, (2) were never exposed, or (3) the features indicative of exposure were obliterated by dolomitization.

3.2 Pervasive Matrix Dolomitization

There are four types of dolomite in the Nisku Formation: gray matrix dolomite, brown porous matrix dolomite, white saddle dolomite, and limpid dolomite cement. The most widespread and most abundant types, the two matrix dolomites, are discussed below.

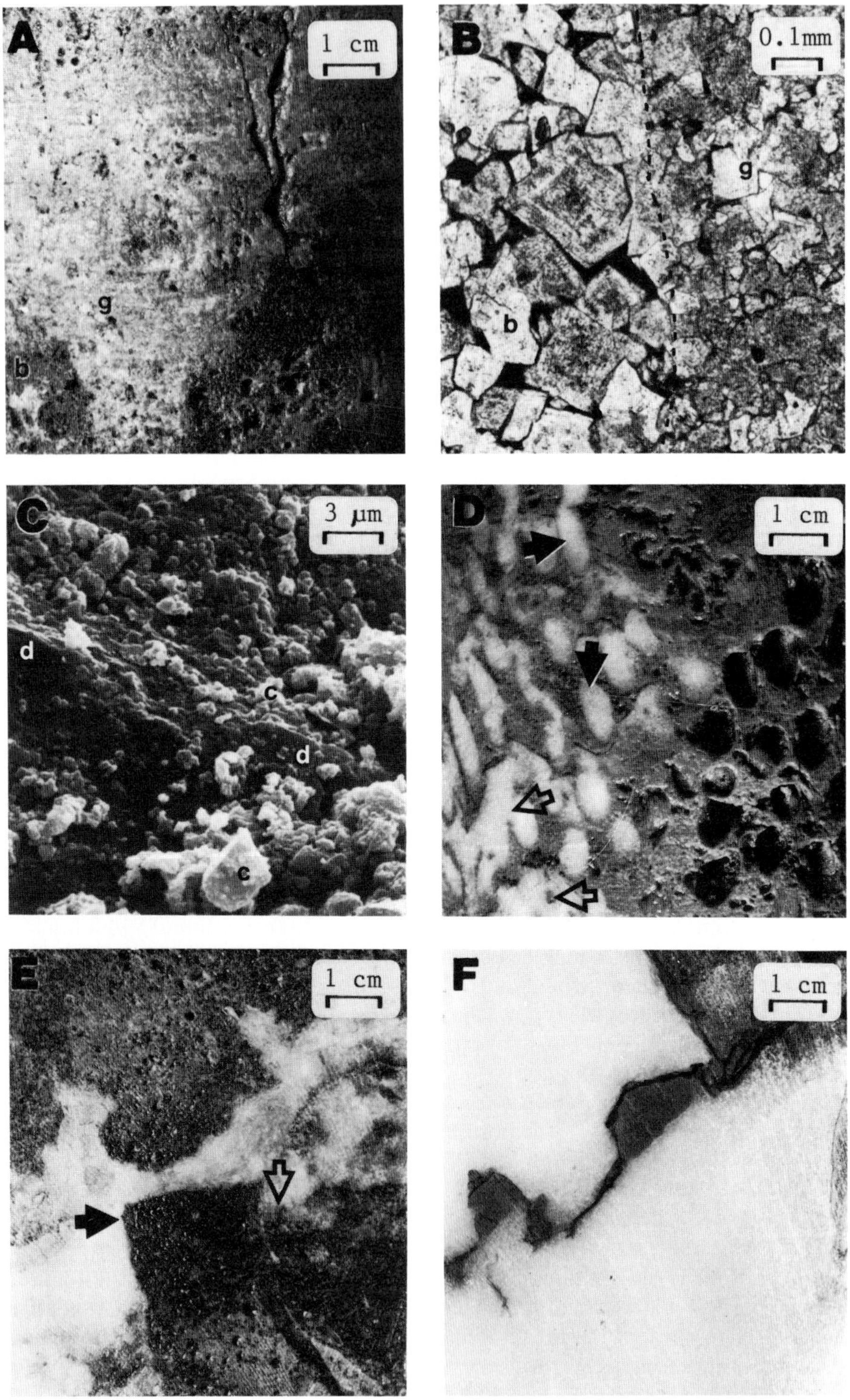

Fig. 10

3.2.1 Petrography

Gray matrix dolomite (*g* in Fig. 10A) is variably gray to gray-green in hand specimen, depending on the amount and color of clay impurities. In most places, large calcite "particles" (e.g., molluscs, fibrous cements etc.), and even delicate uncemented corals, were not affected by this type of selective matrix replacement (Fig. 5E). Brown porous matrix dolomite (*b* in Fig. 10A) is intimately associated with gray matrix dolomite. The contacts are often irregular (Fig. 10A, B).

Gray matrix dolomite is tight in most hand specimens but contains submicroscopic pore throats with remnant micrite grains that are visible only under the SEM (Fig. 10C). On the other hand, brown porous matrix dolomite is sucrosic and commonly has more than 10% intercrystalline porosity (Fig. 10A, B). Both matrix dolomites contain a variety of moldic and vuggy pore types, e.g., coral molds (Fig. 10D).

In thin section, gray matrix dolomite consists of variably cloudy crystals with average diameters of 50 to 100 microns. Where dolomitization was very minor, these crystals occur as euhedral to subhedral rhombohedra scattered within the micrite matrix, and microspar and pseudospar are nearly free of dolomite. Where dolomitization was more complete, the dolomite crystals form tight subhedral to anhedral mosaics (gray matrix dolomite), and microspar and pseudospar contain a loose meshwork of larger dolomite crystals, with average diameters of 100 to 500 microns. In completely dolomitized strata, the micrite matrix is gray matrix dolomite, and the former sparry domains consist of typical brown porous matrix dolomite with larger crystals and intercrystalline porosity (Fig. 10B). Hydrocarbons in these intercrystalline pores stain the rock brown. Stylolites (generation I) both pre- and post-date the two matrix dolomites.

3.2.2 Spatial Distribution

The spatial distribution of both matrix dolomites is heterogeneous. On a regional scale, most of the Nisku in the updip Bigoray area is relatively little dolomitized, whereas almost the entire Nisku, Calmar, and Blue Ridge are completely dolomitized in the downdip Pembina and Brazeau areas. However, notable amounts of limestone occur even within the Brazeau area in several off-reef wells and in the *B*-, *D*-, and *I*-Pools (stippled in Fig. 3). Importantly, the distribution of the matrix dolomites in these pools is asymmetrical; limestones occur almost exclusively in the updip (northeastern) parts.

Fig. 10A – F. Matrix dolomitization and anhydritization. **A** Unpolished core surface, showing typical gray matrix dolomite (*g*) and brown porous matrix dolomite (*b*). Many intervals of the Nisku buildups consist of this lithology. **B** Thin section (plain polarized light) of sample shown in **A**. Black matter between crystals in *b* is oil. **C** SEM of fractured surface in gray matrix dolomite. Remnant calcite (*c*) occurs as anhedral micrite between dolomite (*d* large dark crystals). **D** Unpolished core surface with coral molds in gray matrix dolomite. Anhydrite (white) occurs as cement in coral molds (*filled arrows*) and as a replacement (*open arrows*). **E** Polished slab of typical replacive anhydrite in dolostone. The boundaries anhydrite/dolomite resemble corrosion surfaces. *Arrows* are explained in captions to Fig. 11D,E. **F** Polished slab of massive replacive anhydrite from a 1-m-thick interval. Note "floating" stylolite with clay and dolomite residue

On a more local scale, gray matrix dolomite occurs fairly randomly throughout the sequence in partially dolomitized cores, displaying variable thickness and no vertical or horizontal trends in abundance. Brown porous matrix dolomite is nearly absent in partially dolomitized strata. Both matrix dolomites form massive dolostones exceeding several tens of meters in thickness in the completely dolomitized buildups.

3.2.3 Nature of the Dolomitizing Fluids

Some 140 samples of both matrix dolomites from all parts of the study area were analyzed for 27 trace elements each; of these, 60 samples were analyzed for their stable isotope composition. These data are comprehensively discussed in Machel (1985a); only the essential aspects are presented here.

The Sr-concentrations of both matrix dolomites throughout the study area are between about 80 to 120 ppm. If matrix dolomitization was a bulk solution equilibrium process (which is likely), it took place in diagenetic solutions that had a molar Sr/Ca ratio very different from normal seawater, evaporated seawater, or brackish water (dolomites formed in equilibrium with these solutions would have much higher Sr-contents: e.g., Veizer 1983). Also, the Nisku matrix dolomites did not form in equilibrium with dilute waters that derived most of their Sr from the calcite precursors. Dilute waters dominated by Sr derived from the dissolution of these calcites (which have about 120 to 320 ppm Sr) would precipitate dolomite with less than 10 ppm Sr. Na, Mn, Fe, and other trace elements confirm the interpretation that the dolomitizing fluids were neither normal marine, evaporated marine, or meteoric waters (Machel 1985a).

Two lines of evidence suggest that the dolomitizing fluids were chemically influenced by the illitic shales that surround (Cynthia) and underly (Ireton) the Nisku buildups. (1) Matrix dolomites of buildups on the slope (e.g., the Brazeau *A*-Pool, Fig. 3) have higher *B* (boron) contents than coeval matrix dolomites in the bank-edge reef (15 to 45 ppm versus less than 15 ppm, respectively). The additional *B* in the *A*-Pool was probably derived from the Cynthia shale, because the overlying and underlying strata are mineralogically uniform across the area, and dolomitization via surface-derived fluids would not selectively enrich buildups located on the slope in *B*. (2) Both matrix dolomites contain Sr that is considerably more radiogenic than Nisku or any younger seawater: the dolomite $^{87}Sr/^{86}Sr$ ratios vary from 0.7089 to 0.7107, the Nisku seawater ratio is about 0.7084 (data from Anderson 1985, Burke et al. 1982, respectively). Clearly, much of the Sr in the Nisku matrix dolomites is derived from sources more radiogenic than Nisku (and younger) seawater, most probably from the illitic shales that either surround or underlie the Nisku Formation.

The stable isotopes of both matrix dolomites are fairly uniform across the study area. Most carbon values are close to those of their precursor calcites (between +1 and +2 permil PDB). Most oxygen values are about −6.0 to −5.8 permil PDB, and thus are also nearly identical to those of the precursor calcites. Because dolomitization took place after marine phreatic/shallow burial recrystallization and was partly contemporaneous with stylolitization, the most likely interpretation of the observed oxygen isotope values is that the dolomitizing fluids were of marine parentage with slightly elevated temperatures. Such fluids are likely to be isotopically similar to seawater, or slightly heavier than that (e.g., Land 1980). A temperature range of about 42° to 48 °C during dolomitization is suggested, taking −5.8 permil PDB as representative of both matrix dolomites, about −4 to −3 permil SMOW as representative of Upper Devonian seawater (Popp et al. 1986), and using the temperature equation recommended by Land (1983). An increase in the water oxygen composition by 1 permil would increase these temperatures by about 6 °C.

The direction of fluid flow during dolomitization is indicated by trace element trends. Fifteen trace elements of gray matrix dolomite from 20 wells were plotted in traverses that extend vertically (cross-formational), along and across the paleostrike, and along the structural dip from the Brazeau to the Bigoray area. Trends were found only in vertical traverses. In particular, Mn decreases (Fig. 9) and Sr increases upward in several buildups and non-reefal strata, indicating cross-formational and upward fluid flow during dolomitization (see mathematical flow model in Machel 1985a).

3.2.4 Discussion and Interpretation

The initially selective dolomitization of lime mud explains the irregular distribution of gray matrix dolomite in the partially dolomitized buildups: the sparry domains were relatively resistant to dolomitization. In more advanced cases spar was also replaced, but by larger crystals with fewer nucleation sites. Subsequently, remnant calcite (fossils, cements, pseudospar matrix) was dissolved, forming moldic, vuggy, and sucrosic (brown porous matrix dolomite) porosity. Calcite dissolution did not occur in the partially dolomitized strata but was ubiquitous in the completely dolomitized areas; hence, calcite dissolution appears to be an integral part of the *advanced* stages of the dolomitization process. The causes of calcite dissolution are not clear. The dolomitizing fluids may have been undersaturated with respect to calcite after all Mg and/or CO_3 available for dolomitization had been used up, or the pH may have been lowered by any of a variety of reasons (e.g., Schmidt and McDonald 1979).

Both matrix dolomites probably formed from nearly identical fluids, because their trace element and isotope compositions are nearly identical. Hence, the time difference between mud/micrite and microspar/pseudospar replacement was probably negligible on a geological time-scale. In other words, gray matrix dolomite and brown porous matrix dolomite formed sequentially during one major dolomitization phase. Petrographic and geochemical data provide compelling evidence that matrix dolomitization took place at intermediate burial depths from predominantly advecting (ascending) fluids. Because shallow burial limestone diagenesis and the onset of stylolitization predate dolomitization, a minimum depth of at least a few tens if not hundreds of meters is likely for the onset of matrix dolomitization. Depending on the surface temperature and the geothermal gradient, a burial depth range of perhaps 300 to 1000 m is suggested by the oxygen isotope data. The main hydrologic force for the advecting flow

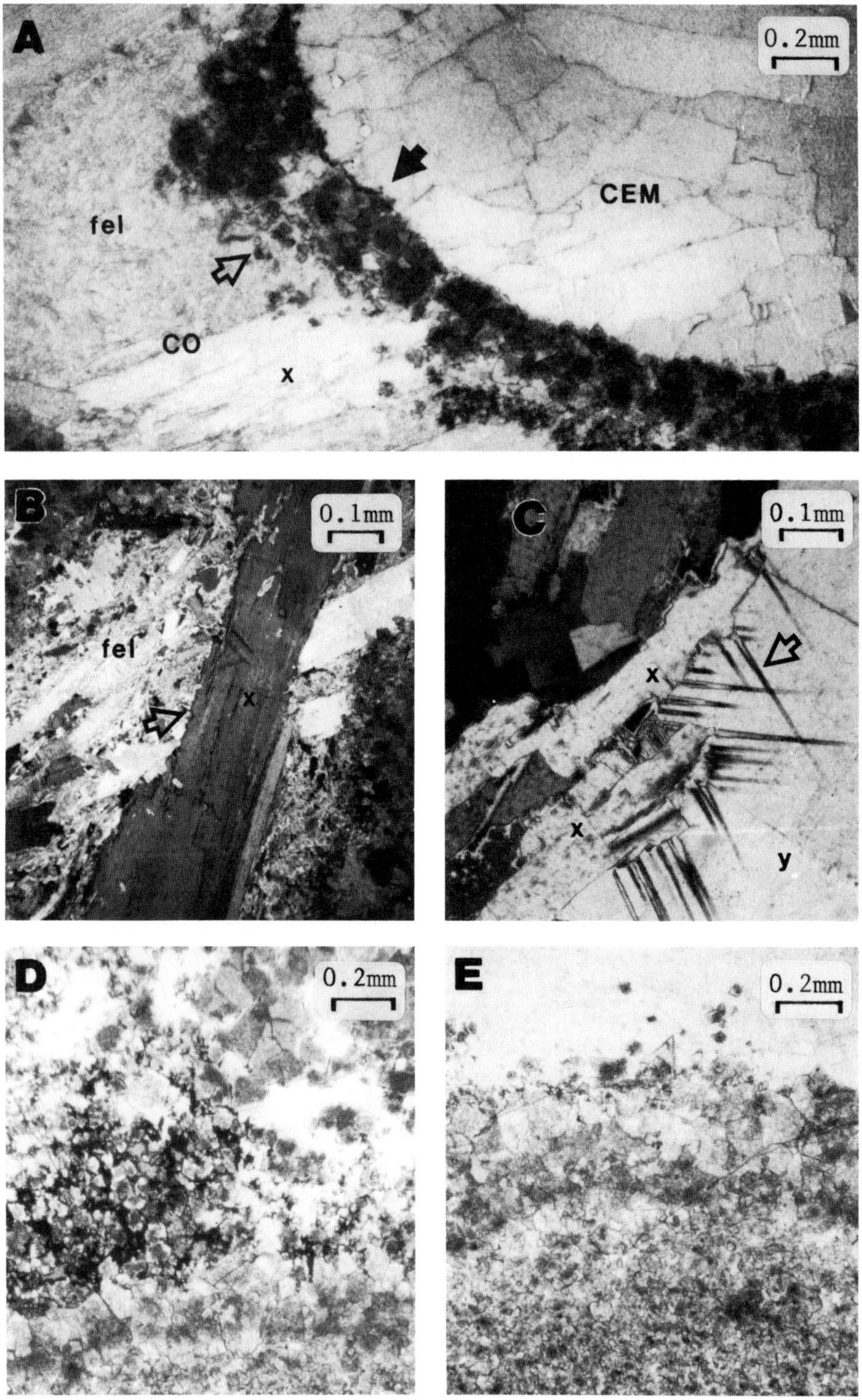

Fig. 11

pattern was probably compaction, because the Nisku Formation was being buried progressively at a steady rate to about 1100 m between Upper Devonian and Upper Mississippian time. At least three factors complicated the general flow pattern. (1) Superimposed on the advective flow pattern were locally variable flow directions due to large permeability variations. Relatively impermeable limestone nodules and beds in reefal and nonreefal strata probably deflected fluid flow, which also explains the absence of well-defined vertical trace element trends in these wells. (2) Partial derivation of dolomitizing fluids from the off-reef areas is indicated by *B*-enrichments of buildups on the slope. If the fluids came exclusively from below, no selective *B*-enrichments would have been possible, since the underlying Ireton shale is uniform across the area. (3) The entire Western Canada Sedimentary Basin was affected by subvertical faulting as early as the Devonian, and throughout the Mesozoic (e.g., Jones 1980). Possibly related to this faulting, the initiation of the regional structural tilt may have partially coincided with dolomitization. It is not unlikely that syncompactional faults funnelled and/or deflected fluid flow. These faults are the most probable reason for the regionally asymmetrical distribution of both matrix dolomites.

It may be argued that the compactional fluids did not contain enough Mg for pervasive matrix dolomitization. However, (a) only a part of the Nisku Formation is pervasively dolomitized, (b) Devonian seawater (and compaction waters derived from it) may have contained much more Mg than Recent or Mesozoic seawater, and (c) the ionically modified Upper Devonian seawater invoked for dolomitization may have obtained additional Mg from the dissolution of older dolomites in the Nisku and/or Ireton Formations, or via the subvertical faults from farther below. The Mg now forming Nisku matrix dolomites may be derived from one or several of these sources.

3.3 Anhydritization

The Nisku Formation contains some 10 to 15 different types of anhydrite in the study area (SA in Fig. 1) and in a region of "primary" (sedimentary) sulfate deposition (NPA in Fig. 1). The "secondary" types that predominate in the study area are discussed below.

Fig. 11. Microfabrics of anhydrites. **A** Thin section (plain polarized light) from the lower left of sample shown in Fig. 10D, showing replacive corrotopic (*CO*) and cement (*CEM*) anhydrite. Note the diffuse boundary between CO and dolomite (*open arrow*) and the sharp boundary between CEM and dolomite (*filled arrow*). CO consists of felted matrix (*fel*) and large porphyrotopes (*x*), CEM consists of subhedral to anhedral laths. **B** Thin section (crossed nicols) of corrotopic anhydrite shown in **A**. Note apparent "corrosion" (*arrow*) of porphyrotope (*x*) by felted anhydrite (*fel*). The porphyrotope formed first within replacive gypsum, then the remnant gypsum was transformed to felted anhydrite. **C** Thin section (crossed nicols) of strain twins (*arrow*). These twins occur in porphyrotope *y* where porphyrotobe *x* impinges on *y*. Strain twins formed when the porphyrotopes grew ("floating" in gypsum). **D** Thin section (plain polarized light) of typical diffuse replacive anhydrite/dolomite boundary (open arrow in Fig. 10E). The anhydrite largely fills intercrystalline pore space and only partially digested dolomite. **E** Thin section (plain polarized light) of sharp replacive anhydrite/dolomite boundary (filled arrow in Fig. 10E). The anhydrite apparently digested the "floating" dolomite crystals almost completely

3.3.1 Types of "Secondary Anhydrite"

In the study area, most anhydrites are milky-white in hand specimens. Some anhydrite occurs as nodular masses in fine-grained non-reefal strata, and as cements in molds and vugs (Fig. 10D) and fractures. Approximately 90 vol% of all anhydrite occurs as irregularly shaped replacive masses in reefal dolostone (Fig. 10E, F, and 11 A – F). Replacive anhydrites cross-cut compactional features, and contain many "floating" remnants of the host rocks. Importantly, these remnants are stylolitized dolostone (Fig. 10F).

Six types of anhydrite were distinguished on the basis of microfabrics. (1) Most of the replacive anhydrite consists of two populations: up to several mm large porphyrotopic crystals, and an irregularly felted mass of smaller (10 to 500 microns) crystals. The felted variety occurs also without the porphyrotopic crystals. Because many porphyrotopic crystals appear intensely corroded by the felted matrix (Fig. 11A, B), this type of anhydrite has been termed corrotopic (preliminarily called corroblastic in Machel 1985a). Other features of the porphyrotopic crystals are strain twins (Fig. 11C), undulous and chessboard-pattern extinction, criss-crossing intergrowths, and partial internal breakage. The contact with the replaced host rock is diffuse, and commonly contains inclusions of dolomite (Fig. 11A). Some porphyrotopic crystals which protrude into the dolostone have square terminations. (2) The second most abundant type of anhydrite is also replacive. The crystals are equidimensional to elongate, and generally form subhedral mosaics. This type of anhydrite began as cement, filling intercrystalline pore spaces (Fig. 11D). Where anhydritization was more complete, most of the dolomite was digested, leaving only scattered "islands" of host rock (Fig. 11E). (3) The third most abundant type of anhydrite is a cement that formed subhedral to anhedral mosaics in molds (Fig. 11A) and vugs. (4) Euhedral to subhedral anhydrite cement crystals with oil-filled, primary fluid inclusions occur in some buildups. (5) Small, replacive anhedral anhydrite crystals with numerous undigested host rock inclusions occur in some limestone beds. (6) Fibrous anhydrite occurs in veins (Machel 1985b).

Types (1), (2), and (3) are the most abundant types in the study area. Corrotopic anhydrite is interpreted to result from replacement of replacive gypsum (Machel 1985a). The other types of anhydrite do not contain features indicative of gypsum precursors.

3.3.2 Spatial Distribution

In the study area, the "secondary" anhydrites occur only where matrix dolomites are extensive (some intervals consist of 30 to 100 vol% anhydrite). Where matrix dolomites are scarce or absent, these anhydrites are also scarce or absent. Furthermore, no "primary" sedimentary sulfate deposits occur in the study area, either in the Nisku, or in the underlying or overlying strata. The closest occurrence of Nisku-equivalent "primary" sulfate deposits (typical shallow-subaqueous, laminated, bedded-massive, and ropy-bedded anhydrites) is about 200 to 250 km east and southeast of the study area (NPA in Fig. 1).

3.3.3 Geochemistry

Some 100 samples of the various types of anhydrite were analyzed for their trace element and stable isotope composition. The data are extensively discussed in Machel (1985a), and only a summary of the results is presented here.

The most important question regarding the "secondary" anhydrites in the study area concerns the source of the sulfate. "Primary" anhydrites from the NPA-region have, on the average, lighter oxygen but about the same range of sulfur isotope ratios as the "secondary" anhydrites (Fig. 12). Apparently, the dissolved sulfate in the fluids which precipitated the "primary" anhydrites (Nisku seawater modified mainly by bacterial sulfate reduction and/or groundwater/river water input?) was different from the dissolved sulfate in those fluids that precipitated the "secondary" anhydrites in the study area. Alternatively, if the source for the "secondary" anhydrites was isotopically similar to the source for the "primary" anhydrites, the reservoir for the "secondary" anhydrites was isotopically fractionated and modified before and/or during emplacement of the anhydrites in the buildups.

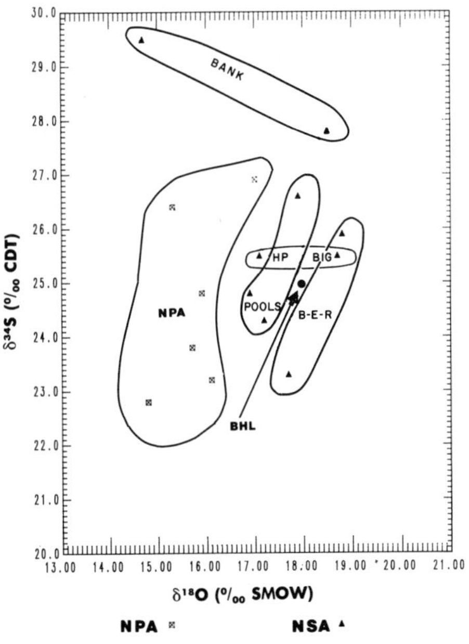

Fig. 12

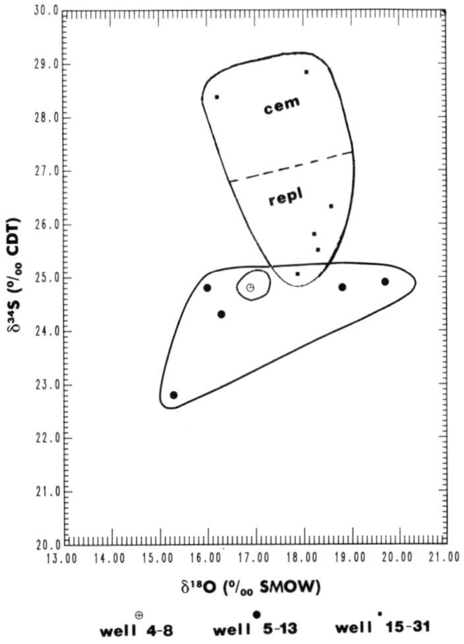

Fig. 13

Fig. 12. Plot of sulfur/oxygen isotope ratios of Nisku "primary" anhydrites (NPA) and "secondary" anhydrites (NSA) from various wells in pools on the slope, in the bank, and in the bank-edge reef. All NSA are from the Brazeau area, except for HP and Big (Pemina *C*-Pool and Bigoray *C*-Pool, respectively). Each data point is the mean value of all data from the respective well. *BHL* is one sample of Beaverhill Lake anhydrite

Fig. 13. Sulfur/oxygen isotope plot of individual samples from three Brazeau pools (see Fig. 3 for location). The data display large and inconsistent spreads, and cements and replacive anhydrites form separate groups (well 15 – 31). Similar patterns occur in most other wells

Three other isotopic features are conspicuous. (1) Petrographically syngenetic anhydrites display considerable isotopic variations within some pools; the greatest variation occurs in the Pembina *C*-Pool, where the sulfur isotopes in the replacive anhydrite range from +13 to +20 permil CDT. (2) The various types of anhydrite form distinct groups in several buildups (e.g., Fig. 13). (3) Corrotopic, felted, and cement anhydrites display isotopic variations of more than 6 permil for both oxygen and sulfur from one pool to another. These features suggest closed-system fractionation, bacterial sulfate reduction, and partial oxygen reequilibration *within* the buildups and/or derivation of sulfate from various sources. In any case, the meticulous isotopic differentiations from one pool to another and within individual pools are possible only in closed systems. Therefore, the buildups no longer had hydrologic communication at the time of anhydritization.

3.3.4 Discussion and Interpretation

There is no doubt that the "secondary" anhydrites in the study area formed at considerable depth, and that most of them replaced dolostone. Unequivocal evidence include (a) stylolites with clay and dolostone residues floating in massive replacive anhydrite, and (b) anhydrite filling molds and vugs that formed during the late stages of and/or after dolomitization. Furthermore, the various types of anhydrite did not form at the same time, as suggested by their chemical and petrographic differences. Corrotopic anhydrite most probably formed as gypsum, which was transformed to its present texture during progressive burial; the other types of anhydrite precipitated directly as anhydrite. This suggests that anhydritization overlapped with the gypsum/anhydrite transformation depth, which is normally in the order of about 600 to 800 m, depending on temperature, pressure, and fluid composition.

Most anhydrites are replacive and occur in extensively dolomitized strata. Hence, a genetic association between dolomitization and anhydritization is likely as, for example, in sabkhas where dolomitization and anhydritization are believed to enhance one another (e.g., Kinsman 1966). Among other factors, precipitation of gypsum will decrease the Ca/Mg ratio so that dolomite is formed instead of gypsum, which in turn increases the Ca/Mg ratio so that more gypsum is formed, and so forth. This process may also take place in the subsurface, implying that anhydritization in the Nisku may have partially coincided with the advanced stages of matrix dolomitization, and may have been caused by the same general flow pattern.

The exact source of the sulfate in the study area remains uncertain. Considering the absolute sulfur isotope values of the "secondary" anhydrites relative to the secular sulfur isotope curve of Claypool et al. (1980), there are three theoretical sources: Ireton formation waters, formation waters from the underlying Beaverhill Lake Formation, and Beaverhill Lake sulfate deposits (anhydrites occur in the Beaverhill Lake Formation close to the study area: BHL in Fig. 1; one sample of BHL anhydrite was analyzed, and is similar to NSA anhydrites:

Fig. 12). The "secondary" anhydrites in the study area may contain sulfate from one or the other, or from all three sources.

4 Conclusions

The diagenetic features discussed constitute a composite paragenetic sequence and formed during distinct, partially overlapping phases which were limited in space and time (Fig. 14).

The diagenetic features in the Nisku buildups are interpreted as having formed during progressive burial, except for molds and vugs (phase 3) and pendant calcite cements (phase 4) which probably formed during a brief period of exposure in some buildups. Phases 1 to 7 comprise the limestone diagenesis, and resulted mainly in cementation and lithification. Stylolites (phase 8) began to form before the onset of matrix dolomitization (phase 9). Chemical compaction continued throughout calcite dissolution (phase 10), and anhydritization (phase 12), and some stylolites of generation II (phase 11) predate anhydritization. The integration of all petrographic and geochemical data suggests that physical compaction and lithification (phase 6) were essentially completed at a burial depth of about 300 m, and anhydritization (phase 12) was essentially completed at about 1000 m. Because the study area reached a burial depth of 1100 m near the end of Mississippian time, the textures discussed in this paper formed relatively early during the burial history. The next distinct phase (oil migration) took place in upper Cretaceous time at much greater depth, and was followed by thermochemical sulfate reduction, and the formation of saddle dolomite, calcite spar, celestite, native sulfur, and pyrobitumen near the maximum burial depth of about 4000 m (Machel 1985a).

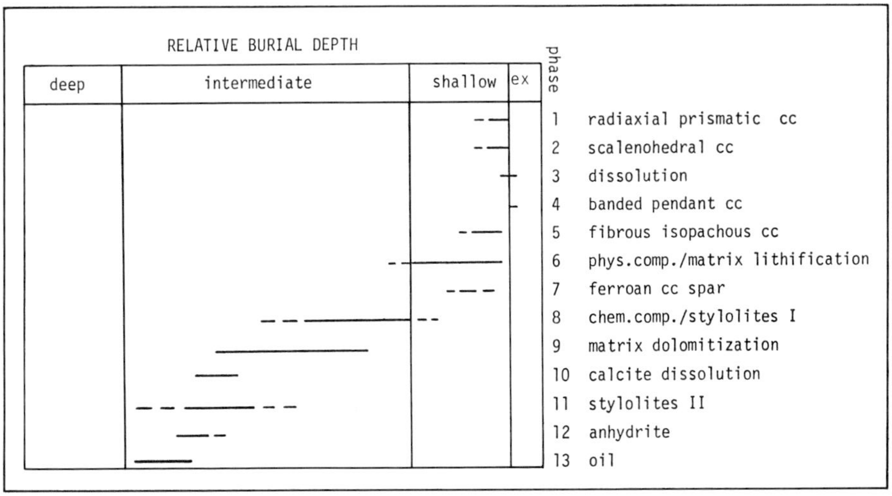

Fig. 14. Composite paragenetic sequence of the Nisku in the study area (*ex* exposed; *cc* calcite)

Acknowledgments. This research was supported by Petro Canada Ltd. (Calgary), NSERC grant No. A2128, and the Basin Research Institute at Louisiana State University. J. H. Anderson, B. H. Purser, M. Scherer, and J. H. Schroeder reviewed the manuscript.

References

Allen JR, Matthews RK (1977) Carbon and oxygen isotopes as diagenetic and stratigraphic tools: surface and subsurface data, Barbados, West Indies. Geology (Boulder) 5:16 – 20

Anderson JH (1985) Depositional facies and carbonate diagenesis of the downslope reefs in the Nisku Formation, Central Alberta, Canada. PhD Thesis, Univ Texas at Austin

Brand U, Veizer J (1980) Chemical diagenesis of a multicomponent carbonate system I: Trace elements. J Sediment Petrol 50:1219 – 1236

Burke WH, Denison RE, Hetherington RE, Koepnick RB, Nelson HF and Otto JB (1982) Variation of seawater $^{87}Sr/^{86}Sr$ throughout Phancerozioc time. Geology (Boulder) 10:516 – 519

Claypool GE, Holser WT, Kaplan IR, Sakai H, Zak I (1980) The age curves of sulfur and oxygen isotopes in marine sulfate and their mutual interpretation. Chem Geol 28:199 – 260

Friedman I, O'Neil JR (1977) Compilation of stable isotope fractionation factors of geochemical interest. In: Fleischer M (ed) Data of geochemistry. US Geol Surv Prof Pap 440K

Jones RMP (1980) Basinal isostatic adjustment faults and their petroleum significance. Bull Can Petrol Geol 28:211 – 251

Kinsman DJJ (1966) Gypsum and anhydrite of recent age, Trucial Coast, Persian Gulf. In: Rau JL (ed) 2nd Symp Salt. Northern Ohio Geol Soc, pp 302 – 326

Land LS (1980) The isotopic and trace element geochemistry of dolomite: the state of the art. In: Zenger DH, Dunham JB, Ethington RA (eds) Concepts and models of dolomitization. Soc Econ Paleontol Miner Spec Publ 28:87 – 110

Land LS (1983) Dolomitization. Am Assoc Petrol Geol Education Course Note Ser 24:1 – 20

Machel HG (1983) Facies and diagenesis of some Nisku buildups and associated strata, Upper Devonian, Alberta, Canada. In: Harris PM (ed) Carbonate buildups – a core workshop. Soc Econ Paleontol Miner Core Workshop No 4:144 – 181

Machel HG (1984) Facies and dolomitization of the Upper Devonian Nisku Formation in the Brazeau, Pembina, and Bigoray areas, Alberta, Canada. In: Eliuk L (ed) Carbonates in subsurface and outcrop. Can Soc Petrol Geol Core Workshop 1984:191 – 224

Machel HG (1985a) Facies and diagenesis of the Upper Devonian Nisku Formation in the subsurface of central Alberta. PhD Thesis, McGill Univ, Montreal

Machel HG (1985b) Fibrous gypsum and fibrous anhydrite in veins. Sedimentology 32:443 – 454

Machel HG (1986) Limestone diagenesis of Upper Devonian Nisku carbonates in the subsurface of central Alberta. Can J Earth Sci (in press)

Pingitore NE (1982) The role of diffusion during carbonate diagenesis. J Sediment Petrol 52:27 – 39

Popp BN, Anderson TF, Sandberg PA (1986) Brachiopods as indicators of original isotopic compositions in Paleozoic limestones. Geol Soc Amer Bull (in press)

Schmidt V, McDonald DA (1979) The role of secondary porosity in the course of sandstone diagenesis. In: Scholle PA, Schluger PR (eds) Aspects of diagenesis. Soc Econ Paleontol Miner Spec Publ 26:175 – 207

Veizer J (1983) Chemical diagenesis of carbonates: theory and application of trace element technique. In: Arthur MA, Anderson TE, Kaplan IR, Veizer J, Land LS (eds) Stable isotopes in sedimentary petrology. Soc Econ Paleontol Miner Short Course No:3 – 1 – 3 – 100

Marine Diagenesis in Devonian Reef Complexes of the Canning Basin, Western Australia

C. KERANS[1], N. F. HURLEY[2], and P. E. PLAYFORD[3]

1 Introduction

Marine diagenesis is widely recognized as a critical process in the development of reef complexes, both ancient and modern (James and Choquette 1983, James and Ginsburg 1979, Krebs 1969, Mountjoy and Krebs 1983, Playford 1980, 1984, Purser 1969, Schroeder 1972, Walls and Burrowes 1985). Marine cementation, together with frame-building reef organisms, is largely responsible for building reefs into rigid, wave-resistant structures with positive relief above the seafloor. Indeed, it has been suggested for some reefs that cementation was the dominant reef-building process and that they should therefore be regarded as inorganic "reefs" (Schmidt 1977).

The Canning Basin reef complexes display spectacular quantities of marine cement which is largely confined to the reef-flat, reef-margin, and reefal-slope limestones. This early cementation played an important role in the growth and post-depositional evolution of the reef complexes. Other processes of marine diagenesis, especially early void sedimentation and aragonite dissolution, were also important in modifying the primary depositional reef fabrics and influencing porosity evolution.

Major consequences of marine cementation in the Canning Basin reefs included early fracturing of the rigid, strongly cemented reef limestones, leading to the development of neptunian dykes and sills and the collapse of sections of the reef margins. Later post-burial diagenetic processes, including dolomitization and pressure-solution, were also controlled to a large degree by the extent of early marine cementation (Playford 1980, 1984).

Diagenesis of the reef complexes has recently been studied, on a regional basis, by Kerans (1985), and, for the Oscar Range reef complex, by Hurley (1986). Kerans' work was undertaken during a 2-year assignment as a post-doctoral research fellow with the Western Australian Mining and Petroleum Research Institute, while Hurley's work formed part of this Ph. D. thesis at the University of Michigan, sponsored by the Geological Survey of Western Australia. Their diagenetic research involved petrographic, cathodoluminescene, and isotopic studies. The analytical isotope work for Kerans' study was conducted by the Baas Becking Geobiological Laboratory, Canberra (Etminan et al. 1984). The present paper will deal only with early marine cementation of the reef

[1] Bureau of Economic Geology, The University of Texas at Austin, Austin, TX 78712, USA
[2] Marathon Oil Company, Denver Research Center, Littleton, CO, USA
[3] Department of Mines, Perth, Western Australia

Reef Diagenesis
Edited by J. H. Schroeder and B. H. Purser

complexes and its consequences, leaving discussion of subsequent burial and meteoric cementation to a later publication.

2 Geological Setting

The Devonian reef complexes outcrop in a belt of rugged limestone ranges about 350 km long following the northern margin of the Canning Basin and adjoining Precambrian rocks of the Kimberley Block (Fig. 1). They have been exhumed by Cenozoic erosion, so that the topography today resembles that of the Devonian seafloor (Fig. 4A). Excellent cross-sections are exposed in river gorges, notably in Windjana Gorge, which is cut by the Lennard River through the Napier Range reef complex (Fig. 4B).

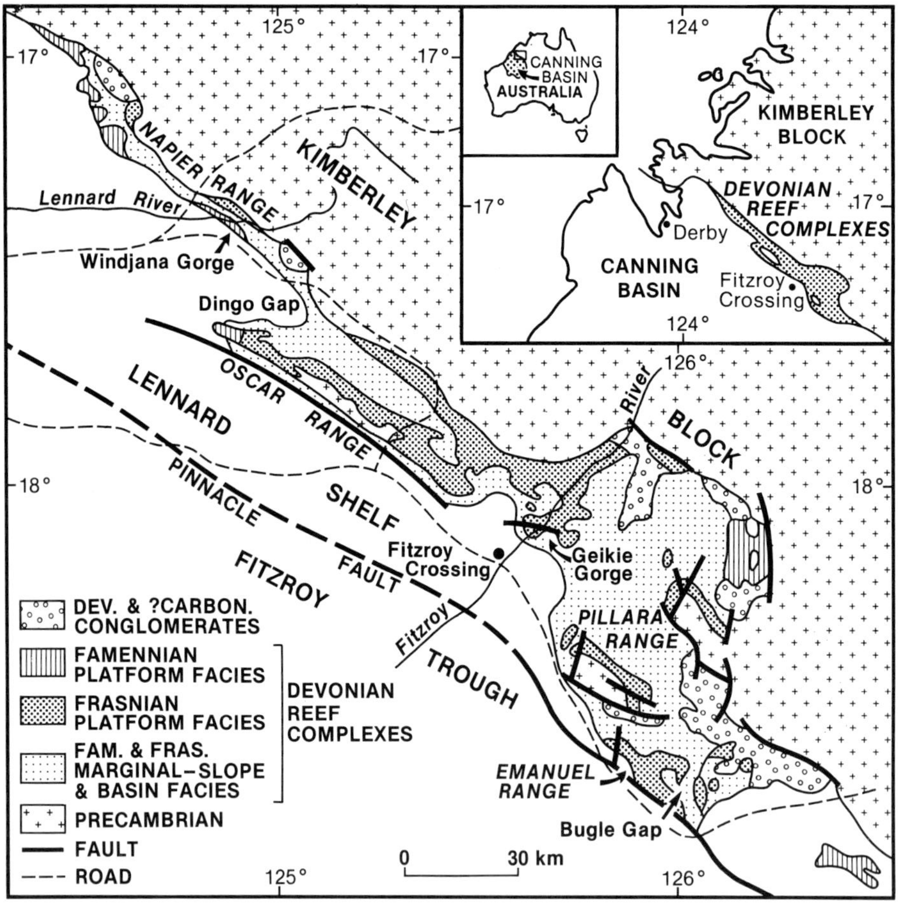

Fig. 1. Locality and generalized geological maps, Canning Basin

The reefs also continue in the subsurface of the Lennard Shelf and south of the Fitzroy Trough, where they are important targets in oil exploration (Playford 1982, Purcell 1984).

The Canning Basin is an intracratonic basin, initiated in the Ordovician as a broadly subsiding downwarp, with greatest subsidence along the northwest-trending Fitzroy Trough, which was initiated during a Middle Devonian aborted-rift phase (Brown et al. 1984, Middleton 1984, Towner and Gibson 1983). The stratigraphy and development of the Devonian reefs has been documented by Playford and Lowry (1966) and Playford (1980, 1981, 1982, 1984). Read (1973a, b) discussed the stratigraphy and depositional environments of the Frasnian backreef deposits (Pillara Limestone), and Hall (1984) described the evolution of the Limestone Billy Hills atoll.

3 Depositional Model

The facies terminology defined by Playford (1980) and used in this paper is shown in Fig. 2, and the stratigraphy and evolution of the reef complexes through time is shown in Fig. 3. Three main facies are recognized in the complexes: platform, marginal-slope, and basin facies. The platform facies is subdivided into the reef-margin, reef-flat, and back-reef subfacies, and the marginal-slope facies into reefal-slope and fore-reef subfacies.

The complexes developed as flat-topped reef-rimmed limestone platforms (Fig. 4A), built by stromatoporoids, corals, and cyanobacteria. They were flanked by dipping marginal-slope deposits (Fig. 4B) composed largely of platform-derived debris, with contributions from indigenous organisms and terrigenous sources. The basin deposits were laid down essentially horizontally, and consist primarily of terrigenous sediments with some platform-derived debris deposited by debris flows and turbidites.

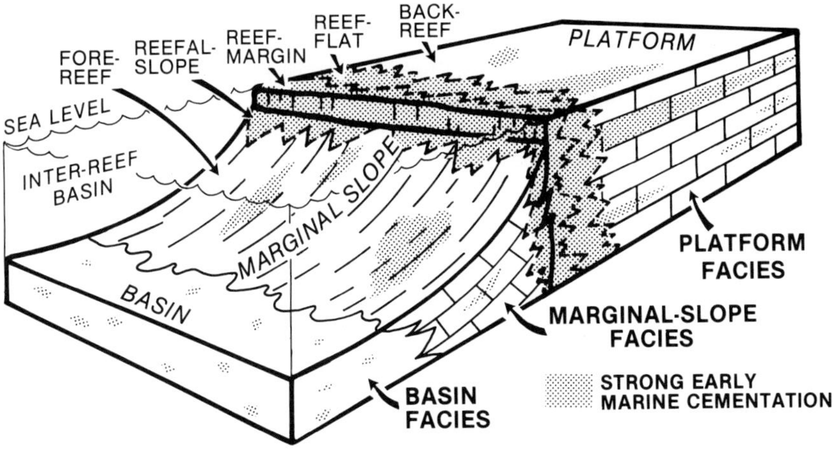

Fig. 2. Generalized morphology, facies and subfacies relationships, and extent of strong early marine cementation in the Devonian reef complexes. (After Playford 1984)

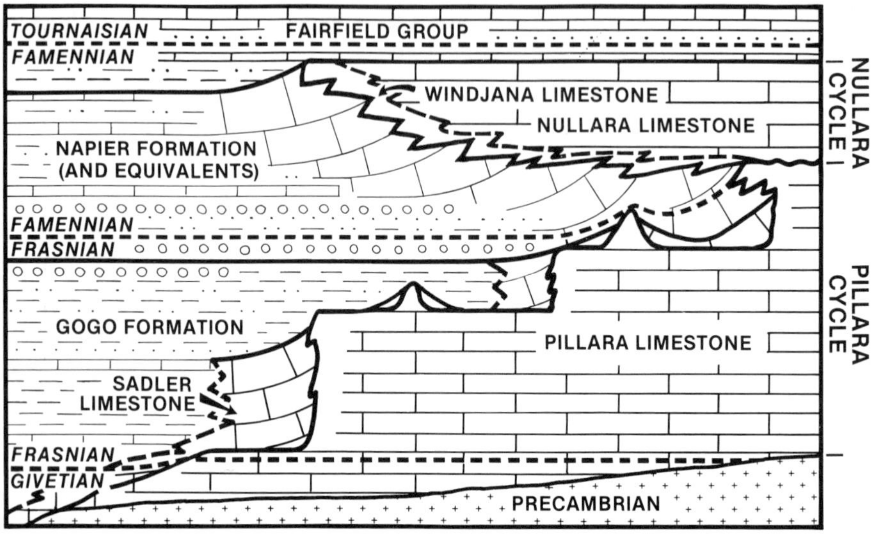

Fig. 3. Stratigraphic nomenclature and development of the reef complexes through time, showing the two major reef cycles. (After Playford 1984)

Reef growth began in the late Givetian, and continued through the Frasnian and most of the Famennian (Fig. 3). The Givetian-Frasnian phase of reef building is referred to as the Pillara Cycle, and the Famennian phase as the Nullara Cycle (Playford 1980). The Pillara Cycle reefs were built by stromatoporoids, corals, and cyanobacteria. However, a global mass extinction of reef-building organisms and other shallow-water metazoans occurred at the end of the Frasnian, so that the cyanobacteria alone remained to build the Famennian reefs of the Nullara Cycle (McLaren 1970, Playford et al. 1984).

Playford (1984) defined six types of platform margin in the complexes: upright-scarp, upright-rollover, advancing, retreating, backstepping, and pinnacle. The oldest (late Givetian) platforms were essentially low-relief banks, lacking prominent reef rims, and with generally retreating or upright-rollover margins. Upright-scarp margins with well-developed reefs first developed in the early Frasnian, followed in the middle to late Frasnian by intervals of abrupt backstepping and the associated development of pinnacle reefs. A brief emergence, with minor erosion of the platforms, occurred at the close of the Frasnian, and was followed by the development of advancing Famennian platforms of the Nullara Cycle, with well-developed reefs around their margins (Windjana Limestone).

Fig. 4. A Aerial view looking north over exhumed Devonian platform atolls in the Bugle Gap area; Laidlaw Range in *right center,* Lloyd Hill in *left center.* **B** The "classic face" in Windjana Gorge, showing flat-lying reef-flat (*right*), massive reef-margin (*center*) and steeply dipping marginal-slope limestones (*left*). **C** Neptunian dyke filled with radiaxial spar and lesser amounts of sediment, stratified parallel to the dyke margins; southern Oscar Range. **D** Large allochthonous block of reef in marginal-slope limestones (Napier Formation) near to the base of the original marginal slope; Napier Range near McSherry's Gap

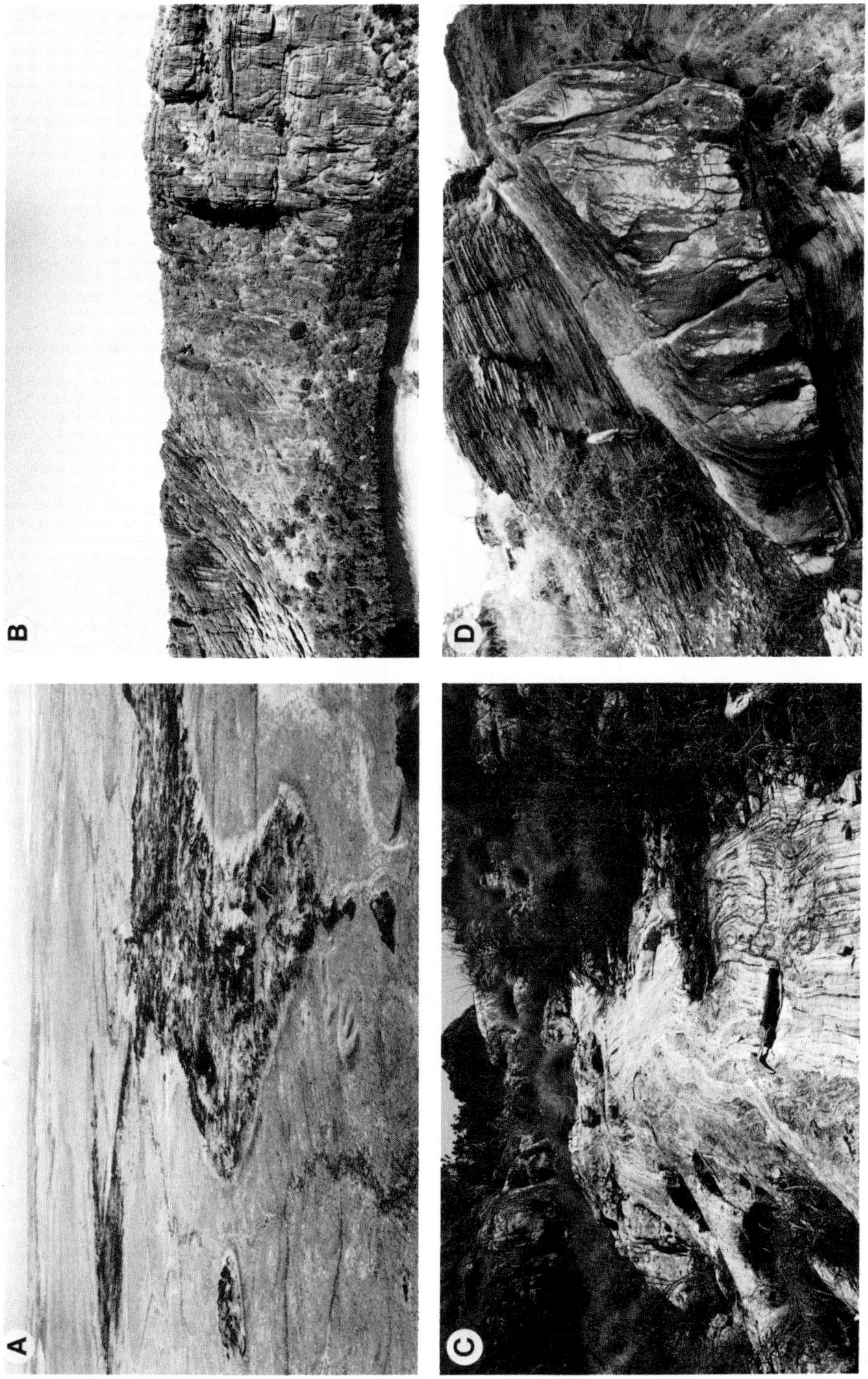

Fig. 4

4 Marine Diagenesis

Marine diagenesis in this paper is taken to include those diagenetic processes that occur on the seafloor (including the intertidal zone) and within the marine-burial realm, extending to those depths below the seafloor where the sediments are still subject to circulating seawater.

Marine diagenesis in the Canning Basin carbonates consisted primarily of extensive calcite cementation and associated void sedimentation. Processes of bioerosion, micritization, and seafloor neomorphism were rare. No aragonitic cements were precipitated, and early dissolution of aragonitic skeletal material was widespread.

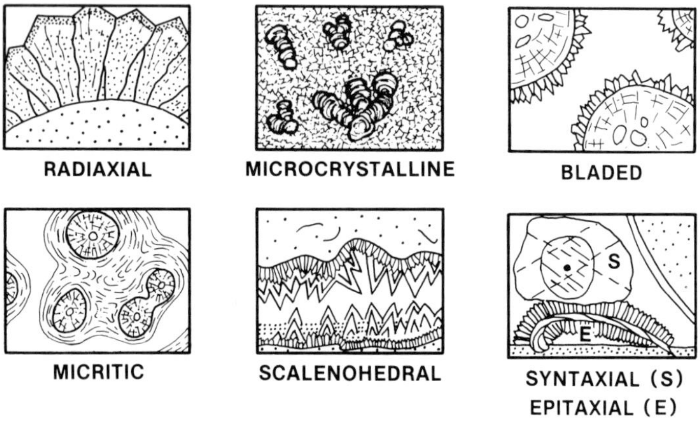

Fig. 5. Types of marine calcite cement recognized in the Devonian reef complexes

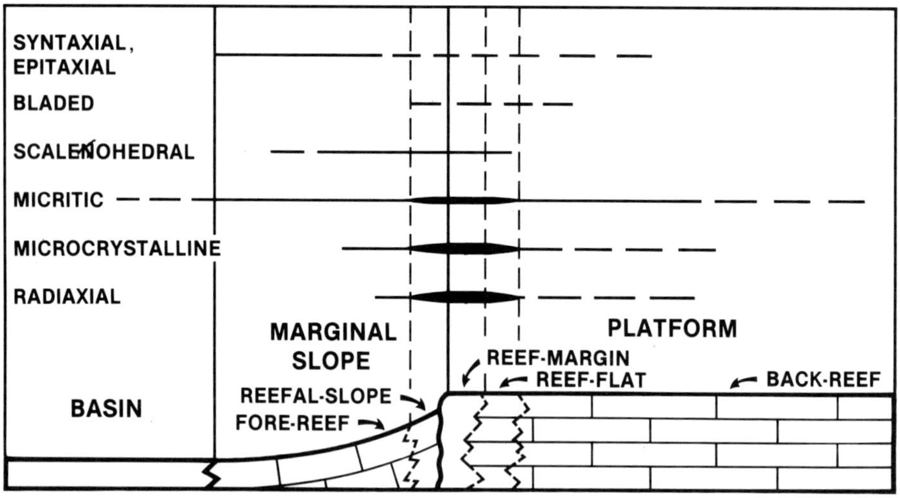

Fig. 6. Distribution of marine cements

4.1 Marine Cementation

Kerans (1985) and Hurley (1986) recognized three principal types of marine cement in the Devonian reef complexes: Radiaxial, microcrystalline, and micritic calcite. Other less important types are bladed, scalenohedral, syntaxial, and epitaxial calcite. These cement types are illustrated in Fig. 5, their distribution within the reef complexes is shown in Fig. 6, and their stable- isotope characteristics in Fig. 7.

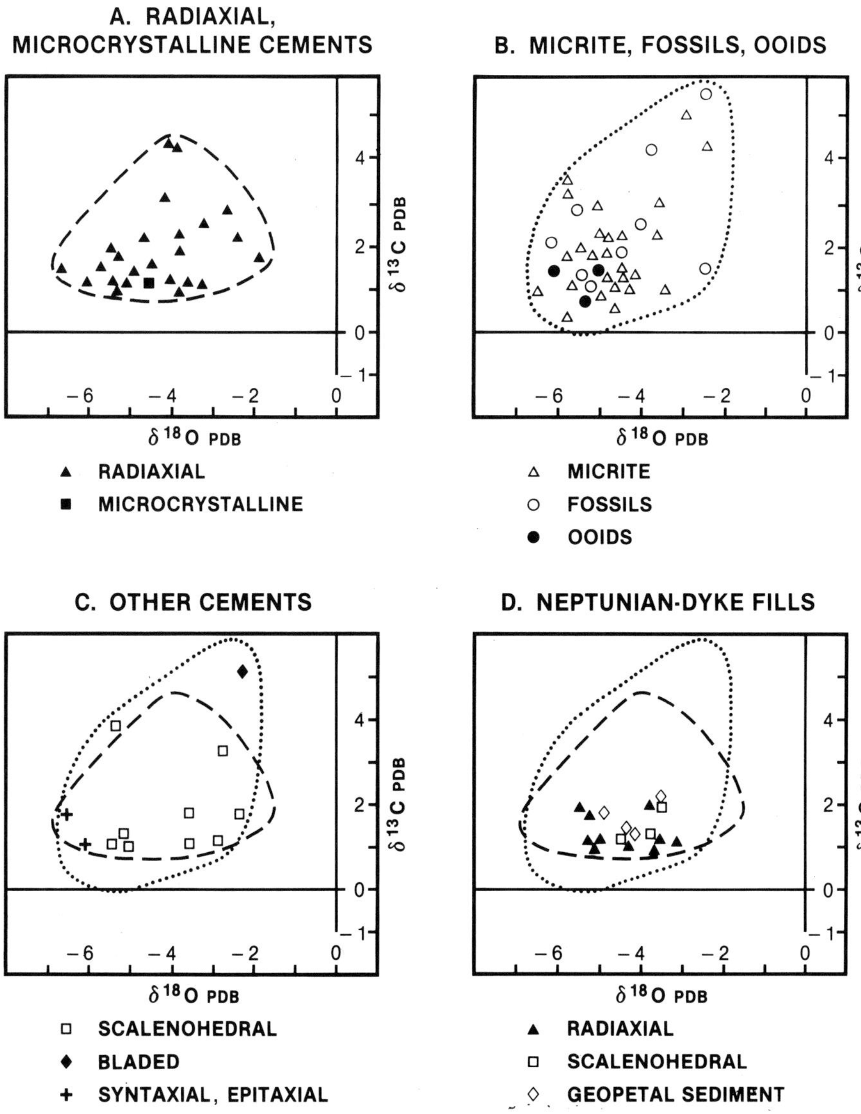

Fig. 7A – C. Stable-isotope compositions of marine cements and their associated sediments. (After Kerans 1985)

Radiaxial, microcrystalline, micritic, and bladed cements. For the most part, were precipitated at or near the seafloor, although they could also form at considerable depths below this surface in neptunian fissures and other cavity systems. Scalenohedral, syntaxial, and epitaxial calcite are regarded as marine-burial cements. Later blocky-calcite cements, which will not be discussed further in this paper, formed in burial and meteoric environments (Kerans 1985, Hurley 1986).

4.1.1 Radiaxial Calcite

Radiaxial calcite is the single most impressive diagenetic feature of the reef complexes, commonly comprising more than 20% and locally more than 50% of reef-margin, reef-flat, and reefal-slope limestones. It forms finely banded fibrous layers filling cavities in the limestones (Figs. 5, 8).

Canning Basin radiaxial spar is characteristically milky white, as a result of high densities of microdolomite inclusions (Fig. 10B) (cf. Lohmann and Meyers 1977), and is believed to have been precipitated originally as high-magnesium calcite. Minor amounts of radial fibrous calcite have been recognized in close association with radiaxial calcite (Kendall 1985). Locally the two are gradational within individual growth bands, and any genetic differences between them must be minor.

The layering in radiaxial crusts is generally isopachous, although it may be thicker in the upper parts of cavities and between protuberances on the cavity walls (Fig. 8A). Individual layers commonly range in thickness from less than a millimeter to several millimeters. However, in large reef cavities, composite thicknesses of several tens of centimeters of radiaxial spar are common, and in wide neptunian dykes total widths of radiaxial spar can amount to many meters (Fig. 4C).

Radiaxial spar was clearly precipitated very early, under marine and marine-burial conditions. Multiple alternations of radiaxial spar and marine internal sediment (including fossil material) are common in cavities (Figs. 8C, 10A), and encrustation of radiaxial cements by the cyanobacteria *Renalcis* and *Frutexites* is known from neptunian fissures and other cavities in the reefs. Fragments of radiaxial spar also occur as clasts in contemporary fore-reef breccias.

Precipitation of radiaxial, microcrystalline, and micritic calcite cement was contemporaneous with growth of the reefal limestones. However, microcrystal-

→

Fig. 8. A Radiaxial cement, in part interlaminated with sediments (cavity peloids) filling a primary cavity in *Stachyodes-Renalcis* reef limestone (Pillara Limestone). The final filling consists of clear blocky (post-burial) spar. Bugle Gap area, sample GSWA 3497. **B** Radiaxial cement partly filling cavities below laminar stromatoporoids in reef limestone (Pillara Limestone). *Renalcis* and microcrystalline cement form protuberances from the upper surface of the cavities. The final filling of the upper cavity consists of clear blocky (post-burial) cement. Windjana Gorge, sample GSWA 3298. **C** Interlaminated radiaxial spar and cavity peloids in reef limestone (Pillara Limestone). Bugle Gap area, sample GSWA 3497. **D** Small neptunian sills and dykes, showing evidence of repeated fracturing and cementation, filled with alternating void sediments, radiaxial calcite, and scalenohedral calcite, in early-cemented marginal-slope limestone (Virgin Hills Formation). Bugle Gap area, sample GSWA 11677B

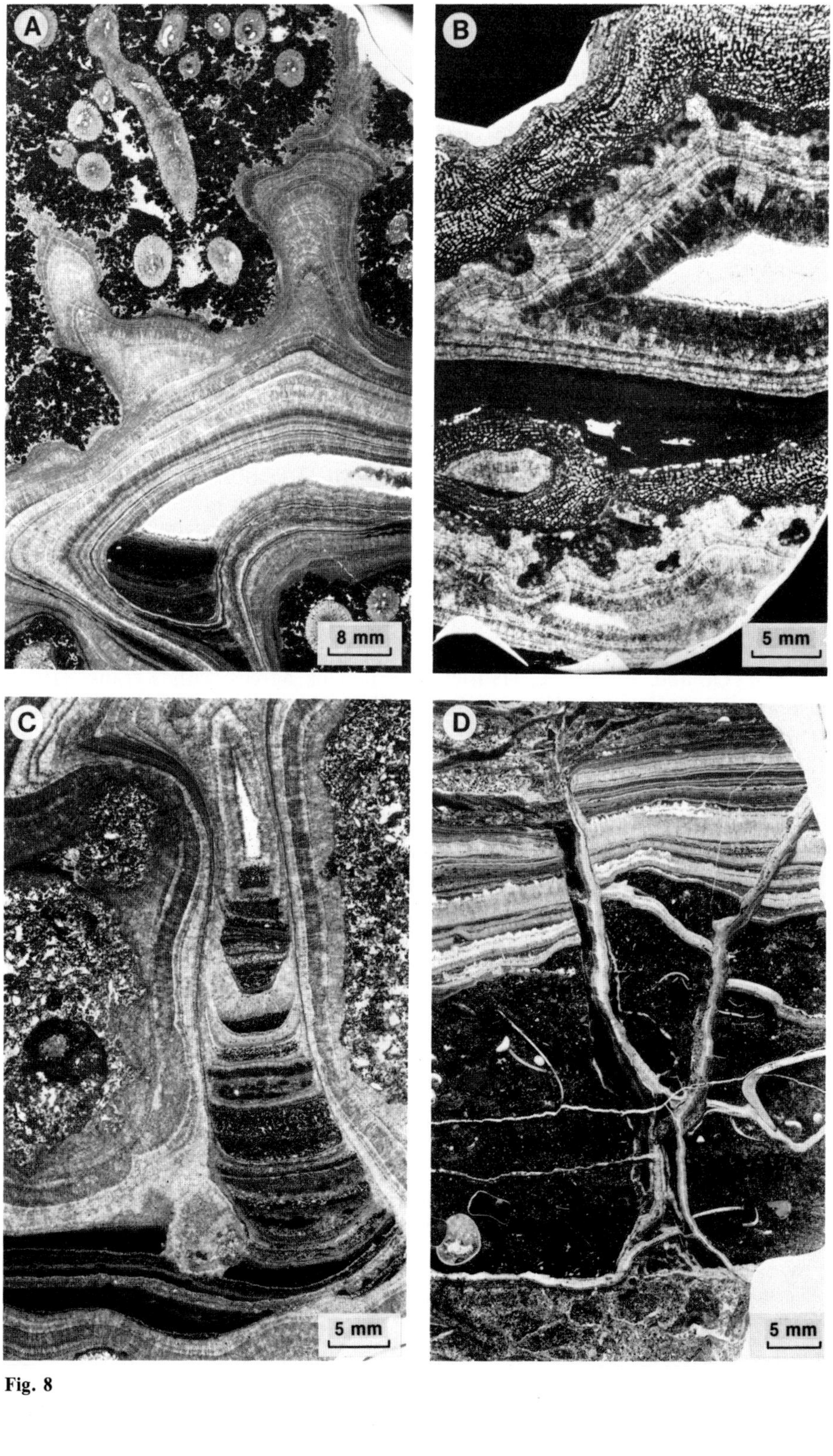

Fig. 8

line and micritic cements normally precede radiaxial cement in the linings of cavities. Radiaxial and scalenohedral spar layers are partly interlaminated in some neptunian dykes and sills (Figs. 8D, 10A), but radiaxial calcite is generally the earlier of the two.

The distribution of radiaxial spar is largely restricted to reef-margin, reef-flat, and reefal-slope subfacies, and to the infillings of neptunian dykes and sills. In passing from the back-reef to the reef, the first appearance of radiaxial cement in cavities generally corresponds to the boundary between the back-reef and reef-flat deposits, but radiaxial spar in neptunian fissures can extend for hundreds of meters into early-cemented parts of the back-reef.

While it is clear that abundant radiaxial spar was precipitated at or near the depositional interface, the deepest limit of precipitation below this surface has not been firmly established. However, radiaxial spar can be shown to have lined neptunian dykes to depths of many tens of meters below the surface, and it must similarly have been precipitated in other intra-reef cavity systems to significant depths. Indeed, it seems likely that radiaxial cements were precipitated more than 100 m below the seafloor in such cavity systems. This observation is consistent with reports of seawater circulation to considerable depths in modern atolls (Saller 1984, Simms 1984, Tator and Hatfield 1975), and of radiaxial cements forming in other ancient marine-burial environments (Halley and Scholle 1985).

Detailed stable-isotope geochemistry of radiaxial spars from platform deposits and neptunian dykes in the Oscar Range shows converging trends, indicating that Late Devonian seawater had an isotopic composition of about $-4.5\,\delta^{18}O$, and $+2.0\,\delta^{13}C$ (Hurley 1986). This agrees with Kerans' (1985) average value of $-4.7\,\delta^{18}O$, $+1.8\,\delta^{13}C$ for various marine components (Fig. 7B), and for radiaxial spar from a variety of settings (average $-4.3\,\delta^{18}O$, $+1.9\,\delta^{13}C$, Fig. 7A).

Radiaxial spar is generally nonluminescent or displays dull background luminescence. However, localized patches of moderate to bright luminescence occur along boundaries between radiaxial spar crystals and in zones that cross-cut growth banding. These luminescent patches typically display low densities of inclusions.

The original mineralogy of radiaxial fibrous spar is interpreted to have been high-magnesium calcite, based on the presence of microdolomite inclusions and the composition of closely comparable cements in Tertiary limestones of Bikini and Eniwetok atolls (Videtich 1985). Kendall (1985) has presented arguments for the primary (unrecrystallized) nature or radiaxial calcites, based on a study of radiaxial calcite filling neptunian sills at McWhae Ridge in the Bugle Gap area of the Canning Basin. While his split-crystal model for growth or radiaxial spars is feasible, the abundant microdolomite, patchy cathodoluminescence, and patchy

Fig. 9. A Spar balls, consisting of concentric layers of radiaxial spar around breccia fragments of earlier spar or country rock, in a neptunian dyke following the Cadjebut Fault, Bugle Gap area. **B** Microcrystalline calcite cementing small *Renalcis* colonies in reef limestone (Windjana Limestone). Oscar Range, sample GSWA 192B. **C** Micritic cement in *Stachyodes-Renalcis* reef limestone (Pillara Limestone), with void-filling sediments and some radiaxial spar (*left*). Bugle Gap area, sample GSWA 3497. **D** Stromatactis structures in polished slab of *cryptalgal-Renalcis* boundstone, reefal-slope subfacies (Napier Formation). Napier Range, sample GSWA 29499A

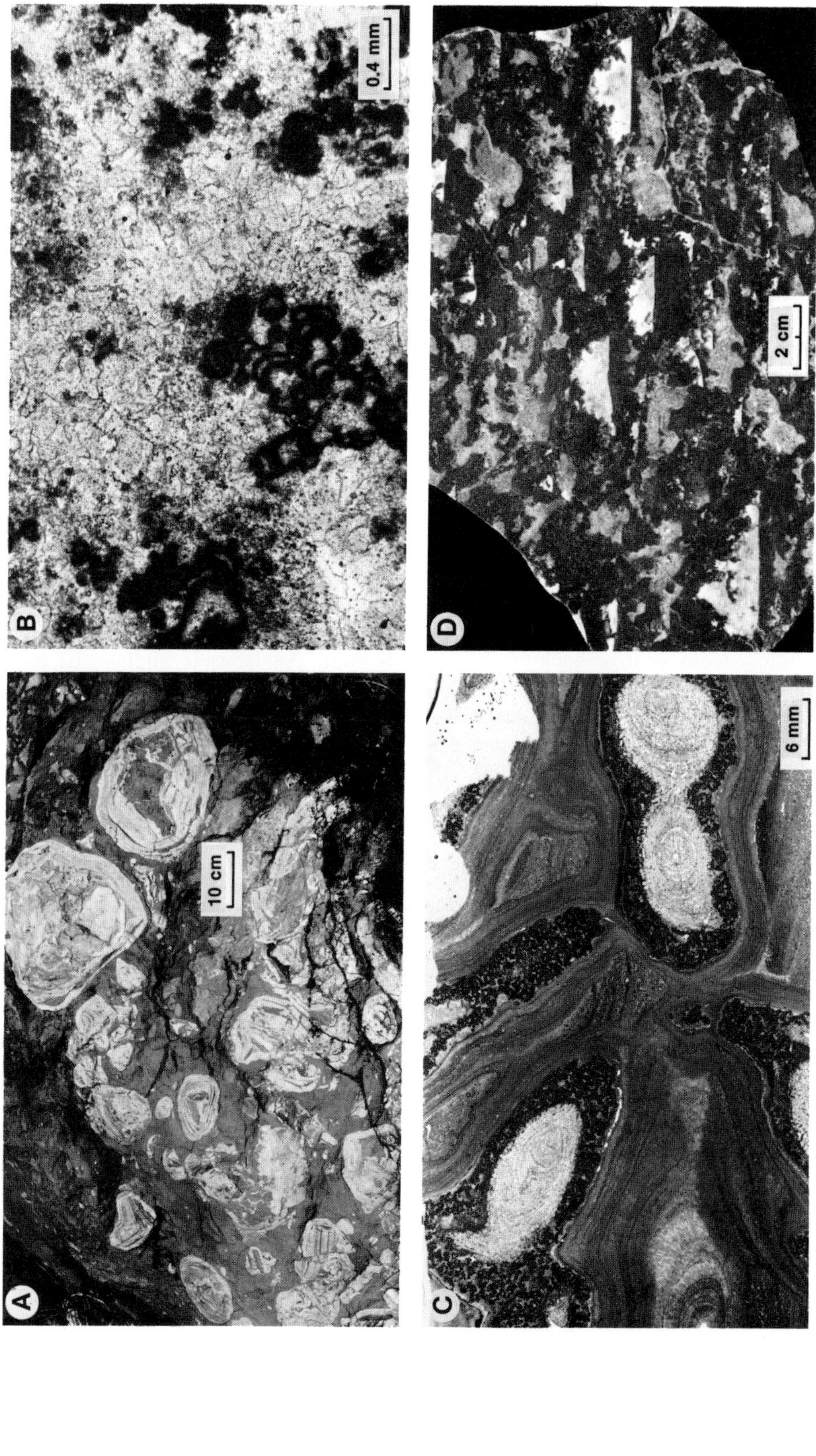

Fig. 9

distribution of inclusions indicate that some dissolution-reprecipitation must nevertheless have occurred.

4.1.2 Microcrystalline Calcite

Microcrystalline calcite makes up major parts of reef limestones constructed by the cyanobacteria *Renalcis* and *Sphaerocodium,* commonly constituting as much as 20% of such reefs by volume, and locally more than 50%. This cement was precipitated contemporaneously with growth of the cyanobacteria (Figs. 5, 9B). It commonly shows early small-scale fracturing and is present in debris blocks within equivalent marginal-slope deposits.

The cement is white and inclusion-rich, giving it a cloudy appearance in thin section, but the nature of the inclusions themselves has not been determined. The crystal size is variable, within the broad range of 0.01 to 0.2 mm, and there is no preferred orientation of the crystal c-axes.

Microcrystalline calcite cementing *Renalcis* forms large masses in parts of the reef subfacies, especially in Famennian reefs of the Nullara Cycle. This cement with associated *Renalcis* also lines stromatactis and other reef cavities. A typical cavity-filling sequence, from earliest to latest, consists of microcrystalline calcite, radiaxial calcite, and scalenohedral calcite (marine cements), and finally clear equant blocky calcite (burial or meteoric cement).

Microcrystalline calcite characteristically lacks luminescence or displays weak homogeneous orange background luminescence. An isotopic analysis of one sample of this cement ($-4.8\ \delta^{18}O$, $+1.2\ \delta^{13}C$) is close to the mean value for radiaxial spar (Fig. 7A). The original crystallinity is uncertain, but the lack of luminescence is consistent with minimal neomorphism, and it is likely that the cement was precipitated as microcrystalline high-magnesium calcite.

4.1.3 Micritic Calcite

Micritic calcite cement can be difficult to distinguish from its detrital counterpart. However, it commonly forms isopachous crusts of cement lining reef cavities (Figs. 5, 9C). It also often displays evidence of early fracturing, followed by precipitation of radiaxial or other well-documented marine spar, showing that it was cemented in the marine environment. Synsedimentary reworking of micritic nodules from the fore-reef slope (Kerans and Playford 1984) and of

Fig. 10. A Radiaxial spar with interlaminae containing cavity peloids, and a single layer of clear scalenohedral spar (*upper part*). *Arrow* marks the area shown in **B** Neptunian sill in marginal-slope facies (Virgin Hills Formation). McWhae Ridge, Bugle Gap area, specimen 80881B. **B** Scanning electron microscope photo of the area arrowed in **A**, showing radiaxial calcite (*above*) with microdolomite inclusions (*arrowed*), and scalenohedral calcite (*below*). **C** Cavity filled with scalenohedral spar and internal sediments, in early-lithified marginal-slope facies (Virgin Hills Formation). Bugle Gap area, specimen 37141C. **D** Goniatite-nautiloid packstone showing dissolution of aragonitic shells, and infilling by radiaxial spar and marine internal sediment, in marginal-slope deposits (Virgin Hills Formation). Bugle Gap area, specimen 37141I

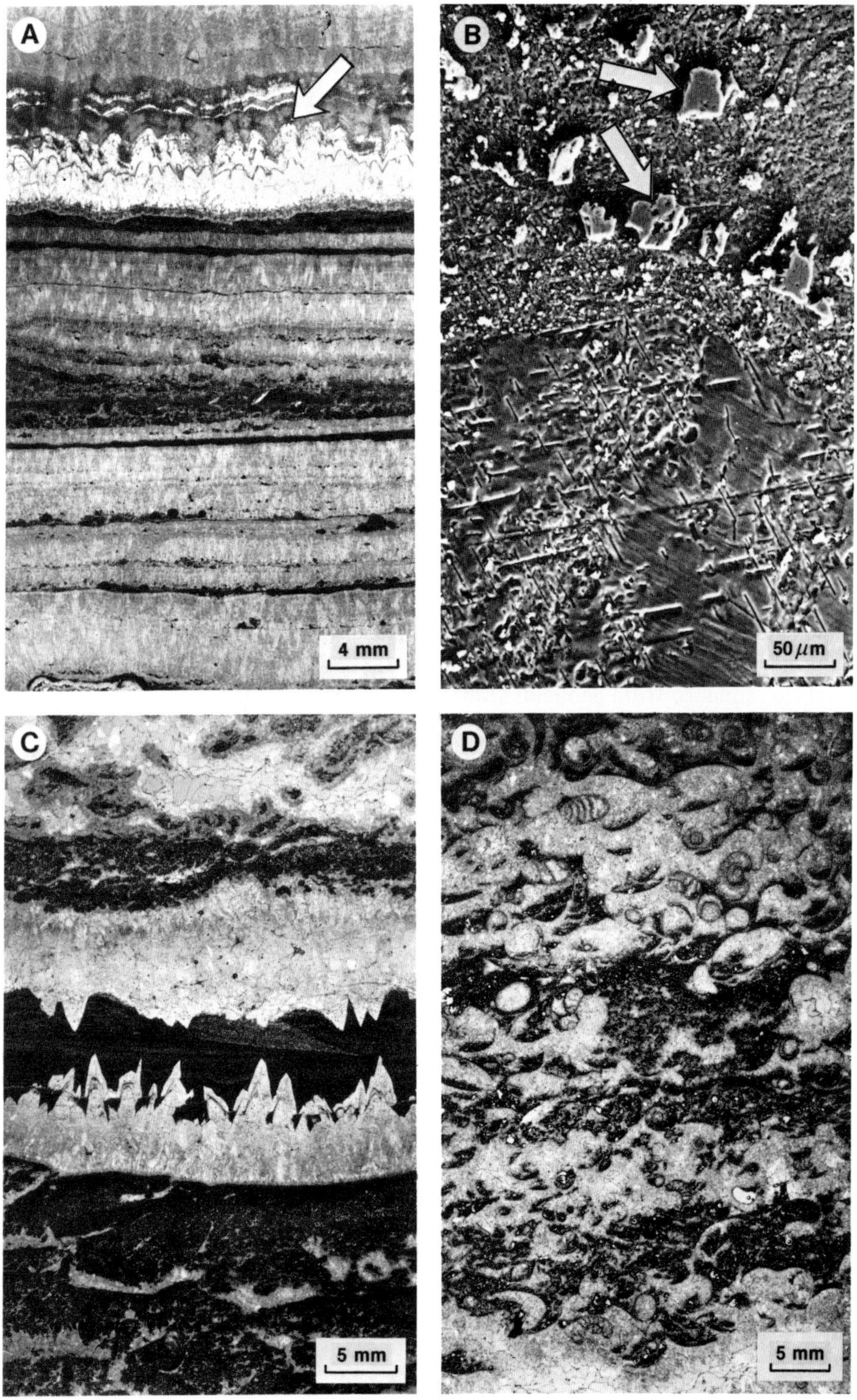

Fig. 10

blocks of micritic cyanobacterial boundstone from the platform testify to the widespread nature of micritic cementation.

The luminescent character of micritic cement is a dull homogeneous "background" orange-red. Isotopic analyses of micrite (both detrital micrite and micritic cement) are given in Fig. 7B, which also shows analyses for fossils (stromatoporoids, corals, crinoids, brachiopods, *Renalcis,* and stromatolitic micrite), and ooids. The isotopic values show a wide scatter, but with no significant separation between micrite, fossils, and ooids. The mean value is $-4.7\ \delta^{18}O$, $+1.8\ \delta^{13}C$, close to the mean for radiaxial spar (Fig. 7A) and Hurley's (1986) determination of the composition of cements precipitated from Devonian seawater.

4.1.4 Scalenohedral Calcite

Scalenohedral calcite cement consists of clear columnar "dogtooth" crystals (Figs. 5, 10A, B, C), or stubby crystal crusts, which are clear (largely inclusion-free) and display unit extinction.

Scalenohedral cement occurs as crusts lining cavity systems in Frasnian reef and reefal-slope limestones, and in neptunian dykes and sills of Frasnian and Famennian age. Scalenohedral cements are later than microcrystalline and micritic cements, and generally later than radiaxial cements. However, in neptunian dykes and sills that are characterized by frequent refracturing, a cyclical sequence from geopetal sediment to radiaxial spar to scalenohedral spar is observed. In one neptunian sill the iron-fixing cyanobacterium *Frutexites* has been observed, encrusting the tips of scalenohedral calcite crystals.

Scalenohedral calcite is nonluminescent and has a mean stable isotopic composition ($-4.3\ \delta^{18}O$, $+1.8\ \delta^{13}C$) comparable to that of radiaxial spar and other marine components (Fig. 7C, D).

Petrographic, luminescent, and stable-isotope relationships indicate the scalenohedral cement was precipitated as low-magnesium calcite from oxidized marine waters in cavity systems at some depth below the seafloor.

A lower nucleation density for scalenohedral spar relative to radiaxial spar indicates that the precipitating fluid for scalenohedral cement was less saturated in calcium carbonate (Given and Wilkinson 1985). In addition, the lack of microdolomite inclusions in scalenohedral cement favors derivation from a relatively magnesium-poor fluid. These differences are probably related to the evolution of seawater passing through the cavity systems, with radiaxial calcite being precipitated first, from normal seawater with strong through-flow, and scalenohedral calcite afterwards, from more mature seawater that had become less saturated in calcium carbonate and more depleted in magnesium, as it moved through the progressively occluded cavity systems.

4.1.5 Bladed Calcite

Bladed calcite cement occurs in Frasnian reefal-slope limestones and early-cemented lenses in Frasnian back-reef limestones. A sample of bladed spar from the back-reef of Geikie Gorge (Fig. 7C) shows distinctly heavy carbon-isotope values (+5.1 $\delta^{13}C$, −2.3 $\delta^{18}O$). These cements were precipitated prior to significant mechanical compaction of the sediments, and must have formed within a few meters of the surface. They are best interpreted as having been precipitated from marine water modified by bacterial fermentation (Irwin et al. 1977).

4.1.6 Syntaxial and Epitaxial Calcite

Syntaxial and epitaxial cements are found principally in marginal-slope deposits, and most are thought to have formed in shallow marine-burial conditions. Epitaxial cements are developed as isopachous crusts on brachiopods, corals, stromatoporoids, tentaculitids, calcispheres, and ooids, all of which have fibrous microstructure of primary calcitic origin. No epitaxial cements are developed on molluscan shells of primary aragonitic composition. Syntaxial cements are developed on pelmatozoan debris.

These cements are all inclusion-poor and non-luminescent, with isotope values (for single samples of each type) of +1.73 $\delta^{13}C$, −6.59 $\delta^{18}O$ (syntaxial cement), and +1.02 $\delta^{13}C$, −6.05 $\delta^{18}O$ (epitaxial cement) (Fig. 7C). An original low-magnesium calcite mineralogy is probable.

4.2 Internal Sedimentation

Wide varieties of internal sediment occur in cavities within the reefal limestones, ranging from structureless lime mudstones, to graded oolitic grainstones, fossiliferous wackestones and packstones, and terrigenous sandstones. Internal sediments occlude much of the primary porosity in many cavity systems, and they are frequently interlaminated with marine cements (Figs. 8A, C, D, 10A, C). The internal sediments are commonly geopetal, forming fossil "spirit levels" that can be used to determine depositional dip and the allochthonous versus autochthonous nature of isolated masses of reefal limestone in marginal-slope deposits (Playford 1981).

Much of the internal sedimentation in primary cavities was derived by downward percolation from the depositional interface. However, many cavities are partly filled by autochthonous cavity peloids (Figs. 8A, 8C, 10A) 0.08 – 0.5 mm in size, which are typically yellow or red in color. They are know only from reef and reefal-slope cavities and neptunian fissures, and may be of bacterial origin (Playford 1984). Fossils in the internal sediments include forms derived from the seafloor above, encrusting cyanobacteria, and troglodytic organisms (especially ostracods) that lived only in cavity systems.

4.3 Aragonite Dissolution

Early dissolution of aragonitic shells (predominantly molluscs) occurred throughout growth of the Devonian reef complexes, in environments from deep-water toe of slope to shallow-water back-reef (Fig. 10D). Several lines of evidence indicate that marine rather than meteoric water was responsible for this dissolution.

Molluscan fossils in the reef complexes are invariably preserved as molds, regardless of the depositional setting, from shallow to deep water. First-generation mold-filling material typically consists or marine internal sediment, identical to that making up the surrounding marine sediment, followed by marine cement, either microcrystalline, radiaxial, or micritic. Some molds show evidence of partial collapse before being cemented.

Aragonite dissolution (Schlager 1974) by seawater in the shallow marine-burial realm best fits the observed sequence of solution of the shell, followed immediately by marine infill, without any evidence of freshwater diagenesis. However, if seawater was responsible for this penecontemporaneous dissolution, the degree of aragonite saturation in Devonian seawater must have been significantly lower than it is today. In modern oceans seafloor aragonite dissolution occurs at depths below 300 – 350 m, but not in shallower waters (Scholle et al. 1983). However, dissolution may have occurred as shallow as 100 – 250 m in the recent past (Balsam 1983, Land and Moore 1980).

It is reasonable to expect that significantly shallower depths of aragonite dissolution would have prevailed in past periods when inorganic aragonite cement was not being precipitated ("submergent stages" of Wilkinson et al. 1985). This was the situtation during growth of the Canning Basin reef complexes, there being no indication of any cementation by aragonite.

An alternative explanation for dissolution of aragonitic shells in the shallow-water limestones would be that it was related to repeated emergence of the Devonian platforms because of fluctuations in sealevel, as has been suggested for shallow platform carbonates elsewhere (Prezbindowski 1985). However, there is no evidence to indicate that the Devonian platforms were repeatedly exposed above sea level, as would be required to explain dissolution of aragonitic shells throughout the sequence. Marine dissolution through all water depths is the more likely explanation.

5 Consequences of Early Marine Cementation

Early marine cementation had profound effects on the development of the reef complexes both during growth and after burial. Some of the principal consequences are summarized diagrammatically in Figs. 11, 12, and 13.

5.1 Reef Morphology

Early marine cementation was fundamentally important in controlling the morphology of the reef complexes. The oldest platforms, of late Givetian and early

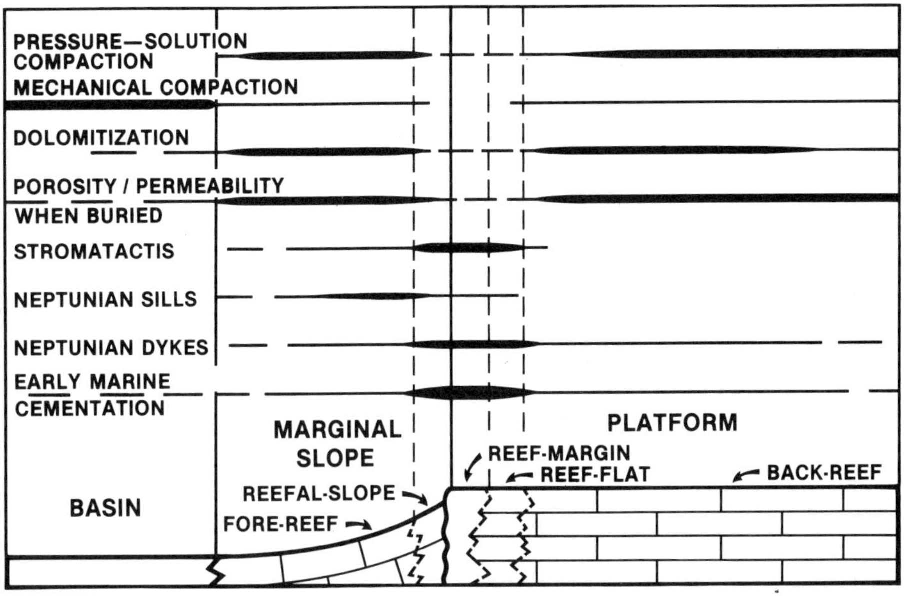

Fig. 11. Distribution of early marine cementation and consequent structures and diagenetic effects

Frasnian age, lacked significant marine cementation of the platform margins. They were unable to build up strong relief above the surrounding seafloor, and were essentially low-relief banks. On the other hand, the succeeding Frasnian and Famennian platforms had well-cemented, wave-resistant reef margins. They maintained precipitous reef scarps elevated above the surrounding basin floors and were flanked by steep marginal-slope deposits. Moreover, the strong early cementation was responsible for sustaining very steep depositional dips, well above the angle of repose, in reefal-slope limestones forming the upper parts of the marginal slopes.

5.2 Porosity Evolution

Some of the highest initial porosities in the complexes were reef cavity systems around the platform margins. However, this porosity was almost entirely destroyed at a very early stage by a combination of void sedimentation and pervasive early marine cementation. Thus, the highest porosity and permeability remaining at the time of burial was commonly in parts of the back-reef and fore-reef subfacies. This porosity and permeability, in turn, was largely destroyed after burial by a combination of mechanical and pressure-solution compaction.

5.3 Neptunian Dykes and Sills

Neptunian dykes and sills formed at an early stage in strongly cemented parts of the Devonian reef complexes, especially in reef-margin, reef-flat, and reefal-slope limestones, but also in cemented parts of the back-reef and fore-reef limestones. Playford (1984) suggested that neptunian fissures formed in these rigid limestones during earthquakes associated with local tectonism, and by slippage along underlying marginal-slope bedding and compaction of underlying basinal muds. Hurley (1986) has further suggested that differential compaction over a buried ridge of Precambrian rocks was responsible for some neptunian fissuring in the Oscar Range reef complex.

The open neptunian fissures were filled soon after they formed by combinations of sedimentation from above, growth of encrusting organisms (especially *Renalcis* and stromatolites) down the walls, internal sedimentation (cavity peloids), and precipitation of marine cement (mainly radiaxial calcite). The cements have isotopic signatures very similar to those of the same types of cement in other reefal cavity systems (Fig. 8D).

Dykes and sills commonly show a complex history of repeated fracturing and infilling. Some dykes are up to 6 km long, 20 m wide, and 80 m deep, but most are only a few tens of meters long, a few centimeters wide, and several meters deep. Most dykes pinch out at depth, but some sole out along bedding planes of underlying marginal-slope deposits. The neptunian fissures generally formed parallel to the platform margins, but some are at right angles to the margins. The maximum measured extension resulting from neptunian fissuring is 6.7% over a width of some 850 m in Nullara and Windjana Limestones near the northwestern end of the Oscar Range (Hurley 1986).

Neptunian dykes are most common along the platform margins in reef-margin, reef-flat, and reefal-slope limestones. However, in the Famennian reef complexes there is also extensive development of neptunian dykes in early-cemented parts of the back-reef subfacies. Neptunian sills are most abundant in marginal-slope deposits, especially in the reefal-slope deposits and early-cemented parts of the fore-reef deposits. They are believed to have resulted from downslope sliding of cemented limestones along curved bedding planes, causing fissures to open up parallel to bedding. Some sills are up to 80 cm thick, but most are only a few centimeters thick.

Spar balls, consisting of concentric layers of radiaxial spar up to 25 cm in diameter (Fig. 9A) are interesting features of neptunian dykes in several localities (Playford 1984). They are thought to have formed as a result of rotation during spar precipitation in the neptunian fissures, caused by strong water movement (probably wave surge).

5.4 Debris Flows and Allochthonous Blocks

Massive debris-flow deposits and huge allochthonous reef blocks are conspicuous features of the marginal-slope deposits in many areas (Figs. 4D, 12). They resulted from collapse of strongly cemented parts of the platform margins along neptunian fissures, probably often initiated by earthquakes. The resulting platform debris either rolled and slid down the marginal slopes as isolated blocks, or gave rise to slides of marginal-slope material which combined with the blocks to form debris flows. Some debris flows also resulted from the sliding and breaking

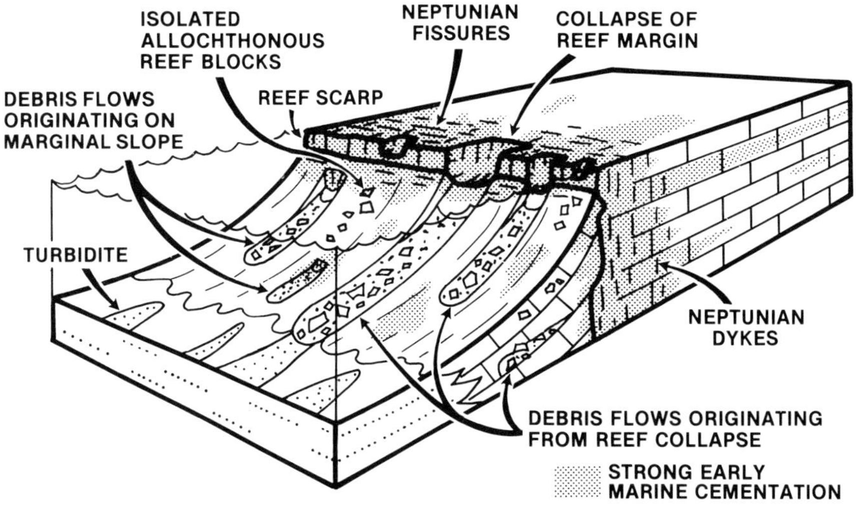

Fig. 12. Block diagram illustrating the origin of debris flows and allochthonous blocks, and the distribution of neptunian dykes, in relation to early marine cementation. Turbidites induced by the debris flows spread over the adjoining basin. (After Playford 1984)

up of early-cemented reefal-slope and fore-reef limestones (Playford 1984, cf. Hopkins 1977).

Most debris flows "froze" near the base of the marginal slopes, but some continued for short distances into the adjoining basins. In many cases the debris flows gave rise to turbidity currents which spread turbidite deposits (Fig. 12) over the basin floors.

5.5 Stromatactis

Stromatactis structures (Fig. 9D) are very common features of early-cemented limestones in the Devonian reef complexes, within the reef-flat, reef-margin, reefal-slope, and fore-reef subfacies. They are cavity fillings consisting largely of marine cement (microcrystalline, micritic, radiaxial, and scalenohedral), with irregular upper surfaces and flat bases, which are usually floored with clastic sediment.

Stromatactis cavity systems made up major pore networks in parts of the reefal limestones, locally constituting as much as 30% by volume. They provided important pathways for seawater flowing through the reefs, thereby facilitating their pervasive cementation.

Although there is general agreement that stromatactis structures are syndepositional, there is no consensus regarding their genesis (Heckel 1972, Bathurst 1982, Bourque and Gignac 1983). However, in the Canning Basin it is clear that stromatactis cavities had diverse origins. The original cavities formed by the "bridging" of reef-building organisms, the decay of soft-bodied organisms (notably sponges), and the dissolution of aragonitic shells (mainly masses of small nautiloids and ammonoids) (Playford 1984, Kerans 1985).

5.6 Dolomitization

Dolomitization is not widespread in the Devonian reef complexes in outcrop, but is more extensive in the subsurface. There is a close inverse relationship between the degrees of early marine cementation and dolomitization. The reef and reefal-slope limestones that were strongly cemented at an early stage show least dolomitization, whereas parts of the back-reef and fore-reef subfacies that were little cemented and retained high porosities and permeabilities when buried are most dolomitized.

These porous rocks apparently allowed the passage of dolomitizing solutions, which are thought to have been basin-derived (Kerans 1985, cf. Mattes and Mountjoy 1980). Neptunian fissures are also believed to have provided pathways for these solutions through the strongly cemented reefs.

5.7 Mechanical Compaction

Mechanical compaction of uncemented parts of the reef complexes began at shallow depths of burial. Maximum compaction occurred in basinal muds, amounting to as much as 70%. Connate water expelled during this compaction is likely to have played a major role in dolomitization of adjoining carbonates, and also in later mineralization and hydrocarbon migration (Playford 1984). There was also some significant mechanical compaction of uncemented carbonates in the fore-reef and back-reef subfacies. On the other hand, the pervasively cemented limestones of the reef margin, reef-flat, and reefal slope were not subject to mechanical compaction.

5.8 Pressure-Solution Compaction

Pressure solution was responsible for major compaction of the platform and marginal-slope carbonates. It was most intense in those deposits that had retained high porosities and permeabilities at the time of burial, especially in fore-reef grainstones and back-reef grainstones and Amphipora biostromes. Such deposits allowed compaction to proceed by solution of calcium carbonate along stylolite seams and its reprecipitation in available void space. This compaction resulted in the destruction of much of the remaining porosity in the reef complexes.

The effect of compaction after burial on the shape of some complexes is illustrated in Fig. 13. The platforms were originally flat-topped, but as a result of strong compaction in the back-reef carbonates and little or no compaction around the strongly cemented platform margins, most Frasnian platforms now have concave shapes (Playford 1980, 1984).

Givetian and Famennian platforms, on the other hand, do not usually have well-developed concave shapes, probably because the Givetian platforms lacked strongly cemented reef rims. Furthermore, the Famennian back-reef deposits were often quite strongly cemented prior to burial, and were therefore not subject to much differential pressure solution.

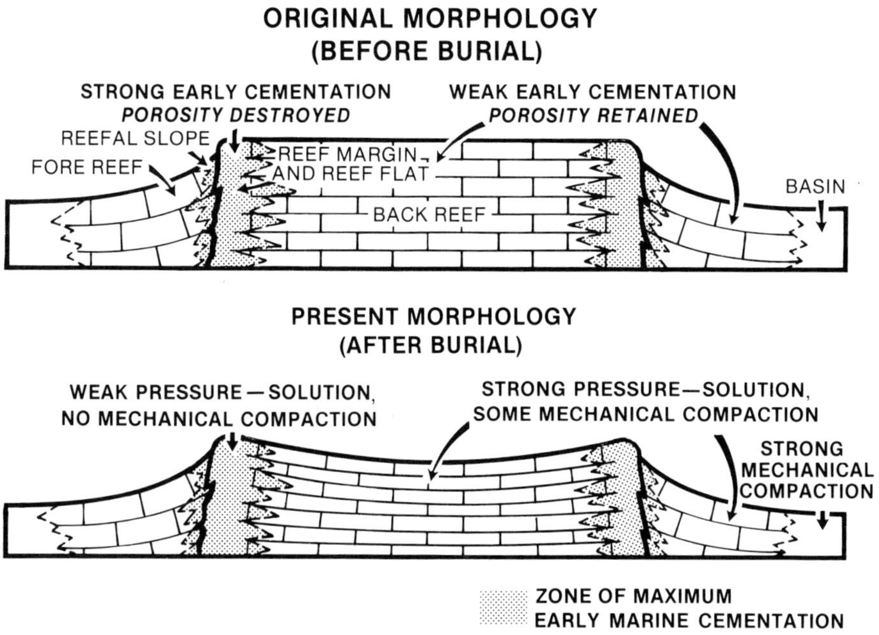

Fig. 13. Diagrammatic cross sections illustrating morphological changes in the reef complexes resulting from mechanical and pressure-solution compaction. (After Playford 1984)

6 Conclusions

The diagenetic history of the Devonian reef complexes was strongly influenced by early marine cementation. This had important consequences both during growth of the complexes and after their burial.

Extensive early marine cementation by radiaxial, microcrystalline, and micritic calcite welded the reef-margin, reef-flat, and reefal-slope subfacies into rigid, wave-resistant masses around the margins of the platforms and upper marginal slopes. Scalenohedral, bladed, epitaxial, and syntaxial marine cements were less extensive, although locally important. Some parts of the fore-reef and back-reef subfacies were subjected to varying amounts of early cementation, especially in back-reef subfacies of the Famennian complexes. Marine cementation extended below the sediment interface into the marine-burial realm, where circulation of seawater was maintained in porous limestones and neptunian fissures.

Typical cementation sequences in cavities show microcrystalline calcite (which generally cements *Renalcis* colonies) or micritic calcite followed successively by radiaxial and scalenohedral calcite. In some cavities radiaxial calcite and scalenohedral calcite are interlaminated.

Other processes of marine diagenesis included early, internal sedimentation and aragonite dissolution. The void-filling sediments were derived from the seafloor above, or were generated as cavity peloids, possibly by bacteria, within the cavity systems. No aragonite cements were precipitated in the reef complexes, and aragonitic shells were subject to early dissolution at shallow depths in the sediments, from sealevel through all water depths.

Internal sediments are commonly interlaminated with marine cements (mainly radiaxal and scalenohedral calcite), and they may include fossils, which were either forms derived from the seafloor above (via the void system), or encrusting forms (cyanobacteria) that grew in place, or troglodytic forms (mainly ostracods) that lived only in cavities.

Major consequences of early marine cementation in the complexes were: (1) The development of precipitous reef scarps around the platform margins, and depositional dips well above the angle of repose in reefal-slope limestones; (2) The destruction by early cementation and void sedimentation of the high initial porosities of the reef and reefal-slope limestones, so that the highest porosities and permeabilities remaining at the time of burial were located in parts of the back-reef and fore-reef subfacies; (3) The development of extensive networks of neptunian dykes and sills in the early-cemented limestones, mainly as a result of synsedimentary seismicity/associated with nearby tectonism; (4) The collapse of early-cemented platform-margin and reefal-slope limestones following neptunian fissuring, leading to the deposition of massive megabreccia debris flows and isolated allochthonous blocks; (5) The widespread development and cementation of stromatactis cavities in reef and reefal-slope limestones; and (6) The general restriction of dolomitization and both mechanical and pressure-solution compaction to those parts of the complexes that were not subject to strong early marine cementation, i.e., to parts of the back-reef and fore-reef subfacies.

References

Balsam WL (1983) Carbonate dissolution on the Muir seamount (Western North Atlantic): interglacial and glacial changes. J Sediment Petrol 53:719 – 731

Bathurst RGC (1982) Genesis of stromatactis cavities between submarine crusts in Palaeozoic carbonate mud buildups. J Geol Soc Lond 139:165 – 181

Bourque PA, Gignac H (1983) Sponge-constructed stromatactis mud mounds, Silurian of Gaspe, Quebec. J Sediment Petrol 53:521 – 532

Brown SA, Boserio IM, Jackson KS, Spence KW (1984) The geologic evolution of the Canning Basin – implications for petroleum exploration. In: Purcell PG (ed) The Canning Basin, WA. Proc Geol Soc Aust and Petrol Expl Soc Aust symp Perth 1984, pp 85 – 96

Etminan H, Lambert IB, Buchhorn I, Chaku S, Murphy GC (1984) Research into diagenetic and mineralising processes, Lennard Shelf reef complexes, WA. In: Purcell PG (ed) The Canning Basin, WA. Proc Geol Soc Aust and Petrol Expl Soc Aust symp Perth 1984, pp 447 – 453

Given RK, Wilkinson BH (1985) Kinetic control of morphology, composition, and mineralogy of abiotic sedimentary carbonates. J Sediment Petrol 55:109 – 119

Hall WDM (1984) The stratigraphy and structural development of the Givetian-Frasnian reef complex, Limestone Billy Hills, western Pillara Range, WA. In: Purcell PG (ed) The Canning Basin, WA. Proc Geol Soc Aust and Petrol Expl Soc Aust symp Perth 1984, pp 215 – 222

Halley RB, Scholle PA (1985) Radiaxial fibrous calcite as early-burial, open-system cement: isotopic evidence from Permian of China. Ann Assoc Petrol Geol Bull 69:261

Heckel PH (1972) Possible inorganic origin for stromatactis in calcilutite mounds in the Tully Limestone, Devonian of New York. J Sediment Petrol 42:7 – 18

Hopkins JC (1977) Production of foreslope breccia by differential submarine cementation and downslope displacement of carbonate sands, Miette and Ancient Wall buildups, Devonian, Canada. In: Cook HE, Enos P (eds) Deep-water carbonate environments. Soc Econ Paleontol Miner Spec Publ 25:155 – 170

Hurley NF (1986) Geology of the Oscar Range Devonian reef complex, Canning Basin, Western Australia. Michigan Univ Dept Geol PhD Thesis

Irwin H, Curtis C, Coleman M (1977) Isotopic evidence for source of diagenetic carbonates formed during burial of organic-rich sediments. Nature 269:209 – 213

James NP, Choquette PW (1983) Diagenesis 6. Limestones – The sea floor diagenetic environment: Geosc Canada 10:162 – 179

James NP, Ginsburg RN (1979) The seaward margin of the Belize barrier and atoll reefs. Int Assoc Sediment Spec Publ 3:1 – 191

Kendall AC (1985) Radiaxial fibrous calcite: a reappraisal. In: Schneidermann N, Harris PM (eds) Carbonate cements. Soc Econ Paleontol Miner Spec Publ 36:59 – 78

Kerans C (1985) Petrology of Devonian and Carboniferous carbonates of the Canning and Bonaparte Basins. West Aust Miner Petrol Res Inst: Rept 12

Kerans C, Playford PE (1984) Scheck breccias from Devonian reef complexes of the Canning Basin, Western Australia. Am Assoc Petrol Geol Bull 68:495

Krebs W (1969) Early void-filling cementation in Devonian fore-reef limestones (Germany). Sedimentology 12:279 – 299

Land LS, Moore CH (1980) Lithification, micritization, and syndepositional diagenesis of biolithites on the Jamaican Island slope. J Sediment Petrol 50:357 – 370

Lohmann KC, Meyers WM (1977) Microdolomite inclusions in cloudy prismatic calcites: a proposed criterion for former high-magnesium calcites. J Sediment Petrol 47:1078 – 1088

Mattes BW, Mountjoy EW (1980) Burial dolomitization of the Upper Devonian Miette buildup, Jasper National Park, Alberta. In: Zenger DH, Dunham JB, Ethington RL (eds) Concepts and models of dolomitization. Soc Econ Paleontol Miner Spec Publ 28:259 – 297

McLaren DJ (1970) Time, life and boundaries. J Pal 44:801 – 815

Middleton MF (1984) Lithospheric processes involved in the formation of the Canning Basin. 7th Aust Geol Convent Geol Soc Aust (Abstr) 12:382 – 383

Mountjoy EW, Krebs W (1983) Diagenesis of Devonian reefs and buildups, western Canada and Europe – a comparison, Z Dt Geol Ges 134:5 – 60

Playford PE (1980) Devonian "Great Barrier Reef" of Canning Basin, Western Australia. Am Asso Petrol Geol Bull 64:814 – 840

Playford PE (1981) Devonian reef complexes of the Canning Basin, Western Australia. Geol Soc Aust field trip guidebook

Playford PE (1982) Devonian reef prospects in the Canning Basin: implications of the Blina oil discovery. Aust Petrol Exp Soc J 22:258 – 271

Playford PE (1984) Platform-margin and marginal-slope relationships in Devonian reef complexes of the Canning Basin. In: Purcell PG (ed) The Canning Basin, WA. Proc Geol Soc Aust and Petrol Expl Soc Aust symp Perth 1984, pp 189 – 214

Playford PE, Lowry DC (1966) Devonian reef complexes of the Canning Basin, Western Australia. West Aust Geol Surv Bull 118

Playford PE, McLaren DJ, Orth CJ, Gilmore JS, Goodfellow WD (1984) Iridium anomaly in the Upper Devonian of the Canning Basin, Western Australia. Science 226:437 – 439

Prezbindowski DR (1985) Burial cementation – is it important? A case study, Stuart City Trend, south central Texas. In: Schneidermann N, Harris PM (eds) Carbonate cements. Soc Econ Paleontol Miner Spec Publ 36:241 – 264

Purcell PG (ed) (1984) The Canning Basin, WA. Proc Geol Soc Aust and Petrol Expl Soc Aust symp Perth 1984

Purser BH (1971) Syn-sedimentary marine lithification of Middle Jurassic limestones in the Paris Basin. Sedimentology 12:205'30

Read JF (1973a) Paleo-environments and paleogeography, Pillara Formation (Devonian) Western Australia. Can Petrol Geol Bull 21:344 – 394

Read JF (1973a) Carbonate cycles, Pillara Formation (Devonian) Canning Basin, Western Australia. Can Petrol Geol Bull 21:38 – 51

Saller, AH (1984) Petrologic and geochemical constraints on the origin of subsurface dolomite, Eniwetok Atoll: an example of dolomitization by normal seawater. Geology (Boulder) 12:217 – 220

Schlager W (1974) Preservation of cephalopod skeletons and carbonate dissolution on ancient Tethyian sea floors. In: Hsu KJ, Jenkyns HC (eds) Pelagic sediments on land and under the sea. Int Assoc Sediment Spec Publ 1:71 – 92

Schmidt V (1977) Inorganic and organic reef growth and subsequent diagenesis in the Permian Capitan reef complex, Guadalupe Mountains, New Mexico. In: Hileman ME, Mazzullo SJ (eds). Soc Econ Paleontol Miner Permian Basin Sec 77-16:93 – 132

Scholle PA, Arthur MA, Ekdale AA (1983) Pelagic environment. In: Scholle PA, Bebout DG, Moore CH (eds) Carbonate depositional environments. Am Assoc Petrol Geol Memoir 33:619 – 692

Schroeder JH (1972) Fabrics and sequences of submarine carbonate cements in Holocence Bermuda cup reefs. Geol Rundsch 61:708 – 730

Simms MA (1984) Dolomitization by groundwater-flow systems in carbonate platforms. Trans Gulf Coast Assoc Geol Socs 34:411 – 420

Tator BA, Hatfield LE (1975) Bahamas present complex geology, part 1. Oil Gas J Oct 1975:172 – 176

Towner RR, Gibson DL (1983) Geology of the onshore Canning Basin, Western Australia. Aust Bur Miner Res Bull 215

Videtich PE (1985) Electron microprobe study of Mg distribution in Recent Mg-calcites and recrystallized equivalents from the Pleistocene and Tertiary. J Sediment Petrol 55:421 – 429

Walls RA, Burrowes G (1985) The role of cementation in the diagenetic history of Devonian reefs, Western Canada. In: Schneidermann N, Harris PM (eds) Carbonate cements. Soc Econ Paleontol Miner Spec Publ 36:185 – 220

Wilkinson BH, Owen RM, Carroll AR (1985) Submarine hydrothermal weathering, global eustacy, and carbonate polymorphism in Phanerozoic marine oolites. J Sediment Petrol 55:171 – 184

Diagenetic History of the Union 8 Pinnacle Reef (Middle Silurian), Northern Michigan, USA

K. R. CERCONE[1] and K. C. LOHMANN[2]

1 Introduction

The subsurface Niagaran pinnacle reef trend of northern Michigan, USA, contains over 500 individual carbonate buildups in a linear belt 15 km wide and 250 km long. This system formed during a period of continuous subsidence in the Michigan Basin (Fischer 1975) and as a result the entire spectrum of reef diagenesis, from near-surface to deep subsurface, can be studied within a framework of known physical and temporal constraints. Correlation of specific diagenetic phases to the burial history of the host carbonates can therefore be used to identify important diagenetic controls such as absolute timing and fluid composition.

Earlier studies of the northern Silurian reef trend have demonstrated the presence of significant regional variation in diagenetic histories (Sears and Lucia 1980, Cercone and Lohmann 1985). In order to limit this variability in the current study, correlation of diagenesis and burial history was restricted to a single pinnacle reef located in Grand Traverse county (Fig. 1). This pinnacle reef was chosen because of the amount of core material available (over 100 m taken from the Shell State Union 1 – 8 well) and because it exhibits an extensive diagenetic history. Previous work on this core has been done by Mesolella et al. (1974) and Petta (1980).

2 Burial History

Niagaran "pinnacle reefs" grew as small (100 m high and 1000 m wide) biohermal mounds on the slopes of the Michigan basin in Middle Silurian time. Reef growth ceased in the Late Silurian when Michigan Basin waters became restricted and evaporites began to be deposited (Briggs et al. 1980). The Union 8 pinnacle reef is currently encased in upper Silurian Salina Group evaporites, the lower members of which (A-0 Carbonate, A-1 Evaporite) form fringing deposits around the base of the pinnacle reef (Fig. 2). Deposition of the initial A-0 Carbonate unit occurred during the onset of restriction, when the Union 8 pinnacle reef could

[1] Indiana University of Pennsylvania, Department of Geoscience, 114 Walsh Hall, Indiana, PA 15705, USA
[2] Department of Geological Sciences, The University of Michigan, Ann Arbor, MI 48109 – 1063, USA

Reef Diagenesis
Edited by J. H. Schroeder and B. H. Purser

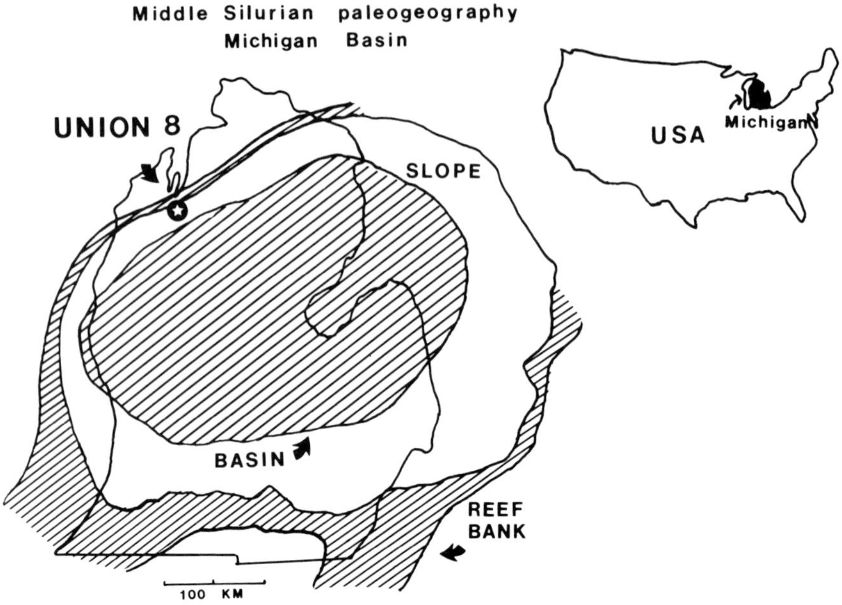

Fig. 1. Location of the Shell State Union 1 – 8 pinnacle reef weel core relative to the Middle Silurian paleogeography of the Michigan Basin. (After Briggs et al. 1980)

have been either submerged in marine water or slightly emergent. Subsequent deposition of the A-1 Evaporite around the base of the pinnacle reef must have coincided with extensive subaerial exposure, since some lithofacies in the A-1 Evaporite have been interpreted as shallow-water deposits (Gill 1977, Nurmi and Friedman 1977). This is the only period when the Union 8 carbonate buildup could have been in contact with meteoric water.

The A-1 Carbonate, which is the first Salina unit to overlie the entire Union 8 reef, marks a return to more normal salinities and higher relative sea-levels in the Michigan Basin. Following deposition of the A-1 Carbonate, the pinnacle reef was buried by Upper Silurian and Devonian strata to a depth of approximately 2000 m, and has remained buried since then at similar or greater depths (Fig. 3). The exact subsidence history of the Michigan Basin during the late Paleozoic is uncertain because of severe erosional truncation. It has been estimated elsewhere that as much as 1000 m of Carboniferous sediment may have been removed from the basin prior to Middle Jurassic time (Cercone 1984), and this figure has been used to complete the curve of Union 8 burial history.

3 Petrography

Samples of reefal carbonate from the Union 8 reef were examined in thin section using plane polarized, cathodoluminescence and ultraviolet (366 nm excitation) fluorescence microscopy. The latter two techniques commonly permit observa-

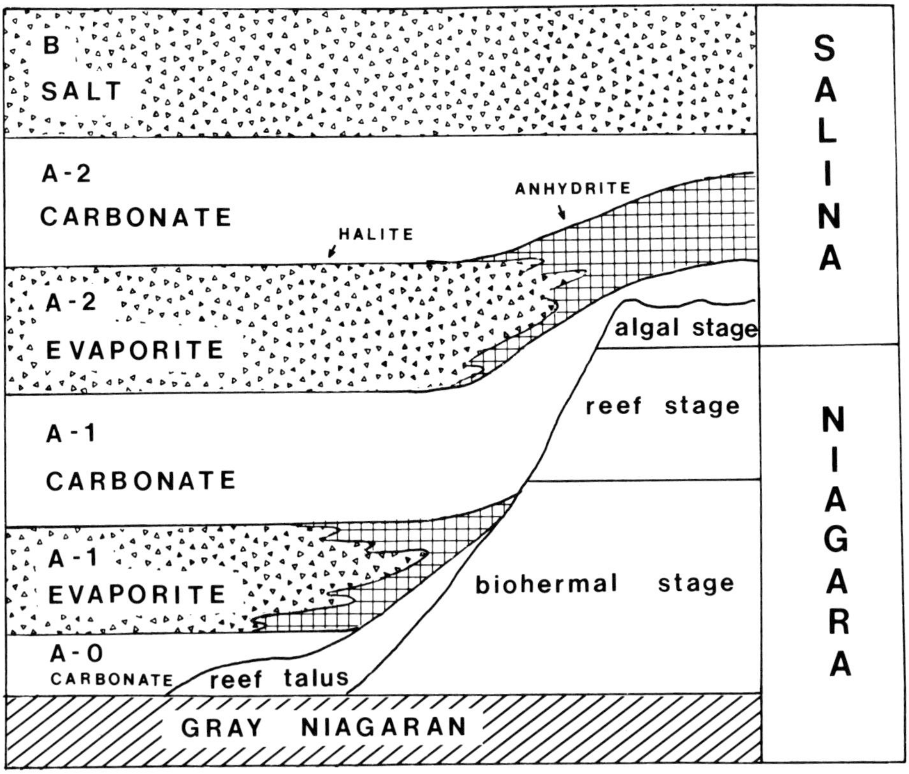

Fig. 2. Schematic stratigraphy of the Middle Silurian Niagaran pinnacle reefs and Upper Silurian A and B members of the Salina Group. All units above the A-0 Carbonate post-date the algal stage of pinnacle reef growth. (After Huh et al. 1977)

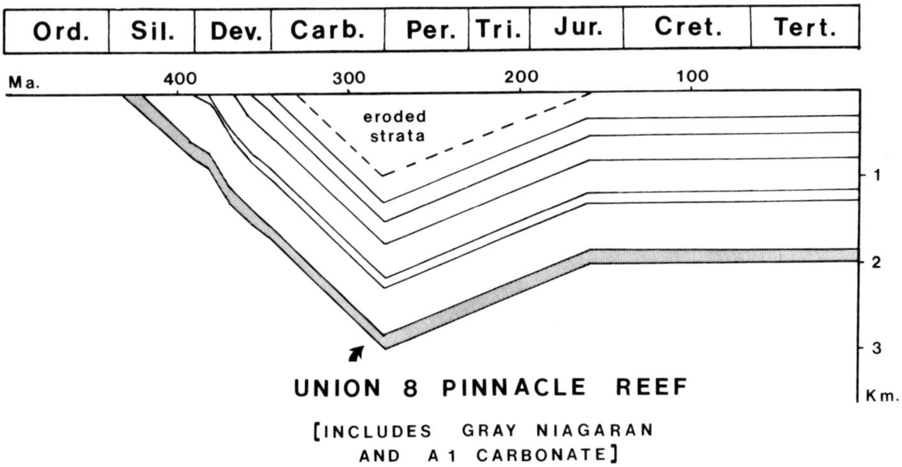

Fig. 3. Burial history of the Union 8 pinnacle reef based on thickness of overlying strata (recorded by drillers' logs). An additional 1000 m of sediment deposition has been assumed during the Carboniferous, followed by 1000 m of erosion in Permian and Triassic time. This assumption is based on the extreme organic maturity and erosional topography of the present Michigan Basin. (Cercone 1984)

tion of fine-scale growth zones caused by variations in trace element incorporation in carbonate minerals. In addition, ultraviolet fluorescence microscopy can detect the presence of liquid hydrocarbons trapped as fluid inclusions in diagenetic minerals (Burruss 1981), thus providing important constraints on diagenetic timing.

The Union 8 pinnacle reef is approximately 130 m thick and consists of crinoidal mudstones at its base, grading upward into crinoidal wackestones and then into coral-stromatoporoid wackestones and packstones. The upper 10 m of the core contain a transition between the Niagaran reef rock and the overlying A-1 Carbonate. Although this contact is abrupt, the rock immediately below it does not appear to be brecciated or affected by karst. The A-1 Carbonate itself consists of dark, stromatolitic micrite that has been completely dolomitized. Several cavernous fracture systems resembling neptunian dikes (5 – 10 cm wide and over 50 m deep) occur in the Union 8 pinnacle reef. These invariably contain a dark dolomitic micrite which strongly resembles the overlying A-1 Carbonate (Petta 1980).

The Union 8 pinnacle reef records a complex sequence of diagenesis (Fig. 4). Throughout the core, primary cavities in micrite are filled with isopachous, inclusion-rich, fibrous calcite (Fig. 5 A) that is occasionally interlayered with marine internal sediment. These calcites, which have been interpreted as former marine cements by Sears and Lucia (1980), are dully luminescent and do not fluoresce under ultraviolet excitation. Two types of replacive dolomite occur in

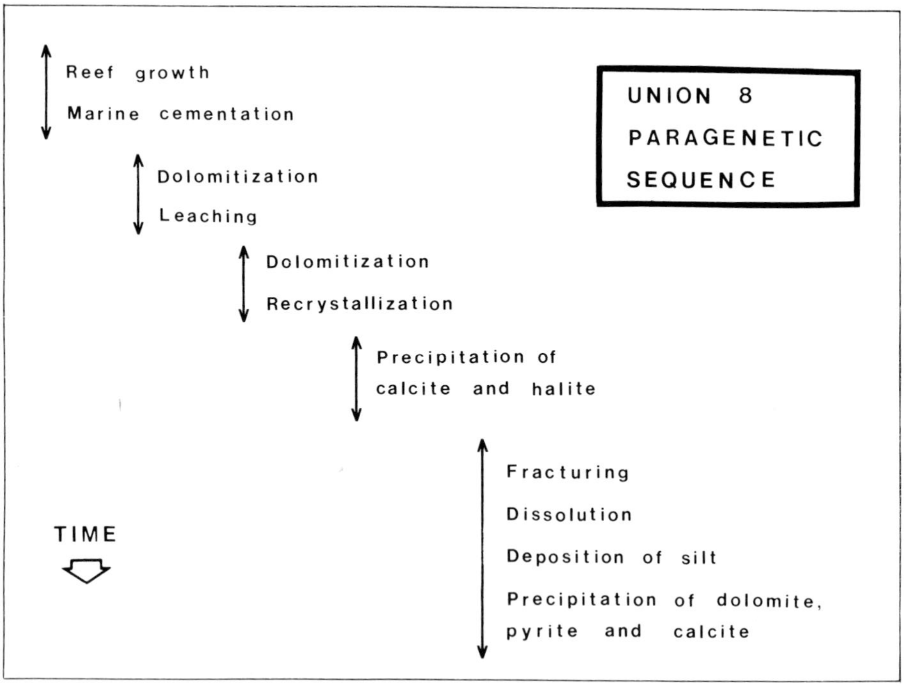

Fig. 4. Summary of diagenetic features observed in the Union 8 pinnacle reef

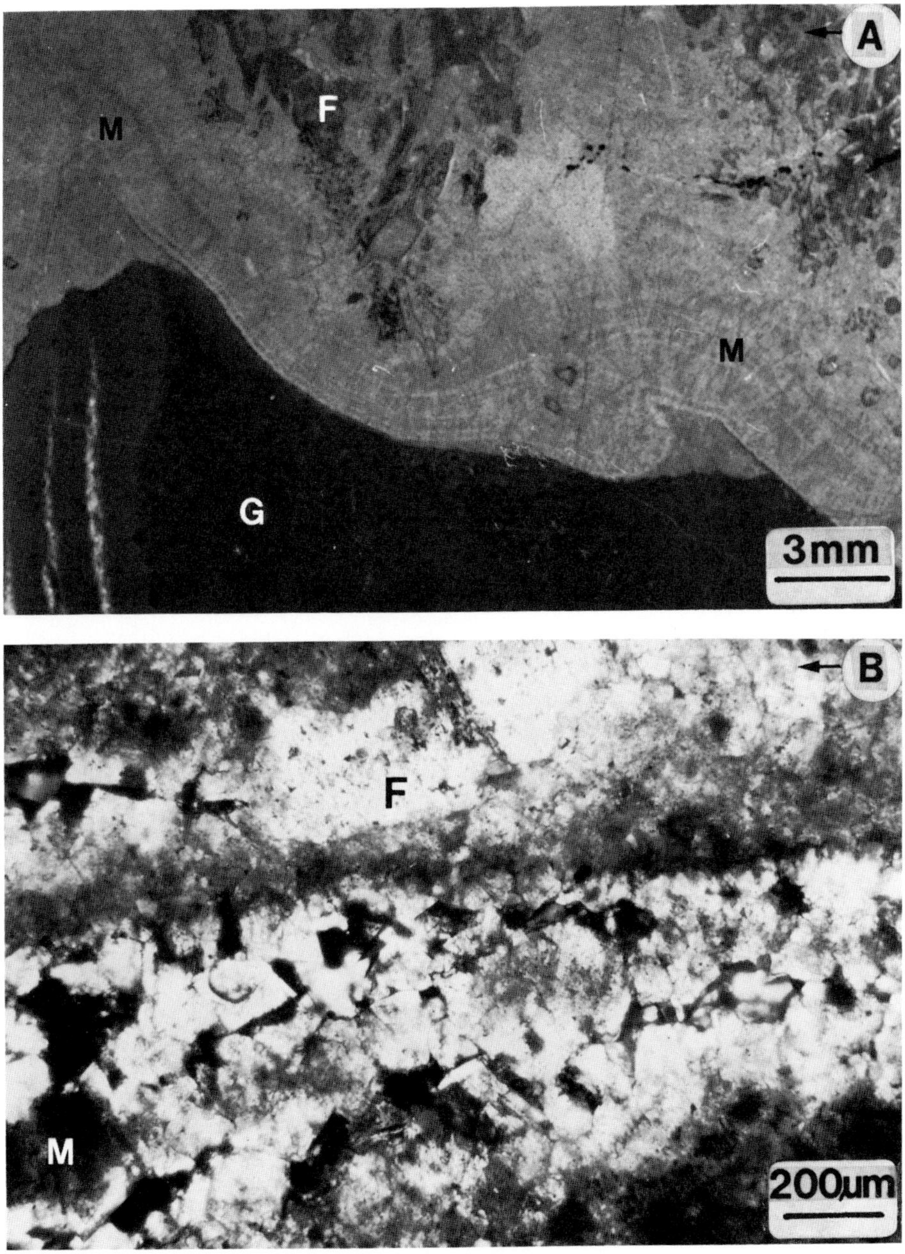

Fig. 5. A Former fibrous marine cement (*M*) in a matrix of bioclastic wackestone (*F*). Note overlap of marine geopetal cavity filling sediment (*G*) which postdates cement deposition. *Arrows* on all photos indicate up directions. Sample from 1953 m (6407 feet). **B** Type I dolomite replacing former marine cement (*M*) and crinoid fragments (*F*) with gross fabric retention. Sample from 1949 m (6396 feet)

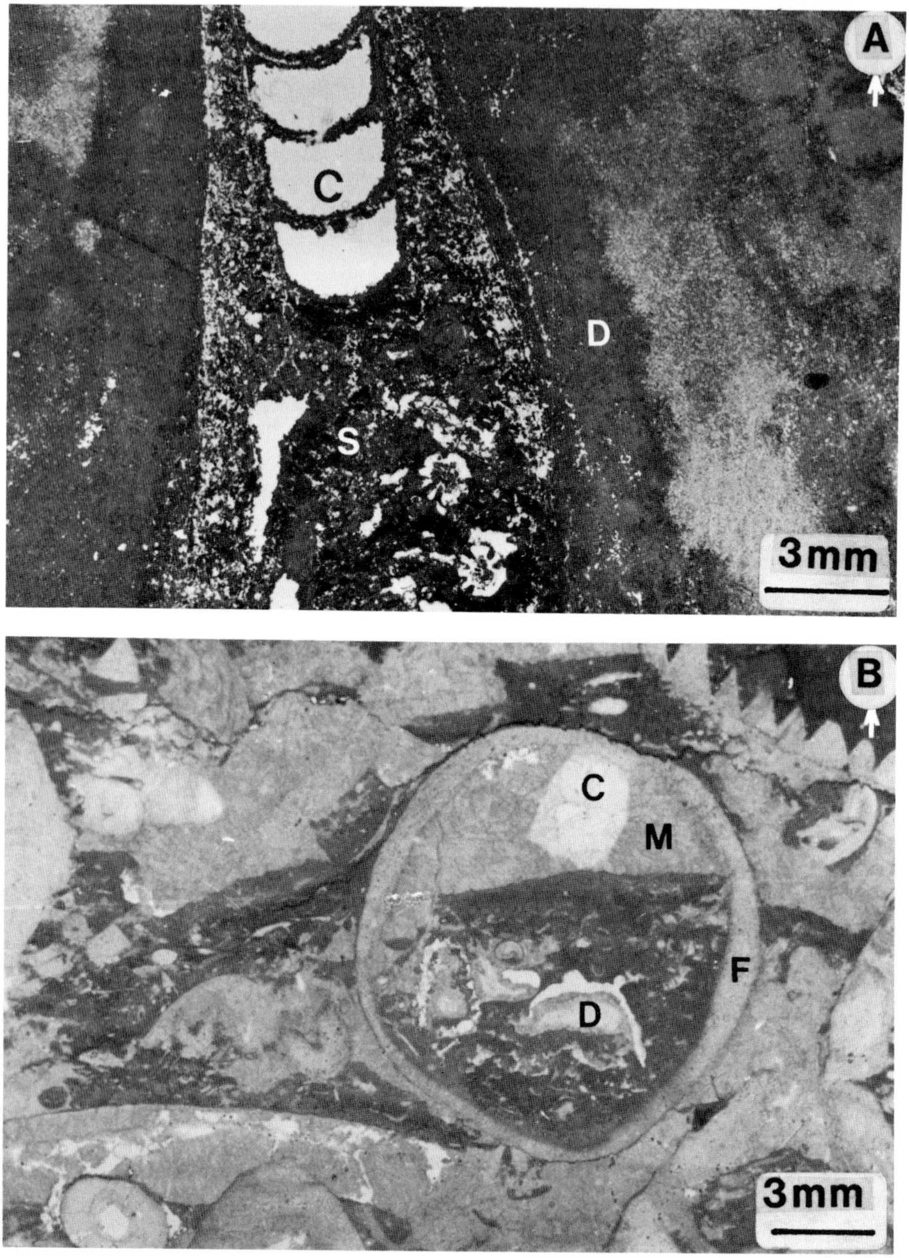

Fig. 6. A Type II dolomite (*D*) associated with a small sediment-filled neptunian fractures. Note leached reef fauna such as corals (*C*) in fracture sediment (*S*). Sample from 1840 m (6034 feet). **B** Early void-filling cement (*C*) filling remnant porosity in fibrous former marine cement (*M*). Note that dolomite II (*D*) has affected the marine geopetal sediment inside the crinoid (*F*) but has not affected the calcite spar. Sample from 1872 m (6129 feet)

the Union 8 pinnacle reef. Type I dolomite is fabric-retentive and in some cases can preferentially affect growth layers in marine cements (Fig. 5B). This dolomite is pervasively developed in the lower part of the buildup (1920 – 1935 m) and is commonly associated with leached zones marked by dissolution of micrite and development of red-stained clay laminae. Type II dolomite is fabric-destructive and occurs only in the upper parts of the pinnacle reef (1860 – 1885 m). It is frequently associated with the presence of sediment-filled neptunian fractures (Fig. 6A) and appears to be identical to the pervasive dolomites that occur in the overlying A-1 Carbonate. Both types of replacive dolomite are nonfluorescent and dully luminescent.

Primary pores in the Union 8 pinnacle reefs are commonly occluded by clear dully luminescent and nonfluorescent equant calcite spar. This void-filling calcite (Fig. 6B) immediately overlies either former marine cement or early replacive dolomite, and never appears to be dissolved or replaced by dolomite. In the upper part of the reef, this spar becomes less common and primary pores are filled with blocky halite. Where the two minerals co-exist, calcite spar underlies halite but appears to contain small (less than 10 μm wide) inclusions of solid halite (Cercone and Lohmann, in review). The two phases are therefore assumed to be genetically related.

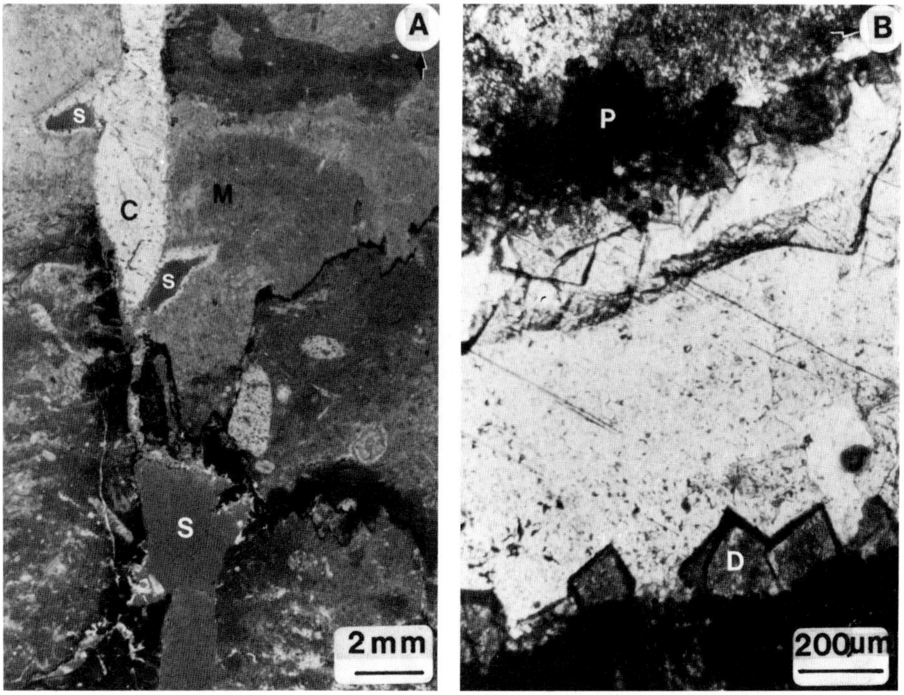

Fig. 7. A Late fracture filled by dolomitic sediment (*S*) before being mineralized by rhombic dolomite, pyrite and calcite (*C*). Note that dolomitic sediment (*S*) also fills the center of a pore in marine cement (*M*) which has been offset by fracturing. Sample from 1956 m (6414 feet). **B** Detail of fracture in **A**. Note disseminated pyrite (*P*) and rhombic dolomite cement (*D*) on fracture walls

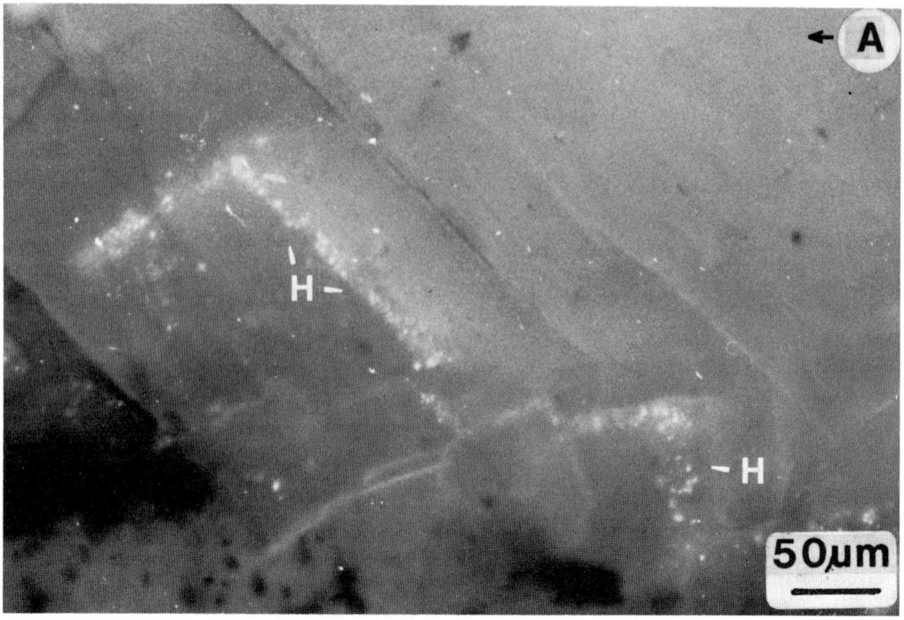

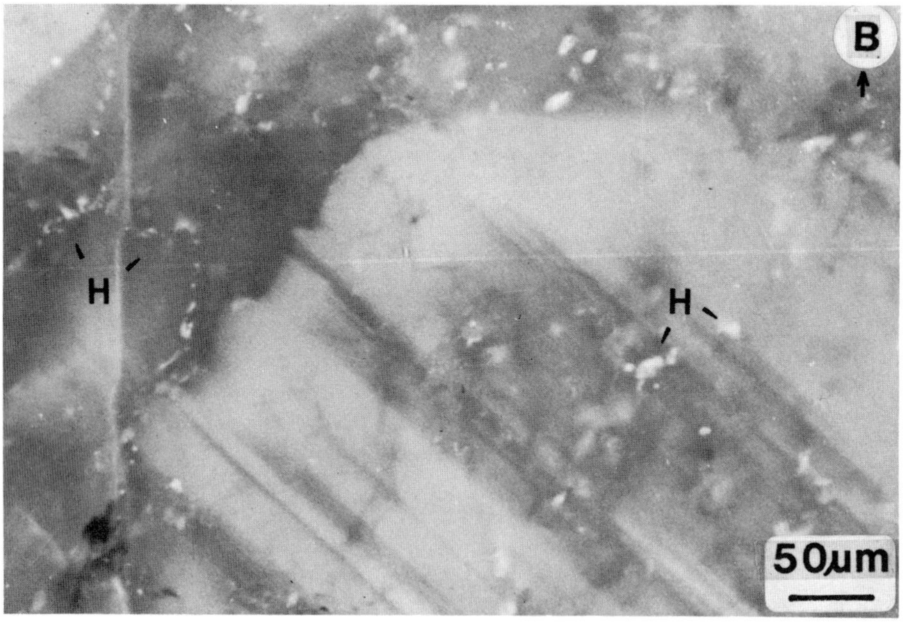

Fig. 8. A Ultraviolet illumination of rhombic dolomite cements shown in Fig. 7. Note fluorescent hydrocarbon inclusions (*H*) trapped along primary growth zones in dolomite rhombohedra. **B** Ultraviolet illumination of late fracture-filling calcite cement shown in Fig. 7. Note striated mineral fluorescence and fluorescent hydrocarbon inclusions (*H*)

Very little secondary porosity is developed in the Union 8 pinnacle reef, with the exception of a small fracture set that occurs in the lowest four meters of the core. These fractures apparently transmitted fluids that were undersaturated with respect to calcite, since solution voids are often formed where fractures intersect marine cements. Many secondary pores are partially filled with unfossiliferous dolomitic sediment (Fig. 7A), which has been interpreted as a residue of halite dissolution within the reef (Cercone and Budai 1985). Fractures and secondary pores are commonly lined by disseminated pyrite and dully luminescent rhombic dolomite cement, and occluded by dully luminescent calcite spar (Fig. 7B). Unlike the early void-filling calcite spar that is associated with halite precipitation, late fracture-filling calcite spar fluoresces in the yellow and white bands when examined under ultraviolet light. In addition, both rhombic dolomite cements and the fracture-filling calcite cements contain numerous fluorescent hydrocarbon-filled fluid inclusions (Fig. 8A, B). Because some of these hydrocarbon inclusions are primary, precipitation of late dolomite cement and late fracture-filling calcite spar must post-date the entry of hydrocarbons into the Union 8 pinnacle reef.

4 Fluid Inclusions

Aqueous inclusions are present in both early void-filling calcite and late fracture-filling calcite from the Union 8 pinnacle reef. Fluid inclusions which were determined to be primary based on their position along growth bands were analyzed in situ using a heating/cooling stage. When a fluid inclusion has been frozen, the melting behavior of the liquid can be used as an indicator of fluid salinity, since freezing point depression depends directly on the moles of solute present. The initial melting point of a frozen fluid inclusion is a eutectic point determined by the type of cations present in solution, whereas the final melting point of the inclusion is a function of its total salinity, usually calculated in weight percent (Crawford 1981). The intermediate melting points, caused by melting of hydrates when ice is the final phase to melt, can be used to estimate the chemical composition of the fluid.

By measuring the temperature of liquid-vapor homogenization point in a two-phase fluid inclusion, the ambient temperature of the host rock at the time of entrapment can be estimated. This calculation can be made only by assuming that the fluid was trapped as a single phase and that no leakage, internal solution and reprecipitation, or entrapment of carbon dioxide and methane has occurred. Since the vapor pressure of water is a function of total pressure, an estimate of burial depth at the time of inclusion entrapment is needed in order to correct inclusion homogenization temperatures to true temperatures of entrapment (Hollister 1981). In the case of the Union 8 pinnacle reef such an estimate cannot be made *a priori* and homogenization temperatures thus represent only the minimum temperature at which fluid inclusions could have been trapped.

The initial and final melting temperatures and homogenization temperatures of fluid inclusions from diagenetic calcite in the Union 8 pinnacle reef are listed

Table 1. Uncorrected heating and freezing measurements on primary fluid inclusions from early and late diagenetic calcite spar in the Union 8 pinnacle reef

Fluid inclusion analysis		
First melt	Final melt	Homogenization
Void-filling calcite (°C)		
−54	−50	+196
−51	−50	+207
−48	−27	–
Fracture-filling calcite (°C)		
−62	−57	+83
–	−35	+89
–	−30	+108

in Table 1. Homogenization temperatures of fluid inclusions from early void-filling calcite exceed 180 °C, a much higher temperature than could ever have been attained during the reef's burial history (Cercone 1984). Although it is possible that very hot fluids precipitated this calcite spar, it is more likely that the fluid inclusions within it have been affected by entrapment of methane or carbon dioxide, or by leakage and contamination of the liquid phase due to aquathermal pressuring. Neither a valid salinity nor a valid temperature of formation can therefore be determined for this diagenetic phase.

Fluid inclusions are abundant in late fracture-filling calcite, but many are either secondary (microfracture traces), hydrocarbon-filled or simply too small to observe the formation of hydrates and ice. When frozen, primary fluid inclusions in this phase began to melt at temperatures below −55 °C, which could correspond within experimental error to the eutectic point of the $MgCl_2-CaCl_2-NaCl-H_2O$ system (−57 °C) or to that of an even more complex brine, possibly with minor amounts of strontium, barium or other metals in solution. In some inclusions, final melting occurred immediately after initial melting, probably indicating that a vitreous (noncrystalline) ice had formed on freezing. In other inclusions, final melting of a phase that appeared to be ice occurred between temperatures of −30° to −40 °C in the presence of halite. If hydrates were present in these inclusions, their melting points were not observed and thus the relative amounts of Ca, Mg, and Na in the liquid could not be estimated. Final melting below −30 °C, however, indicates that total salinities of the fluids are high and that brines are Ca-Mg enriched (Crawford 1981). Primary fluid inclusions from pyrite-associated fracture-filling calcite spar homogenized at temperatures ranging from +83° to +122 °C with no expandable gas phases (carbon dioxide or methane) found upon crushing in anhydrous glycerin. A temperature of +80 °C will be assumed as the minimum possible ambient temperature of the Union 8 pinnacle reef during precipitation of this diagenetic phase.

5 Carbon and Oxygen Stable Isotopes

Samples of several Union 8 pinnacle reef components including former fibrous marine cements, fossil fragments, micrites, replacive dolomites, and calcite spars were analyzed for stable isotopic composition. Samples for analysis were obtained using a microscope-mounted drill with a 0.5 mm bit; each sample consisted of approximately 0.2 mg of powdered carbonate. Samples were roasted at 380 °C under vacuum for 1 h to remove volatile components. Calcite components were dissolved in anhydrous phosphoric acid at 50 °C in an extraction line connected directly to the inlet of a VG 602E ratio mass spectrometer. Dolomitic samples were dissolved in anhydrous phosphoric acid at 50 °C in separate off-line vessels to insure digestion. All $\delta^{13}C$ and $\delta^{18}O$ enrichments were measured relative to a lab reference gas calibrated to PDB; data were then corrected for ^{17}O according to the procedure of Craig (1957). Precision and calibration of analyses were determined by running a laboratory standard (NBS 20) at the beginning and end of every analytic session; measured precision was 0.11 permil for oxygen and 0.08 permil for carbon.

The isotopic compositions of all samples analyzed are shown in Fig. 9. At approximately 1890 m in the Union 8 pinnacle reef, there is a marked change in the isotopic distribution of reefal components. Below this depth, all carbonate components have isotopic compositions that range from -5.0 to $-8.0‰$ $\delta^{18}O$ (PDB) and $+1.5$ to $+1.7‰$ $\delta^{13}C$ (PDB). The ^{18}O-enriched end-members of this trend consist of former marine cements and micrites, components which were originally composed of metastable aragonite and magnesian calcite. The enriched isotopic composition of these phases probably results from partial preservation

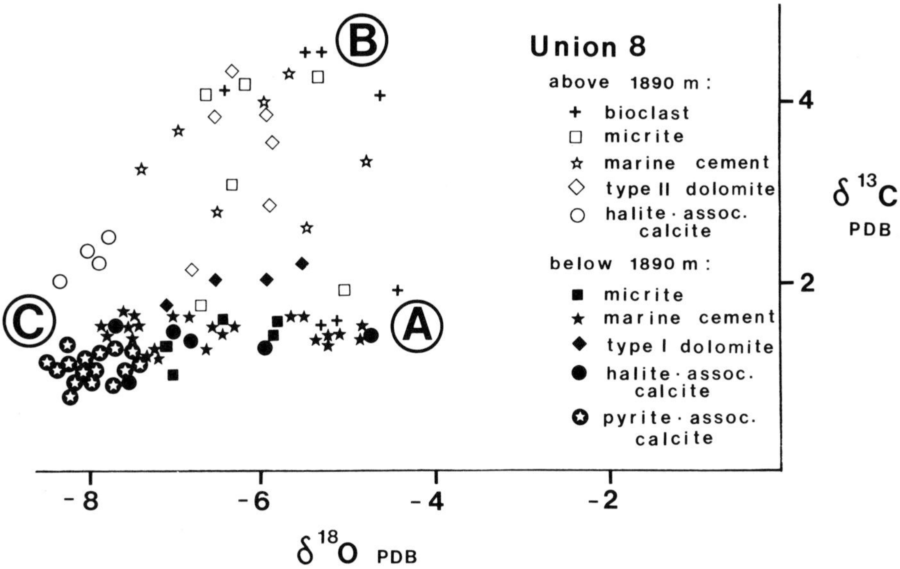

Fig. 9. Cross-plot of stable oxygen and carbon isotopic data gathered from the Union 8 pinnacle reef. See text for discussion

of the original isotopic composition of carbonate precipitated from Silurian seawater. Other studies of Silurian carbonates have proposed a value of $-4.0‰$ $\delta^{18}O$ and $+1.5‰$ $\delta^{13}C$ for the original composition of Silurian marine carbonate (Lohmann 1983). Late fracture-filling calcite spar (found only in the base of the pinnacle reef) plots near the ^{18}O-depleted end of the lower-reef field, with slightly more ^{13}C-depleted compositions. Type I dolomite plots near the center of the field, along with other marine cements and micrites. In zones of high porosity (near neptunian dikes), Type I dolomites exhibit a slight ^{13}C enrichment.

Reef components sampled from strata above 1890 m exhibit isotopic compositions ranging from -5.0 to $-8.0‰$ $\delta^{18}O$ (PDB) and from $+2.0$ to $+4.5‰$ $\delta^{13}C$ (PDB). These components form a roughly triangular field, one corner of which (labeled A) probably represents partial preservation of the same original marine composition observed in the lower part of the pinnacle reef. The triangle's apex (B) reflects the presence of early diagenetic carbonates such as type II dolomite, micrite and recrystallized fossil components that exhibit enriched ^{13}C compositions relative to original marine values. The third corner of the triangle (labeled C) reflects the presence of other diagenetic components such as early void-filling calcites that are depleted in ^{18}O relative to original marine values. Salina A-1 dolomites plot near point B. The scatter of marine cements and micrites within this triangular field can sometimes be attributed to a physical admixture of recrystallized marine carbonate with type II dolomite or early void-filling calcite. In many cases, however, intermediate samples can be shown by petrographic and X-ray analysis to consist of entirely recrystallized material.

6 Discussion

Based on the petrographic, fluid inclusion and stable isotopic data presented above, the diagenetic sequence recorded in the Union 8 pinnacle reef can now be correlated to the burial history of the host carbonates in order to estimate diagenetic timing and identify possible sources of diagenetic fluids. These conclusions are summarized in Fig. 10.

6.1 Type I Dolomitization

The occurrence of type I dolomite at the base of the reef and its common association with oxidized and leached zones suggests that it may have formed during exposure of the pinnacle reef to the atmosphere and to meteoric fluids. The burial history of the Union 8 pinnacle reef thus restricts it in age to the very early Late Silurian (A-1 Evaporite time). Sears and Lucia (1980) concluded that early dolomitization in the northern pinnacle reef trend resulted from mixing of dilute meteoric water and seawater at the base of a large phreatic lens. This model requires that much of the pinnacle reef above the pervasively dolomitized base be exposed to dilute water for a geologically significant period of time. Both the extremely arid climate needed to preserve halite, which was accumulating

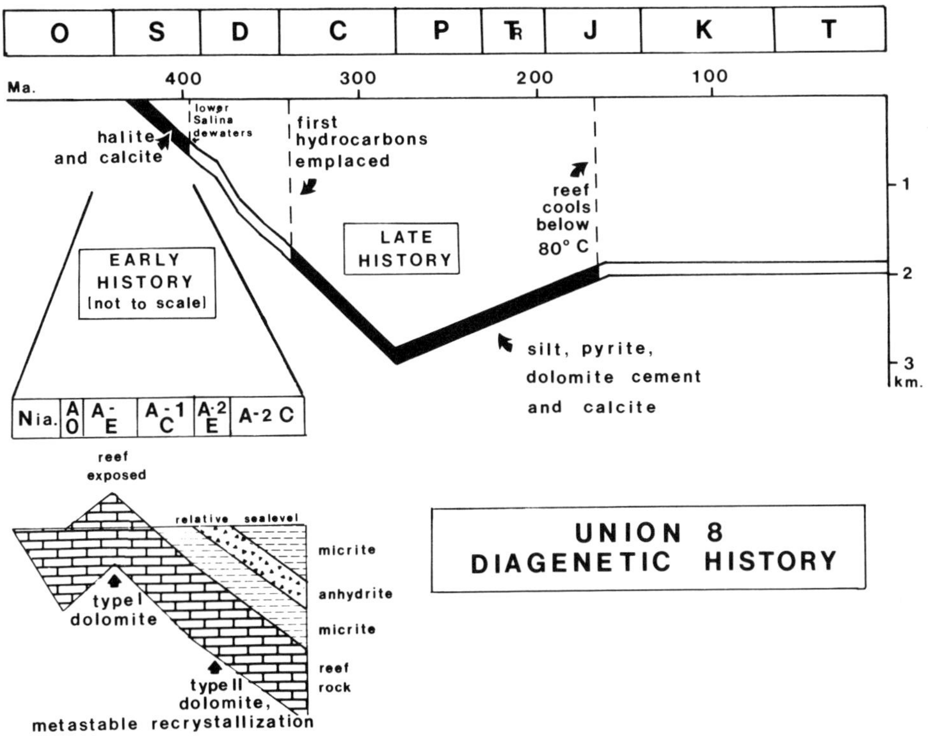

Fig. 10. Summary of the correlations between diagenetic history and burial history for the Union 8 pinnacle reef. Enlarged portion of the burial history curve is schematic and not to scale. The absolute timing of early diagenetic events is hindered by the lack of faunal control in the Salina Group. See text for discussion

throughout the period of reef exposure, and the small size of pinnacle reefs argue against the formation of a large-scale dilute phreatic system. It seems more likely that Type I dolomite in the Union 8 pinnacle reef formed in a small schizohaline phreatic lens that developed only at the base of the carbonate buildup during A-1 Evaporite deposition. Although the intermediate isotopic composition of Type I dolomite appears to be more compatible with a freshwater/seawater mixing zone than with a thin and sometimes hypersaline lens, the correlation of slight ^{13}C enrichment (a signature typical of the A-1 Carbonate) with zones of high porosity suggest that Type I dolomite has been recrystallized and its isotopic composition altered by later and possibly warmer diagenetic fluids. No definite conclusions concerning the chemistry of Type I dolomite parent fluids can therefore be drawn from its isotopic composition.

6.2 Type II Dolomitization

The fact that Type II dolomite is both petrographically and isotopically identical to the replacive dolomite that occurs in the A-1 Carbonate implies that both dolomites formed together following deposition of the A-1 Carbonate. Sears and Lucia (1980) attributed the second replacive dolomite in the pinnacle reef trend to regionally extensive brine reflux during A-2 Evaporite sabkha deposition in early Late Silurian time (Fig. 10), a model which is supported by isotopic data from the Union 8 pinnacle reef. Anaerobic decomposition in organic-rich rocks has been shown to produce diagenetic carbonates enriched in ^{13}C (Irwin et al. 1977). The ^{13}C-enriched composition of Type II dolomites therefore appears to document the passage of parent brines downward from the A-2 Evaporite through the organic-rich A-1 Carbonate.

6.3 Recrystallization of Metastable Carbonates

Based on analogy with other modern and ancient carbonate systems, it seems probable that micrites, fossil components and former fibrous marine cements in the Union 8 pinnacle reef were originally composed of either aragonite or magnesian calcite. Former metastable components in both the upper and lower parts of the reef trend towards a similar Silurian marine isotopic signature, suggesting that they originally had similar isotopic compositions and that their current separation into two distinct groups results from recrystallization in the presence of different fluids. In the upper part of the Union 8 pinnacle reef, entirely calcitic micrites and marine cements have isotopic compositions similar to Type II dolomite, suggesting that A-2 Evaporite refluxing brines were agents of recrystallization as well as dolomitization. The conclusion that metastable components remained in their unaltered state in the upper part of the Union 8 pinnacle reef until A-2 Evaporite time provides further support for the absence of a chemically reactive freshwater lens during earlier A-1 Evaporite time.

In the lower part of the reef, the rather ambiguous invariant carbon isotopic trend exhibited by former metastable components is more difficult to interpret. It is possible that some recrystallization occurred in the schizohaline phreatic lens that caused early (Type I) dolomitization. It is also possible that recrystallization of metastable components occurred after halite-saturated brines had entered the lower part of the Union 8 pinnacle reef, during deposition of laterally adjacent evaporites. In this case, unfavorable kinetic controls may have caused recrystallization to occur slowly in a nearly closed chemical system affected only by temperature increases due to burial. Recrystallization of metastable components in the lower part of the Union 8 pinnacle reef can therefore not be given a precise minimum or maximum age.

6.4 Precipitation of Blocky Halite and Associated Early Calcite Spar

These two void-filling diagenetic phases appear to be related to an influx of evaporite-derived brines into the Union 8 pinnacle reef. Both phases overlie Type II dolomite and must therefore post-date shallow burial of the pinnacle reef by the A-1 carbonate in early Late Silurian time. The most likely sources of halite over-saturated brines in the subsurface are halite deposits undergoing pressure solution. Pore-filling halite in some northern Michigan pinnacle reefs has in fact been correlated by bromide content and chlorine isotopic composition to halite in the A-1 and A-1 Evaporites (McCollough 1975, R. Kaufman, unpublished data). According to Sonnenfeld (1984), most salt deposits lose their interstitial water during the first 300 m of burial, yielding a maximum age of late Silurian for both blocky halite and associated pore-filling calcite spar (Fig. 10). Due to its high solubility, later dissolution and/or recrystallization of halite is possible and in fact may be documented by the presence of unfossiliferous dolomitic sediments in secondary pores of the Union 8 reef. Combined with the absence of valid fluid inclusion data in early void-filling calcite, this uncertainty precludes any estimate of the chemistry of original parent fluids for these two early burial diagenetic phases.

6.5 Precipitation of Pyrite, Rhombic Dolomite Cement and Associated Late Calcite Spar

Several lines of evidence indicate that the mineralization of secondary pores by pyrite, dolomite and calcite occurred late in the burial history of the Union 8 pinnacle reef. Firstly, both rhombic dolomite cements and late fracture-filling calcite spar contain primary liquid hydrocarbon inclusions and must therefore post-date the entry of liquid hydrocarbons into the Union 8 pinnacle reef. These hydrocarbons have been determined by other studies (Ells 1967, Gardner and Bray 1985) to have been derived largely from local or basin-center Salina A-0 and A-1 Carbonates. Previous Lopatin modeling of these units indicates that neither was sufficiently mature to expel liquid hydrocarbons until Carboniferous time, at the very earliest (Cercone 1984). Secondly, pyrite-associated calcite spar contains primary fluid inclusions that homogenize at minimum temperatures of +80 °C. Based on the burial history of the Union 8 pinnacle reef and assuming a geothermal gradient of 35 °C/km in the Paleozoic decreasing to today's average of 25 °C/km in Mesozoic time (Cercone 1984), this temperature of formation requires an age of formation between Carboniferous and Jurassic (Fig. 10). Because true temperatures of inclusion entrapment were probably higher, the true age of calcite formation probably corresponds to the time of maximum reef burial (Pennsylvanian-Permian). This age is only applicable to the calcite spar in which inclusions were trapped. Underlying pyrite and dolomite may be older than this phase (i.e., Mississippian), but cannot be younger.

The chemistry of the fluid which precipitated late mineralizing phases in the Union 8 pinnacle reef is more problematic than its timing. The isotopic composition of late fracture-filling calcite spar does not differ significantly from some earlier diagenetic phases such as recrystallized marine cements present in the lower part of the pinnacle reef. If the fluid which precipitated calcite spar was allochthonous, as its association with dissolution and extensive iron sulfide mineralization suggests, it does not seem likely that passage through open fractures could have caused the required equilibration with pinnacle reef host rock. The Union 8 pinnacle reef is completely encased in impermeable evaporite deposits with the exception of the overlying Salina A-1 Carbonate and the underlying Gray Niagaran units. Although Gray Niagaran strata have not been cored below the Union 8 reef, carbonate components taken from these rocks elsewhere have isotopic compositions similar to those of the lower Union 8 pinnacle reef, whereas components from the A-1 Carbonate isotopically resemble the upper Union 8 reef (Cercone and Lohmann 1985). The isotopic composition of late fracture-filling calcite therefore suggests that its parent fluid must have migrated for long distances in, and equilibrated with, the Gray Niagaran before entering the Union 8 pinnacle reef.

The fact that pyrite-associated, fracture-filling calcite is the only diagenetic or depositional phase in the Union 8 pinnacle reef which displays inorganic (mineral) fluorescence provides further support for its derivation from allochthonous fluids, as does the presence of extremely saline, calcium-enriched brines trapped as inclusions in this phase. Brines with similar compositions occur in many sedimentary basins and it has been concluded by Graf (1982) that such brines result from shale-filtration due to large-scale cross-formational flow. If this theory is correct, the presence of such fluids in diagenetic phases dated to the late Paleozoic and early Mesozoic implies that basin-wide fluid flow could have been occurring in the Michigan Basin since the onset of terrigenous clastic deposition in the Carboniferous.

7 Conclusions

By combining the burial history of a single Middle Silurian pinnacle reef with detailed petrographic, fluid inclusion and stable isotopic analysis, constraints can be placed on both the absolute timing and fluid chemistry of the reef's diagenetic sequence. Most reef diagenesis occurred in early Late Silurian time as a result of subaerial exposure, sabkha brine reflux and evaporite compaction. Following the emplacement of hydrocarbons during deeper burial of the reef, an influx of allochthonous brines caused mineralization of secondary porosity. This late event occurred between Carboniferous and Jurassic time and may have been related to the onset of large-scale cross-formational fluid flow in the Michigan Basin.

It is interesting to note that throughout its diagenetic evolution, the Union 8 pinnacle reef evolved from an extremely open hydrochemical system to one that became progressively much more isolated. During the second episode of replacive dolomitization that affected this pinnacle reef, its entire upper portion underwent either dolomitization or significant recrystallization to calcite. In both cases, the carbonate diagenetic product phases exhibit a 2 to 3 permil enrichment in ^{13}C over the original marine reactant phases, an isotopic signature which was apparently carried into the reef by fluids percolating through the organic-rich overlying unit. Since the carbon isotopic reservoir in diagenetic fluids is many times smaller than the carbon isotopic reservoir in carbonate host rocks, this wholescale shift in composition of the reef rock implies that a large water-rock ratio was characteristic of this diagenetic event. At this time, the Union 8 reef was encased in evaporites which had not yet undergone compaction and were probably more than 50% fluid themselves. In the nearly 400 million years following evaporite compaction, however, allochthonous diagenetic fluids appear to have reached the Union 8 reef only once, and never to have caused a large shift in the isotopic composition of the reef-rock. It can be concluded that the timing of carbonate diagenesis relative to evaporite compaction constitutes an important fluid constraint in the pinnacle reef system of northern Michigan, and may also do so in other evaporite-related carbonate systems.

Acknowledgments. All isotopic analyses cited in this paper were performed in the University of Michigan's Stable Isotopic Laboratory and all core materials were provided courtesy of the University of Michigan's Subsurface Laboratory. Scientific discussions with J. M. Budai, L. A. Gonzalez, V. Pedone, and R. K. Given were greatly appreciated. Technical and logistical support was provided by D. Cercone, E. M. Cercone, R. Regitz, and M. A. Eaverly. This research was supported by grants from the Marathon Oil Company, the Geological Society of America and the American Association of Petroleum Geologists.

References

Briggs LI, Gill D, Briggs DZ, Elmore RD (1980) Transition from open marine to evaporite deposition in the Silurian Michigan Basin. In: Nissenbaum A (ed) Hypersaline brines and evaporitic environments. Elsevier, New York, pp 253 – 270

Burruss RC (1981) Hydrocarbon fluid inclusions in studies of sedimentary diagenesis. Min Assoc Can Short Course Handbook 6:138 – 156

Cercone KR (1984) Thermal history of the Michigan Basin. Am Assoc Petrol Geol Bull 68:130 – 136

Cercone KR, Budai JM (1985) Formation of geopetal diagenetic sediment by evaporite solution: evidence that not all crystal silt is vadose. Soc Econ Paleontol Miner Midyear Meet 2:17

Cercone KR, Lohmann KC (1985) Early diagenesis of Middle Silurian pinnacle reefs, northern Michigan. Mich Bas Geol Soc Spec Pap 4:109 – 130

Craig H (1957) Isotopic standards for carbon and oxygen and correction factors for mass spectrometric analysis of carbon dioxide. Geochim Cosmochim Acta 12:133 – 149

Crawford ML (1981) Phase equilibria in aqueous fluid inclusions. Min Assoc Can Short Course Handbook 6:75 – 100

Ells GD (1967) Michigan's Silurian oil and gas fields. Mich Geol Surv Rep Invest 2, p 49

Fischer AG (1975) Origin and growth of basins. In: Fischer AG, Judson S (eds) Petroleum and global tectonics. Princeton Univ Press, Princeton, pp 47 – 79

Gardner WC, Bray EE (1985) Oils and source rocks of Niagaran reefs (Silurian) in the Michigan basin. Am Assoc Petrol Geol Stud Geol 18:33 – 44

Gill D (1977) Salina A-1 sabkha cycles and the late Silurian paleogeography of the Michigan Basin. J Sediment Petrol 47:979 – 1017

Graf DL (1982) Chemical osmosis, reverse chemical osmosis, and the origin of subsurface brines. Geochim Cosmochim Acta 46:1431 – 1448

Hollister LS (1981) Information intrinsically available from fluid inclusions. Min Assoc Can Short Course Handbook 6:1 – 12

Huh JM, Briggs LI, Gill D (1977) Depositional environments of pinnacle reefs, Niagara and Salina Groups, northern shelf, Michigan Basin. Am Assoc Petrol Geol Stud Geol 5:1 – 21

Irwin H, Curtis C, Coleman M (1977) Isotopic evidence for source of diagenetic carbonates formed during burial of organic-rich sediment. Nature 269:209 – 213

Lohmann KC (1983) Unravelling the diagenetic history of carbonate reservoirs: integration of petrographic and geochemical techniques. Dallas Geol Soc Short Course, p 41

McCollough CN (1975) Origin of pore-filling salt in the Niagaran reefs of northern Michigan. Am Assoc Petrol Geol Bull 59:1737

Mesolella KJ, Robinson JD, McCormick LM, Ormiston AR (1974) Cyclic deposition of Silurian carbonates and evaporites in the Michigan Basin. Am Assoc Petrol Geol Bull 58:34 – 62

Nurmi RD, Friedman GM (1977) Sedimentology and depositional environments of basin-center evaporites, lower Salina group (upper Silurian), Michigan Basin. Am Assoc Petrol Geol Stud Geol 5:23 – 52

Petta TJ (1980) Silurian pinnacle reef diagenesis – northern Michigan. Soc Econ Petrol Miner Core Workshop 2:32 – 42

Sears SO, Lucia FJ (1980) Dolomitization of northern Michigan Niagara reefs by brine refluxion and freshwater/seawater mixing. Soc Econ Petrol Miner Spec Publ 28:215 – 235

Sonnenfeld P (1984) Brines and evaporites. Academic Press, New York, p 613

Diagenesis of Silurian Bioherms in the Klinteberg Formation, Gotland, Sweden

P. FRYKMAN[1]

1 Introduction

Most work on diagenesis of biohermal rocks on Gotland has focused on the early cement features (Watts 1981, 1982, Nield 1981). Except for cementation studies of megapolygonal bedrock structures by Laufeld et al. (1978), and a detailed cementation study by Frykman (1985), little has been published on the general aspects of the diagenetic processes affecting the sequence on Gotland.

At a single locality in the Klinteberg Formation, meteoric diagenesis has been invoked from the cement stratigraphy alone (Frykman 1985). An expansion of the investigation showed that similar cementation phenomena are present throughout the biohermal facies of the Klinteberg Formation, and that the cement stratigraphy could be correlated with other localities.

The present study illustrates how traditional petrographic investigations, supplemented by cathodoluminescence studies and geochemical analyses of stable isotopes, may elucidate the diagenetic history of a biohermal facies association.

2 Stratigraphy and Environmental Setting

The exposed Silurian sequence on the island of Gotland in the Baltic comprises approximately 500 m of shallow marine carbonate sediments. The succession of Late Llandovery to Late Ludlow age was developed in an intracratonic situation at tropical latitudes. An introduction to the stratigraphy and depositional environments was given by Riding (1981) and Laufeld and Bassett (1981).

The sequence is relatively undisturbed, with a NE-SW strike, and dips at 2° – 3° to the SE. The Klinteberg Formation spans the Wenlock-Ludlow boundary and is mainly composed of carbonates. The outcrop extends across the island (Fig. 1) and the unit attains a maximum thickness of 70 m.

The present investigation has been carried out in the southwestern area, where facies include small bioherms and interbiohermal crinoid-rich sediments. Northeast of the area where bioherms occur, the formation is dominated by bedded crinoidal grainstones. Further northeast, this facies grades into argillaceous finer-grained wacke- to packstones with peloidal beds and abundant

[1] Institute of Historical Geology and Palaeontology, Øster Voldgade 10, 1350 Copenhagen K, Denmark

Reef Diagenesis
Edited by J. H. Schroeder and B. H. Purser

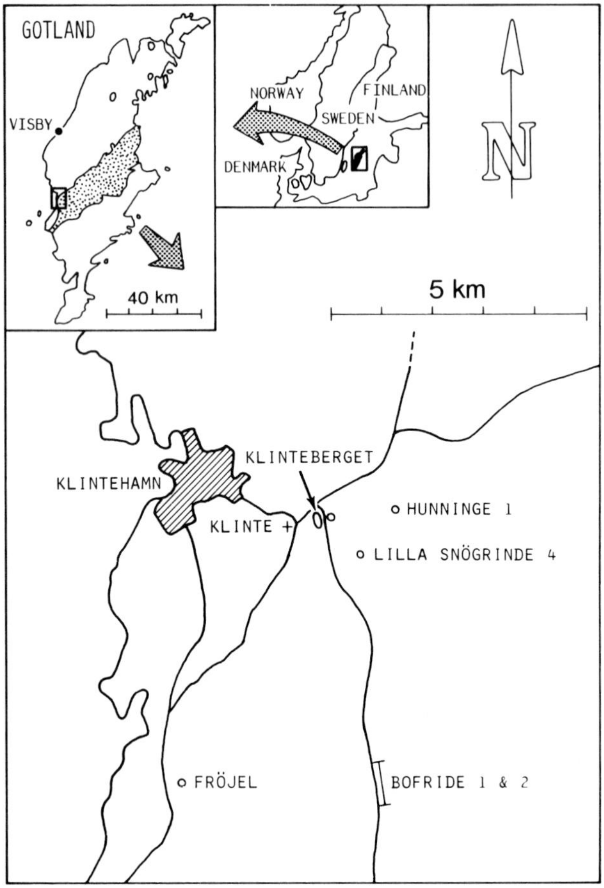

Fig. 1. Map showing outcrop of Klinteberg Formation (*stippled area*)

oncolitic algae. These deposits in the northeastern area of the outcrop represent relatively more protected environments.

Recent work has indicated that the Klinteberg Formation is diachronous (Martinsson 1967, Laufeld and Jeppsson 1976).

The Silurian sequence on Gotland was deposited in the shallow Balto-Scandian basin situated on the Precambrian shield area toward the western margin of the North European craton. The area was flooded during the Early Silurian. To the west of the shield the Scandinavian Caledonide orogen developed, which increasingly influenced sedimentation during the Silurian (Laufeld and Bassett 1981, Riding 1981).

3 Description of Bioherms

At the main locality at Klinteberget (Fig. 1) only small patches of autochthonous framework are found in association with rudstones and grain- to wackestones.

Similar relationships are visible at the other localities. The overall appearance of the Klinteberg bioherms is comparable to that at Wenlock Edge, England, of approximately the same age (Wenlock) (Scoffin 1971, Riding 1981).

The largest biohermal accumulation in the Klinteberg Formation is approximately 9 m thick, but generally the bioherms attain thicknesses of only a few meters. All are of lenticular shape. The fauna is dominated by stromatoporoids, tabulate corals, and colonial rugose corals.

The stromatoporoid assemblage is dominated by flat to laminar forms, but domal specimens also occur. The larger skeletons rarely construct a framework but occur as accumulations. Much of the larger skeletal material is not in growth position, but appears to be overturned (Fig. 2A). This material clearly has not been moved far from its growth site as it shows little sign of intensive abrasion through transport, and is therefore interpreted as essentially autochthonous.

The biohermal accumulations are associated with adjacent facies of bioclastic sands and gravel in a complex facies mosaic, and the boundaries are often clearly defined at individual localities (Fig. 2B). The bioclastic sands and gravel are often evenly bedded and some parts show cross-bedding and erosional contacts (Fig. 2C). The sediments are composed of crinoidal debris supplemented by stromatoporoid/coral fragments and bryozoans. Many beds are separated by argillaceous laminae up to 10 mm thick.

The matrix in the biohermal accumulations consists dominantly of crinoid debris, but in some areas appears to be slightly more argillaceous than the adjacent bedded bioclastic sediment. Fine-grained, finely laminated internal sediment occurs in the biohermal rocks, at one locality filling voids up to 8 cm in diameter.

Bioerosion is sparse and only few macroborings of *Trypanites* type occur in the stromatoporoid skeletons. Microboring and micrite envelopes are seen on some grains.

4 Methods

Investigations with petrographic microscope in combination with cathodoluminescence (CL) have been of great value in this study. With CL the clear calcite cements exhibit detailed zonation which can be used for outlining the cement stratigraphy.

Staining of thin sections was carried out with potassium ferricyanide to test for ferroan calcite and dolomite, using the method of Dickson (1966).

Stable isotopes have also been analyzed, in all cases on separate phases in the rocks in samples of 2 mg size. These samples were processed with phosphoric acid and analyzed on a Finnigan MAT 250 mass-spectrometer, and the isotope ratios are expressed against the PDB standard.

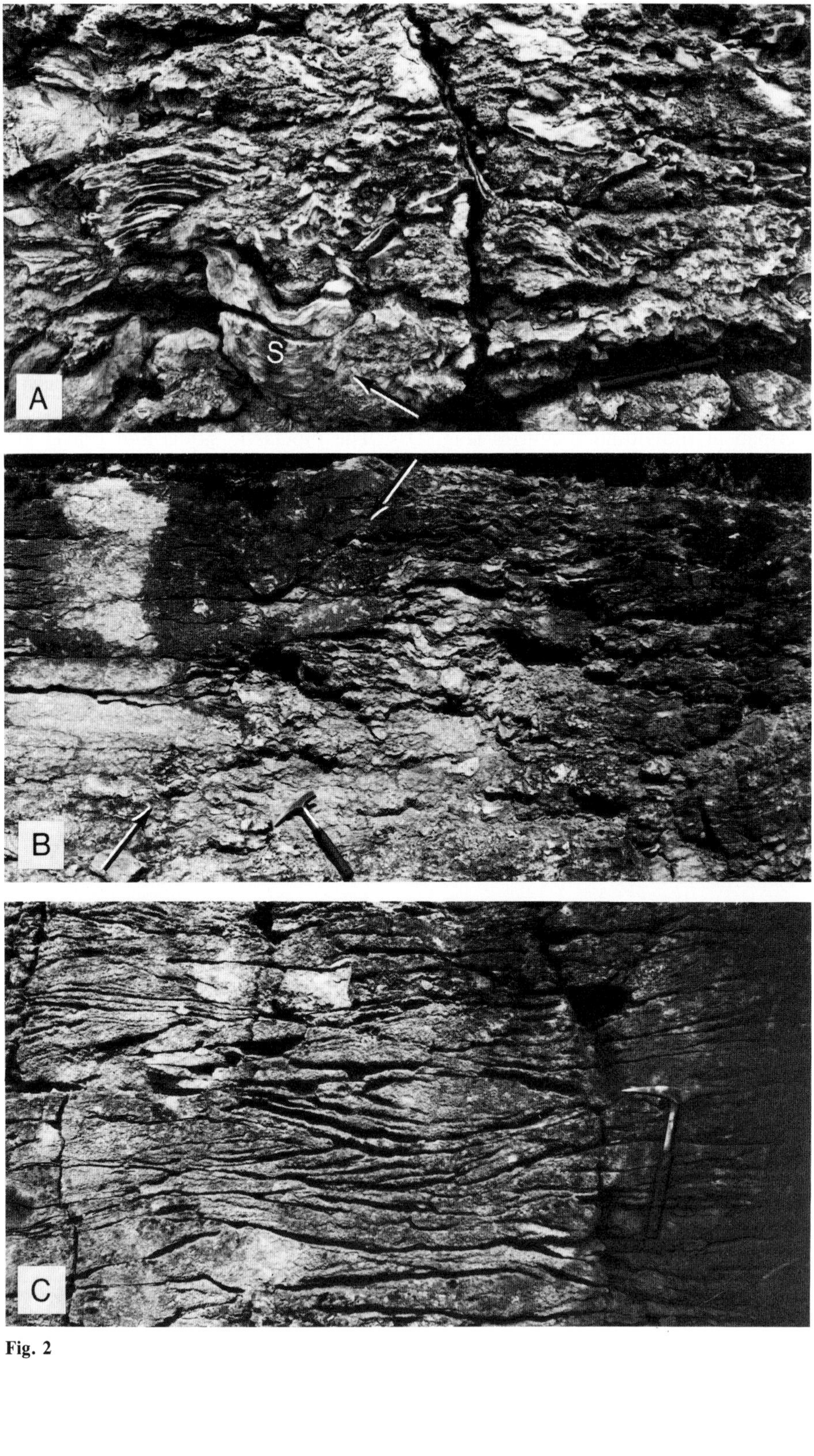

Fig. 2

5 Diagenetic Setting

Repeated meteoric influence in the Gotland sequence during the Silurian is suggested by descriptions of intertidal deposits and karstic phenomena (Laufeld et al. 1978, Watts 1981, 1982, Jeppsson 1982, Cherns 1982).

Burial of the sequence seems to have been moderate, and temperatures have never exceeded 100 °C, as indicated by the excellent preservation of microfossils (Laufeld 1974). Conodonts have received no thermal alteration according to Jeppsson (1983); he even proposed a new class "0" within the alteration index for conodonts for the very well-preserved Gotland material.

The succession on Gotland was buried beneath the Upper Silurian-Devonian (Carboniferous?) sequence, which now outcrops off-shore south of Gotland (Flodén 1980). Late- and Post-Paleozoic uplift and erosion contributed significantly to the present outcrop pattern which was completed by the Quaternary glacial erosion.

6 Diagenesis of Skeletal Material

The fabric and mineralogy of calcareous invertebrate skeletons have been reviewed in detail by, e.g., Bathurst (1975), and recently studies of selected groups from the Lower Cambrian have been carried out by James and Klappa (1983). The full spectrum of fossil types will not be described here, as focus is placed on two types of skeleton; crinoids that are the most abundant, and brachiopods which contribute to the interpretation of the diagenesis. All skeletal material investigated is nonferroan.

6.1 Crinoids

Crinoid debris, including root-systems and larger stem-parts, are abundant.

6.1.1 Petrography

The crinoid skeleton is believed originally to have been of HMC, having a highly porous stereome network and behaving as a monocrystalline unit under the light microscope (Bathurst 1975, Meyers and Lohmann 1978, James and Klappa 1983). The crinoidal fragments nearly always have an epitaxial cement overgrowth in optical continuity with the crinoidal monocrystalline skeleton.

Fig. 2. A Rudstone from biohermal accumulation. Notice overturned stromatoporoid (*s*). Pencil at lower right for scale. **B** Boundary (*arrowed*) between stromatoporoid rudstone (*center* and *right*) and bedded grainstones (*left*) in the biohermal facies complex. Hammer for scale. **C** Grainstone facies with trough cross-bedding. Hammer for scale

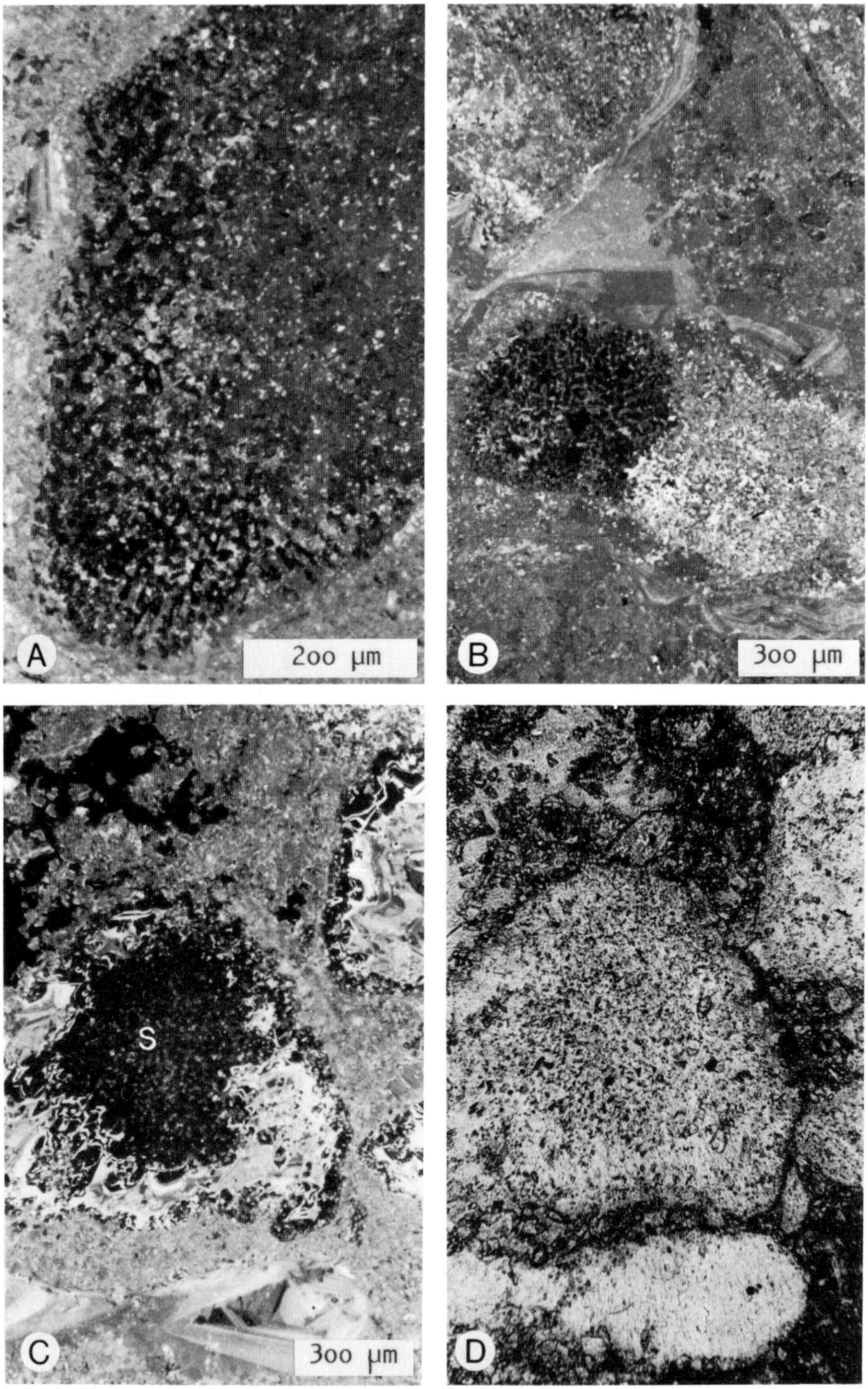

Fig. 3

In many cases investigation with CL clearly outlines the stereome structure because of contrasting skeleton/porefill luminescence. The contrasts range from bright/nonluminescent to different shades of dull luminescence (Fig. 3A, B). In all cases the crinoids are calcite, containing no iron, and have various amounts of inclusions giving a dusty appearance (Fig. 3B). Rare dolomite rhombs (15 – 20 μm) can be recognized within the crinoid grains with CL. Diagenesis of the crinoid grains is not consistently homogeneous, and the same grain may contain parts showing different aspects of diagenesis (Fig. 3A, B).

Dissolution of crinoid skeletons is recorded only in a few cases (Fig. 3C, D), where the pore space subsequently is filled with spar or internal sediment.

6.1.2 Geochemistry

Echinodermata are known to be a group that exerts a vital effect on the isotope content of their skeleton (Weber 1968). The fractionation is reported to be in the order of 3.5‰ towards lighter $\delta^{18}O$ (Brand 1981) for crinoid material. The crinoid material analyzed here falls within -6.0 to -7.2‰ $\delta^{18}O$ (Fig. 4A).

6.1.3 Interpretation

The crinoid particles are petrographically identical to both recent and other fossil examples which are shown or interpreted to be original HMC (Bathurst 1975, Meyers and Lohmann 1978, James and Klappa 1983, Leutloff and Meyers 1984). In the Gotland material the effects of early dissolution and partial diagenetic obliteration of the stereome structure, combined with the presence of inclusions, confirm a HMC precursor mineralogy.

6.1.4 Discussion

The diagenetic transformation of HMC echinoderm skeletons is a subject of much discussion. Echinoderms have a highly porous reticulate ultrastructure, and the pores are thought to be filled with cement at a very early stage in diagenesis. As the crinoids have a high initial porosity the isotope analysis will include also the diagenetic calcite filling the pores, making interpretations difficult.

Diagenesis of the crinoid skeletal particles in the Gotland material seems complicated, and no general scheme for the alteration and cementation can be

Fig. 3. A (CL). Crinoid grain exhibiting stereome structure in *left* half, nearly obliterated in *right* part by diagenesis. **B** (CL). Crinoid grain with dull epitaxial cement of zone 4. Stereome structure in nonluminescent and dull calcite in *left* half, nearly obliterated in *right* half consisting of bright and dull calcite. **C** (CL). Partly dissolved crinoid grain where the central part has stereome structure (*s*) in nonluminescent and dull calcite. The dissolved part is filled by cement zones 2, 3 and 4 (nonluminescent, bright and dull). **D** (normal light). Same view as **C**. Inclusion material in the crinoid grain is preferentially preserved in the part showing stereome structure

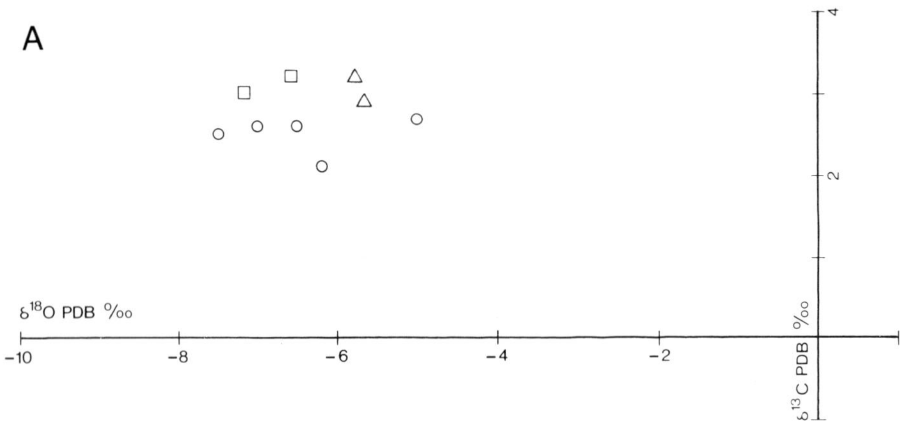

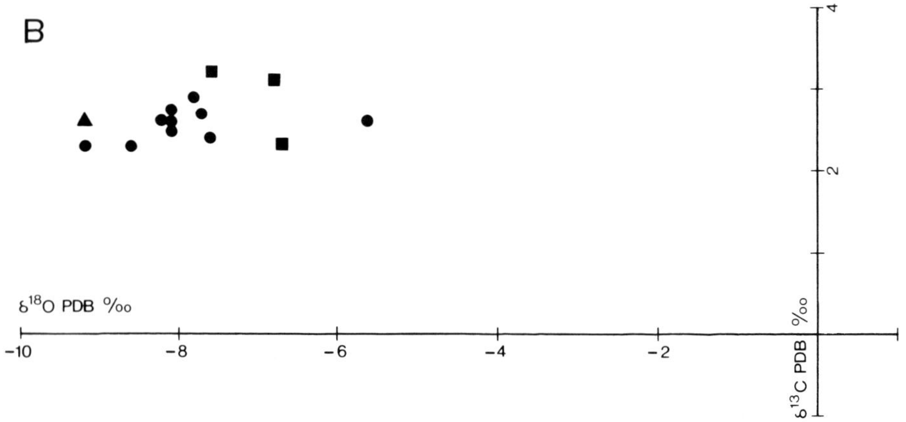

Fig. 4. A Plot of oxygen and carbon isotopic composition of fossil components in biohermal facies in Klinteberg Formation. *Circles* are crinoidal material. *Squares* are stromatoporoids and *triangles* are articulate brachiopods. **B** Plot of oxygen and carbon isotopic composition of calcite spar and internal sediment in biohermal facies in Klinteberg Formation. *Filled circles* are analyses of clear calcite cement filling voids and fractures. *Filled squares* are analyses of internal geopetal sediment, *filled triangle* is analysis of clear calcite spar filling primary voids in tabulate coral

presented. Recrystallization of the crinoid skeleton probably occurred at the initial stage of the introduction of meteoric water. Some crinoid particles were dissolved before precipitation of cement zone 2 (Fig. 3C); this is also ascribed to meteoric influence.

6.2 Brachiopods

6.2.1 Petrography

The brachiopods observed in the Silurian material are mostly articulate forms showing a characteristic fibrous, laminated fabric (Fig. 5A). Phosphatic inarticulates occur in minor quantities.

Some of the articulate brachiopods have a slightly dirty micritic rim and, with CL, tubular microborings are seen to penetrate into the marginal part of the shell. This is especially clear when the shell is nonluminescent (Fig. 5B). The shells are rarely homogeneously nonluminescent, but show thin bright or dull laminae (Fig. 5B). Some brachiopod shells show lamination and a dull luminescence, often overgrown by cloudy bladed cement (Fig. 5C).

6.2.2 Geochemistry

The shell material analyzed is nonluminescent and has preserved original skeletal texture.

The isotopic composition of the shell material averages −5.8‰ $\delta^{18}O$, 3.1‰ $\delta^{13}C$ (Fig. 6).

6.2.3 Interpretation

The nonluminescent brachiopod material with preserved original skeletal texture is interpreted to have suffered relatively little recrystallization, and exchange of elements with the pore water is expected to be small. Lowenstam (1961) has shown that shell material from recent and fossil brachiopods carries no vital effect in the isotope composition.

The isotopic composition (−5.8‰ $\delta^{18}O$, 3.1‰ $\delta^{13}C$) is therefore suggested to be close to the isotopic composition of carbonate precipitated in equilibrium with Wenlock-Ludlow sea water at this locality.

6.2.4 Discussion

The pattern of preservation and interpretation of the stability of the brachiopod skeletal material is consistent with investigations of Devonian brachiopods by Popp et al. (1981).

In Gotland material both nonluminescent and luminescent brachiopod shells occur, thus suggesting variable stability. The incorporation of Mn presumably takes place during diagenesis, since the primary content of Mn in brachiopods is seemingly very low (Al-Aasm and Veizer 1982).

When considering the significance of the isotopic signatures the stability obviously is important, but this skeletal material generally is expected to have preserved the original isotopic composition. Evaluation usually assumes only slight or no vital effect on the isotopic composition of the original skeleton, as indicated by Lowenstam (1961).

The isotope composition can be compared to values selected to represent a marine carbonate isotope composition by James and Choquette (1983) from Silurian material presented by Sears and Lucia (1980). These show values of approximately −4.0‰ $\delta^{18}O$; 1.4‰ $\delta^{13}C$ isotopic composition. Analyses of Silurian material by Brand and Veizer (1980) show a wide range in compositions

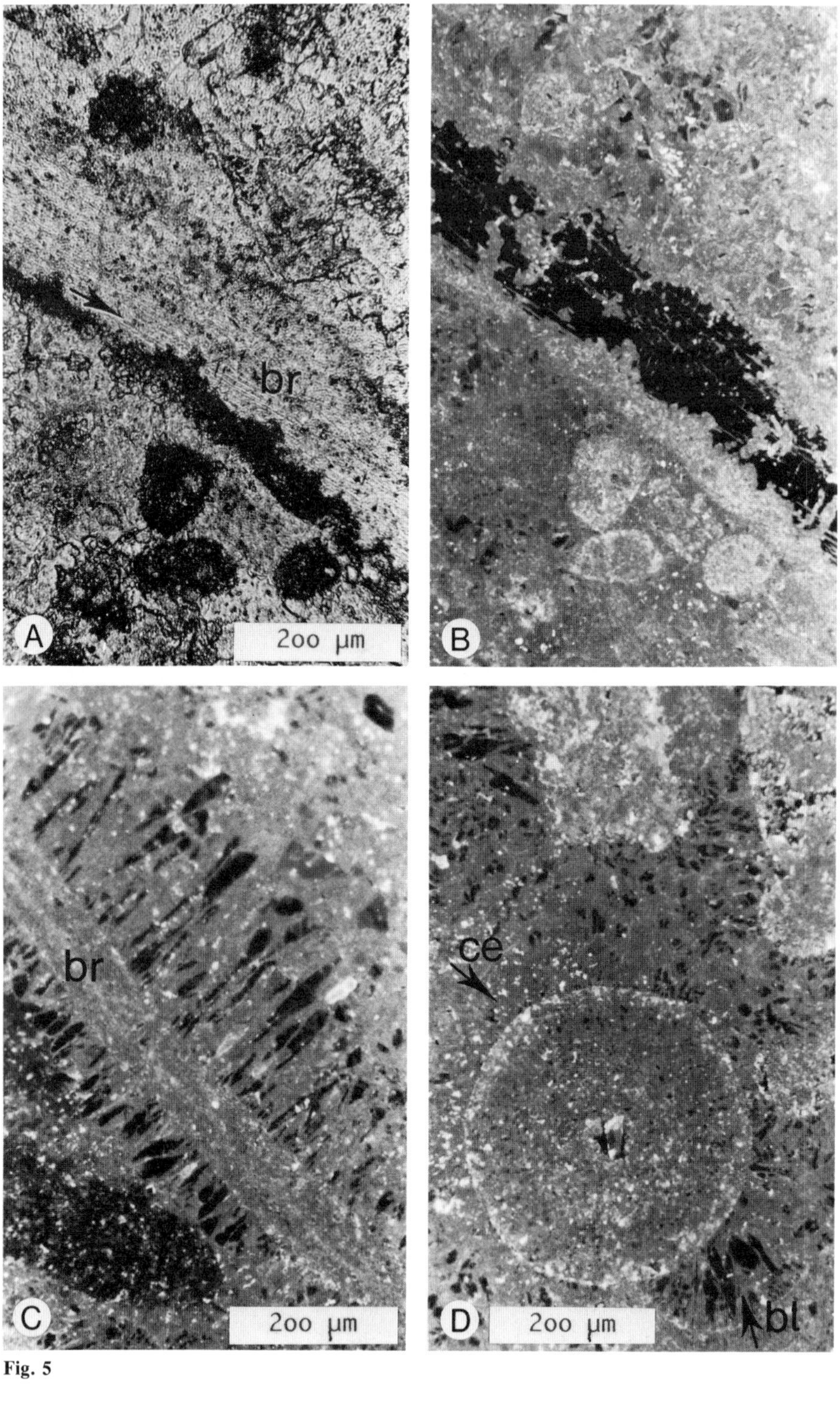

Fig. 5

but their heaviest $\delta^{18}O$ value is -5.0‰. All these values deviate somewhat from the composition found in this investigation. Some of the deviation might be ascribed partly to different precipitation temperatures in the different settings, but there seems no doubt that isotope values for carbonate in equilibrium with seawater were more negative in the Silurian than in the Recent.

7 Diagenesis of Cements

Cement fills nearly all pores and voids in the biohermal facies and associated rocks. Both cloudy and clear calcite occur with different crystal morphologies. Cloudy bladed cement is most important within the biohermal facies itself, but also occurs in the associated bedded grainstones. In crinoid-dominated grainstones, epitaxial calcite cements dominate.

Significantly, all cement types investigated in this study proved to be nonferroan when tested with potassium ferricyanide.

Zonation seen with CL, especially in the epitaxial calcite cement, is used for establishing the cement stratigraphy. The four main cement zones are defined by their general CL characteristics and are numbered zones 1 to 4.

7.1 Cloudy Bladed Cements

7.1.1 Petrography

This type of cement occurs mostly on shell fragments such as brachiopods and bryozoans (Fig. 5C). On crinoids, only rare examples of bladed cement occur (Fig. 5D) and these are seen to be contemporaneous with cloudy epitaxial cement.

With CL the cloudy bladed cement nearly always exhibits a very dark, dull luminescence. In this way some of the blades show the classical shape of a roman sword, attaining lengths from 20 to 100 μm (Fig. 5C). Most of the individual blades contain fewer, brightly luminescing inclusions than the cloudy epitaxial cement (Fig. 5D). All bladed cements are nonferroan.

7.1.2 Interpretation

Based on their morphology and inclusion content, bladed cements are interpreted as being former marine HMC cements. The individual blades are characterized

Fig. 5. A (normal light). Brachiopod grain (*br*) with faint oblique laminated fabric (*arrow*). *Left* margin of shell has dirty micritic rim. Peloids occur in the matrix. **B** (CL). Same view as **A** showing nonluminescent brachiopod shell with a few bright laminae. Microborings have penetrated the shell. **C** (CL). Brachiopod grain (*br*) overgrown by cloudy bladed cement with individual blades outlined by differences in luminescence. **D** (CL). Crinoid grain with both cloudy bladed cement (*bl*) and cloudy epitaxial cement (*ce*). The cloudy epitaxial cement contains bright inclusions which are less common in the bladed cement

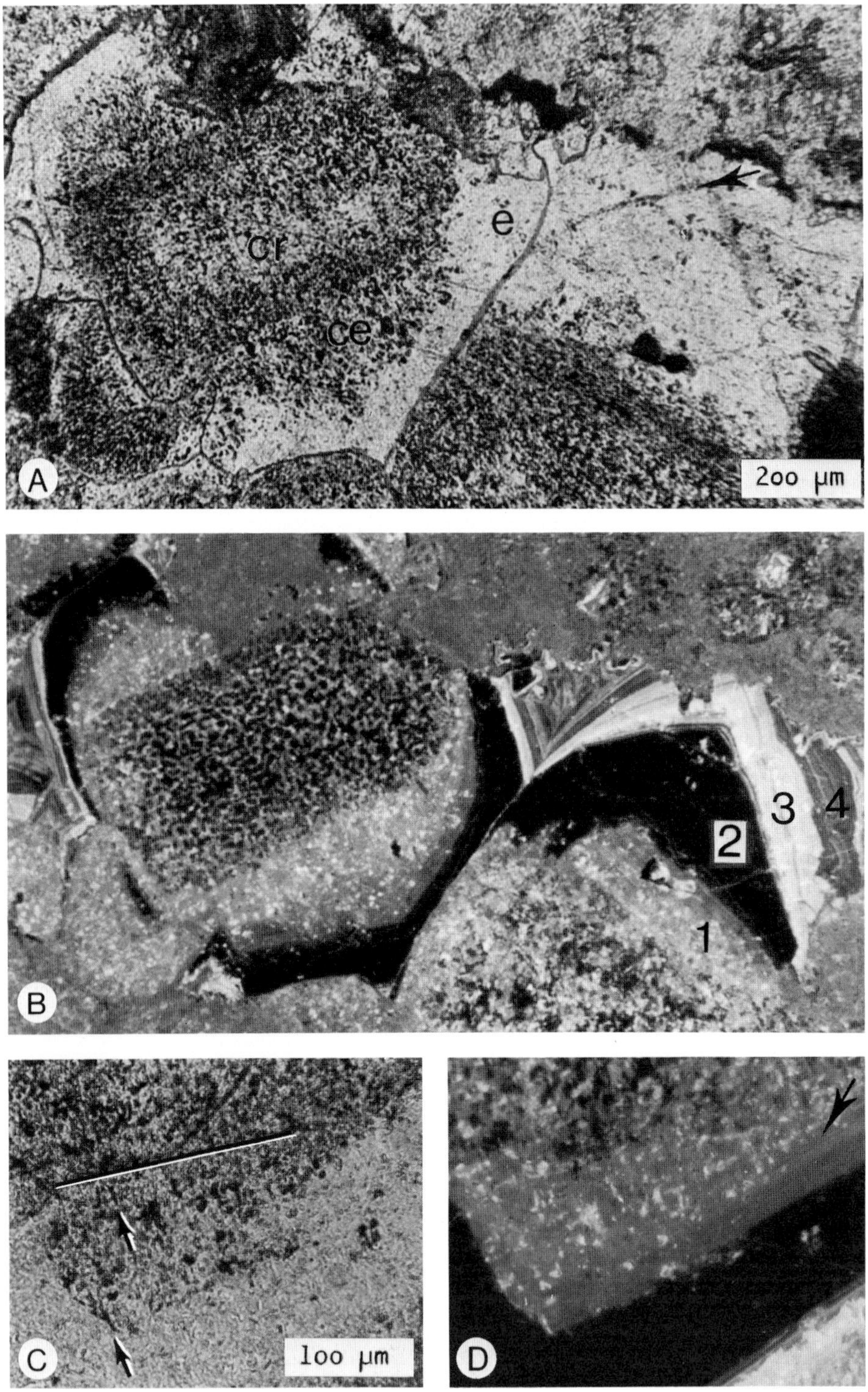

Fig. 6

by increasing width from the substrate, and by rhombohedral terminations, and they closely resemble the morphology of recent bladed HMC marine cement illustrated by James et al. (1976) from Belize reefs, by Schroeder (1973) from Pleistocene Bermuda reefs, and Land and Moore (1980) from Jamaican reefs.

7.1.3 Discussion

A wide spectrum of original mineralogies has been proposed for bladed cements of marine origin. LMC is suggested by Lindström (1979), Wilkinson et al. (1982) and Sandberg (1983), based mainly on textural and morphological criteria. Original dolomitic mineralogy has been suggested for some Precambrian bladed cements by Tucker (1983). However, in recent literature, HMC is commonly suggested as being the precursor for bladed marine cements (Davies 1977, Lohmann and Meyers 1977, Walls et al. 1979, Marshall and Ashton 1980, Marshall 1981).

Cements having fibrous and acicular textures are sometimes interpreted as having had aragonite precursors, as supported by analysis of trace elements such as Sr (Davies 1977, Mazzullo 1980). Fibrous texture however, seems to be shared between many aragonite and HMC cements, and close examination is needed to detect differences. Cements with acicular habit have not been observed in this investigation, and might possibly indicate the absence of aragonite cements in this setting.

The dominant role of bladed and epitaxial cement, both interpreted as original HMC cements, is somewhat in line with Sandberg's (1983) hypothesis outlining a variation of mineralogy of marine precipitates through geologic time. However, Sandberg suggests that LMC cement was the dominant precipitate during the Silurian, and that aragonite cements were lacking during this period; the Silurian is considered to be a "Calcite Sea" (Sandberg 1983).

7.2 Cloudy Epitaxial Cement

7.2.1 Petrography

Crinoid fragments are common substrates for cloudy epitaxial cements which form the main part of cement zone 1 in the cement stratigraphy (Fig. 6A, B).

The cloudy nature of the cement is due to small inclusions of 0.5 – 5 μm. Some of these, analyzed with microprobe, contain up to 40 mol % $MgCO_3$, and

Fig. 6. A (normal light). Crinoid grain (*cr*) with cloudy epitaxial calcite cement (*ce*) and overgrowth of clear epitaxial calcite (*e*). The clear cement includes a dust-line (*arrowed*), coinciding with the boundary between cement zones 2 and 3. **B** (CL). Same view as **A** showing cement zones 1 – 4 (*numbered*). Stereome structure clearly preserved in crinoidal grain to the *left*. **C** (normal light). Cloudy epitaxial cement with inclusions showing faint parallel arrangement and some elongate inclusions (*arrowed*). Boundary to crinoid grain marked with *line*. **D** (CL). Same view as **C** showing faint subzonation in outer part of zone 1 cement (*arrow*) and well-defined boundary to cement zone 2. Some inclusions can be recognized in both **C** and **D**

therefore seem comparable to microdolomite inclusions as described by Meyers and Lohmann (1978). Some of the larger inclusions are elongate and thus are not comparable to microdolomite; these show a faint parallel pattern aligned with the optic axis in the epitaxial overgrowth crystal (Fig. 6C, D).

Most of the inclusions can also be recognized with CL, most commonly as bright yellow spots and streaks. The bright yellow appearance of the inclusions is consistent even when the general luminescence of the cement is bright, dull or nonluminescent.

The cloudy epitaxial cement predates blocky calcite and internal sediment. Terminations of the cloudy epitaxial calcite can be difficult to trace, but some examples clearly have a euhedral crystal face termination (Fig. 6B). In other cases the outline is irregular. CL reveals that the earliest part of the inclusion-free calcite has luminescence comparable to that of the inclusion-rich calcite, but has a faint subzonation (Fig. 6D). Despite the contrast in inclusion density, the subzoned dull, clear calcite is included in cement zone 1, because of the close association with this zone and the very well defined boundary to cement zone 2 as seen with CL (Fig. 6D).

In some examples the cloudy spar shows an apparent pendant habit; however, the cement also grows irregularly on the upper surfaces on some grains (Fig. 6B).

7.2.2 Interpretation

Being the first generation of cement on the crinoids, and because of the content of microdolomite inclusions, this cement is interpreted as having a HMC precursor.

The cloudy epitaxial cement (zone 1) is interpreted to be recrystallized before zone 2 of clear calcite precipitated. This is suggested by the very well defined and euhedral morphology of the boundary between the two zones.

7.2.3 Discussion

Rims of cloudy epitaxial cement on echinodermal material were described by Meyers (1974), Meyers and Lohmann (1978) from Mississippian limestones in New Mexico, by Grover and Read (1983) from Middle Ordovician carbonates, Virginia, and by Cantrell and Walker (1985). In these studies, some of the cloudy epitaxial cements were interpreted as marine cements, and others as meteoric marine mixing zone cements (Meyers and Lohmann 1978).

Cloudy epitaxial cements were illustrated, by Heckel (1983, Fig. 14) and James and Klappa (1983, Fig. 6b) from Pennsylvanian and Lower Cambrian rocks respectively, but these examples were not commented upon in the text. The occurrence of this type of epitaxial cement therefore could be more widespread than generally indicated.

According to James and Choquette (1983), no epitaxial HMC has been reported from modern environments except for the development of epitaxial HMC crystallites on cores of echinoderm material in HMC ooids described by

Marshall and Davies (1975). One exception concerns the description by Richter (1974, p. 62) of small epitaxial HMC crystals (10 μm) growing on the surface of echinoderms in recent submarine environments.

The alternative interpretation of cloudy epitaxial calcite as having originated in a meteoric-marine mixing zone, as suggested by Meyers and Lohmann (1978) is considered unlikely for the Gotland cements. The described mixing zone cements have a distinct zonation and varying inclusion density (Meyers and Lohmann 1978) which is not a characteristic of the Silurian cements.

To date, no internal sediment of obvious marine nature has been seen to overlie the inferred marine cloudy epitaxial cement, and this admittedly weakens the present interpretation.

The presence of elongate inclusions within the epitaxial cement might suggest a former bladed texture of the epitaxial overgrowth. However, the terminal faces of cloudy epitaxial cements are clearly euhedral in some cases, and growth as a monocrystalline unit is evident. The elongate inclusions might originate also within a single crystal during growth (Kendall 1985).

The difference in luminescence characteristics between the crystal and the inclusions remains unexplained.

Recrystallization of cloudy epitaxial cement probably predated meteoric cementation by clear calcite of zone 2. Leutloff and Meyers (1984) argued that slight meteoric cementation can be partly responsible for an enhanced retention of microdolomite inclusions during neomorphism of the overgrown HMC material.

7.3 Clear Prismatic Cement

7.3.1 Petrography

Clear prismatic cement crystals 150 – 200 μm are seen in shelter voids under skeletal fragments (Fig. 7A), as a prismatic epitaxial overgrowth on cloudy bladed cement, in skeletal voids (Fig. 7B), and as epitaxial continuation of prismatic structure in certain brachiopod shells (Fig. 7C, D). With CL the cement shows varying luminescence and sometimes reveals fine subzonation corresponding closely to the zonation seen in the epitaxial overgrowth on the crinoids (Fig. 7A, B, D).

7.3.2 Interpretation

The limpidity and preservation of growth stages are interpreted as indicating a LMC origin having escaped recrystallization during diagenesis.

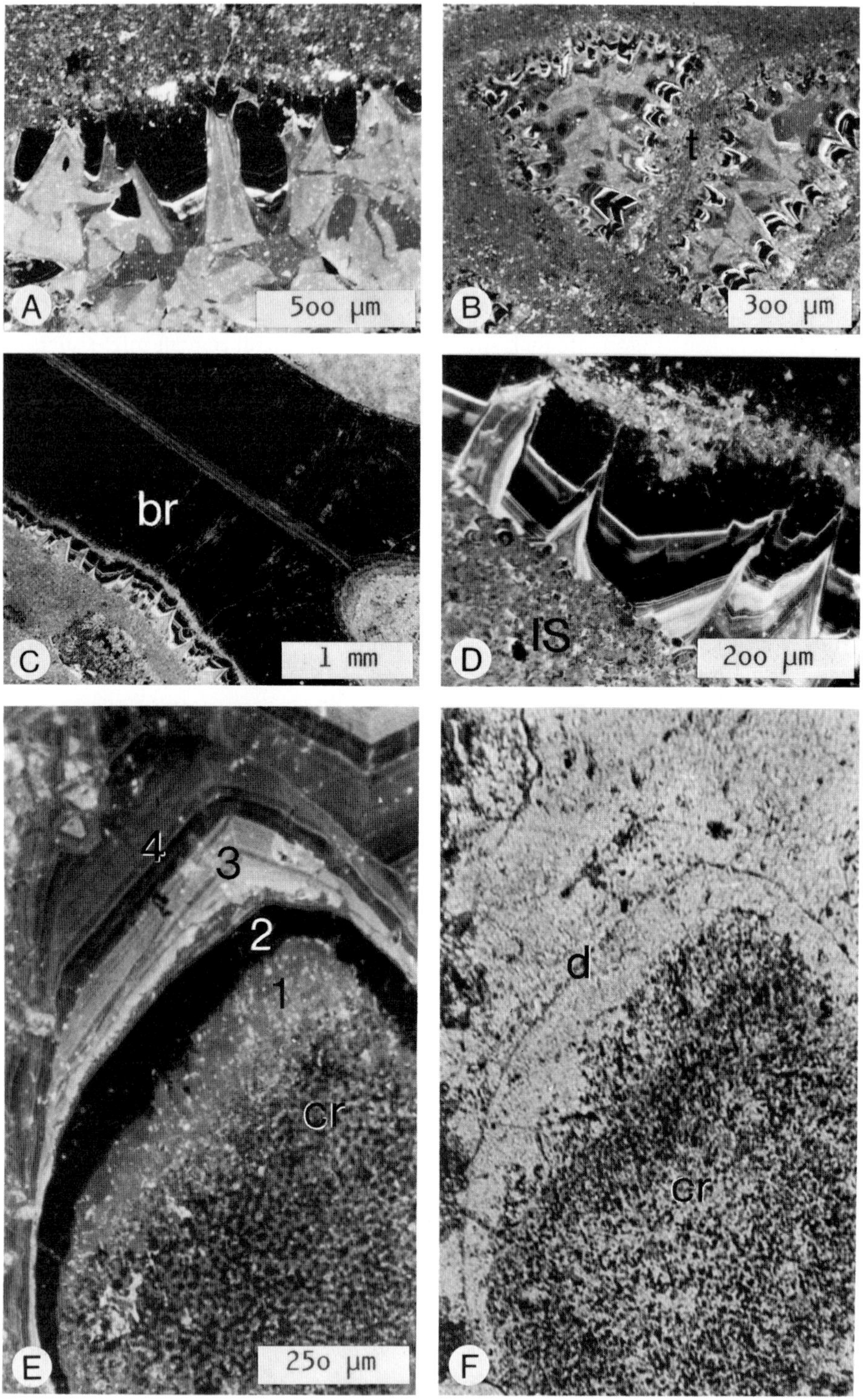

Fig. 7

7.4 Clear Epitaxial Cement

7.4.1 Petrography

On crinoids the epitaxial overgrowth with clear calcite forms rinds up to 500 μm thick, and the very distinct zonation seen with CL in many samples is the basis for outlining the cement stratigraphy (Fig. 7E, F).

Three main zones of clear calcite have been established: zones 2, 3 and 4 are nonluminescent, bright yellow and dull orange respectively (Fig. 7E).

The boundary between zone 2 and the earlier cloudy cement or the skeletons is mostly well defined by the great contrast in inclusion density, and additional contrast in luminescence. The nonluminescent cement zone 2 often forms the innermost zone of clear spar. This zone commonly includes a few thin, brightly luminescing lines marking earlier growth stages. Repetitions and intergrowth of zone 2 and zone 3 of bright luminescent spar are also recorded (Fig. 8A). Zone 2 alone can be relatively thick on some grains, reaching 200 μm. In some samples zone 2 is overlain and restricted by internal sediment (Fig. 8B), and on other grains the boundary between zones 2 and 3 is marked by a dust-line (Figs. 6A, 7F).

The succeeding, brightly luminescent, zone 3 is commonly finely subzoned (Fig. 7E). The dull cement with subzonation is designated zone 4 and forms the final generation filling the last pore spaces (Fig. 7E).

Clear epitaxial calcite cement occurs in all lithologies associated with the bioherms. However, the cement zones 2 and 3 are irregularly distributed.

7.4.2 Geochemistry

The analyzed samples have been selected as the clear calcite spar filling voids and fractures, and cloudy cements have been avoided if possible. The samples are dominated by clear epitaxial calcite and blocky spar.

The isotopic composition ranges from −5.5‰ to −9.3‰ $\delta^{18}O$ (Fig. 4B) and thereby shows the lightest oxygen values recorded in the present material.

7.4.3 Interpretation

The limpidity and preservation of growth stages as seen with CL argue for a diagenetically unaltered LMC mineralogy. The light isotope composition is

Fig. 7. A (CL). Primatic clear calcite cement in shelter void below fossil fragment. Zones 2, 3, and 4 are recognized in the prismatic crystals, zone 4 also forming the blocky cement filling the last pore space. **B** (CL). Prismatic clear cement overgrowing cloudy cement on skeletal walls (*t*) of a tabulate coral. Zonation and repetition of zones are seen in the prismatic crystals. **C** (CL). Prismatic clear cement crystals growing epitaxially on prismatic structure in brachiopod shell (*br*). **D** (CL). Detail of **C**. Prismatic cement crystals show zonation and repetitions of zones. Internal sediment (*IS*) overlies zone 2 cement. **E** (CL). Cement stratigraphy on crinoid grain (*cr*). Cement zones *numbered*. Both zone 3 and 4 are finely subzoned. **F** (normal light). Same view as **E**. Boundary between crinoid (*cr*) and cement zone 1 is difficult to outline in normal light. Dust-line (*d*) separates cement zones 2 and 3

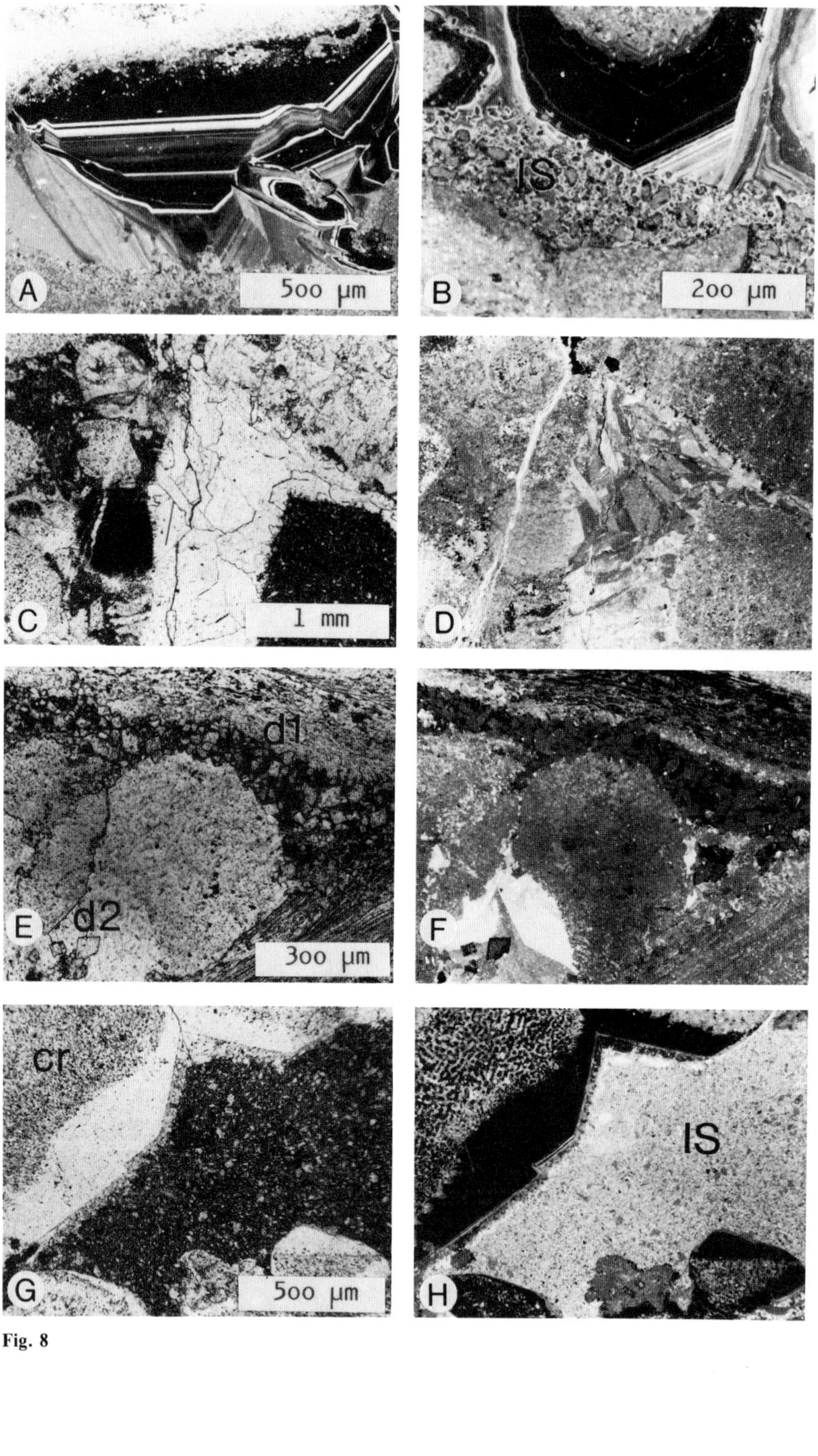

Fig. 8

interpreted as either derived from meteoric influence by depleted water, or from burial temperature effect.

7.4.4 Discussion

Epitaxial overgrowth on echinoderms of clear calcite is normally considered to form only in meteoric phreatic settings (Longman 1980, James and Choquette 1984).

The development of epitaxial overgrowths as so-called solution coronas around the crinoid fragments, as described by Walkden and Berry (1984), is not considered in this study because the grainstone lithologies and euhedral crystal morphologies indicate precipitation in primary void spaces.

7.5 Clear Blocky Cement

7.5.1 Petrography

This clear, nonferroan cement fills the central part of void spaces, solution vugs and fractures, having crystals up to more than 500 µm. The luminescence is bright orange or dull orange, and subzonation occurs frequently (Figs. 7 A, B; 8 C, D).

7.5.2 Geochemistry

The clear blocky calcite is included in the samples of clear cement showing the light oxygen signatures (Fig. 4 B), the lightest being $-9.3‰\ \delta^{18}O$.

7.5.3 Interpretation

The limpidity and preservation of growth stages as seen with CL argue for a diagenetically unaltered LMC mineralogy. Based on luminescence characteristics, some of this cement is interpreted as having formed contemporaneously with zone 4 in the clear epitaxial calcite.

The blocky cement occurring in fractures is interpreted as a relatively late post-compactional cement.

Fig. 8 A (CL). Epitaxial clear calcite showing repetions of the zonation. **B** (CL). Epitaxial clear calcite zone 2 overlain and stopped by internal sediment (*IS*). **C** (normal light). Fracture with clear blocky calcite. **D** (CL). Same view as **C** showing blocky calcite with slightly varying luminescence in fracture. **E** (normal light). Dolomite (*d1*) as replacement probably of internal sediment under brachioppod shell. Isolated dolomite rhombohedra (*d2*) occur in cement zone 4. **F** (CL). Same view as **E** showing dolomite rhombohedra included in the dull calcite cement. **G** (normal light). Slightly argillaceous internal sediment terminating clear epitaxial cement on crinoid grains (*cr*). **H** (CL). Same view as **F**. Cement zone 2 is overlain by internal sediment (*IS*)

8 Dolomite

8.1 Petrography

The dolomite in these rocks is euhedral and sucrosic (Fig. 8 E, F). Dolomitization has often affected selectively the internal sediment or argillaceous limestone matrix in the rock. Dolomite rhombohedra also occur intergrown with the clear blocky calcite cement in the pore fillings of zone 4 (Fig. 8 E, F). The crystals range in size from 20 to 100 μm and luminesce dark red. Small dolomite rhombohedra occur within crinoid fragments, also having dark red luminescence.

8.2 Interpretation

Two types of dolomite seem to be present; a replacement type occurring as the dolomitized internal sediment, and a cement type is intergrown with the clear blocky calcite cement of zone 4. The last type is interpreted as derived during the burial phase, possibly in a mixing zone.

Dolomitization in a marine meteoric mixing zone has been invoked by Kaldi and Gidman (1982) and Jones et al. (1984). The mixing zone is interpreted here to be a result of upward retreat of the meteoric regime in the buried sequence during a transgressive period.

9 Internal Sediment

9.1 Petrography

In biohermal lithologies, internal sediment has infiltrated even the smallest voids and shelter pores and forms geopetal configurations. Most internal sediment seems to be of the crystal silt type (Dunham 1969) containing no recognizable skeletal material. Lamination is rarely seen, but the upper layer of some internal sediment appears to be slightly coarser than the underlying sediment. This inverse gradation has been noted also by Aissaoui and Purser (1983) in internal sediment.

The luminescence of internal sediments is generally bright or dull, with few nonluminescent grains.

Cement zone 2 is overlain and terminated by the internal sediment in some samples (Fig. 8 B, G, H). In pores where internal sediment is absent, the boundary between cement zones 2 and 3 is marked by a dust-line (Figs. 6 A, 7 F).

9.2 Interpretation

The internal sediment, separating two phases of meteoric cementation, is interpreted as derived by infiltration of eroded material and crystal fragments. A dust-line formed contemporaneously in some cements. There is no positive evidence of deposition in a vadose zone. The infiltration seems to coincide with a slight geochemical change in some locations, which is recorded by the change from nonluminescent zone 2 to brightly luminescent zone 3 cement.

10 Summary of Diagenetic History

Three main stages of diagenesis have affected the bioherms:

1. synsedimentary marine cementation,
2. near-surface meteoric diagenesis,
3. shallow burial diagenesis.

10.1 Phase 1 – Synsedimentary Cementation

The bioherms and associated sediments in the Klinteberg Formation have been subjected to extensive marine cementation. Although no hard-grounds or macroborings attest to sea-floor lithification, intraclasts cemented by bladed cements are present and strongly suggest the marine cementation.

Early cementation also occurs in the crinoidal grainstones deposited adjacent to the biohermal complexes. Both bladed cements and cloudy, zone 1 epitaxial calcite are probably marine precipitates. The cementation of crinoids by marine cement involved both filling of the stereome pores and growth of epitaxial cement.

Furthermore, no major compactional phenomena are seen such as broken thin shells or crushed void-supporting fossils; this supports the suggestion of early marine cementation.

Recrystallization of the crinoids and the marine cement in the marine or mixed waters is a possibility, since recrystallization of HMC material can occur in the marine environment (Richter and Füchtbauer 1978). As suggested earlier, the cloudy cement of zone 1 probably recrystallized prior to precipitation of the succeeding zone 2 of clear calcite. The neomorphism of zone 1 took place either in the marine phreatic environment or in the initial stages of meteoric influence, i.e., in the diagenetic phase 2.

10.2 Phase 2 – Near-Surface Meteoric Diagenesis

Dissolution phenomena and possibly also neomorphism of HMC material are the first signs of the penetrating meteoric waters. Dissolution has preferentially affected the crinoid grains in the samples investigated and dissolution predates cement zone 2.

The sequence of luminescence in the zones of clear calcite from nonluminescent to bright to dull is interpreted to reflect the evolving redox conditions in the pore waters: a transition from oxidizing to increasingly reducing character of the waters may result in the sequence described (Frank et al. 1982). Taking this sequence of luminescence as a basis, the nonluminescent zone 2 indicates precipitation from oxidizing meteoric water. Zone 2 cement filled part of primary pore space, crinoidal stereome pores, and dissolution cavities.

In some samples the cloudy cement of zone 1 has a nonluminescent character instead of being dull orange, but still with brightly luminescing inclusions. This is interpreted as recrystallization during the influence of meteoric oxidizing water giving rise to the nonluminescent character of the cement.

Cement zone 2 in some examples is overlain by internal sediment, recording a change in pore water conditions. Some internal sediment is comparable to examples of crystal silt interpreted as having been derived from a vadose zone by Dunham (1969); however, neither the sediments nor the cements show positive criteria for growth under vadose conditions.

The nonluminescent cement of zone 2 is often interlayered with bright luminescent cement zone 3. The change from nonluminescent to bright can be brought about by only minor variations in the redox conditions in the pore water (Frank et al. 1982). The cyclic pattern is interpreted as reflecting minor repeated episodes of stagnation in the pore water, leading to periodically more reducing conditions.

Precipitation of zones 2 and 3 occurred in this phase; cement zone 4 might range from this into the next phase.

10.3 Phase 3 – Shallow Burial Diagenesis

The Klinteberg Formation was subsequently buried by the overlying deposits. Data on burial depth are sparse, but low thermal alteration of microfossils indicate burial to a maximum of 2 – 3 km.

Owing to extensive marine and near-surface meteoric cementation, compaction features are very rare, but cementation, stylolitization, and fracturing occurred as the final diagenetic stages including fracture filling by clear, blocky, calcite spar.

Selective dolomitization occurred in internal sediment and some parts of the argillaceous carbonate matrix within the bioherms. Scattered dolomite rhombohedra occur intergrown with the last generation of void-filling cement. The dolomite is interpreted as having formed during relatively shallow burial in the marine meteoric mixing zone.

The isotope analyses of clear calcite cements show a trend towards light $\delta^{18}O$. This is compatible with influence from meteoric water combined with a burial temperature effect.

11 Summary and Conclusions

The present composition of the Silurian limestones on Gotland is the result of three main stages of diagenesis occurring (a) at the sea floor, (b) in a meteoric setting, and (c) during shallow burial diagenesis. The three diagenetic stages include complex and nonuniform processes or sequences of processes, which have affected the bioherms in the Klinteberg Formation.

1. The bioherms lack true framework, but are rather dense biogenic accumulations of essentially autochthonous material, much of it, however, overturned. Very little bioerosion can be traced in the skeletal material. The sedimentary facies are all indicative of shallow water.
2. Marine cementation dominated by HMC was widespread. Both bladed cements and cloudy epitaxial cements helped to lithify the sediment in the biohermal accumulation. No aragonite cement seems to have been present, in contrast to modern reefal environments. Sandberg's (1983) "calcite sea" seems to have been a "HMC sea".
3. Penetrating meteoric waters in parts of the sequence had important diagenetic impact on the sediments containing metastable phases. Recrystallization, dissolution and precipitation of cement occurred in a complex pattern. Cementation was interrupted by an episode of internal sedimentation.
4. The trend in the oxygen isotopes towards depletion of the last diagenetic phases ($-9.2‰\ \delta^{18}O$) can be explained with meteoric influence combined with burial temperature fractionation of the diagenetic and precipitated components.
5. Shallow burial diagenesis involved fracturing and further cementation. Dolomite formed in minor parts of the sediment probably during influence of mixed marine-meteoric waters. The mixed waters may have arrived during a transgressive period, causing retreat of the meteoric regime.

Acknowledgments. For support and discussions on Gotland, I thank Sven Laufeld and Björn Sundquist, Uppsala. Jan Aagaard and Henrik Egelund helped prepare the illustrations. Richard Bromley kindly improved the language and suggested revisions. The two editors of this volume are thanked for their critical reviewing. Isotope analyses were performed at the Stable Isotope Laboratory at the Geological Institute of Copenhagen where instrumentation is funded by the Danish Science Research Council (Grant no. 11-9372). Tommy Cederberg helped during the analysis program. This paper forms a contribution to the IGCP Project Ecostratigraphy on Gotland, and the Swedish group is thanked for permission to use the Allekvia Field Station on Gotland.

References

Aissaoui DM, Purser BH (1983) Nature and origin of internal sediments in Jurassic limestones of Burgundy (France) and Fnoud (Algeria). Sedimentology 30:273 – 283

Al-Aasm IS, Veizer J (1982) Chemical stabilization of low-Mg calcite: an example of brachiopods. J Sediment Petrol 52:1101 – 1109

Bathurst RGC (1975) Carbonate sediments and their diagenesis. 2nd edn. Elsevier, Amsterdam, 658

Brand U (1981) Mineralogy and chemistry of the Lower Pennsylvanian Kendrick fauna, eastern Kentucky II. Stable isotopes. Chem Geol 32:17 – 28

Brand U, Veizer J (1981) Chemical diagenesis of a multicomponent carbonate system II. Stable isotopes. J Sediment Petrol 51:987 – 997

Cantrell DL, Walker KR (1985) Depositional and diagenetic patterns, ancient oolite Middle Ordovician, eastern Tennessee. J Sediment Petrol 55:518 – 531

Cherns L (1982) Palaeokarst, tidal erosion surfaces and stromatolites in the Silurian Eke Formation of Gotland, Sweden. Sedimentology 29:819 – 833
Davies GR (1977) Former magnesian calcite and aragonite submarine cements in Upper Paleozoic reefs of the Canadian Arctic. Geology (Boulder) 5:11 – 15
Dickson JAD (1966) Carbonate identification and genesis as revealed by staining. J Sediment Petrol 36:491 – 505
Dunham RJ (1969) Early vadose silt in Townsend mound (reef), New Mexico. In: Friedman GM (ed) Depositional environments in carbonate rocks. Soc Econ Paleontol Miner Spec Publ 14:139 – 181
Flodén T (1980) Seismic stratigraphy and bedrock geology of the central Baltic. Stockholm Contrib Geol 35:240 pp
Frank JR, Carpenter AB, Oglesby TW (1982) Cathodoluminescence and composition of calcite cement in the Taum Sauk Limestone (Upper Cambrian), southeast Missouri. J Sediment Petrol 52:631 – 638
Frykman P (1985) Subaerial exposure and cement stratigraphy of a Silurian bioherm in the Klinteberg Beds, Gotland, Sweden. Geol Fören Stockh Förhandl 107:77 – 88
Grover G Jr, Read JF (1983) Paleoaquifer and deep burial related cements defined by regional cathodoluminescence patterns, Middle Ordovician carbonates, Virginia. Am Assoc Petrol Geol Bull 67:1275 – 1303
Heckel PH (1983) Diagenetic model for carbonate rocks in midcontinent Pennsylvanian eustatic cyclothems. J Sediment Petrol 53:733 – 757
James NP, Choquette PW (1983) Limestones – the sea floor diagenetic environment. Geosci Canada 10:159 – 179
James NP, Choquette PW (1984) Limestones – the meteoric diagenetic environment. Geosci Canada 11:161 – 194
James NP, Klappa CF (1983) Petrogenesis of Early Cambrian reef limestones, Labrador, Canada. J Sediment Petrol 53:1051 – 1096
James NP, Ginsburg RN, Marszalek DS, Choquette PW (1976) Facies and fabric specificity of early subsea cements in shallow Belize (British Honduras) reefs. J Sediment Petrol 46:523 – 544
Jeppsson L (1982) Third European conodont symposium (ECOS III) Guide to excursion. Publ Inst Mineral Palaeontol Quat Geol, Univ Lund, Sweden 239:1 – 32
Jeppsson L (1983) Silurian conodont faunas from Gotland. Fossils Strata 15:121 – 144
Jones B, Lockhart EB, Squair C (1984) Phreatic and vadose cements in the Tertiary Bluff Formation of Grand Cayman Island, British West Indies. Bull Can Petrol Geol 32:382 – 397
Kaldi J, Gidman J (1982) Early diagenetic dolomite cements: examples from the Permian Lower Magnesian Limestone of England and the Pleistocene carbonates of the Bahamas. J Sediment Petrol 52:1073 – 1085
Kendall AC (1985) Radiaxial fibrous calcite: a reappraisal. Soc Econ Paleontol Miner Spec Publ 36:59 – 77
Land LS, Moore CH (1980) Lithification, micritization and syndepositional diagenesis of biolithites on the Jamaican Island slope. J Sediment Petrol 50:357 – 369
Laufeld S (1974) Silurian Chitinozoa from Gotland. Fossils Strata 5:1 – 130
Laufeld S, Bassett MG (1981) Gotland: the anatomy of a Silurian carbonate platform. Episodes 2:23 – 27
Laufeld S, Jeppsson L (1976) Silicification and bentonites in the Silurian of Gotland. Geol Fören Stockh Förhandl 98:31 – 44
Laufeld S, Sundquist B, Sjöström H (1978) Megapolygonal bedrock structures in the Silurian of Gotland. Sver Geol Unders C 759:1 – 26
Leutloff AH, Meyers WJ (1984) Regional distribution of microdolomite inclusions in Mississippian echinoderms from southwestern New Mexico. J Sediment Petrol 54:432 – 446
Lindström M (1979) Diagenesis of Lower Ordovician hardgrounds in Sweden. Geol Palaeontol 13:9 – 30
Lohmann KC, Meyers WJ (1977) Microdolomite inclusions in cloudy prismatic calcites: a proposed criterion for former high magnesium calcites. J Sediment Petrol 47:1075 – 1088
Longman MW (1980) Carbonate diagenetic textures from nearsurface diagenetic environments. Am Assoc Petrol Geol Bull 64:461 – 487
Lowenstam HA (1961) Mineralogy, $^{18}O/^{16}O$ ratios, and strontium and magnesium contents of Recent and fossil brachiopods and their bearing on the history of the oceans. J Geol 69:241 – 260

MacKenzie FT, Pigott JD (1981) Tectonic controls of Phanerozoic sedimentary rock cycling. J Geol Soc Lond 138:183 – 196
Marshall JD (1981) Zoned calcites in Jurassic ammonite chambers: trace elements, isotopes and neomorphic origin. Sedimentology 28:867 – 887
Marshall JD, Ashton M (1980) Isotopic and trace element evidence for submarine lithification of hardgrounds in the Jurassic of eastern England. Sedimentology 27:271 – 289
Marshall JF, Davies PJ (1975) High-magnesium calcite ooids from the Great Barrier Reef. J Sediment Petrol 45:285 – 291
Martinsson A (1976) The succession and correlation of ostracode faunas in the Silurian of Gotland. Geol Fören Stockh Förhandl 89:350 – 386
Mazzullo SJ (1980) Calcite pseudospar replacive of marine acicular aragonite, and implications for aragonite cement diagenesis. J Sediment Petrol 50:409 – 422
Meyers WJ (1974) Carbonate cement stratigraphy of the Lake Valley Formation (Mississippian) Sacramento Mountains, New Mexico. J Sediment Petrol 44:837 – 861
Meyers WJ, Lohmann KC (1978) Microdolomite-rich syntaxial cements: proposed meteoric marine mixing zone phreatic cements from Mississippian limestones, New Mexico. J Sediment Petrol 48:475 – 488
Nield EW (1981) The inception and development of reef growth in the Visby Formation (Sheinwoodian/late Llandovery) of Gotland, Sweden. Thesis, Univ Wales, Cardiff
Popp BN, Sandberg PA, Anderson TF (1981) Coordinated textural isotopic and elemental analysis of constituents in some Middle Devonian limestones. Geol Soc Am Abstr Programs, p 531
Richter DK (1974) Zur subaerischen Diagenese von Echinidenskeletten und das relative Alter Pleistozäner Karbonatterrassen bei Korinth (Griechenland). N Jahrb Geol Palaeontol Abh 146:51 – 77
Richter DK, Füchtbauer H (1978) Ferroan calcite replacement indicates former magnesian calcite skeletons. Sedimentology 25:843 – 860
Riding R (1981) Composition, structure and environmental setting of Silurian bioherms and biostromes in northern Europe. Soc Econ Paleontol Miner Spec Publ 30:41 – 83
Sandberg PA (1983) An oscillating trend in Phanerozoic non-skeletal carbonate mineralogy. Nature 305:19 – 22
Schroeder JH (1973) Submarine and vadose cements in Pleistocene Bermuda reef rock. Sediment Geol 10:179 – 204
Scoffin TP (1971) The conditions of growth of the Wenlock reefs of Shropshire (England). Sedimentology 17:173 – 219
Sears SO, Lucia FJ (1980) Dolomitization of northern Michigan Niagara reefs by brine refluxion and freshwater/seawater mixing. In: Zenger DH, Dunham JB, Ethington RL (eds) Concepts and models of dolomitization. Soc Econ Paleontol Miner Spec Publ 28:215 – 237
Tucker ME (1983) Diagenesis, geochemistry, and origin of a Precambrian dolomite: the Beck Spring Dolomite of eastern California. J Sediment Petrol 53:1097 – 1119
Walkden GM, Berry JR (1984) Syntaxial overgrowths in muddy crinoidal limestones: cathodoluminescence sheds new light on an old problem. Sedimentology 31:251 – 267
Walls RA, Mountjoy EW, Fritz P (1979) Isotopic composition and diagenetic history of carbonate cements in Devonian Golden Spike reef, Alberta, Canada. Bull Geol Soc Am 90:963 – 982
Watts N (1981) Sedimentology and diagenesis of the Högklint reefs and their associated sediments, Lower Silurian, Gotland, Sweden. Thesis, Univ Wales, Cardiff
Watts NR (1982) Early carbonate fabrics in Silurian reefs on Gotland, Sweden. Am Assoc Petrol Geol Bull 66:640
Weber JN (1968) Fractionation of the stable isotopes of carbon and oxygen in calcareous marine invertebrates – the Asteroidea, Ophiuroidea and Crinoidea. Geochim Cosmochim Acta 32:33 – 70
Wilkinson BH, Janecke SU, Brett CE (1982) Low-magnesian calcite marine cement in Middle Ordovician hardgrounds from Kirkfield, Ontario. J Sediment Petrol 52:47 – 57

Conclusions

The Diagenesis of Reefs: A Brief Review of Our Present Understanding

B.H. PURSER[1] and J. H. SCHROEDER[2]

1 Introduction

It is fairly obvious, both from published literature and from contributions to this Volume, that reefs are particularly susceptible to diagenetic modifications and, because of their relatively well-defined limits, are exceptionally favorable sites for the study of diagenetic processes and sequences. The scales of any given study, however, are extremely variable, ranging from ultrastructure of skeletal components measured in micrometers to the regional framework measured in tens or even hundreds of kilometers. Depending on purpose and interest, any given study concentrates on some portion of that range: one studying the forest without knowing the trees, the other concentrating on the leaves with little interest in trees and none in the forest. This Volume includes studies at all scales. Comprehensive studies covering the full range of scales and applying the various tools of the trade are rare. The work or the late W. Krebs and his students on Devonian reefs of the Rheinisches Schiefergebirge and that of E. Mountjoy and his team in the Canadian Rockies (Mountjoy and Krebs 1983) constitute two of the very few examples.

In view of the inherent difficulty in producing such comprehensive works, the global aspects of reef diagenesis must be pieced together mostly from less complete, fragmentary case studies. Nevertheless, at this stage it appears reasonable to attempt an evaluation of the state of our ignorance. We are prepared to take a certain amount of risk when speculating a little, notably concerning models and processes, for this may provoke criticism and thus lead to improvement of our understanding of reef diagenesis and, perhaps, of carbonate diagenesis in general.

2 The Sedimentary and Biological Controls of Reef Diagenesis

Perhaps the main reason for the frequent and varied nature of diagenesis in reefs is their particular sedimentary and biological composition.

[1] Laboratoire de Pétrologie Sédimentaire et Paléontologie, Bâtiment 504, Université de Paris-Sud, 91405 Orsay Cedex, France
[2] Institut für Geologie und Paläontologie, Technische Universität Berlin, Hardenbergstr. 42, 1000 Berlin 12, FR Germany

Reef Diagenesis
Edited by J. H. Schroeder and B. H. Purser

2.1 Reef Morphology

Because reefs, by definition, rise above the sea floor, they tend to form barriers which deflect or block ocean currents. Unlike flat sediment bottoms, they are susceptible to penetration by seawater and therefore to renewal of their interstitial fluids. This is clearly demonstrated by Buddemeier and Oberdorfer (this Vol.), who calculated (for Eniwetak atoll) a turnover of approximately 5 m^3 $year^{-1}$ m^{-3} of permeable reef, this volume probably being infinitely higher in fractures and coarse rudites. However, in spite of considerable relief, not all reefs are notably affected by diagenesis, renewal of interstitial fluids depending on current directions and intensity.

2.2 Reef Stability

Again, reefs, by definition, are rigid bodies whose stable substrates and high porosity permit sustained crystal growth. This situation contrasts with sand bodies whose mobile grains are less favorable for early cementation; as Shinn (1969) has pointed out, the growth of cement requires time, during which the substrate must be stable, remobilization disrupting or eroding incipient cement.

2.3 Reef Oceanography

Most Quaternary, and many older reefs, have developed near sea level. As such they are particularly subject to the strongest marine currents and therefore to the rapid renewal of interstitial waters. Because the reef surface is near sea level, interstitial waters are susceptible to change in quality, notably when the mass is exposed during minor drops in sea level. Finally, modern (and probably many ancient) reefs grow preferentially along breaks in sea-floor slope and are thus situated close to deep, oceanic waters. As such, they are affected by oceanic currents, including upwelling effects capable of delivering the ions necessary for crystal growth. Excellent examples of these phenomena are given by James and Ginsburg (1979) at Belize, and by Marshall (this Vol.) for the Great Barrier reef. However, not all reefs appear to grow in environments favorable for diagenesis. As suggested by Marshall, the deeper reefs of the Great Barrier are less affected than those which have lived at sea level for a considerable period. Lagoonal patch reefs, because they are protected from oceanic currents, tend to be less affected by early marine cementation. Wells drilled on the small (500-m-wide) emergent patch reef of Mba within the lagoon at Noumea (New Caledonia) encountered only loose coralgal gravels (Aïssaoui, personal communication).

2.4 Internal Morphology, Sedimentary Textures, and Skeletal Microstructures

The cavernous nature of the reef structure, today readily apparent to any diver floating in leisurely fashion down the fore-reef slope, has been known since the

19th century. For example, Walther (1891) described the considerable space within the reef structure, partly filled by internal sediments, but part of which remains open and serves as conduits for water both in living and in raised reefs. The presence of the cavernous architecture is demonstrated by daily tidal fluctuations in boreholes, notably Funafuti (Cullis 1904) and at Bikini atoll (Saller 1984). On uplifted atolls in the Loyalty Group (New Caledonia), where the primary cave system may have been widened by karst processes, tidal fluctuations of 1 m occur some 5 km from the coast (Purser, personal observation). Clearly, these water movements are highly significant for diagenetic processes.

On a more local scale, the considerable variation in sedimentary texture and related permeability results in a heterogeneous hydrodynamic regime whose variations are discussed by Buddemeier and Oberdorfer in this Volume. Coarse, permeable rudstone fabrics are often cemented with centimeter-thick crusts of marine calcite cement.

On an even smaller scale, intraskeletal diagenesis is related closely to the microstructure of the skeleton. Thus, coexisting corals or sponges within a given environment exhibit varying degrees of dissolution and/or neomorphism clearly demonstrated in Quaternary corals of Barbados by Constantz (this Vol.), who relates dissolution to the more porous microstructures, especially when these are composed of the smallest aragonite crystals, as in the Astrocoeniid corals. Other families, notably the Faviinids, composed of larger, more densely packed crystals, are much less affected. The latter, however, are more susceptible to neomorphic inversion to calcite. Similar conclusions are formulated by Dullo (this Vol.) concerning the Pleistocene reef terraces on the Saudi Arabian coast of the Red Sea, and by Schroeder for Paleocene corals of the Western Desert of Egypt. Because the permeability of coral skeletons is highly variable, their internal voids are affected to varying degrees by intraskeletal waters. This may result in varying cement sequences within adjacent corals and even within a given coral colony, a point made by Schroeder, and in Permian sponges by Scherer (both this Vol.).

The variable susceptibility of coral and other skeletons to diagenetic modification implies that care must be taken when determining diagenetic sequences; distinction must be made between the more important fundamental effects and those resulting from particular substrate controls.

2.5 Skeletal Mineralogy

The varying susceptibility of carbonate skeletons to diagenetic modification is, of course, well known. However, this variability involves somewhat more than aragonite and calcite polymorphs; variable skeletal microstructure (see preceding paragraph) may be equally important in determining their susceptibility to dissolution. The fact that a skeleton is aragonitic does not imply, *a priori*, that it will dissolve in the presence of meteoric waters, there being many examples where corals are well preserved in emergent reef terraces, for example, at Barbados (James 1974, Constantz, this Vol.), Hogsty Reef in the Bahamas (Pierson and Shinn 1985) and the Permian of Tunisia (Scherer, this Vol.), to name but a few.

In some cases preservation is related seemingly to dry climates or to the early precipitation of intraskeletal aragonitic cements, this primary aragonite tending to be relatively stable.

Skeletal mineralogy, together with high internal porosity and permeability, may be major factors determining early dolomitization. At Mururoa atoll, the first traces of dolomite are often located within secondary voids created by the dissolution of aragonite skeletons.

2.6 Organic Metabolism

That carbonate precipitation may be favored by photosynthetic activity is well known. Thus, the intimate association of early marine cements and algal crusts within the Triassic Wetterstein reefs of the Calcareous alps has led Henrich and Zankl (this Vol.) to propose an algal origin for certain calcite cements. A microbial influence has been suggested by Monty (1982) within the Devonian reefs of Belgium. It is also possible that microbial activity and decomposition of organic matter may modify pore-water geochemistry, leading to undersaturation with respect to calcite, perhaps favoring dolomitization, as suggested by Buddemeier and Oberdorfer (this Vol.).

In summary, we have listed six major sedimentary factors which, to varying degrees, influence the nature and intensity of diagenesis in reefs. Any study of diagenesis should clearly bear in mind these important factors.

3 Hydrodynamics and Reef Diagenesis

Diagenesis is closely related to water movement, this being especially evident in pre-burial diagenesis; extensive dissolution requires flushing, important cementation needs many pore volumes of water, and dolomitization requires that Mg ions be transported into the sediment and Ca removed, etc. As already noted, reefs are frequently characterized by movement of interstitial waters, this not necessarily being the case for sediments deposited on horizontal bottoms. Reefs, together with islands, are potentially favorable sites for measuring the overall hydrodynamic budget. However, in spite of their highly favorable attributes, the number of studies relating hydrology and diagenesis are very limited; hydrodynamics in reefs is still in its infancy. Most would agree that an overall understanding of diagenesis must be based on hydrodynamic models but, unfortunately, there are many discouraging technical and financial difficulties.

3.1 Hydrodynamics Relating to Oceanic Water Movement

Again, most of the "facts" are theoretic. One of the few exceptions is that of Buddemeier and Oberdorfer, who have quantified tidal movements at Davies reef (Great Barrier), showing that disharmonic oscillations should produce water movement between the lagoon and the open ocean, much of it passing through the reef mass. Tidal oscillation at Funafuti, Eniwetak, and in the uplifted atolls of the Loyalty Group, has been mentioned already.

Oceanic currents, either regional or wind-driven, force seawater through reefs, at least in their shallower parts. Water enters the reef system mainly via coarse permeable conduits, this probably leading to heterogeneous cementation of the mass. However, as one approaches the external flanks of the reef, marine cementation is more pervasive, probably reflecting the greater effects of oceanic water movement. This phenomenon is expressed along the oceanic sides of the barrier reef at Belize (James and Ginsburg 1979), parts of the Great Barrier (Marshall, this Vol.), at Mururoa atoll (Aïssaoui et al., this Vol.) and along the flanks of the Devonian barrier reef of the Canning Basin (Kerans et al., this Vol.). If oceanic currents determine, in part, the intensity of marine diagenesis, it is logical to expect an asymmetric distribution of marine cements around islands or pinnacles, these being best developed on the windward sides of the buildups. This seemingly is the case at Stanley reef in the Great Barrier (Marshall, this Vol.) and, possibly, at Mururoa atoll. Cementation may possible be related to upwelling, as noted by James and Ginsburg (1979) and by Marshall (this Vol.). However, there are no precise data supporting this.

An increasingly popular theory relating water movement to thermal convection is proposed by Saller (1984) and by Samaden et al. (1985) for Bikini and by Aïssaoui et al. for Mururoa. Related to lateral variations in temperature, the movement of oceanic waters into the reef body may explain, for example, the supply of Mg necessary for dolomitization within the buildup and for the mixing of marine and meteoric waters whose chemical qualities may also favor dolomitization.

However, there would appear to be important problems in cementing the major part of wide platforms by oceanic drive alone; water forced into the reef from the exterior must displace water within the pile. If movement is centripetal, this requires that water move upward, a fact that has never been measured. This upward movement will be severely limited by impermeable stratiform horizons.

3.2 Meteoric Aquifers and Mixing Zones

As already noted, reefs are inherently susceptible to meteoric diagenesis, confirmed by the presence of nonmarine calcitic cements in most Quaternary and older reefs. Limiting our remarks, for the moment, to the hydrodynamic aspects of a freshwater body, it is clear that the efficiency of this system will depend on the lateral and vertical dimensions of the emerged area. It will also depend on climate. If there is centrifugal movement of freshwater relative to vertical relief, then clearly this will favor both dissolution and mixing. Furthermore, a thick aquifer implies a relatively important zone of mixing around the lateral margins of the aquifer, as implied by Aïssaoui et al. (this Vol.) to explain dolomitization in atolls.

The classical Ghyben-Herzberg model, while basically acceptable, probably is severely modified by the heterogeneous nature of reef carbonates. As noted by

Buddemeier and Oberdorfer (this Vol.), the aquifer will probably be truncated by the input of marine waters along the more permeable horizons; the freshwater lense probably has a very complicated geometry whose irregularities and thickness depend in part on the permeabilities of the reef mass.

Interestingly, most dolomites are currently being explained in terms of mixing zones. These interpretations are probably valid in many cases, but where mixing is also being suggested for relatively small reef bodies, it should be recalled that an aquifer requires a minimum area of emergence and that it is hardly reasonable to propose mixing zones for small reefs unless these are associated with regional emergence.

Eustatic movement obviously will modify the dynamics of aquifers leading eventually to the formation of karst. This aspect has been stressed by Henrich and Zankl (this Vol.), who demonstrate that shallow burial dolomitization in Triassic Wetterstein reefs may be due to meteoric input during sea-level lows. Further, it is plausible to assume that reef bodies in a given reef complex which are situated at different depths will be to a greater or lesser extent (or not at all) affected by a drop in sea level. In this way considerable spatial variations in diagenesis may arise (Koch and Schorr, this Vol., Pisera 1985).

3.3 Refluxing Brines

This classical model, proposed for only 2 of the 14 examples of dolomitization discussed in this Volume, should nevertheless be retained. Modern reefs, including atolls, may have closed lagoons whose waters may have abnormally high or low salinities, Clipperton Island being an example (Bourrouilh 1985). Differences in density with respect to adjacent oceanic waters may favor a reflux system. Many fossil buildups with peripheral reef systems have evaporite-filled lagoons, the Permian Capitan system of New Mexico being one the better-known examples.

3.4 Late Diagenesis Relating to Expulsion of Connate Waters

Hydrodynamic movement becomes increasingly slow with depth and this causes some difficulties in explaining the origins of late dolomites and calcites. A number of authors have proposed the expulsion of formation waters from surrounding marine shales in order to satisfy these contraints, notably with respect to the diagenesis of Paleozoic reefs. Examples include those of Mattes and Mountjoy (1980) and Machel (this Vol.) for the Devonian reefs of Alberta, and Kerans et al. (this Vol.) for the Devonian reefs of the Canning Basin, NW Australia.

In summary, there are at least four basic hydrological systems which explain, with varying degrees of conviction, the early and/or late diagenetic products in reefs.

4 Fluid Composition and Early Cements

It is hardly surprising that all contributions to this Volume include descriptions of early cement phenomena. These fabrics vary in nature and in quantity, attaining maximum development on the oceanic flanks of both Quaternary reefs such as Mururoa atoll, and Belize (James and Ginsburg 1979), and in certain fossil buildups such as the Permian reefs of England (Tucker and Hollingworth, this Vol.) and, especially, the Devonian barrier reef of the Canning Basin NW Australia (Kerans et al., this Vol.). In these examples, early marine cements form crusts often exceeding 10 mm in thickness, the development of which must depend on one or other favorable hydrodynamic factor discussed in the preceding paragraphs.

4.1 Marine Cementation

In Recent and Quaternary reefs, the marine origin of various fibrous or micritic calcites and aragonites is readily demonstrated (Land and Goreau 1970, Macintyre 1977, etc). Generally best developed in coarse rudstones, probably in response to active water circulation, these cements may produce very hard, rigid masses whose physical properties possibly play an important role in reef construction. That these cements, at least in part, form very close to the sea floor is demonstrated by frequent internal marine sediments and by boring of the cements by marine organisms, as demonstrated in the Great Barrier by Marshall (this Vol.), in the Triassic reefs of Bavaria by Henrich and Zankl (this Vol.) and in the Devonian reefs of the Canning Basin (Kerans et al., this Vol.). However, not all marine cements are spectacular, nor are they invariably formed in coarse permeable sediments; Scherer (this Vol.) shows that fibrous aragonite (still preserved as such) may grow preferentially in skeletons, an observation supported by many workers. In relatively fine sediments, both aragonite and magnesian calcite are often best developed, both as diagenetic peloids and as micrite (Lighty 1985). In these less permeable systems, it is difficult to imagine rapid water circulation. Therefore, marine cementation would appear to be dependent not only on active hydrodynamics (as demonstrated by the thick crusts formed in rudstones), but also by other mechanisms related to water composition, microbial activity, or to substrate control.

In both Recent and ancient reefs, fibrous or micritic Mg calcite is the predominant marine cement, interpretation of initial mineralogy generally being based (in fossil reefs) on the presence of microdolomite inclusions within the neomorphosed cement (Lohmann and Meyers 1977). Curiously, the two polymorphs (aragonite and Mg calcite) are often interstratified within a given void, suggesting that temperature is not the main factor determining the formation of one or other mineral.

Because various marine cements are discussed and well illustrated in the recent volume edited by Schneidermann and Harris (1985), no attempt is made to review the petrographic nature of cements. There are, however, two exceptions which merit comment.

The first concerns radiaxial calcite spar, one of the most frequent and thickest early cements, especially in Paleozoic reefs. This well-known fabric, characterized by elongate crystals having undulatory extinction, appears to have several origins. Initially interpreted by Kendall and Tucker (1973) as a recrystallization product of a precursor fibrous cement – an interpretation confirmed in this Volume by Tucker and Hollingworth for certain Permian radiaxial cements – it is now clear that bladed spars with undulose extinction are also of primary origin. As demonstrated by Kendall (1985) and by several contributions to this Volume, compound crystal structure results from crystal splitting; bundles of fibrous low Mg calcite are common in certain reef bodies where their genesis is still poorly understood. The example illustrated in Fig. 1, from the uplifted atoll of Ouvea (New Caledonia) is composed of low Mg calcite.

The second unusual cement concerns botryoidal aragonite (Fig. 2), a spectacular, fan-like growth which may attain 5 cm in radius in certain Quaternary fringing reefs of the Red Sea (Aïssaoui 1985). Occurring both in Recent (James and Ginsburg 1979) and in fossil reefs (see Tucker and Hollingworth, this Vol.), it was initially described as "fan druse" by Shinn (1969). Generally developed within large cavities (>1 cm), notably in the oceanic flanks of reefs, this cement occurs to depths of 150 m below sea level at Mururoa atoll.

It seems to be particularly well developed in certain Permian reef slopes in England (Tucker and Hollingworth, this Vol.) and in Permian reefs in S. Tunisia, where it is still partly preserved as aragonite (M'Rabet, personal communication). The factors favoring the formation of this unusual diagenetic product are not known.

4.2 Mixing-Zone Cementation

Theoretically, mixed marine and meteoric waters are undersaturated with respect to calcite, and thus tend to dissolve carbonate, good examples being described by Hanshaw and Back (1980) from the coast of Yucatan. Undersaturation with respect to calcite is also indicated by measurements of mixed interstitial waters at Eniwetak atoll by Buddemeier and Oberdorfer (this Vol., Fig. 9). The diagenetic effects of mixing will depend on the relative proportions of the two waters, and on other geochemical factors. The term "mixing" has relatively little meaning in terms of diagenetic consequence unless accompanied by precise geochemical data. There appears to be very little information available concerning the natural effects of "mixed" waters, although this is often inferred from the isotopic qualities of dolomite (see Ward and Halley 1985), and calcite.

4.3 Meteoric Cementation

Virtually all Quaternary and older reefs are affected by early meteoric cementation relating to repeated periods of emergence. The sparitic or micritic, low magnesian calcites which result are amply described (Schneidermann and Harris 1985) etc. However, sparitic cements also occur, possibly in limited quantities, in

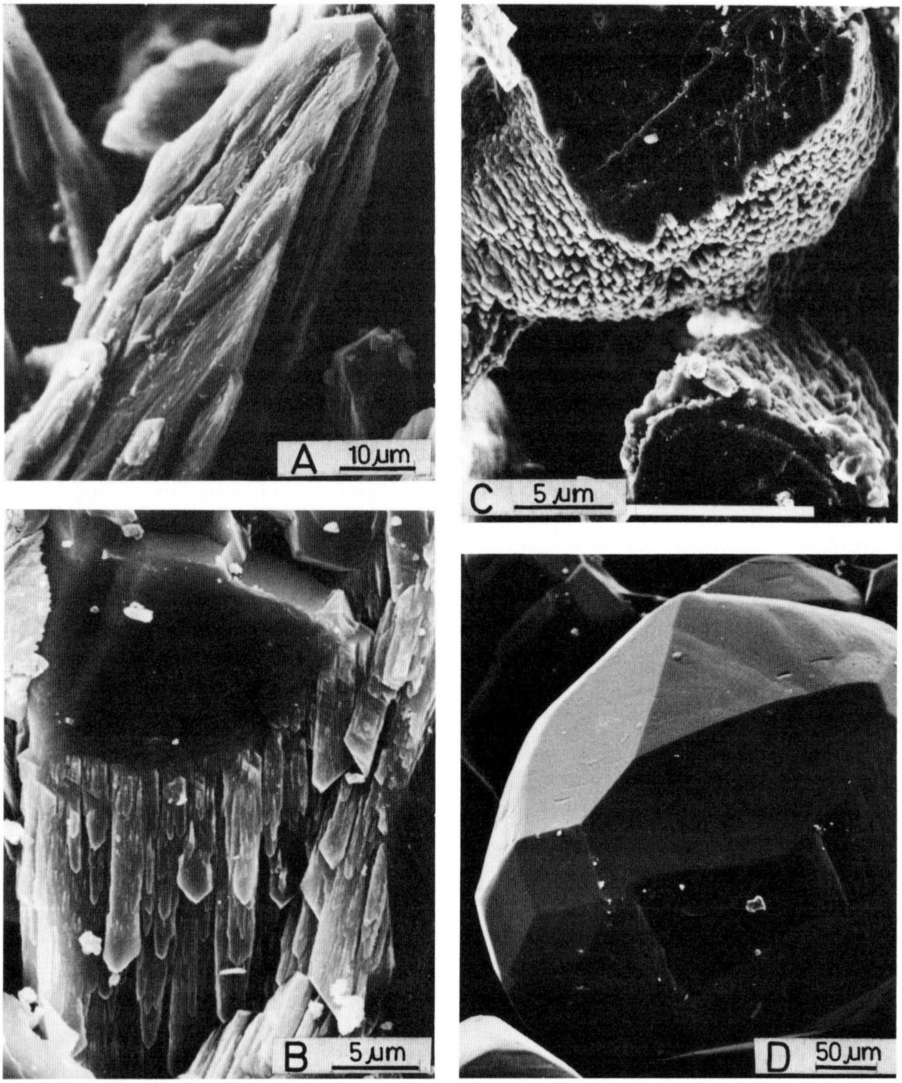

Fig. 1A – D. Composite low magnesium calcite cement and varying crystal morphologies. **A** Elongate, composite crystal of low magnesium calcite showing beginning of development of terminal faces (Holocene, Ouvea Atoll, New Caledonia; SEM Aissaoui). **B** Detail of composite calcite crystal with well developed rhombohedral terminal faces (Holocene Ouvea Atoll, New Caledonia; SEM Aissaoui). **C** Coarse sparitic crystal showing cleavage face and composite fabric visible on natural surface (Holocene Poé, New Caledonia; SEM Aissaoui). **D** Low magnesium calcite crystal exhibiting several sets of rhombohedral and scalenohedral faces as well as a basal pinacoid (Miocene, NE Sudan; SEM Schroeder)

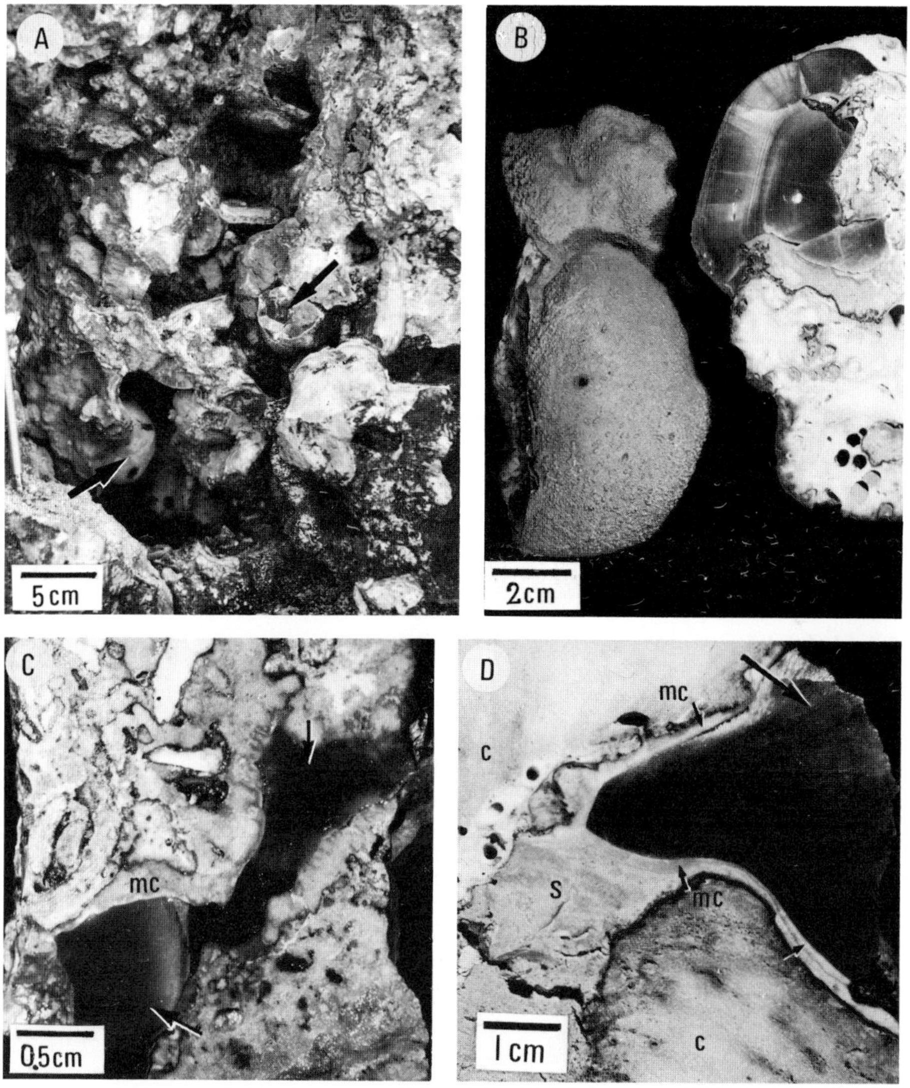

Fig. 2 A – D. Botryoidal aragonite cements in cavernous Holocene reef terrace (+3 m), Egyptian Coast of the Red Sea. **A** Natural outcrop showing botryoids (*arrows*) growing into constructional cavity. **B** Botryoid from cavity (**A**) showing natural surface (*1*) and polished surface (*2*). **C** Polished surface showing two botryoids (*arrows*) developed on magnesian calcite cement (*mc*). **D** Polished surface showing relationship between botryoid and substrate; *c* coral; *s* internal sediment; *mc* magnesian calcite cement; *large arrows* indicate botryoid (fan druse). Photos D. Aïssaoui

submarine environments (Schlager and James 1978). Inversely, meteoric, low magnesian calcite cements, may be relatively fibrous and thus resemble classical marine fabrics. Fibrous LMC is well developed within karst, both at Mururoa atoll (see Aïssaoui et al., this Vol.) and in uplifted atolls of the Loyalty group (Fig. 1), this cement often having a characteristic brown color.

Literature (Schroeder 1985) and personal research suggests, as for marine cements, that the fabric of sparry calcite cement is highly variable, both in size and morphology of individual crystals (Fig. 1), the causes of which are rarely discussed.

In summary, modern and fossil reefs exhibit considerable petrographic and mineralogical variations, whose respective genesis still remains to be elucidated; there exist a multitude of different "fibrous marine cements" and a comparable variety of "sparry LMC, meteoric cements", very few of which are described precisely with respect to chemical composition and crystallography, or related with precision to the chemical quality of the parental waters.

5 Skeletal Diagenesis and Dissolution

Neomorphic inversion and dissolution of aragonitic skeletons, although typical of most carbonates, may be more frequent in reefs because of their susceptibility to emergence and thus to the evolution of their interstitial waters.

5.1 Skeletal Neomorphism

The inversion of skeletal aragonite to a coarser calcite mosaic, although generally resulting in the loss of initial organic microstructure, is not invariably fabric-destructive; corals and sponges may retain much of their primary structure, in spite of calcitization, as shown by Scherer (this Vol.). This preservation is seemingly favored by the early precipitation of aragonite cement, which reduces skeletal porosity, notably around the periphery of the colony. Subsequent inversion to calcite destroys neither the fibrous fabric of the cement nor that of the enclosed coral skeleton. Therefore, the preservation of initial skeletal or crystal structure in calcitic rocks cannot be regarded as absolute proof of an initial calcitic mineralogy, as has been proposed by Sandberg (1975, 1983) and by Loreau (1982).

5.2 Dissolution of Skeletons

As already noted, the susceptibility of skeletons to dissolution depends not only on their primary mineralogy, but also on their microstructure and morphology, an important point demonstrated by Constantz and by Schroeder (this Vol.). While most dissolution probably expresses the passage of aggressive "mixed" or meteoric waters, some dissolution may be related to the lowering of pH following decomposition of organic matter, as suggested by Miller (this Vol.) concerning the development of secondary voids in the Waulsortian reefs of Britain. This type of dissolution may explain certain types of secondary porosity in reefs and other facies which lack karst or other obvious discontinuity surfaces. If dissolution results, in part, from the degradation of organic matter, it is clear that dissolution of skeletal carbonate should not be used *a priori* as a criterium for meteoric water diagenesis.

That dissolution may be independent of emergence has also been suggested by James and Ginsburg (1979) concerning the Holocene barrier reef at Belize, by Saller at Eniwetak atoll (1984) and by Kerans et al. (this Vol.) for the Devonian barrier reef in the Canning Basin, NW Australia. These and other authors suggest that aragonitic skeletons may dissolve at relatively shallow depths below sea level (+300 m) due to the aggressive properties of seawater at this depth. Relatively shallow marine dissolution of aragonite skeletons, although of considerable potential interest, remains to be demonstrated. Recognition is complicated by the lowering of sea level (and freshwater influx) during the Quaternary, and by the possibility of dissolution by other processes discussed in the preceding paragraph.

5.3 Dissolution of Cements

Early aragonitic or high magnesian calcite cements and calcian dolomites have variable susceptibility to dissolution with respect to each other and to their skeletal substrates. The observations of Scherer (this Vol.) on Permian material confirms that aragonite cement, in some cases, may be more resistant to dissolution than the aragonite skeletal substrate.

The selective dissolution of fibrous magnesian calcite marine cements is demonstrated by Aïssaoui et al. (this Vol.) at Mururoa atoll. The planar voids resulting from selective dissolution may be filled subsequently with dolomite cement. Thus, these authors have observed in reefs a phenomenon previously reported only from Kenyan beachrock (Schroeder 1979), where isopachous pores were formed by partial dissolution of high magnesian palisade and low magnesian bladed cements. In some cases the pores formed by dissolution of high magnesian calcite were filled by later, low magnesian cement. As shown by Aïssaoui et al. (this Vol., Figs. 11, 12), partial dissolution and later precipitation of new cements in the secondary pores results in a stratigraphic sequence of cements which is partially inverted.

In summary, it is clear that dissolution is highly selective, even for a given mineral, i.e., aragonite or calcite. Furthermore, for reasons already elaborated concerning skeletal molds, it would seem that dissolution of cements cannot be regarded as absolute proof of emergence.

6 Dolomitization and Dedolomitization

Reefs appear to be more susceptible to dolomitization than other sedimentary facies, with the obvious exception of sabkhas; of the 19 contributions in this Volume, 14 include dolomite. This susceptibility to dolomitization most probably results from the combination of favorable factors including their frequently evolving interstitial waters partly due to sea-level fluctuation, their permeable nature, the presence of metastable minerals and, possibly, the presence of organic (microbial) substances.

6.1 Petrography and Mineralogy of Reef Dolomites

In common with dolomites in general, reef dolomites are highly variable and multiphased, many reefs including both early and late (burial) dolomites whose distinction is based both on isotopes and on fluid inclusions (see Cercone and Lohmann, this Vol.). Most early dolomites are calcium-rich. However, the Ca/Mg ratios in Quaternary dolomites are variable, even within a given crystal, as shown by Ward and Halley (1985) in partially dolomitized Pleistocene reefs in Yucatan. Although the precise factors controlling the Ca/Mg ratio are uncertain, these may relate to salinity (Füchtbauer and Goldschmidt 1965), the relatively high Ca/Mg ratios noted by Ward and Halley seemingly coincide with a "mixed" parental waters. However, the situation is not entirely clear; Aïssaoui et al. (this Vol.) note that high Ca/Mg ratios are characteristic of early dolomites situated around the oceanic periphery of Mururoa atoll. The near-stoichiometric composition characteristic of many dolomites in fossil reefs, as noted by Koch and Schorr (this Vol.) may be due to progressive equilibration.

The well-known petrographic variability of dolomites is confirmed in this volume. However, surprisingly little attempt is made to relate this variability to the overall geometry of the dolomite body, one important exception being that of Aïssaoui et al. (this Vol.). These authors demonstrate that dolomite cement, in common with the marine calcites and aragonites, is better developed toward the periphery of the platform.

6.2 Dolomitization Models

Because reefs generally have clearly defined limits they are particularly favorable for "modeling". However, mainly because of limited outcrop or well control, little attempt has been made to map dolomite facies.

The hydrodynamic systems responsible for dolomitization of reefs are based mainly on the geochemical nature of the parental fluids which are deduced mainly from isotopic signatures of the dolomites; the "dolomite model" thus is highly conceptual, generally with very limited geometric support.

Of the 14 examples of dolomitized reefs cited in this Volume, 5 are interpreted as having been the product of a "mixing" or freshwater system. While this "model" certainly lacks originality, it nevertheless finds confirmation in reef studies. Not only are the isotopic compositions of many reef dolomites relatively light ($\delta^{18}O$ -3 to -7), but the dolomite is almost invariably associated with dissolution phenomena. Theoretic geochemistry suggests that certain mixtures of marine and meteoric waters favor dolomitization (Badiozamani 1973). That this potentially favorable situation may exist in nature is confirmed by the data offered by Buddemeier and Oberdorfer (this Vol., Fig. 9); brackish interstitial waters at Eniwetak atoll are undersaturated with respect to calcite, but are oversaturated with respect to dolomite.

The mechanism responsible for the "magic mixing" are multiple. In pericontinental situations, such as Yucatan Peninsula, Ward and Halley (1985) suggest seaward flow of aquifers, an observation confirmed by local groundwater studies by Hanshaw and Back (1980). On atolls, such as Eniwetak (Saller 1984) and Mururoa (Aïssaoui et al., this Vol.), a centripetal drive related to oceanic thermal convection is proposed.

Mixed-water dolomitization, however, is not without its problems. Although numerous lagoonal and adjacent interstitial waters are brackish, modern sediments are rarely dolomitized. The near-surface sediments of Pacific atolls, the Great Barrier Reef and Caribbean islands are notoriously devoid of dolomite, in spite of the presence of brackish waters. It would appear that mixed waters are not the sole ingredient for dolomitizing fluids; dolomitization probably depends not only on the absolute proportions of these solutions, but also on the efficiency of the mixing mechanism and, possibly, on the presence of other "catalyzers" such as organic material.

6.3 Dissolution of Dolomite ("Dedolomitization")

Although surface dolomites are rare in Holocene reefs, secondary voids created by dissolution of precursor rhombohedra are frequent, notably in Pacific atolls where they have been noted at Mururoa (Aïssaoui et al., this Vol.) and at the uplifted atolls of the Loyalty group (New Caledonia). Partial dissolution of Pleistocene dolomite is also demonstrated by Ward and Halley (1985) in coastal Yucatan, where the more soluble parts of individual rhombohedra are shown to be calcium-rich. In somewhat older Pleistocene dolomites both at Mururoa and in certain Caribbean dolomites studied by Pierson (1982), the central parts of the crystal are dissolved. All these examples confirm the relatively soluble nature of early, calcian dolomites. More importantly, they also indicate that interstitial waters have evolved following dolomitization, probably becoming more meteoric.

Several contributions to this Volume also describe dedolomite phenomena in older reefs, including those occurring in the subsurface. M'Rabet et al. mention dedolomites from Cretaceous rudist mounds in Tunisia, Koch and Schorr describe subsurface dedolomite from Jurassic subsurface reefs in S. Germany, and Tucker and Hollingworth note dedolomite phenomena in the Permian reefs in Britain. Eogenetic dedolomitization has also been demonstrated in Jurassic oolite from the Paris Basin (Purser 1985). These early dedolomitization fabrics may indicate that the mixing model, while in part favoring dolomitization, may evolve subsequently into a "dedolomitizing model" as the mixing zone is deplaced seaward perhaps during lowering of relative sea level.

7 The Evolution of Early Reef Diagenesis

7.1 Diagenetic Sequences

Clearly, interstitial waters and their related diagenetic products evolve. This leads to diagenetic sequences from which is derived the notion of "diagenetic stratigraphy". Virtually every contribution to this Volume, and many other publications, document this well-known phenomenon. Possibly the most detailed examples of diagenetic evolution are figured in the "petrogenetograms" of Schroeder, Koch and Schorr, and Scherer (this Vol.). The studies reveal, however, that the principle of diagenetic sequence must be regarded more critically; the sequence within a given sample is not necessarily typical for the entire rock unit. As shown by Enos and by Schroeder (this Vol.), considerable differences often exist, even within adjacent voids. These differences probably reflect subtle differences in inter- or intraskeletal permeability, as noted by Constantz and by Dullo (this Vol.).

Reef studies indicate that for a given diagenetic stage, care must be taken to separate the general (regional to semiregional) diagenetic effects from their local variants relating to substrate or micro-environmental differences. They further suggest, with respect to temporal succession, that the products of distinct diagenetic environments, such as cements of characteristic composition or fabric, should be distinguished from minor environmental fluctuations ("noise") reflected, for example, by zonation recognized by staining, cathodoluminescence or microprobe.

7.2 Evolution of Surface Diagenesis Through Geological Time

It has been suggested by Sandberg (1975, 1983) and by Mackenzie and Pigott (1981) that the mineralogy of marine cements and ooids has evolved through geological time. This fundamental notion is somewhat complicated by the study of reef diagenesis. Paleozoic reefs, notably those of Permo-Trias age, often exhibit cements whose geochemical and petrographic properties strongly suggest original aragonitic mineralogies (see Scherer, this Vol.), in conformity with the "Sandberg principle". Devonian reefs, notably those of Alberta (Mattes and Mountjoy 1980 and Machel, this Vol.) and in the Canning Basin (Kerans et al., this Vol.) are characterized by precursor magnesian calcite cements, also in agreement with Sandberg's observations. However, Quaternary and Recent reefs contain many magnesian calcite cements, while contemporaneous oolites are virtually all aragonitic. It would seem therefore that the factors determining the mineralogy of marine cements, while partly relating to the global evolution of PCO_2 or Mg/Ca ratios in seawater, depend also on other important factors.

Finally, one may also wonder whether the overall importance of marine cementation has evolved through geological time. Although lacking the statistical base necessary for valid conclusions, it would seem that Paleozoic reefs nearly always exhibit well-developed marine cements, as strikingly demonstrated by several contributions to this Volume. Mesozoic reefs, on the contrary, frequently have only moderately developed marine cements (see contribution Koch and Schorr, this Vol.). The numerous Jurassic reefs of the Paris Basin have virtually no early cements (observation B.H.P.). However, a notable exception is demonstrated by Enos (this Vol.) in Cretaceous rudist buildups in Mexico.

Holocene reefs demonstrate that the degree of marine cementation is highly variable and depends partly on local hydrological factors and permeability; reefs in oceanic settings appear to be very susceptible to marine cementation, especially around their peripheries. The evolution of the mineralogy and the importance of marine cements as a whole, in reefs as in other sediments, is complicated by the mixing of long-term regional and local environmental effects.

8 Burial Diagenesis in Reefs

Although pre-burial diagenesis generally is the predominant phase in many reefs, virtually all fossil reefs exhibit varying amounts of calcite or dolomite whose light oxygen isotopes or fluid inclusions, including hydrocarbon (Cercone and Lohmann, this Vol.) suggest crystal growth at subsurface temperatures. Certain sparry calcites or dolomites clearly post-date fractures or stylolites (Mattes and Mountjoy 1980).

8.1 Late Calcite Cementation

There is a distinct petrographic difference between most eogenetic calcites and those formed during burial; the latter almost invariably form coarse sparry mosaics whose average crystal size generally exceeds the somewhat finer sparry calcites formed under near-surface conditions. Late fibrous calcites seem to be very rare. Geochemically, these burial calcites may be characterized by increasing ferrous iron content. Thus, Kerans et al. (this Vol.) and Frykman (1985, and this Vol.) demonstrate that early burial calcites are succeeded by later burial, ferroan cement, this evolution expressing the change from oxidizing to reducing conditions with increasing depth.

8.2 Late Burial Dolomitization

Of the 14 examples of dolomitized reefs discussed in this volume, 8 include a late, burial phase. This late dolomite is generally characterized by light oxygen isotopes, by undulatory extinction and by relatively large crystals with curved surfaces (saddle dolomite of Radke and Mathis 1980). Burial diagenesis is most frequently evoked to explain dolomitization of Devonian reefs, both in Canada (Illing 1959, Mattes and Mountjoy 1980, Machel, this Vol.) and in the Canning Basin (Kerans et al., this Vol.). For some unknown reason, burial dolomitization is rarely proposed concerning the diagenesis of Mesozoic or Tertiary reefs.

8.3 Origins of Parental Fluids

While the formation of near-surface calcites and dolomite does not pose major problems, in the subsurface, on the contrary, the supply of ions necessary for sparitization and dolomitization is not easy to explain. Calcite and, possibly, moderate amounts of dolomite, may result from pressure-dissolution processes (Wanless 1979). However, a major source of Mg required for massive dolomitization is rarely demonstrated convincingly by the proponents of burial dolomitization. The expulsion of marine formation waters during burial has been suggested by several authors. McHargue and Price (1982) have reviewed the role of marine clay deposits as source of ions for the dolomitization of limestones, while Mattes and Mountjoy (1980) invoke this process for dolomitization in Devonian reefs and Grover and Read (1983) for certain stages of burial diagenesis of Ordovician ramp carbonates including buildups. While this source may well be effective in some cases, it requires specific clay compositions and alteration processes and also specific facies relationships, and thus seems not to be generally applicable.

In summary, the mesogenetic origins of certain sparry calcite and dolomites, although clearly demonstrated, pose a considerable problem concerning the source of ions, notably Mg. If indigenous formation waters do not constitute a sufficient source, one should perhaps reconsider artesian supply (Fig. 3 D), notably for the formation of shallow burial calcite and dolomite, as has been suggested by Henrich and Zankl and by Scherer (both this Vol.).

9 Porosity Evolution in Reefs

Many fossil reefs are well-known oil reservoirs, whose porosity and permeability depend essentially on the combined effects of both early and burial diagenesis; much porosity occurs as dissolution voids and as intercrystalline pores in dolomite. Primary porosity is severely modified, if not totally destroyed, by pore-filling cement.

9.1 Porosity and Early Diagenesis

The initial sedimentary porosity in reefs is high, as shown by Buddemeier and Oberdorfer (this Vol.) and the greater part of the reef mass is potential reservoir. However, the rapid evolution of this initially favorable situation is clearly demonstrated by most contributions in this Volume. Porosity destruction, notably by massive marine cementation around the oceanic flanks of both Recent and fossil reefs, is demonstrated by James and Ginsburg (1979) at Belize, by Aïssaoui et al. (this Vol.) at Mururoa atoll and by Marshall (this Vol.) in parts of the Great Barrier reef. A similar situation exists on the basinal flank of Miocene reefs in the Red Sea (Aïssaoui et al., this Vol.) and on the slopes of the Devonian reefs in the Canning Basin by Kerans et al. (this Vol.).

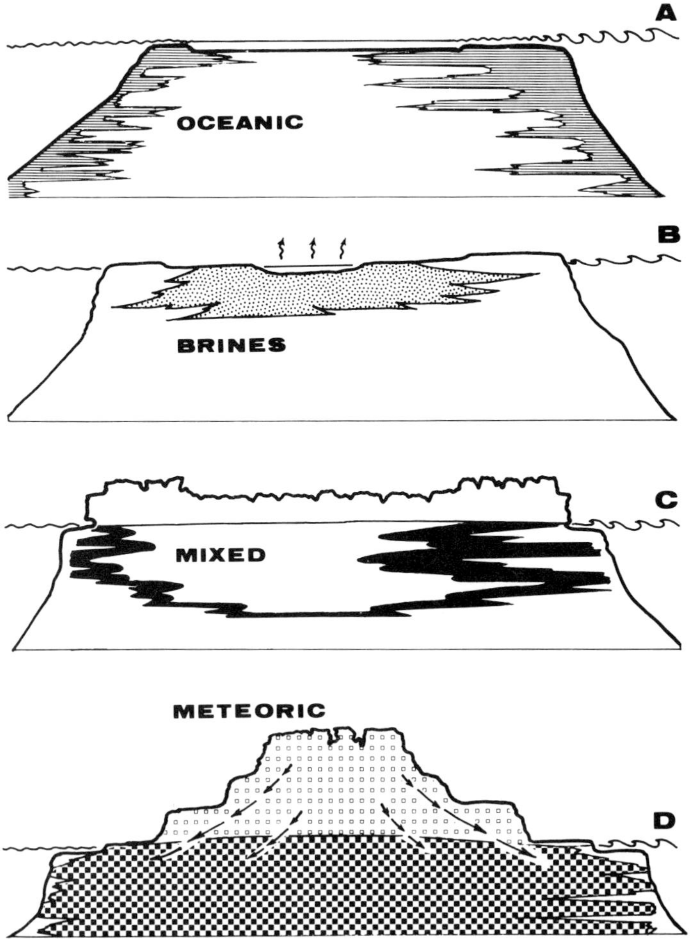

Fig. 3 A – D. Simplistic models showing the relationships between the geometry of a given water body and its hydrodynamic regime. **A** Centripetal oceanic drive by wave, tide, oceanic currents or by thermal convection, favors input around the periphery of the buildup. The irregular form of the body will reflect the asymmetric nature of the external drive (windward/leeward) and the variable permeability of the platform constituents. **B** Refluxing waters will depend on the density difference between the lagoonal and oceanic waters; the drive may be centrifugal if lagoonal waters are highly saline, or centripetal if these waters are diluted. **C** Mixed waters, relating to emergence, will be best developed on the lateral flanks of the Ghyben-Herzberg lense because of oceanic input. **D** Artesian, meteoric systems will depend on the relative sea-level, size of the emerged area, climate and permeability of the mass. High relief may favor relatively deep penetration of meteoric or mixed waters causing shallow burial diagenesis

Initial porosity is also modified both by meteoric dissolution and by brecciation, as demonstrated by Henrich and Zankl in the Triassic Wetterstein reefs of Bavaria (this Vol.). These processes are most effective in the fore-reef.

Porosity and permeability also depend on the diagenetic fabric of the muds and silts trapped within the reef. Because these may be volumetrically important as, for example, in Cretaceous rudist buildups, the petrophysical qualities of this matrix may strongly influence the reservoir potential of the reef. Surprisingly little information is available concerning the diagenetic evolution of reef micrite, one exception being that of M'Rabet et al. (this Vol.), who demonstrate an important early microsparitization of micrites in rudist reefs in Central Tunisia. Clearly, much research remains to be done in this domain.

In many fossil reefs, reservoir potential is related closely to dolomitization, which may be multiphased. Dolomitization, almost invariably, is related to dissolution phenomena which are expressed on various scales. Secondary molds obvious in cores and thin sections also occur on as micrometer-sized voids peripheral to many dolomite crystals (Aïssaoui et al., this Vol., Fig. 10). Initial porosity in pure dolomite depends to a considerable degree both on these dissolution phenomena and, especially, on the pre-dolomite, sedimentary porosity. However, this initially favorable situation deteriorates with the progressive development of overgrowth cement on dolomite crystals and the development of dolospar cements in residual voids. In common with early marine calcite cementation, this excessive development of pore-destroying dolomite, at least at Mururoa atoll, is best developed toward the periphery of this oceanic platform. Thus, the highest reservoir potential is located within the central part of the buildup.

9.2 Porosity and Burial Diagenesis

Mesogenetic sparites and dolomites, clearly described by Cercone and Lohmann, Machel and Scherer, and other contributions to this Volume, show that burial diagenesis, both in reefs and in carbonates in general, is porosity-destructive. Part of this pore-filling cement may be derived from compacting marine shales, as already noted.

Burial diagenesis is also caused by pressure dissolution and by mechanical compaction, whose effects are negative. In reefs, and in carbonates in general, stylolitization is best developed in those parts of the reef which have been little affected by pre-burial cementation. As this appears to be the case within the central, lagoonal parts of certain fossil reefs, compaction may be highest in these areas. This is confirmed within certain Devonian reefs of Alberta by Mossop (1972), where stylolitization has reduced thickness of lagoonal facies and thus exaggerated the initial peripheral relief of the platform.

In summary, the initial porosity in reefs evolves in a centripetal manner; early peripheral marine cementation grades laterally into late compaction and cementation of lagoonal sediments. The residual porosity in a given reef will therefore depend on the relative importance of these two major phases. Thus, within the Miocene reefs of the Red Sea, which have suffered only minor burial, residual porosity is best preserved within the lagoonal parts of the platform. In older reefs the situation may be reversed when burial compaction is excessive.

10 Conclusions

To "conclude the conclusions", the authors would like to make two points.

10.1 The Relative Importance of Early Diagenesis in Reefs

This Volume, and many other published studies, clearly show that most reefs are affected to varying degrees by early diagenesis. In particular, early cement plays at least three major roles. Firstly, in consolidating the pre-existing substrate, it may determine the localization of certain reefs, as has been suggested by Tucker and Hollingworth (this Vol.). Secondly, cements are volumetrically important and, in certain Paleozoic reefs such as the Permian Capitan reef of New Mexico, may locally exceed the sedimentary component in volumetric importance, as noted by Schmidt (1977). Thirdly, cements tend to impart a certain rigidity to the reef, their presence creating a wave-resistant mass with steep slopes, these being the essential attributes of reefs. In certain cases "stratigraphic reefs" may be formed by the diagnetic consolidation of essentially detrital carbonate to create "lithoherms" (Neumann et al. 1977).

10.2 Models of Reef Diagenesis

By studying a clearly defined rock unit such as reefs, it was hoped that a broader view of diagenetic patterns would emerge; as noted in the *Introduction*, this was one of the major objectives. Predictably, it has not been fully attained. Trends *are* emerging, especially concerning the peripheral concentration of early marine cements. However, partly because of limited outcrop and subsurface control, diagenetic phases, although generally arranged in chronological order, are rarely mapped.

Currently, there exists a multitude of "diagenetic microfacies", many already defined by Cullis (1904) at Funafuti atoll. Most of these facies are related to marine, mixing or meteoric waters, but generally in rather vague terms. Even in Quaternary reefs there are very few attempts to correlate the mineral phase with its parental solutions. This, of course, is rarely possible because of the age of the reef. Diagenesis is currently in the situation analogous to that of sedimentary facies in the 1960's which were then described and catalogued. By now most sedimentary "microfacies", however, are placed with relative ease within an overall sedimentary model.

Knowledge of the broader picture of reef diagenesis, and probably diagenesis as a whole, is severely handicapped by our lack of precise data concerning hydrodynamic models, the study of which is costly and technically difficult; sedimentary models are founded mainly on studies of surface phenomena; diagenetic equivalents generally require drilling. These difficulties should not, however, be overestimated. That diagenetic facies can be correlated and mapped is demonstrated by the application of relatively simple techniques, notably cathodoluminescence. Clearly, the geometry of any given diagenetic product

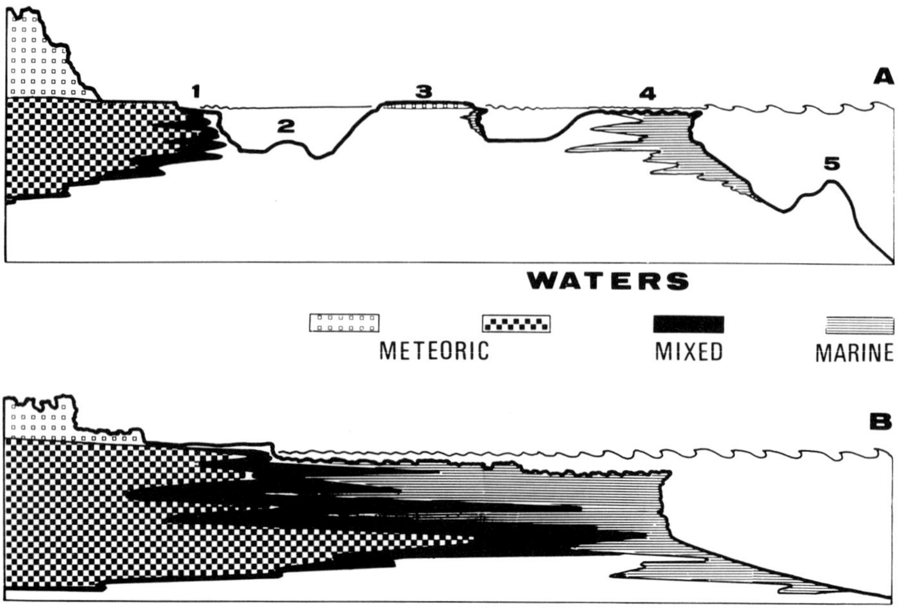

Fig. 4 A, B. Simplistic sketch showing possible geometric relationships of various water masses within a given profile. **A** Profile showing the relationship between reef-type and its potential diagenesis. *1* a fringing reef bordering high relief is strongly affected by meteoric and/or mixed waters; *2* submerged patch reef protected from oceanic currents probably has little diagenetic modification; *3* lagoonal island may have a thin brackish lense and minor marine cementation on its seaward margins; *4* barrier reef strongly affected by oceanic currents will probably be affected by marine cementation; *5* deep oceanic reef, little affected by oceanic currents, will probable be little affected by diagenesis. **B** Profile showing the relationships between various water-types within a given reef. NB. These examples are based mainly on contributions to this Volume

(e.g., dolomite) will depend partly on the parental hydrodynamic system. Although the principles are fairly well known, the authors have sketched these (Figs. 3, 4) in the hope that the rapid amelioration of this admittedly simplistic base will lead to an improved, overall view of reef diagenesis.

References

Aïssaoui DM (1985) Botryoidal aragonite and its diagenesis. Sedimentology 32:345 – 361

Aïssaoui DM and Purser BH (1985) Reef diagenesis: cementation and dissolution at Mururoa atoll, French Polynesia. Proc 5th Int Coral reef Congress, Tahiti

Badiozamani K (1973) The dorag dolomitization model. Application to the Middle Ordovician of Wisconsin. J Sediment Petrol 43:965 – 984

Bourrouilh F (1985) Sédimentation phosphatée actuelle dans le lagon confiné de l'île de Clipperton, (Océan Pacifique). Datations, sédimentologie et géochimie. Sci Géol Mém 77, Univ de Strasbourg

Cullis CG (1904) The mineralogical changes observed in the cores of the Funafuti borings, in Bonnay TG (ed), The Atoll of Funafuti, Rept. of the Coral Reef Committee of the Royal Society (London), 392 – 420

Frykman P (1985) Subaerial exposure and cement stratigraphy of a Silurian bioherm in the Klinteberg Beds, Gotland, Sweden. Geol Foren Stockh Forhandl 107:77 – 88

Füchtbauer H, Goldschmidt H (1965) Beziehungen zwischen Calciumgehalt und Bildungsbedingungen der Dolomite. Geol Rundsch 55:29 – 40

Grover G, Read JE (1983) Paleoaquifer and deep burial related cements defined by regional cathodoluminescent patterns, Ordovician carbonates, Virginia. Am Assoc Petrol Geol Bull 67:1275 – 1303

Hanshaw BB, Back W (1980) Chemical mass-wasting of the Northern Yucatán Peninsula by groundwater dissolution. Geology (Boulder) 8:222 – 224

Illing LV (1959) Deposition and diagenesis of some Upper Paleozoic carbonate sediments in Western Canada. 5th World Petroleum Congr. New York, Proc. I:23 – 52

James NP (1974) Diagenesis of scleratinian corals in the subaerial vadose environment. J Paleontol 48:785 – 799

James NP, Ginsburg RN (1979) The seaward margin of Belize barrier and atoll reefs. Int Assoc Sedimentol Spec Publ 3:1 – 191

Kendall AC (1985) Radiaxial fibrous calcite: a reappraisal. In: Schneidermann N, Harris PM (eds) Carbonate cements. Soc Econ Paleontol Miner Spec Publ 36:58 – 78

Kendall AC, Tucker ME (1973) Radiaxial fibrous calcite: a remplacement after acicular carbonate. Sedimentology 20:365 – 389

Land LS, Goreau TF (1970) Submarine lithification of Jamaican reefs. J Sediment Petrol 40:457 – 462

Lighty RG (1985) Preservation of internal reef porosity and diagenetic sealing of submerged Early Holocene barrier reef, Southeast Florida Shelf. In: Schneidermann N, Harris PM (eds) Carbonate cements. Soc Econ Paleontol Miner Spec Publ 36:123 – 151

Lohmann KC, Meyers WJ (1977) Microdolomites in cloudy prismatic calcites: a proposed criteria for former high magnesium calcites. J Sediment Petrol 47:1078 – 1088

Loreau JP (1982) Sédiments aragonitiques et leur genèse. Mémoire Museum Nat. Hist. Natur. de Paris, série C, XLVII

Mackenzie FT, Pigott JD (1981) Tectonic controls of Phanerozoic sedimentary rock cycling. J Geol Soc (Lond) 138:183 – 196

Macintyre IG (1977) Distribution of submarine cements in a modern Caribbean fringing reef, Galeta Point, Panama. J Sediment Petrol 47:503 – 516

Mattes BW, Mountjoy EW (1980) Burial dolomitization of the Upper Devonian Miette buildup, Jasper National Park, Alberta. In: Zenger DH, Dunham JB, Ethington RL (eds) Concepts and models of dolomitization. Soc Econ Paleontol Miner Spec Publ 28:259 – 297

McHargue TR, Price RC (1982) Dolomite from clay in argillaceous shale-associated marine carbonates. J Sediment Petrol 52:873 – 886

Monty C (1982) Microbial spars, 11th Internat. Congress Sedimentology, Abstracts of papers 26

Mossop GD (1972) Origin of the peripheral rim, Redwater Reef, Alberta. Bull Can Petrol Geol 20:238 – 280

Mountjoy EW, Krebs W (1983) Diagenesis of Devonian reefs and buildups, Western Canada and Europe – a comparison. Z Dt Geol Ges 134:5 – 60

Neumann AC, Kofoed JW, Keller GH (1977) Lithoherms in the Straits of Florida. Geology 5:4 – 10

Pierson BJ (1982) Cyclic sedimentation, limestone diagenesis and dolomitization in Upper Cenozoic carbonates of the Southeastern Bahamas. PhD Thesis, University of Miami

Pierson BJ, Shinn EA (1985) Cement distribution and carbonate mineral stabilization in Pleistocene limestones of Hogsty Reef, Bahamas. In: Schneidermann N, Harris PM (eds) Carbonate cements. Soc Econ Paleontol Miner Spec Publ 36:153 – 168

Pisera A (1985) Paleoecology and lithogenesis of the Middle Miocene (Badenian) algal-vermetid reefs from the Roztocze Hills, Southeastern Poland. Acta Geol Pol 35:89 – 155

Purser BH (1985) Dedolomite porosity and reservoir properties of Middle Jurassic carbonates in the Paris Basin, France. In: Roehl PO, Choquette PW (eds) Carbonate petroleum reservoirs. Springer, Berlin Heidelberg New York Tokyo, pp 341 – 356

Purser BH, Aïssaoui DM (1985) Reef diagenesis: dolomitization and dedolomitization at Mururoa Atoll, French Polynesia. Proc. 5th Int Coral Reef Congress, Tahiti

Radke BM, Mathis RL (1980) On the formation and occurrence of saddle dolomite. J Sediment Petrol 50:1149 – 1168

Saller AH (1984) Diagenesis of Cenozoic limestones on Eniwetak Atoll. PhD Thesis, University of Louisiana.

Samaden G, Dallot P, Roche R (1985) L'atoll d'Eniwetak: système géothermique insulaire à l'état naturel. La Houille Blanche 2:143 – 151
Sandberg PA (1975) New interpretations of Great Salt Lake ooids and of ancient non-skeletal carbonate mineralogy. Sedimentology 22:497 – 537
Sandberg PA (1983) An oscillating trend in Phanerozoic non-skeletal carbonate mineralogy. Nature 305:19 – 22
Schlager W, James NP (1978) Low magnesium calcite limestones forming at the deep-sea floor, Tongue of the Ocean, Bahamas. Sedimentology 25:675 – 702
Schmidt V (1977) Inorganic and organic reef growth and subsequent diagenesis in the Permian Capitan reef complex, Guadeloupe Mountains, Texas, New Mexico. In: Hileman ME, Mazzulo SJ (eds), Soc Econ Paleontol Mines Permian Basin Section, Pub. 77-16, 93-132
Schneidermann N, Harris PM (eds) (1985) Carbonate cements. Soc Econ Paleontol Miner Spec Publ 36:379
Schroeder JH (1979) Carbonate diagenesis in Quaternary beachrock of Uyombo, Kenya: sequences of processes and coexistence of heterogeneous products. Geol Rundsch 6:894 – 919
Schroeder JH (1985) Sparry calcite cement in Miocene corals from Khor Eit, NE Sudan. Proc. 5th Intern Coral Reef Congress, Tahiti 3:283 – 288
Shinn EA (1969) Submarine lithification of Holocene carbonate sediments in the Persian Gulf. Sedimentology 12:109 – 144
Walther (1891) Die Adamsbrücke und die Korallenriffe der Palkstrasse. Petermanns Mitteilungen, Ergänzungsheft No. 102:1 – 40
Wanless HR (1979) Limestone response to stress: pressure solution and dolomitization. J Sediment Petrol 49:437 – 462
Ward WC, Halley RB (1985) Dolomitization in a mixing zone of near-seawater composition, Late Pleistocene, Northeastern Yucatán Peninsula. J Sediment Petrol 55:407 – 420

Subject Index

10µm

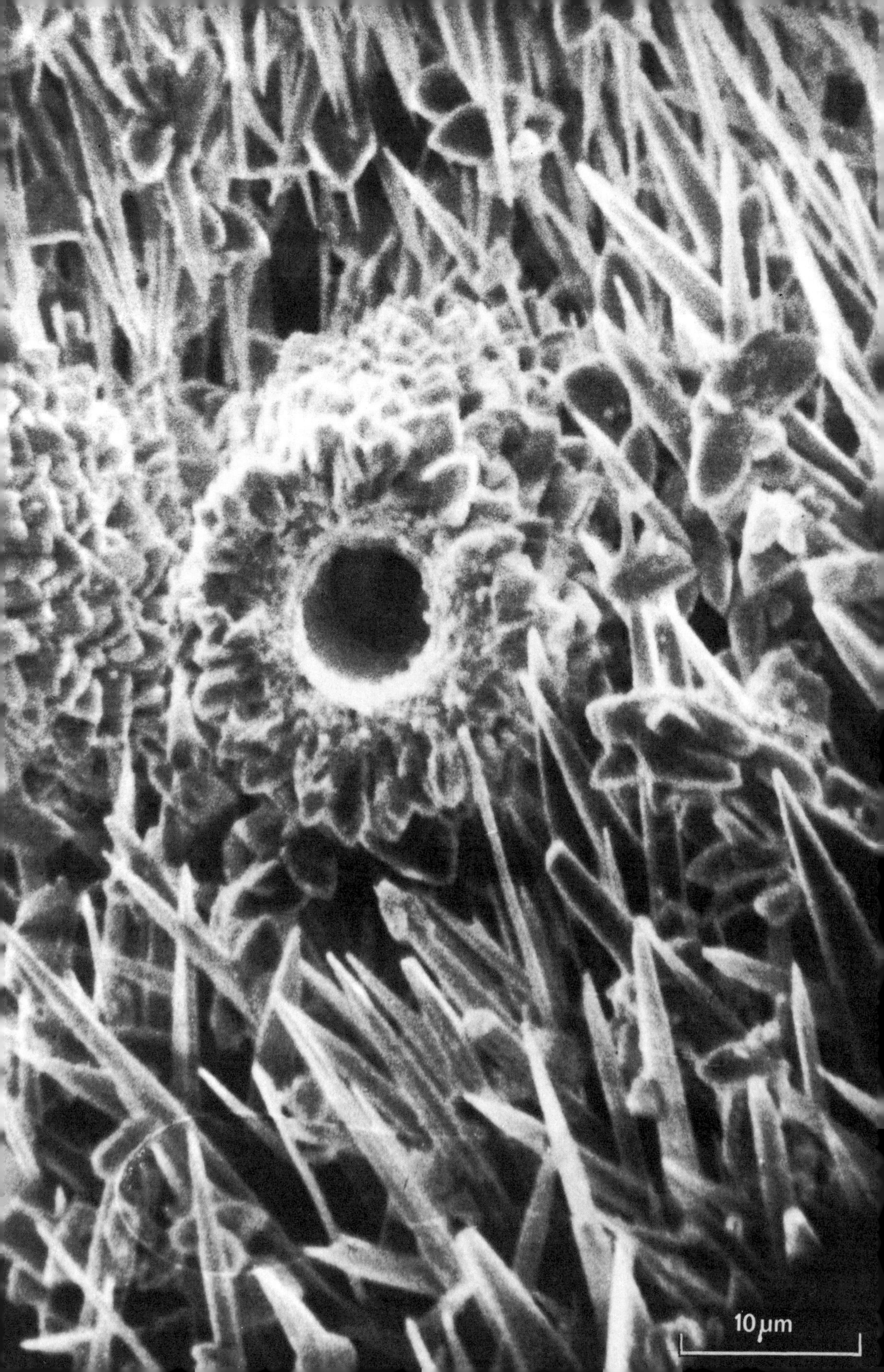
10 µm